Applied Signal Processing

Sadasivan Puthusserypady

now
the essence of knowledge

Published, sold and distributed by:
now Publishers Inc.
PO Box 1024
Hanover, MA 02339
United States
Tel. +1-781-985-4510
www.nowpublishers.com
sales@nowpublishers.com

Outside North America:
now Publishers Inc.
PO Box 179
2600 AD Delft
The Netherlands
Tel. +31-6-51115274

ISBN: 978-1-68083-978-4
E-ISBN: 978-1-68083-979-1
DOI: 10.1561/9781680839791

Suggested citation: Sadasivan Puthusserypady. (2021). *Applied Signal Processing*. Boston–Delft: Now Publishers

Table of Contents

Dedication

I dedicate this book to my mother, Thanka Kumaran and (late) father, Kumaran Puthusserypady. I hope that my achievements have made you proud.

Preface

"Applied Signal Processing" is a textbook mainly for students, who are studying in the Medicine and Technology, Electrical Engineering, Electronics Engineering, Computer Engineering and Biomedical Engineering Disciplines. The main objective of this book is to provide the students with a solid basis for the analysis and processing of analog and digital signals emanating from either deterministic or stochastic systems. This book is the outcome of more than two decades of my experience in teaching signal processing. For more than ten years, I have been teaching 'Applied Signal Processing' course to Medicine and Technology students at the Technical University of Denmark (Department of Health Technology). It introduces the basic concepts, definitions, analysis and implementation of signal processing techniques as well as signal processing systems. Since the digital signal processing techniques have evolved from its analog counterpart, the book begins by explaining the concepts in analog signal processing first and then moves towards digital signal processing. This will help the students to gain a general overview of the whole subject and establish links between the various concepts. In addition, understanding the classical analog signal processing techniques would certainly make it easy for the students to understand the digital signal processing counterpart. The book uses a number of good illustrations to explain the fundamental concepts in both analog and digital signal processing along with examples and numerous exercise problems.

The book is divided into seventeen chapters that covers the fundamentals of most signal processing concepts. In the introduction chapter, some basic signals as well as descriptions of some of the most commonly referred types of biomedical signals, such as the electroencephalogram (EEG), electrocardiogram (ECG), electromyogram (EMG), etc., are provided. In Chapter 2, the fundamental concepts of power and energy as well as their evaluations for different types (periodic/aperiodic) of signals is provided. Fourier series representation of periodic signals as well as its important properties are provided in Chapter 3. In Chapter 4, the Fourier representation of aperiodic analog signals, that is, the Fourier transform is explained. The derivation of Fourier transform and its important properties are also explained here. As complex signals are the building blocks for the synthesis of different types of signals, the fundamentals of complex signals is dealt with in Chapter 5. This chapter also introduces the concept of Hilbert transform and analytic signals. The fundamentals of analog systems, especially that of linear time invariant (LTI) systems is provided in Chapter 6. The process of generating digital signals from analog signals by sampling and quantization techniques are introduced in the next Chapter. In Chapter 8, the fundamentals of the z-transform are introduced for the analysis of discrete-time signals. Discrete-time Fourier transform (DTFT) as well as its relationships to Laplace transform and z-transform are explained in Chapter 9. The fundamentals of digital systems are described in detail in Chapter 10. Some of the most important implementation schemes for the digital systems (both finite impulse response (FIR) systems as well as infinite impulse response (IIR) systems) are provided in the next Chapter. The discrete Fourier transform (DFT) techniques are developed in Chapter 12 followed by the concept and derivation of fundamental algorithms for fast Fourier transform (FFT) in Chapter 13. Design methods for digital filters (both FIR and IIR) are provided in Chapter 14. A detailed description of random/stochastic processes and characterization methods for random signals are provided in Chapter 15. In Chapter 16, the fundamentals of analog and digital modulation schemes as well as detection methods are explained. The final chapter introduces the power spectrum estimation methods, which could be applied to signals of both analog and digital types. The effect of windowing, spectral leakage, and spectral smoothing are also explained in this chapter.

While the main focus of this book is on the fundamentals of digital signal processing, the understanding of these topics greatly enriches our confident use and further development of the design and analysis of digital systems for various engineering and medical applications. This book will also equip students to further their studies/research in advanced topics in signal processing.

Acknowledgements

Working on this book manuscript has been a massive challenge, involving lots of intense work over the past several months. Nonetheless, it has been a great experience and I would like to thank all the people who have supported and encouraged me during this process.

I am grateful to the Technical University of Denmark (DTU), especially the Department of Health Technology and the Department of Electrical Engineering, for having given me the opportunity to pursue my interest in teaching, research and publishing scholarly articles. I am thankful to Prof. Jakob Eyvind Bardram, (my division head, Digital Health, DTU Health Tech) for all his support in publishing this book. I would like to express my sincere gratitude to my former division head, Prof. Jørgen Arendt Jensen for introducing me to the 'Applied Signal Processing' course and sharing his resources in teaching this course. I have been fortunate to associate with many outstanding students at DTU, who took my 'Applied Signal Processing' course during the past 11 years. Their constructive feedback has helped me to improve the writing of this book. I am indebted to many of my teaching assistants, who have unconditionally supported and helped me in running the course for more than a decade at DTU. Their invaluable comments and suggestions have helped me improve the book manuscript considerably.

Finally, I would like to sincerely thank my family, my wife (Deena Puliyasseri) and my son (Vaisakh Puthusseryppady) who have always encouraged me to chase my dreams and have given me enormous support throughout all these years. Without your encouragement and support, this would not have been possible.

Notations, Symbols

Notations

$[\cdot]$ indicates discrete valued independent variable, e.g., $x[n]$.
$(\cdot)$ indicates continuous valued independent variable, e.g., $x(t)$.

- Complex numbers

$\|c\|$	Magnitude of complex quantity c.
$\arg\{c\}$	Phase angle of complex quantity c.
$\text{Re}\{c\}$, $\text{Real}\{c\}$	Real part of c.
$\text{Im}\{c\}$, $\text{Imag}\{c\}$	Imaginary part of c.
c^*	Complex conjugate of c.

- Complex functions

$\text{ARG}\{c(\omega)\}$	Principal value of the phase ($-\pi < \text{ARG}\{c(\omega)\} \leq \pi$).
$\arg\{c(\omega)\}$	Unwrapped phase (no jumps of 2π).

- Lower-case functions denote time-domain quantities, e.g., $x(t)$, $w[n]$.
- Upper-case functions denote frequency or transform domain quantities.

$X[k]$	Discrete Fourier transform coefficients for $x[n]$.
$X[k]$	Fourier series coefficients for $x(t)$.
$X(e^{j\omega})$	Discrete-time Fourier transform of $x[n]$.
$X(\Omega)$	Fourier transform of $x(t)$.
$X(z)$	$z-$transform of $x[n]$.

- Boldface lower-case symbols denote vector quantities, e.g., $\mathbf{q}$.
- Boldface upper-case symbols denote matrix quantities, e.g., $\mathbf{A}$.
- Subscript s indicates continuous-time representation of a discrete-time signal.

$x_s(t)$ continuous-time representation for $x[n]$.
$X_s(\Omega)$ Fourier transform of $x_s(t)$.

- Arctan refers to the four quadrant inverse tangent function and produces a value between $-\pi$ and π radians.

Principal Symbols

j or i	Square root of -1.
T_s	Sampling interval of T_s in seconds.
N	Fundamental period for discrete-time signals in samples.
Ω	(Angular) frequency for continuous-time signals in radians/second.
ω	(Angular) frequency for discrete-time signals in radians.
Ω_p	Fundamental (angular) frequency for continuous-time periodic signal in radians/second.
ω_p	(Angular) frequency for discrete-time periodic signal in radians.
f	(Cyclic) frequency of signals (continuous/discrete-time) in Hz.
f_p	Fundamental (cyclic) frequency of periodic signals (continuous/discrete-time) in Hz.
f_s	Sampling (cyclic) frequency ($= \frac{1}{T_s}$) in Hz.
$\tilde{f}$	Normalized (cyclic) frequency ($= \frac{f}{f_s}$).
$u(t), u[n]$	Step function of unit amplitude.
$\delta(t), \delta[n]$	Unit impulse.
$*$ or $\otimes$	Denotes convolution operation.
$\odot$	Circular convolution.
$\text{sinc}(u)$	$\frac{\sin(\pi u)}{\pi u}$.
$\gamma_{xx}(\tau)$	Auto-correlation function (ACF) of analog random signal, $x(t)$.
$\gamma_{xy}(\tau)$	Cross-correlation function (CCF) between two analog random signals, $x(t)$ and $y(t)$.
$r_{xx}(\tau)$	ACF of periodic/aperiodic analog signal, $x(t)$.
$r_{xy}(\tau)$	CCF between two periodic/aperiodic analog signals, $x(t)$ and $y(t)$.
$\Gamma_{xx}(f)$	Power spectra of analog random signal, $x(t)$.
$\Gamma_{xy}(f)$	Cross power spectra of two analog random signals, $x(t)$ and $y(t)$.
$S_{xx}(f)$	Power (energy) spectra of periodic/aperiodic analog signal, $x(t)$.
$S_{xy}(f)$	Cross power (energy) spectra of two periodic/aperiodic analog signals, $x(t)$ and $y(t)$.
$\gamma_{xx}[m]$	ACF of discrete-time random signal, $x[n]$.
$\gamma_{xy}[m]$	CCF between two discrete-time random signals, $x[n]$ and $y[n]$.
$r_{xx}[m]$	ACF of periodic/aperiodic discrete-time signal, $x[n]$.

$r_{xy}[m]$	CCF between two periodic/aperiodic discrete-time signals, $x[n]$ and $y[n]$.
$\Gamma_{xx}(\tilde{f})$	Power spectra of discrete-time random signal, $x[n]$.
$\Gamma_{xy}(\tilde{f})$	Cross power spectra of two discrete-time random signals, $x[n]$ and $y[n]$.
$S_{xx}(\tilde{f})$	Power (energy) spectra of periodic/aperiodic discrete-time signal, $x[n]$.
$S_{xy}(\tilde{f})$	Cross power (energy) spectra of two periodic/aperiodic discrete-time signals, $x[n]$ and $y[n]$.
m_x	Ensemble Mean of the random signal, $x(t)$.
$C_{xx}(\tau)$	Covariance function of the random signal, $x(t)$.
$\tilde{\gamma}_{xx}(\tau)$	ACF of ergodic random signal, $x(t)$.

Signal Representation and Notation

A summary of various notational conventions used in Digital Signal Processing (DSP) for representing signals and spectra is provided below.

Units

In this book, time 't' is always the physical unit of time in seconds (s). The time 'n', 'm', or 'k' are in units of samples (counting integer numbers having no units). Time 't', is a continuous real variable. The physical time 't' corresponding to time 'n' samples is given by,

$$t = nT_s,$$

where, T_s is the sampling interval in seconds.

For frequencies, we have two physical units: (i) cycles/second (Hertz (Hz)), and (ii) radians/second (rad/sec). One cycle equals 2π radians, which is 360°. Therefore, f Hz is the same frequency as $2\pi f$ rad/sec.

For example, a periodic signal with a period of p seconds has a frequency of $f_p = \frac{1}{p}$ Hz, and a radian frequency of $\frac{2\pi}{p}$ rad/sec. The sampling rate, f_s, is the reciprocal of the sampling period T_s, i.e.,

$$f_s = \frac{1}{T_s}\text{Hz}.$$

Glossary

A

ACF - *Autocorrelation Function* - . 341, 342, 351, 352, 357, 359, 362–368, 370, 372

ADC - *Analog-to-Digital Converter* - . 123, 130, 133, 135, 136, 138, 174, 175, 244

B

BIBO - *Bounded Input Bounded Output* - . 180, 183

C

CCF - *Cross-correlation Function* - . 350, 359, 360, 365–368

D

DAC - *Digital-to-Analog Converter* - . 136, 139, 175

dB - *Decibel* - . 119, 467

DFT - *Discrete Fourier Transform* - . 155, 212, 217, 242, 251, 253, 255, 256, 258–282, 297, 372, 408, 413, 418, 419, 429, 440, 443, 444, 449, 451, 452

DTFT - *Discrete-Time Fourier Transform* - . 146, 155–160, 163–167, 183, 191–193, 251–253, 255, 261, 439–443, 447

E

ECG - *Electrocardiogram* - . 2, 3, 14, 123, 173, 200, 342

ECoG - *Electrocortocogram* - . 2

EEG - *Electroencephalogram* - . 2, 3, 11, 13, 14, 173, 362, 372

EGG - *Electrogastrogram* - . 2

EMG - *Electromyogram* - . 2, 15, 16

EOG - *Electrooculogram* - . 2, 14, 15

ERG - *Electroretinogram* - . 15

F

FFT - *Fast Fourier Transform* - . 269, 270, 276, 277, 279–281, 418, 429, 430, 454

FIR - *Finite Impulse Response* - . 145, 193, 198, 211–216, 218, 220–225, 227, 229, 233–235, 237, 239, 283, 285–292, 294, 296–301, 304, 308, 324, 325, 438

FS - *Fourier Series* - . 34–36, 38–46, 159

FT - *Fourier Transform* - . 2, 57–59, 61, 62, 64–67, 70, 71, 78, 103, 105, 106, 112, 113, 116–118, 159, 374, 409

H

Hz - *Hertz* - . 139, 169, 266, 281, 467

I

IIR - *Infinite Impulse Response* - . 193, 211, 212, 226–228, 230–235, 237, 239, 283, 309, 311, 312, 314, 317, 319, 324

L

LTI - *Linear Time Invariant (system)* - . 78, 113, 114, 141, 180–183, 185–188, 190, 191, 193, 196, 205, 206, 208, 209, 211, 228, 245–247, 283, 284, 310, 347, 366–369, 388, 390

R

ROC - *Region of Convergence* - . 142–146, 148, 151, 152, 164, 166, 167, 183, 190, 206

Z

ZT - *z-transform* - . 163, 164, 166, 167, 175, 188, 189, 192–194

DOI: 10.1561/9781680839791.ch1

Chapter 1

Introduction

Signal processing is an interdisciplinary subject which has application in almost all scientific fields. There are mainly two subfields to signal processing: (i) Analog signal processing and (ii) Digital signal processing. Analog signal (continuous in time and amplitude) processing is the processing performed on analog signals to extract various types of information from the signal. Typical examples which uses analog signal processing include crossover filters in loudspeakers, 'Bass', 'Treble' and 'Volume' controls on stereos, old televisions, old radios, audio tapes, old land-line telephones, video cassette recorders, etc. Digital signal (discrete in time and amplitude) processing, on the other hand, is the use of digital processing on signals to perform a wide variety of signal processing operations by computers or more specialized digital signal processors to extract the relevant information. Digital signal processing applications include audio and speech processing, sonar, radar and other sensor array processing, spectral density estimation, statistical signal processing, digital image processing, data/image compression, video/audio coding, signal processing for telecommunications, control systems, biomedical engineering, and seismology, etc., to list a few.

The enormous advancement in technology, especially in digital technology during the last couple of decades has revolutionized the way technology has been exploited in many scientific fields and digital health field in biomedical engineering (medicine and technology) is one of the most benefited. In digital health, digital signal processing plays a significant role to research and innovate technologies by working on the various digital bio-signals to revolutionize healthcare and life science for the society, in order to enhance the efficiency of healthcare delivery and make personalized and precise treatment for individuals.

Biomedical signals are the observations of physiological activities of organisms. It ranges from gene and protein sequences, to neural and cardiac rhythms, to tissue and organ images. Examples of some of the most commonly used biomedical signals are:

- Electrocardiogram (ECG) – are recordings of the electrical potentials induced by the presence of time-varying electrical activity in cardiac muscles,
- Electromyogram (EMG) – are recordings by placing electrodes in, on, or near a muscle and amplifying the electric potential that results from the generation and propagation of action potentials along muscle fibers,
- Eelectroretinogram (ERG) – are recordings of the electrical responses of various cell types in the retina,
- Electrooculogram (EOG) – are recordings of the electrical potentials associated with the eye-ball and eye-lid movements,
- Electrogastrogram (EGG) – are recordings of the electrical signals that travel through the stomach muscles,
- Electroencephalogram (EEG) – are recordings of the electrical activities of the brain by placing electrodes on the scalp, and
- Electrocorticogram (ECoG) – are recordings of the signals directly from the cortex.

Signal processing essentially is the manipulation of signals for the purpose of (i) extracting information (amplitude, frequency, or phase) from the signal, (ii) extracting information about the relationships of two (or more) signals, or (iii) producing alternative representation (time or frequency domain) of the signal. The motivations for processing these signals include: (i) removing unwanted signal components (noise/artefacts) that are corrupting the signal of interest, (ii) rendering it in a more obvious or more useful form, and (iii) predict future values of the signal in order to anticipate the behaviour of its source (modelling). When these processing methods are applied to the types of signals listed above or similar, it becomes biomedical signal processing. Whatever may be the types/sources of signals, the basic processing methods are similar as in one is interested to estimate the basic signal characteristics such as its amplitude (power), frequency, and phase.

This book introduces the basic tools and techniques required for the design and implementation of systems that process analog and digital signals. This book also lays a strong foundation for advanced studies in biomedical engineering, medicine and technology, and digital signal processing. The main topics covered in this book help us to: (i) Use common Fourier transform (FT) pairs and properties to determine the FT of complex analog and digital signals, (ii) Relate spectra of periodic and aperiodic analog and digital signals and plot these using correct physical units such as Hertz and Volt, (iii) Determine the auto- and cross-correlation functions of analog and digital random signals, (iv) Analyze zero-pole diagrams to determine the causality and stability of linear time-invariant systems, (v) Use the z-transform to calculate the impulse response and the transfer function of a linear time-invariant system, (vi) Determine and analyze the quantization errors in analog-to-digital conversion, (vii) Design simple low-pass, high-pass, band-pass, band-stop and notch filters, using windowing, frequency sampling, optimization methods, as well as filter transformation rules, and (viii) Find the power spectra of random/deterministic signals using non-parametric and parametric spectral estimation methods. Knowledge in these topics will enable anyone, for example, to apply signal processing techniques to signals emanating from biological systems (for example, ECG, EEG, etc.) and design procedures to estimate parameters such as the heart rate, blood velocity and profile of blood flow, etc.

In this chapter, we discuss some of the very fundamental aspects in the study of signals and signal processing.

1.1 What is a Signal?

A signal can be considered to be the result of observing a physical phenomenon and converting it into its equivalent electrical format. Thus, formally, a signal is a function of one or more variables that conveys the information on the nature of a physical phenomenon (i.e., current or voltage as a function of e.g., *time*). Consider the following practical scenarios:

- A microphone converts our *speech into an electrical signal.* This signal can be used to record our speech or for making our speech heard to a large group of people through an amplifier and loudspeaker.
- A temperature sensor connected to the boiler in a chemical plant measures the temperature continuously and converts the *temperature readings into an electrical signal.* This signal can be used to display the temperature for monitoring by the plant engineer or can be recorded for later use.
- A security camera placed at a building site captures the images of all the objects that come under its view and converts these *images into electrical signals* and stores them in a database for later review.
- The *ECG signals* recorded by placing electrodes on the chest of a patient/subject convey the information on the condition of his/her heart.

In short, *any time-varying phenomenon which conveys some information is a signal.* The usefulness of the information conveyed by a signal depends on the application

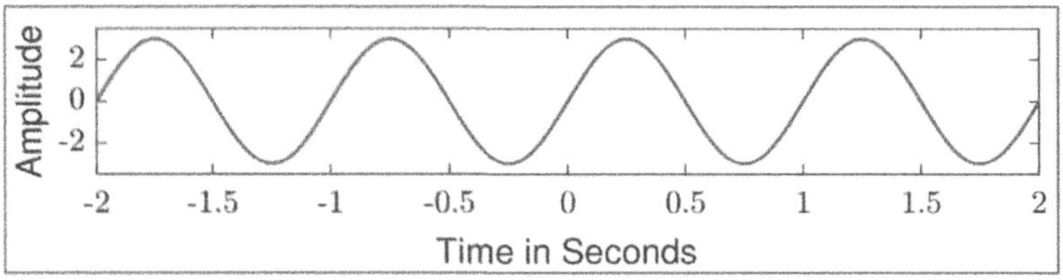

Figure 1.1. Sinusoidal signal.

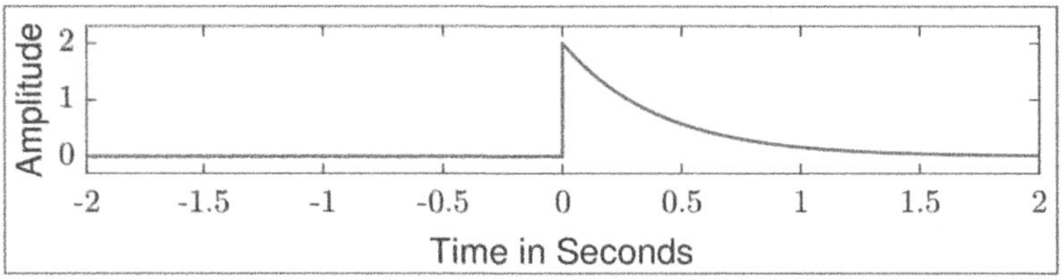

Figure 1.2. Exponentially decaying signal.

under consideration. Thus, in general, *a signal can be considered as a function of time*. We can also have signals that are functions of quantities other than time, such as distance, temperature, height, etc. To specify a signal, *we need to specify the value of the signal at all the instants of time under consideration*. Consider some typical example signals below, that one encounter in signal processing.

1. A *sinusoidal signal* (Fig. 1.1) is mathematically described by,

$$x(t) = A\sin(2\pi f_p t + \phi) \tag{1.1}$$

   where A denotes the amplitude (e.g., $A = 3$), f_p denotes the frequency in Hertz or cycles/second (e.g., $f_p = 50Hz$), and ϕ denotes the phase of the sinusoid (e.g., $\phi = \frac{\pi}{4}$ radians, which is equivalent to[1] 45 degrees). Thus, $x(t)$ gives the value of the signal at time instant t and this value can be calculated using the expression on the right-hand side (RHS) of Eq. (1.1) for any value of t (if the values of A, f_p and ϕ are given).
2. An *exponentially decaying signal* (Fig. 1.2) can be defined mathematically by,

$$y(t) = \begin{cases} Ae^{-\alpha t} & \text{for } t \geq 0 \\ 0 & \text{for } t < 0 \end{cases} \tag{1.2}$$

   where A denotes the value of the signal at $t = 0$ and α is a positive number ($\alpha = 2.1$ in this case) that controls the speed with which the signal value

1. Note: [angle in degrees] = [angle in radians] $\times 180/\pi$.

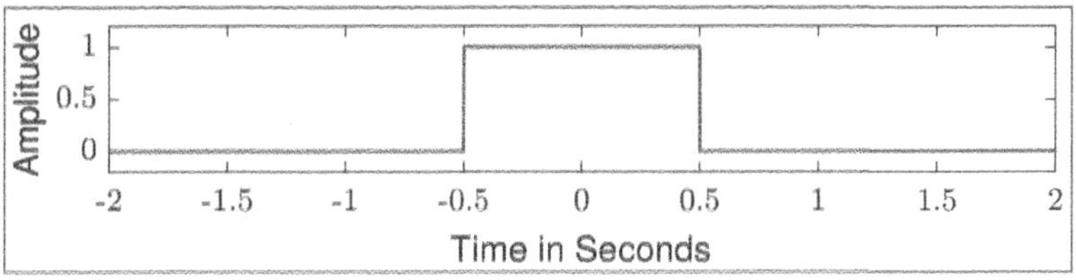

Figure 1.3. Unit rectangular pulse.

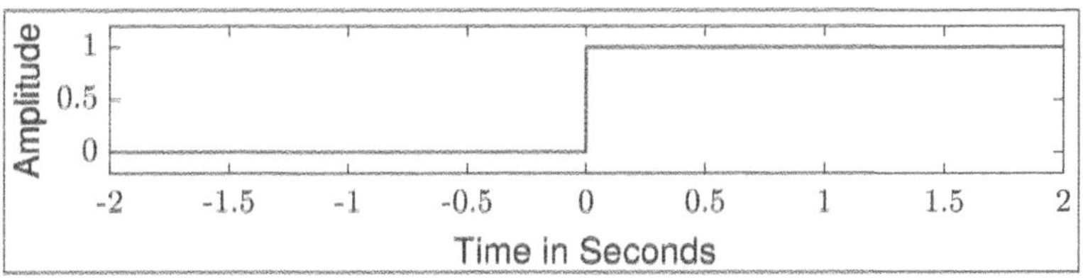

Figure 1.4. Unit step signal.

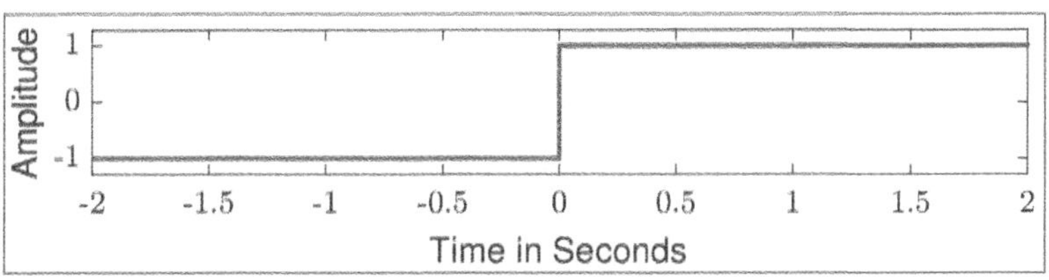

Figure 1.5. Signum function.

decays towards zero. Larger the value of α, faster it reaches towards zero and vice versa.

3. The *unit rectangular pulse signal* (Fig. 1.3) is given by,

$$\text{rect}(t) = \begin{cases} 1 & \text{for } -0.5 \leq t \leq +0.5 \\ 0 & \text{otherwise} \end{cases} \tag{1.3}$$

4. The *unit step (function) signal* (Fig. 1.4) is given by,

$$u(t) = \begin{cases} 1 & \text{for } t \geq 0 \\ 0 & \text{for } t < 0 \end{cases} \tag{1.4}$$

5. The *signum function signal* (Fig. 1.5) is given by,

$$\text{sgn}(t) = \begin{cases} +1 & \text{for } t \geq 0 \\ -1 & \text{for } t < 0 \end{cases} \tag{1.5}$$

6. A *triangular pulse signal* (Fig. 1.6) is given by,

$$y(t) = \begin{cases} 1+t & \text{for } -1 \leq t < 0 \\ 1-t & \text{for } 0 \leq t < +1 \\ 0 & \text{otherwise} \end{cases} \tag{1.6}$$

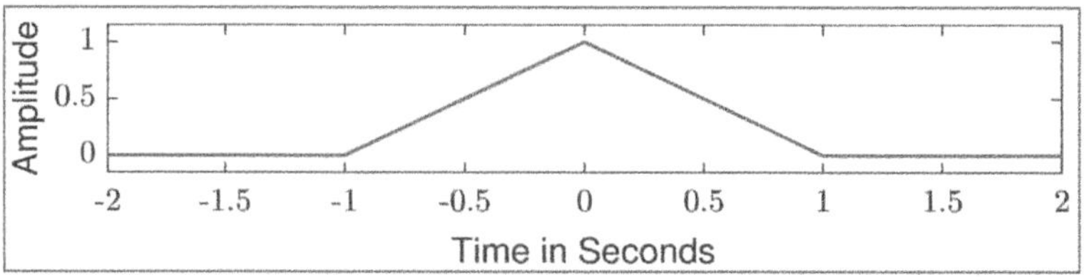

Figure 1.6. Triangular pulse.

§ *Plot the time-shifted versions $u(t-2)$ and $u(t+2)$ of $u(t)$ and compare these plots with $u(t)$ in Fig. 1.4. Comment on your observations.*

1.2 Classification of Signals

There are several ways by which signals can be classified.

1.2.1 Analog or Digital Signals

This classification is based on the values allowed for the time 't' and signal '$x(t)$'. A strictly analog signal has its time and amplitude continuous, whereas a digital signal has its time and amplitude discrete.

- Continuous-Time Signals: A signal is said to be a 'continuous-time signal' if it is defined (or, if it exists) at all the instants of time during any interval under consideration. Examples are the signals given in Eqs. (1.1)–(1.6) above.
- Discrete-Time Signals: A signal is said to be a 'discrete-time signal' if it is defined only at discrete set of instants of time rather than *all* the time instants.

We can generate discrete-time signals from continuous-time signals by, for example, *sampling the signals* at regular instants of time given by kT_s, where k is an integer and T_s is the time duration between any two adjacent samples. Thus, the discrete-time version of the sinusoidal signal in Eq. (1.1) can be given by,

$$\begin{aligned} x[k] &= x(kT_s) = [x(t) \quad \text{at} \quad t = kT_s] \\ &= A\sin(2\pi f_p kT_s + \phi), \\ &\quad \text{for } k = \ldots, -2, -1, 0, 1, 2, 3, \ldots. \end{aligned} \tag{1.7}$$

Figure 1.7 illustrates both the continuous (blue) and discrete-time (red) signals corresponding to the sinusoidal signal defined in Eq. (1.1).

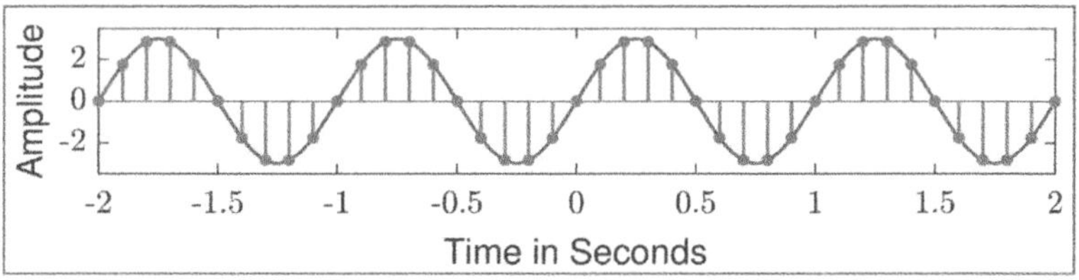

Figure 1.7. Sinusoidal signal (continuous-time and discrete-time).

We use the notation $x[k]$ to represent a discrete-time signal[2] since this signal is defined only at the discrete-time instants, kT_s. Since this notation, $x[k]$, does not explicitly show what the time interval T_s is between adjacent samples, we may provide this information by specifying the *'sample rate'* in the form of *'number of samples per second'*. Thus, if f_s is the sample rate, then we have $T_s = 1/f_s$.

Another example of such a signal is the *discrete-time unit step signal* given by,

$$u[k] = \begin{cases} 1 & \text{for } k = 0, 1, 2, 3, \ldots \\ 0 & \text{for } k < 0 \end{cases} \tag{1.8}$$

§ *Since most of the real-world signals are continuous-time signals, why should we be interested in the study of discrete-time signals?*

§ *Are there real-world signals that are discrete-time in nature?*

- Continuous-Value Signals: A signal is said to be a 'continuous-value signal' if its value is allowed to be anywhere in a given range. For example, the value of the sinusoidal signal in Eq. (1.1) varies between the range $-A$ and $+A$, and the value of the signal in Eq. (1.2) varies between 0 and $+A$. So, these are continuous-value signals.
- Discrete-Value Signals: A signal is said to be a 'discrete-value signal' if its value is restricted to a discrete set of values rather than all the values in a given range. An example for a discrete-value signal can be constructed as follows.

  Let B_k for $k = 0, 1, 2, \ldots$. represents a sequence of binary data. That is, B_k can take only two values and let these values be either '1' or '0'. Thus, this sequence of binary data form a discrete-time signal given by,

$$z[k] = B_k, \tag{1.9}$$

where the signal exists at time instants kT_s (here, $1/T_s$ specifies the *data rate*, i.e., *number of bits per second*) and its value at any such instant is either '1'

2. Other notation used for representing discrete-time signals is x_k.

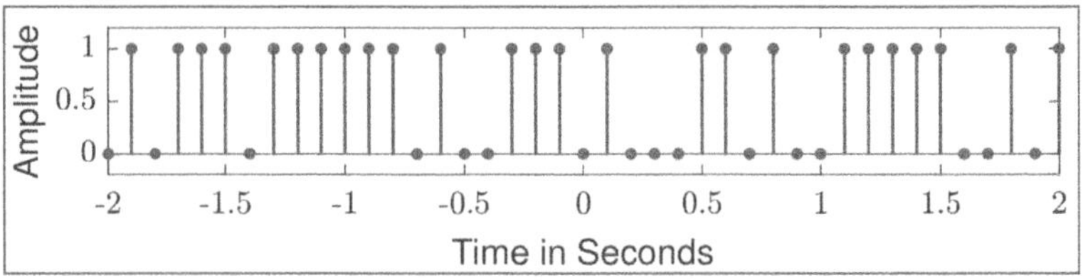

Figure 1.8. Discrete-time discrete-value signals.

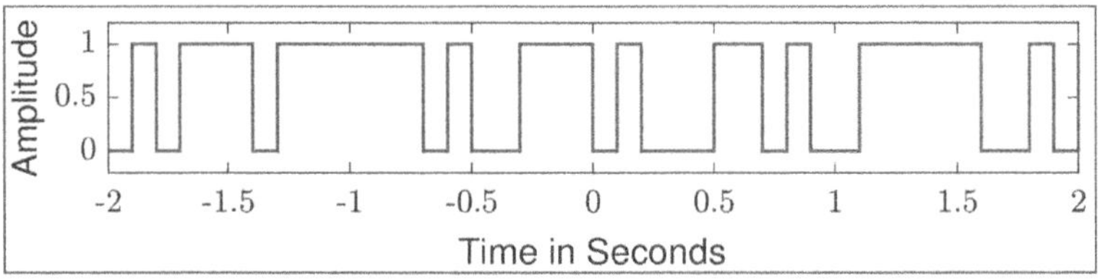

Figure 1.9. Continuous-time discrete-value signals.

or '0'. In other words, $z[k]$ is a *discrete-time discrete-value signal.* An example of such a signal is shown in Fig. 1.8. You may note that values of the signals are either '1' or '0' and it exists at discrete time instances.

- Continuous-time Discrete-value Signals: We can also have *continuous-time discrete-value signals*. For example, we can convert the discrete-time discrete-value signal $z[k]$ in Eq. (1.9) into a continuous-time discrete-value signal as follows:

$$\tilde{z}(t) = 0.9B_k \qquad \text{for } kT_s \leq t < (k+1)T_s. \tag{1.10}$$

Here, the multiplication factor, 0.9 is just an arbitrary scalar constant. In other words, during the time interval $kT_s \leq t < (k+1)T_s$, we get

$$\tilde{z}(t) = \begin{cases} 0.9 & \text{if } B_k = 1, \\ 0 & \text{if } B_k = 0. \end{cases} \tag{1.11}$$

Thus, we see that $\tilde{z}(t)$ is a continuous-time discrete-value signal and it takes only two values '0.9' and '0' at any time instant t. This signal $\tilde{z}(t)$ is illustrated in Fig. 1.9 above.

§ *Let the binary data sequence B_k be given by* [0, 1, 1, 0, 0, 0, 1, 0, 0, 1, 0] *for* $k = 0, 1, 2, \ldots, 10$. *Using the definitions of signals given in Eqs. (1.9) and (1.10), draw the resulting 'discrete-time discrete-value signal' $z(k)$ and 'continuous-time discrete-value signal' $\tilde{z}(t)$.*

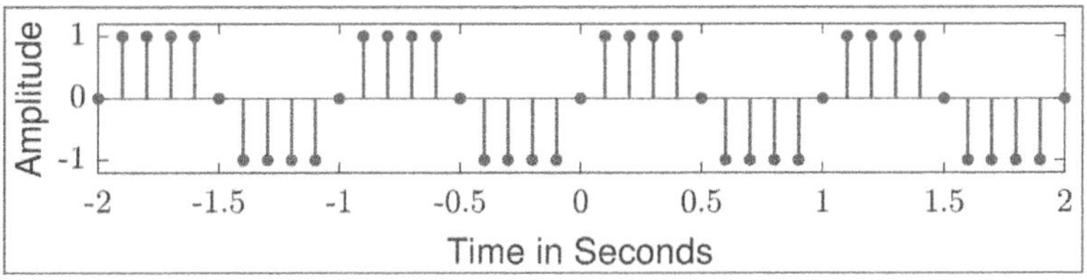

Figure 1.10. Discrete-time discrete-value signals.

- Discrete-time Discrete-value Signals: We can also generate discrete-value signals through the process of *quantization*. For example, consider the discrete-time continuous-value sinusoidal signal $x[k]$ given in Eq. (1.7). We can convert this into a discrete-value signal using the following quantization operation, assuming the amplitude to be $A = 3$:

$$\tilde{x}[k] = \begin{cases} -1 & \text{if } -3 \leq x[k] < -1 \\ 0 & \text{if } -1 \leq x[k] < +1 \\ +1 & \text{if } +1 \leq x[k] \leq +3 \end{cases} \tag{1.12}$$

Thus, we see that $\tilde{x}[k]$ is a *discrete-time discrete-value signal* and it takes only one of the three values $\{-1, 0, 1\}$ at any time instant kT_s (or, *time index* k). The corresponding signal is illustrated in Fig. 1.10.

In general, *continuous-time continuous-value signals are also known as* analog signals and *discrete-time discrete-value signals are known as* digital signals.

§ *Examine the kind of signal processing interface you will need to set-up to store your speech signal in a computer.*

1.2.2 Periodic and Aperiodic Signals

A periodic signal repeats itself with some period T_p in the time domain. That is, if $x(t)$ is a periodic signal, it must obey the following relation for any integer k:

$$x(t + kT_p) = x(t) \qquad \text{for all } t. \tag{1.13}$$

The sinusoidal signal given in Eq. (1.1) is an example of a periodic signal. Its period is given by $T_p = 1/f_p$.

A periodic signal can be completely specified by specifying its period and defining the signal during one period. For example, a *triangular periodic signal* of period 2 seconds can be defined by defining one period during the time interval $0 \leq t < 2$ as follows:

$$x(t) = \begin{cases} t & \text{for } 0 \leq t < +1, \\ 2 - t & \text{for } +1 \leq t < +2. \end{cases} \tag{1.14}$$

§ *Plot the triangular periodic signal defined in Eq. (1.14).*

The signals that are not periodic are called *non-periodic or aperiodic signals*. The exponentially decaying signal considered in Eq. (1.2) and the signal $\tilde{z}(t)$ given by Eq. (1.10) are examples of aperiodic signals.

1.2.3 Deterministic and Random Signals

Deterministic signals are the ones whose values can be predicted exactly, and there is no uncertainty about the value of these signals at any given time. Such signals can be usually described by mathematical expressions. For example, the sinusoidal signal given by Eq. (1.1), the exponentially decaying signal given by Eq. (1.2) and the triangular periodic signal given by Eq. (1.14) are examples of deterministic signals.

A random signal is one whose values cannot be predicted exactly because of some uncertainty caused by physical phenomena. We may not be able to exactly describe random signals using mathematical expressions. Some examples of random signals are as follows.

- The noise that is picked up by a radio/TV receiver.
- The output of a coin-tossing experiment.
- The signal received by a mobile phone is random due to the random (and time-varying) nature of the channel.
- The number of passengers arriving at any given time at an international airport.

1.2.4 Real and Complex Signals

A *real signal* is one whose values are real numbers. All the signals we considered so far in this chapter are real signals.

A *complex signal* is one whose values are complex numbers. A typical example is the *complex sinusoid* defined as,

$$x(t) = Ae^{j(2\pi f_0 t+\phi)}, \tag{1.15}$$

where $j = \sqrt{-1}$. The parameters A, f_0, and ϕ denote the amplitude, frequency and phase, respectively, of the complex sinusoid. Using the Euler's identity

$$e^{jz} = \cos(z) + j\sin(z), \tag{1.16}$$

we can express the complex sinusoid in Eq. (1.15) in the form

$$x(t) = A\cos(2\pi f_0 t + \phi) + jA\sin(2\pi f_0 t + \phi). \tag{1.17}$$

On the other hand, using this identity, we can express real sinusoids as[3]

$$A\cos(2\pi f_0 t + \phi) = A\left(\frac{e^{j(2\pi f_0 t+\phi)} + e^{-j(2\pi f_0 t+\phi)}}{2}\right), \tag{1.18}$$

$$A\sin(2\pi f_0 t + \phi) = A\left(\frac{e^{j(2\pi f_0 t+\phi)} - e^{-j(2\pi f_0 t+\phi)}}{2j}\right). \tag{1.19}$$

Noting that $e^{-j(2\pi f_0 t+\phi)} = e^{j[2\pi(-f_0)t-\phi]}$ and $1/j = -j = e^{-j\pi/2}$, we can conclude the following:

> § *A real 'cos' sinusoid of amplitude* A*, frequency* f_0 *and phase* ϕ *can be obtained by summing two complex sinusoids of amplitudes* $\{A/2, A/2\}$*, frequencies* $\{f_0, -f_0\}$ *and phases* $\{\phi, -\phi\}$*. For the same amplitudes and frequencies, we get a real 'sin' sinusoid if the phases are* $\{\phi - \pi/2, -\phi + \pi/2\}$*. Thus, complex sinusoids act as building blocks for real signals. This means that real signals can be constructed by combining complex sinusoids.*

Different views of the complex sinusoid in Eq. (1.15) with $\phi = 0$ is shown in Fig. 1.11. In the figure, $a(t)$ is the instantaneous magnitude, and $\theta(t)$ is the instantaneous angle/phase of the complex sinusoid. The real component, $A\cos(2\pi f_0 t)$, is the projection of the complex sinusoid onto the real axis and the imaginary component, $A\sin(2\pi f_0 t)$, is the projection of the complex sinusoid onto the imaginary axis in a complex plane. This is illustrated in the 3D view of the complex sinusoid (phasor) in the figure.

1.3 Typical Real World Biomedical Signals

There are hundreds or thousands of types of biomedical signals, that one can think of, from a living organism. In this section, however, some of the most commonly referred types of biomedical signals are briefly described.

1.3.1 Electroencephalogram (EEG)

It is a recording of the electrical activity of the brain by placing electrodes on the scalp. The recorded waveforms reflect the cortical electrical activity generated by the firing of brain cells neurons. The amplitude of EEG activities are quite small, measured in microvolts to millivolts. The main frequency contents of EEG waveforms are Delta (0.5–4 Hz), Theta (4–8 Hz), Alpha (8–12 Hz), and Beta (12–35 Hz).

3. Note: $\cos(z) = \frac{[e^{jz}+e^{-jz}]}{2}$ and $\sin(z) = \frac{[e^{jz}-e^{-jz}]}{2j}$.

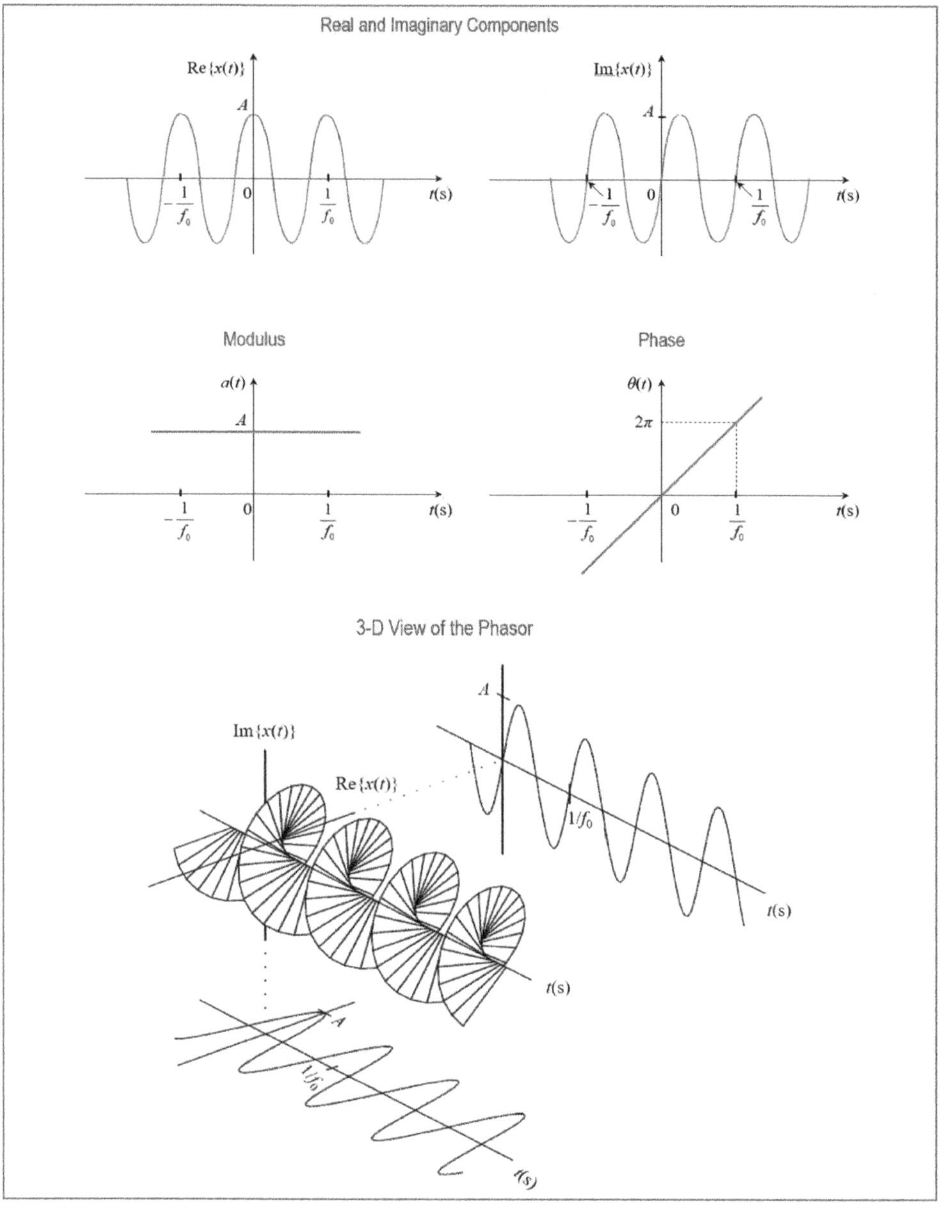

Figure 1.11. Different views of the phasor $Ae^{j2\pi f_0 t}$.

Delta waves: It tends to be the highest in amplitude and the lowest in frequency. It is the dominant rhythm in infants and in stages 3 and 4 of sleep in adults. It may occur focally with subcortical lesions and in general distribution with diffuse lesions, metabolic encephalopathy hydrocephalus or deep midline lesions. It is

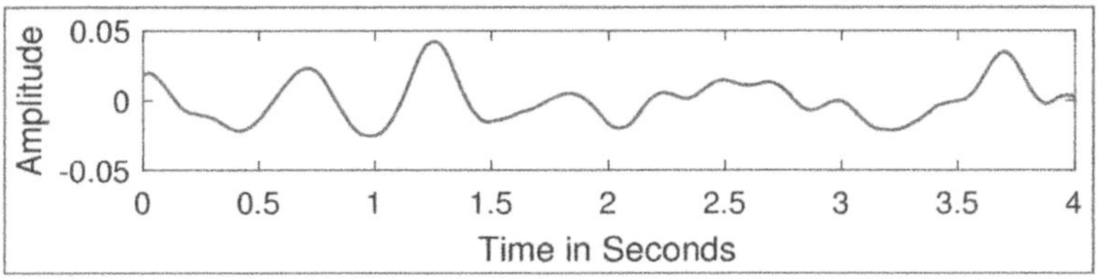

Figure 1.12. EEG delta wave.

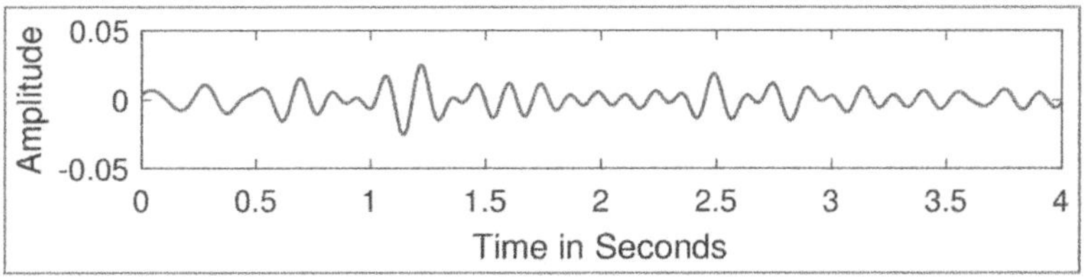

Figure 1.13. EEG theta wave.

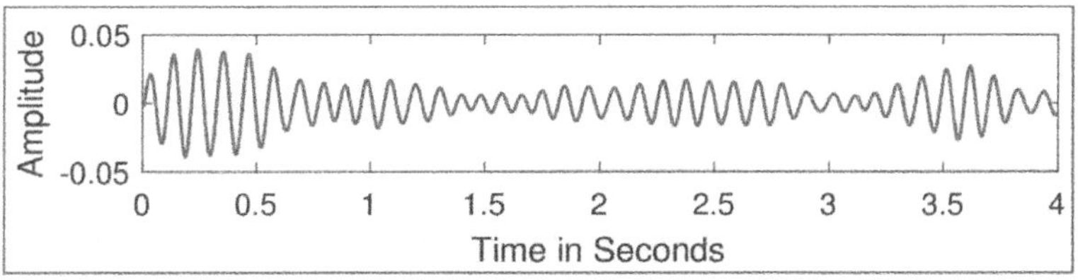

Figure 1.14. EEG alpha wave.

usually most prominent frontally in adults and posteriorly in children. A delta wave segment for four seconds is shown in Fig. 2.1.

Theta waves: It is classified as "slow" EEG activity. It is perfectly normal in children and in sleep but abnormal in awake adults. It can be seen as a manifestation of focal subcortical lesions. It can also be seen in generalized distribution in diffuse disorders such as metabolic encephalopathy or some instances of hydrocephalus. In Fig. 1.13, a small segment of a typical theta wave is provided.

Alpha waves: Figure 1.14 illustrates a segment of alpha EEG waves. Is usually best seen in the posterior regions of the head on each side, being higher in amplitude on the dominant side. It appears when closing the eyes and relaxing, and disappears when opening the eyes or alerting by any mechanism (thinking, calculating). It is the major rhythm seen in normal relaxed adults. It is present during most of life especially during teenage years and after.

Beta waves: As you may see from Fig. 1.15, it is the "fast" EEG activity. It has higher frequency and lower amplitude. It is usually seen on both sides of the brain in symmetrical distribution and is most evident frontally. It may be absent or reduced

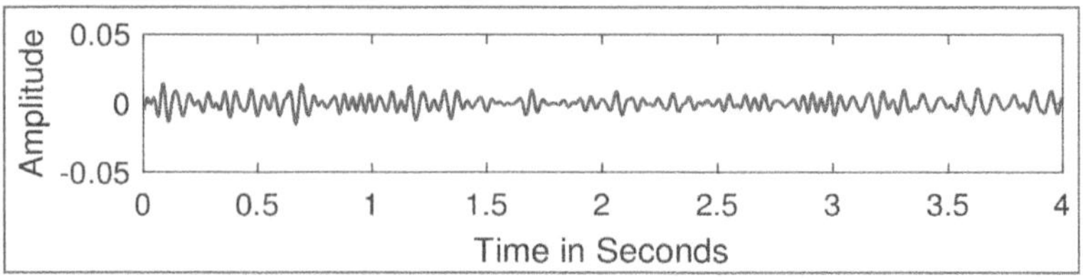

Figure 1.15. EEG beta wave.

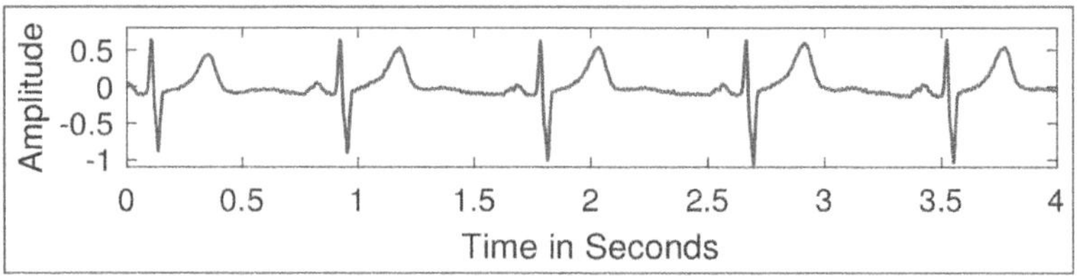

Figure 1.16. ECG signal.

in areas of cortical damage. It is generally regarded as a normal rhythm. It is the dominant rhythm in patients who are alert or anxious or have their eyes open.

1.3.2 Electrocardiogram (ECG)

It is the recording of the electrical activities of the heart based on skin surface electrode measurements. An example of a typical ECG waveform for a normal adult is shown in Fig. 1.16. ECG offers a low-cost non-invasive investigation of the heart, commonly used for the detection of heart arrhythmia and conduction disturbances. It can also be used to to investigate the heart morphology to discover hypertrophy or myocardial infraction.

1.3.3 Electrooculogram (EOG)

It is a technique for measuring the corneo-retinal standing potential that exists between the front and the back of the human eye. Primary applications are in ophthalmological diagnosis and in recording eye movements.

To measure eye movements, pairs of electrodes are typically placed either above and below the eye or to the left and right of the eye. If the eye moves from centre position towards one of the two electrodes, this electrode "sees" the positive side of the retina and the opposite electrode "sees" the negative side of the retina. Consequently, a potential difference occurs between the electrodes. Assuming that the resting potential is constant, the recorded potential is a measure of the eye's position. An example of eye-blink EOGs for ten seconds long is illustrated in Fig. 1.17.

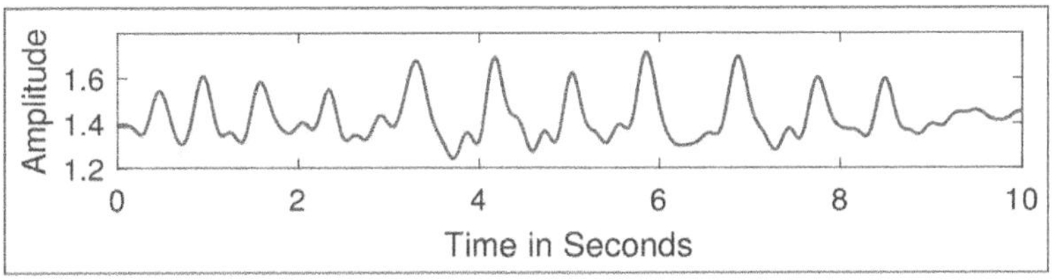

Figure 1.17. EOG (eye blinks) signal.

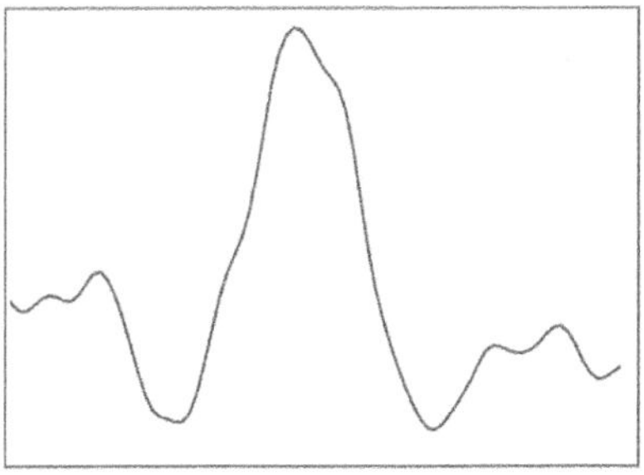

Figure 1.18. A typical ERG signal.

1.3.4 Electroretinogram (ERG)

It measures the electrical responses of various cell types in the retina, including the photo receptors, inner retinal cells, and the ganglion cells. Electrodes are placed on the surface of the cornea or on the skin beneath the eye to measure retinal responses. Retinal pigment epithelium (RPE) responses are measured with an EOG test with skin-contact electrodes placed near the canthi. During a recording, the patient's eyes are exposed to standardized stimuli and the resulting signal is displayed showing the time course of the signal's amplitude (voltage). Signals are very small, and their amplitudes are in the range of typically nano volts to micro volts. The ERG is composed of electrical potentials contributed by different cell types within the retina, and the stimulus conditions (flash or pattern stimulus, whether a background light is present, and the colours of the stimulus and background) can elicit stronger response from certain components. A typical ERG signal is shown in Fig. 1.18.

1.3.5 Electromyogram (EMG)

This is an electrodiagnostic technique for evaluating and recording the electrical activity produced by skeletal muscles. An EMG detects the electric potential generated by muscle cells when these cells are electrically or neurologically stimulated. These signals can be analysed to detect medical abnormalities, activation level, or recruitment order, or to analyse the biomechanics of human or animal movements. In computer science, EMG is also used as middle-ware in gesture recognition

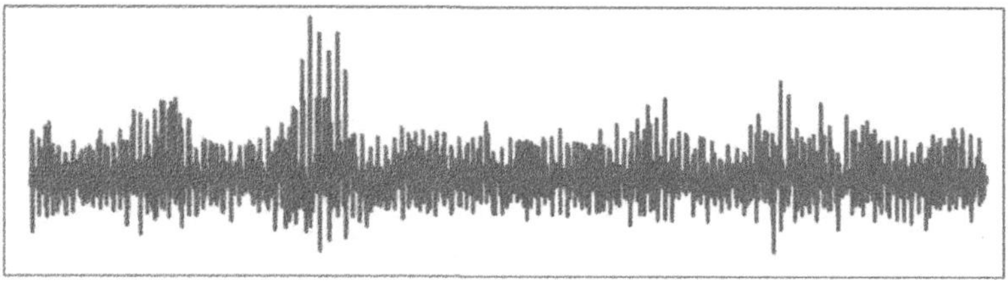

Figure 1.19. A typical EMG signal segment.

towards allowing the input of physical action to a computer as a form of human-computer interaction. Figure 1.19 shows an illustration of a typical EMG signal.

As mentioned earlier, the most important characteristics of a signal are the power, frequency and phase. In the next chapter, the definition of power and energy of signals and the ways to estimate them are briefly described.

1.4 Concluding Remarks

The following points can be deduced from the discussions we have had in this chapter.

- Definitions of some of the most common types of signals were provided with illustrations.
- Ways of classifying signals were provided.
- Most of the *practical real-world signals* are aperiodic and random in nature.
- *Complex sinusoids act as building blocks* in the study of real signals.
- Examples of typical biomedical signals were provided.

DOI: 10.1561/9781680839791.ch2

Chapter 2

Power and Energy

According to physics, power is defined as the rate of doing work, it is the work done in unit time. Energy, on the other hand, is a quantitative property that must be transferred to an object in order for it to perform some work. Hence, energy is defined as the strength to do any kind of physical activity. Thus, energy is the ability to do work. Here, in this chapter, the following points are discussed in detail:

- Definition and meaning of energy and power for signals, and computations of these quantities for typical examples of signals.
- Detailed understanding of sinusoidal (real and complex) signals.

2.1 Power and Energy of Signals

Consider the simple electric circuit shown in Fig. 2.1 below. Here a resistance of $R\ \Omega$ is connected to a voltage source $x(t)$. A current flows through the circuit, which makes the resistance to dissipate some power in the form of heat.

Here, $x(t)$ is the voltage source connected to a resistor of resistance $R\ \Omega$. Then, we know that, the instantaneous power $(p(t))$ delivered to the resistor is given by $p(t) = \frac{|x(t)|^2}{R}$. If $R = 1\ \Omega$, we get the *instantaneous power of the signal* $x(t)$ *at time* t as,

$$p(t) = |x(t)|^2. \tag{2.1}$$

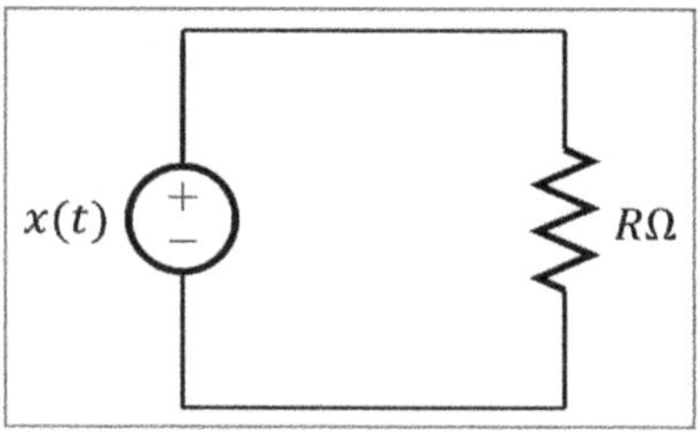

Figure 2.1. A resistor connected to a voltage source.

For real-valued signals, the instantaneous power is given by $x^2(t)$. However, in Eq. (2.1), we use $|x(t)|^2$ (instead of $x^2(t)$) so as to account for complex-valued signals. Since *power is energy per unit time*, we get the *energy (e) of the signal $x(t)$ over the time interval $t_1 < t < t_2$* as,

$$e_{t_1 \to t_2} = \int_{t_1}^{t_2} p(t)dt = \int_{t_1}^{t_2} |x(t)|^2 dt. \tag{2.2}$$

Consequently, we get the *average power of the signal $x(t)$ over the time interval $t_1 < t < t_2$* as

$$p_{t_1 \to t_2} = \frac{e_{t_1 \to t_2}}{t_2 - t_1} = \frac{1}{t_2 - t_1} \int_{t_1}^{t_2} |x(t)|^2 dt. \tag{2.3}$$

Extending the definition in Eq. (2.2) to the complete time interval $-\infty < t < \infty$, we can get the *total energy (or, energy) of the signal $x(t)$* as,

$$e = \lim_{T \to \infty} \int_{-T/2}^{T/2} |x(t)|^2 dt. \tag{2.4}$$

Here, for the sake of convenience and without losing generality, we have used $t_1 = -\frac{T}{2}$ and $t_2 = \frac{T}{2}$, where T denotes an arbitrary time interval. Similarly, we can get the *average power (or, power) of the signal $x(t)$* as

$$p = \lim_{T \to \infty} \frac{1}{T} \int_{-T/2}^{T/2} |x(t)|^2 dt. \tag{2.5}$$

§ *If a signal $x(t)$ is periodic with period T_p, show that its power can be computed using one of the following expressions,*

$$p = \begin{cases} \frac{1}{T_p}\int_0^{T_p} |x(t)|^2 dt, \ \ or \\ \\ \frac{1}{T_p}\int_{-T_p/2}^{T_p/2} |x(t)|^2 dt, \ \ or \\ \\ \frac{1}{T_p}\int_{t_0}^{t_0+T_p} |x(t)|^2 dt. \end{cases} \tag{2.6}$$

and it is equivalent to Eq. (2.5). Here, t_0 is any arbitrary time point. Show also that its energy is infinity.

§ *Examine the power and energy of the signals given in Eqs. (1.1)–(1.6).*

2.2 Energy Signals and Power Signals

Signals are also classified according to their energy and power. *Signals with non-zero finite power are known as power signals.* Similarly, *signals with non-zero finite energy are known as energy signals.*

§ *Show that the energy of a power signal should be infinity, and the power of an energy signal should be zero.*

§ *Can a signal be simultaneously a power signal and energy signal?*

§ *Classify the signals given in Eqs. (1.1)–(1.6) into energy or power signals.*

We shall now use the expressions for the power and energy defined earlier to determine the power and energy of real and complex sinusoidal signals.

2.2.1 Single Real Sinusoid

Let us first consider the case of a real sinusoid given by,

$$x(t) = A\cos(2\pi f_0 t + \phi).$$

where, A, f_0, and ϕ, are the amplitude, frequency and phase, respectively of the sinusoid. We get the instantaneous power as,

$$\begin{aligned} p(t) = |x(t)|^2 &= A^2\cos^2(2\pi f_0 t + \phi), \\ &= \frac{A^2}{2}[1 + \cos(4\pi f_0 t + 2\phi)], \end{aligned} \tag{2.7}$$

as $\cos^2(z) = \frac{1}{2}[1 + \cos(2z)]$. Without loss of generality, we can take the time interval T in Eqs. (2.4) and (2.5) as $T = mT_0$ where m is an integer and $T_0 = \frac{1}{f_0}$ is the period of the sinusoid. Since $\cos(4\pi f_0 t + 2\phi)$ is periodic with period $T_0/2$, and the positive and negative half-cycles of $\cos(4\pi f_0 t + 2\phi)$ have equal areas, we get

$$\int_{-T/2}^{T/2} \cos(4\pi f_0 t + 2\phi)dt = 0. \tag{2.8}$$

To understand Eq. (2.8), let $T = mT_0$. Then, we get (since $f_0 T_0 = 1$)

$$\begin{aligned}
\int_{-T/2}^{T/2} \cos(4\pi f_0 t + 2\phi)dt &= \frac{\sin(4\pi f_0 t + 2\phi)}{4\pi f_0}\Bigg|_{-mT_0/2}^{mT_0/2}, \\
&= \alpha \sin(2\pi f_0 m T_0 + 2\phi) - \alpha \sin(-2\pi f_0 m T_0 + 2\phi), \\
&= \alpha \sin(2m\pi + 2\phi) + \alpha \sin(2m\pi - 2\phi), \\
&= \alpha \sin(2\phi) - \alpha \sin(2\phi) = 0,
\end{aligned}$$

where $\alpha = \frac{1}{4\pi f_0}$. This can also be easily seen (intuitively) by the fact that the areas over the positive half-cycles and negative half-cycles are equal for the function $\cos(4\pi f_0 t + 2\phi)$ over an interval that is an integer multiple of its period.

Therefore, we get *the power and energy of a real sinusoid* of amplitude A (any non-zero frequency and any phase) as

$$p = \lim_{T\to\infty} \frac{1}{T}\int_{-T/2}^{T/2} \frac{A^2}{2}dt = \frac{A^2}{2} \tag{2.9}$$

$$e = \lim_{T\to\infty} \int_{-T/2}^{T/2} \frac{A^2}{2}dt = \infty. \tag{2.10}$$

§ *Show that the power and energy of the signal* $x(t) = A\sin(2\pi f_0 t + \phi)$ *are also given by Eqs. (2.9) and (2.10).*

2.2.2 Single Complex Sinusoid

Next, let us consider the case of a complex sinusoid given by

$$x(t) = Ae^{j(2\pi f_0 t + \phi)}. \tag{2.11}$$

We get the instantaneous power as

$$p(t) = |x(t)|^2 = A^2 \tag{2.12}$$

as $\left|e^{jz}\right|^2 = \cos^2(z) + \sin^2(z) = 1$. Therefore, we get the power and energy of a complex sinusoid of amplitude A (any frequency and any phase) as

$$p = \lim_{T\to\infty} \frac{1}{T} \int_{-T/2}^{T/2} A^2 dt = A^2 \tag{2.13}$$

$$e = \lim_{T\to\infty} \int_{-T/2}^{T/2} A^2 dt = \infty. \tag{2.14}$$

2.2.3 Sum of Complex Sinusoids

Let us consider the signal obtained by summing two complex sinusoids as,

$$x(t) = A_1 e^{j(2\pi f_1 t+\phi_1)} + A_2 e^{j(2\pi f_2 t+\phi_2)}, \tag{2.15}$$

where $\{A_1, A_2\}$, $\{f_1, f_2\}$ and $\{\phi_1, \phi_2\}$ represent the amplitudes, frequencies and phases, respectively, of the two sinusoids. Then, we get the instantaneous power as,[1]

$$\begin{aligned} p(t) &= |x(t)|^2 = x(t)x^*(t) \\ &= \left[A_1 e^{j(2\pi f_1 t+\phi_1)} + A_2 e^{j(2\pi f_2 t+\phi_2)}\right] \\ &\quad \times \left[A_1 e^{-j(2\pi f_1 t+\phi_1)} + A_2 e^{-j(2\pi f_2 t+\phi_2)}\right] \\ &= A_1^2 + A_1 A_2 \left[e^{j(2\pi(f_1-f_2)t+\phi_1-\phi_2)} + e^{-j(2\pi(f_1-f_2)t+\phi_1-\phi_2)}\right] + A_2^2 \\ &= A_1^2 + 2A_1A_2\cos(2\pi(f_1 - f_2)t + \phi_1 - \phi_2) + A_2^2. \end{aligned} \tag{2.16}$$

Note that $\cos(2\pi(f_1 - f_2)t + \phi_1 - \phi_2)$ is periodic with period $T_0 = \frac{1}{|f_1-f_2|}$. Therefore, by choosing $T = mT_0$ (m being an integer), we get,

$$\int_{-T/2}^{T/2} \cos(2\pi(f_1 - f_2)t + \phi_1 - \phi_2)dt = 0. \tag{2.17}$$

Therefore, we get *the power of the sum of two complex sinusoids of different frequencies* $\{f_1, f_2\}$ *and amplitudes* $\{A_1, A_2\}$ (and any phase) as,

$$p = \lim_{T\to\infty} \frac{1}{T} \int_{-T/2}^{T/2} A^2 dt = A_1^2 + A_2^2. \tag{2.18}$$

As before, the corresponding energy of $x(t)$ is infinity.

---

1. Here, $x^*(t)$ denotes complex conjugate of $x(t)$.

§ *Examine the power of the sum of two complex sinusoids given in Eq. (2.15) in the case where* $f_1 = f_2$. *Assume that* $A_1 \neq A_2$ *and* $\phi_1 \neq \phi_2$.
§ *Express a real sinusoid of frequency* f_0 *as the sum of two complex sinusoids with frequencies* f_0 *and* $-f_0$. *Then, using Eq. (2.16), show that the power of a real sinusoid of amplitude* A *is equal to* $\frac{A^2}{2}$.

We can easily *generalize the above results to determine the power of a signal consisting of the sum of* M *complex sinusoids* with different frequencies. Let the signal be given by,

$$x(t) = \sum_{i=1}^{M} A_i e^{j(2\pi f_i t + \phi_i)}, \tag{2.19}$$

where A_i, f_i and ϕ_i represent the amplitude, frequency and phase, respectively, of the ith sinusoid (for $i = 1, 2, \ldots, M$). Then, we get the power of $x(t)$ as,

$$p = \sum_{i=1}^{M} A_i^2 = A_1^2 + A_2^2 + \cdots + A_M^2. \tag{2.20}$$

2.2.4 Sum of Real Sinusoids

Consider the signal obtained by summing two real sinusoids as

$$x(t) = A_1 \cos(2\pi f_1 t + \phi_1) + A_2 \cos(2\pi f_2 t + \phi_2), \tag{2.21}$$

where $\{A_1, A_2\}$, $\{f_1, f_2\}$ and $\{\phi_1, \phi_2\}$ represent the amplitudes, frequencies and phases, respectively, of the two sinusoids, with $f_1 \neq f_2$. Using Euler's identity, we can write $x(t)$ as,

$$\begin{aligned} x(t) &= \alpha_1 e^{j(2\pi f_1 t + \phi_1)} + \alpha_1 e^{j(2\pi(-f_1)t - \phi_1)}, \\ &\quad + \alpha_2 e^{j(2\pi f_2 t + \phi_2)} + \alpha_2 e^{j(2\pi(-f_2)t - \phi_2)}, \end{aligned} \tag{2.22}$$

where $\alpha_1 = A_1/2$ and $\alpha_2 = A_2/2$. Since f_1, $-f_1$, f_2 and $-f_2$ are distinct frequencies, we can use Eq. (2.20) to obtain the power of $x(t)$. Thus, we get *the power of the sum of two real sinusoids of different frequencies* $\{f_1, f_2\}$ and amplitudes $\{A_1, A_2\}$ (and any phase) as,

$$p = \alpha_1^2 + \alpha_1^2 + \alpha_2^2 + \alpha_2^2 = \frac{A_1^2}{2} + \frac{A_2^2}{2}. \tag{2.23}$$

As before, we can generalize the above results to determine the power of a signal consisting of the sum of M real sinusoids with different frequencies. Let the signal be given by,

$$x(t) = \sum_{i=1}^{M} A_i \cos(2\pi f_i t + \phi_i) \tag{2.24}$$

where A_i, f_i and ϕ_i represent the amplitude, frequency and phase, respectively, of the ith sinusoid (for $i = 1, 2, \ldots, M$). Then, we get the power of $x(t)$ as,

$$p = \sum_{i=1}^{M} \frac{A_i^2}{2} = \frac{A_1^2}{2} + \frac{A_2^2}{2} + \cdots + \frac{A_M^2}{2}. \tag{2.25}$$

§ *Derive Eq. (2.23) without using Euler's identity.*[2]

An important step in the study and/or analysis of signals is to express the signals in the form of sum of many complex sinusoids and to identify which frequencies have dominant contribution. This task is called the *spectrum estimation*. The fundamental tool that facilitates spectrum estimation is *Fourier transform* and its variants. Fourier series – the variant of Fourier transform when applied to periodic signals is discussed in the next chapter.

2.3 Concluding Remarks

Based on the discussions above, the following conclusions can be derived:

- The *power of a signal consisting of a sum of sinusoids* (real or complex) is given by the sum of the powers of the individual sinusoids, if the frequencies of these sinusoids are different.
- The *phases of the sinusoids* have no effect on the power if the frequencies of the sinusoids are different. If the frequencies of any two sinusoids are equal, then their phases will have an effect on the total power [11].

2. For this, first show that the instantaneous power of $x(t)$ can be obtained as,

$$p(t) = A_1^2 \frac{1 + \cos(4\pi f_1 t + 2\phi_1)}{2} + A_2^2 \frac{1 + \cos(4\pi f_2 t + 2\phi_2)}{2} + A_1 A_2 \left[\cos(2\pi(f_1 + f_2)t + \tilde{\phi}_1) + \cos(2\pi(f_1 - f_2)t + \tilde{\phi}_2)\right], \tag{2.26}$$

where $\tilde{\phi}_1 = \phi_1 + \phi_2$ and $\tilde{\phi}_2 = \phi_1 - \phi_2$. In doing this derivation, the following identities will be useful: $2\cos(A)\cos(B) = \cos(A + B) + \cos(A - B)$ and $2\sin(A)\cos(B) = \sin(A + B) + \sin(A - B)$. Then, using an appropriate choice of integration interval T, argue that the integration of all the 'cos(·)' terms in Eq. (2.26) will become zero.

Exercises

2.1. The delta function $\delta(t)$ in Fig. P2.1 is a "generalized" function and is frequently used in signal processing. Find its energy.

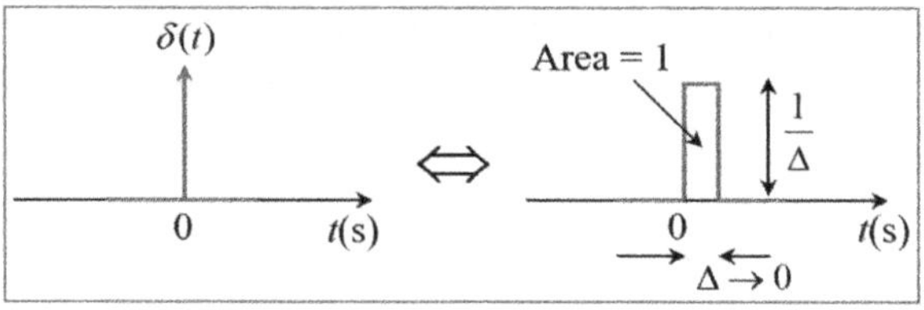

Figure P2.1. Delta function.

2.2. The process of modulation shown in Fig. P2.2 below is frequently used to "up convert" the frequency of a signal to a higher value so that it can be more efficiently transmitted over a channel. Are the energy of $x(t)$ and $y(t)$ the same?

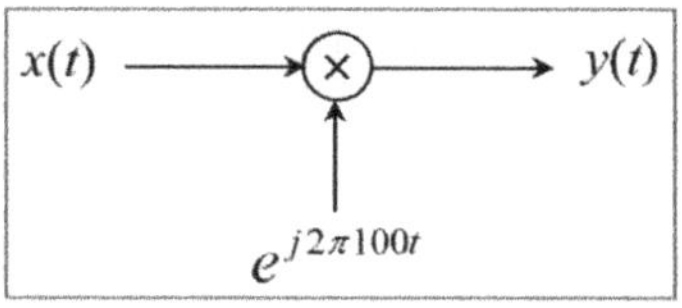

Figure P2.2. Modulation scheme.

2.3. Convert the following complex numbers into polar form, that is, in the form of $s = \rho e^{j\phi}$. Here, ρ is the magnitude of s and ϕ is the phase of s.

(a) $s_1 = 4 + j5$
(b) $s_2(t) = a(t) + jb(t)$
(c) $s_3(t) = 10 \cos\left(20\pi t + \frac{\pi}{4}\right)$

2.4. Convert the following complex numbers into rectangular form, that is, in the form of $s = x + jy$. Here, x is the real part of s and y is the imaginary part of s.

(a) $s_1 = 2e^{j8}$
(b) $s_2 = 2ke^{j\pi k/3}$
(c) $s_3(t) = \Re\left[10e^{j(20t+5)}\right]$, where $\Re$ denotes the real part of a complex number.

2.5. Determine whether or not each of the following signals is periodic. If the signal is periodic, determine its fundamental period.

(a) $x_1(t) = \cos(4t) + \sin(6t)$

(b) $x_2(t) = \cos t + \sin\left(\sqrt{2}t\right)$

(c) $x_3(t) = e^{j[(\pi/2)t-1]}$

2.6. Determine whether the following signals are energy signals, power signals or neither.

(a) $x_1(t) = e^{-at}u(t),$ where $a > 0$ and $u(t)$ is the unit step function

(b) $x_2(t) = tu(t).$

(c) $x_3(t) = a\cos(\omega_0 t + \phi)$

DOI: 10.1561/9781680839791.ch3

Chapter 3

Fourier Series

The study of signals and systems using sinusoidal representations is known as *Fourier analysis* (named after the French mathematician Jean Baptiste Joseph Fourier, who developed Fourier series). In this chapter, we discuss *the representation of periodic signals in terms of sinusoids*. This lays the foundation for *spectrum estimation*, which is one of the most important topics in the study and applications of signals and systems. Since each sinusoid corresponds to a single frequency, the *representation of signals using sinusoids leads us to frequency-domain understanding and interpretation of signals.*

Frequency-domain representation of signals is very necessary as well as useful in practical applications and theoretical studies. For example, we should know what are the frequencies present in a signal, before we can design a communication system to reliably transmit this signal. An important advantage of frequency-domain representation is that it is often easy to see certain signal characteristics when we examine the signal in this domain, compared to the time-domain. Figure 3.1 illustrates this for a signal consisting of sinusoids and noise.

In Fig. 3.1, the signal, $x(t)$, is synthesized by combining three sinusoids of different amplitudes and frequencies along with white Gaussian noise[1] as shown in

1. Will be described in a later chapter of this book.

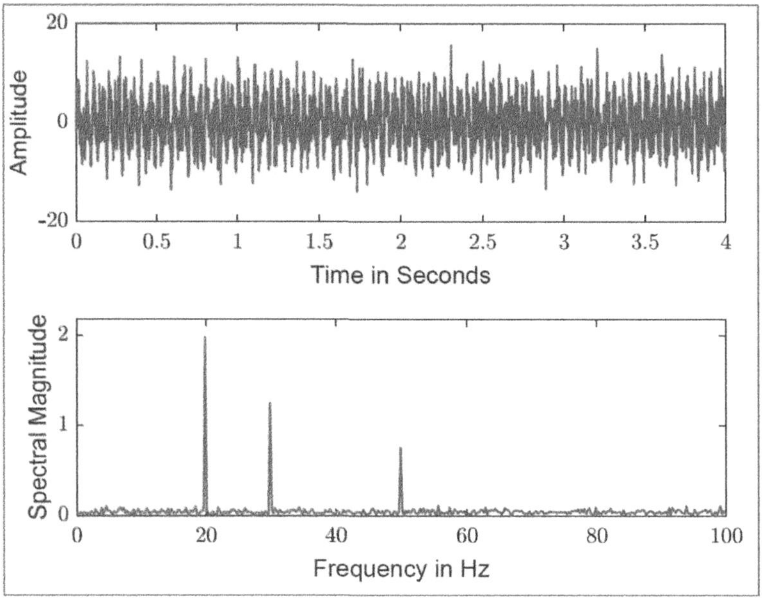

Figure 3.1. Time plot of sum of three real sinusoids (at frequencies 20 Hz, 30 Hz, 50 Hz) and noise (top) and its frequency spectrum (bottom) in the frequency domain.

the following equation:

$$x(t) = A_1 \sin(2\pi f_1 t) + A_2 \sin(2\pi f_2 t) + A_3 \sin(2\pi f_3 t) + n(t), \tag{3.1}$$

where A_1, A_2, A_3 are respectively 4, 2.5 and 1.5. The corresponding three frequencies, f_1, f_2, f_3 are 20Hz, 30Hz and 50Hz. $n(t)$ in Eq. (3.1) is the white Gaussian noise added to the sinusoids. The signal (top; time domain) and its frequency spectrum[2] (bottom; frequency domain) are shown in Fig. 3.1. It can be observed that while it is not possible to infer either the number of sinusoids or their frequencies from the time-domain plot, the frequency-domain spectrum clearly conveys this information. The relative amplitudes of the sinusoids are also clearly visible from the frequency domain spectrum. The background noise spectrum also can be clearly seen in the frequency domain and it spans the complete frequency range. This illustration clearly explains the importance of Fourier representation of signals.

In this chapter, the following points will be discussed in detail.

- The role of sinusoids as basic building block for synthesizing signals.
- Representation of periodic signals using sinusoids (real or complex), leading to the concept of *frequency-domain representation of signals.*

2. Distribution of frequencies in the signal.

- The *definition and interpretation of spectrum*, and interpreting signals based on their spectra.
- Various *properties of Fourier series representation*, and their use in practice.
- *Limitations* of Fourier series representation.

3.1 Introduction to Fourier Series

Consider the vector $\vec{r}$ shown in Fig. 3.2. It connects the origin $O(0, 0)$ to the point $P(x, y)$, in the $X - Y$ plane. Here, x and y are the X-coordinate and Y-coordinate of the point P. The vector $\vec{r}$ can be expressed as,

$$\vec{r} = x \cdot \vec{x}_u + y \cdot \vec{y}_u, \tag{3.2}$$

where $\vec{x}_u$ and $\vec{y}_u$ are unit vectors (i.e., vectors of length 1.0) along X and Y axes, respectively. In other words, given any vector $\vec{r}$ connecting the origin $O(0, 0)$ to a point '$P(x, y)$' in the $X - Y$ plane, we can *express this vector as a sum of scaled unit vectors along the X and Y axes*; the scale factors (i.e., x and y) can be obtained by determining the components of $\vec{r}$ along X and Y axes. Thus, we are able to *represent all vectors in the $X - Y$ plane in terms of the two unit vectors $\vec{x}_u$ and $\vec{y}_u$.*

§ *A special feature of this representation is that the vectors $\vec{x}_u$ and $\vec{y}_u$, which we have chosen as fundamental components, are perpendicular (or, orthogonal) to each other.*

This representation is convenient because *manipulation and interpretation of vectors can now be done in terms of the standard vectors $\vec{x}_u$ and $\vec{y}_u$, which are easy to handle and interpret.*

Fourier series does a representation of the type explained above for the case of periodic signals. In place of the unit vectors $\vec{x}_u$ and $\vec{y}_u$, Fourier series uses complex

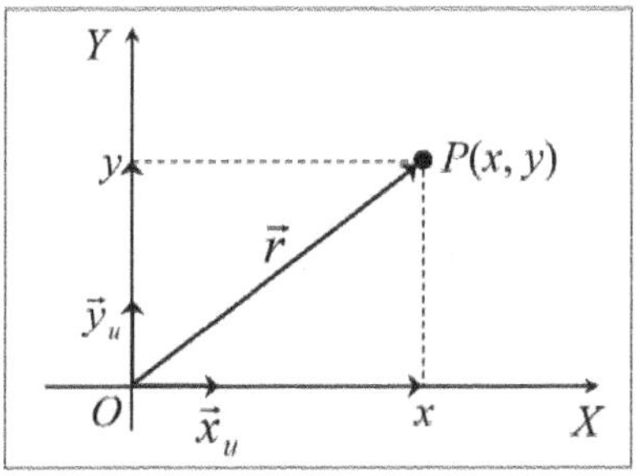

Figure 3.2. Vector $\vec{r}$ in rectangular coordinates representation.

sinusoids as the basic/fundamental signals for representing periodic signals.[3] Since a complex sinusoid $e^{j\Omega_p t}$ is associated with a specific frequency Ω_p ($2\pi f_p$; f_p is the cyclic frequency of the sinusoid), we can generate any arbitrary signal by adding many complex sinusoids of different frequencies, with appropriate amplitude and phase for each sinusoid. In particular, we note the following:

§ *For a periodic signal $x(t)$ with period T_p (or, frequency $f_p = 1/T_p$), the complex sinusoids that are required to represent $x(t)$ are only those at the frequencies kf_p for $k = 0, \pm 1, \pm 2, \ldots$.*

3.2 Why Fourier Series?

It is good for us to ask ourselves the following questions:

§ *Why do we want to express signals using Fourier series?*
§ *How does the knowledge of Fourier series benefit us in practical applications involving signals and systems?*

Let us consider the case of having to transmit a signal $x(t)$ (e.g., speech, video, data files, outputs of some sensors, etc.,) from one place to another. We expect the transmission medium (e.g., telephone cables, optical fiber cables, wireless medium, etc.,) to deliver the signal faithfully to the receiver. We can not check this by simply looking at the time-domain signal $x(t)$. One easy way to check this is to examine the frequencies present in $x(t)$, which can be done by using a tool such as Fourier series. By examining the values of the Fourier series coefficients X_k, we can know the maximum frequency (with non-negligible amplitude) present in the signal. Let the maximum frequency be 100 KHz (kilo Hertz) in a particular application. Then, *we should look for a suitable transmission medium that will not attenuate frequencies up to 100 KHz.* Thus, we can ensure that the chosen transmission medium faithfully transports the signal to the destination.

This example shows that the *knowledge of frequency-domain representation of signals can be very helpful in practical applications.* Even though most of the useful

3. We will study later that complex sinusoids of different frequencies are orthogonal to each other. Mathematically, this means the following:

$$\int_{-\infty}^{\infty} A_1 e^{j(2\pi f_1 t + \phi_1)} A_2 e^{-j(2\pi f_2 t + \phi_2)} dt = 0 \qquad \text{if } f_1 \neq f_2. \tag{3.3}$$

practical signals are non-periodic in nature, the study of periodic signals forms the foundation for the study of non-periodic signals.

3.3 Fourier Series: Definition and Interpretation

Let $x(t)$ be a periodic signal with period T_p (or, frequency $f_p = 1/T_p$). Then, based on the discussion at the end of Section 3.1, we get the Fourier series representation for $x(t)$ as,

$$\begin{aligned} x(t) &= \sum_{k=-\infty}^{\infty} X_k e^{jk\Omega_p t}, \qquad (3.4) \\ &= \cdots + X_{-2}e^{-j2\Omega_p t} + X_{-1}e^{-j\Omega_p t} + X_0 \\ &\quad + X_1 e^{j\Omega_p t} + X_2 e^{j2\Omega_p t} + \cdots . \end{aligned}$$

where, X_ks are the Fourier coefficients and $k\Omega_p$s are harmonics of the fundamental frequency Ω_p. Comparing Eq. (3.4) with Eq. (3.2), we find that the complex sinusoids $e^{jk\Omega_p t}$ take the place of the unit vectors $\vec{x}_u$ and $\vec{y}_u$, and the coefficients X_ks take the place of the scale factors x and y.

We have the following remarks on the Fourier series representation given in Eq. (3.4).

- *Complex sinusoids are used as component signals* to express the given periodic signal $x(t)$.
- *Frequencies of the complex sinusoids are given by* $k\Omega_p$ *or* $2\pi k f_p$ where $f_p = 1/T_p$ and $k = 0, \pm1, \pm2, \ldots$.
- Ω_p is called the *fundamental frequency* (Note: $k\Omega_p$ with $k = 1$) and $k\Omega_p$ is called the kth *harmonic frequency.*
- For all periodic signals with a given period T_p, we can state the following:
  - *The complex sinusoids that constitute the Fourier series are the same for all the signals* and are given by $e^{jk\Omega_p t}$ for $k = 0, \pm1, \pm2, \ldots$.
  - But *the set of coefficients* $\{X_k\}$ *will be different for each signal.*
- The coefficients X_k are obtained as follows: Consider,

$$\begin{aligned} \int_0^{T_p} x(t)e^{-jm\Omega_p t}dt &= \int_0^{T_p} \left[\sum_{k=-\infty}^{\infty} X_k e^{jk\Omega_p t}\right] e^{-jm\Omega_p t}dt \\ &= \sum_{k=-\infty}^{\infty} X_k \int_0^{T_p} e^{j(k-m)\Omega_p t}dt, \qquad (3.5) \end{aligned}$$

where m is an arbitrary integer. Since $\Omega_p T_p = 2\pi f_p T_p = 2\pi$, and $e^{jn2\pi} = 1$ for integer n, we get

$$\int_0^{T_p} e^{j(k-m)\Omega_p t} dt = \frac{e^{j2\pi(k-m)} - 1}{j\Omega_p(k-m)} = \begin{cases} 0 & \text{if } \ k \neq m \\ T_p & \text{if } \ k = m. \end{cases} \tag{3.6}$$

Substituting Eq. (3.6) in Eq. (3.5), we get

$$\int_0^{T_p} x(t)e^{-jm\Omega_p t} dt = T_p X_m \Rightarrow X_m = \frac{1}{T_p}\int_0^{T_p} x(t)e^{-jm\Omega_p t} dt. \tag{3.7}$$

This implies that

$$X_k = \frac{1}{T_p}\int_0^{T_p} x(t)e^{-jk\Omega_p t} dt. \tag{3.8}$$

The interval of integration should be one period and this period can be taken to be anywhere on the time axis. That is either $[0, T_p]$, or $[-T_p/2, T_p/2]$, etc. In general, it can be taken as $[t_p, t_p + T_p]$ *where* t_p *is any arbitrary real number.*

- Because X_k is a complex number, we can express it as

$$X_k = |X_k|e^{j\theta_k}, \tag{3.9}$$

where $|X_k| = \sqrt{X_{k,r}^2 + X_{k,i}^2}$ is the magnitude of the complex quantity and $\theta_k = \tan^{-1}\left(\frac{X_{k,i}}{X_{k,r}}\right)$ is the phase of the complex quantity. Here, $X_k = X_{k,r} + jX_{k,i}$ with $X_{k,r}$ and $X_{k,i}$ being the real and imaginary parts, respectively, of X_k. Thus, $|X_k|$ *and* θ_k *give the magnitude and phase, respectively, of the complex sinusoid* $X_k e^{jk\Omega_p t}$ *at frequency* $k\Omega_p$.

- The coefficient X_k gives the strength (i.e., amplitude and phase) of the kth harmonic $e^{jk\Omega_p t}$ in the given signal $x(t)$.

§ *The set of coefficients* $\{X_k\}$ *gives the frequency composition of the given signal. That is, by observing the values of* X_k, *we can know which all frequencies (or, complex sinusoids) are required to construct the signal* $x(t)$ *and what is the amplitude and phase of each sinusoid.*

In other words, we can say that

§ *Fourier series gives a frequency-domain representation of periodic signals.*

In summary, if $x(t)$ is periodic and satisfies the Dirichlet conditions,[4] it can be represented in a Fourier series as in Eq. (3.4), where the coefficients X_k are specified by Eq. (3.8). So, the frequency analysis of continuous-time periodic signals are summarized below:

- The synthesis equation:

$$x(t) = \sum_{k=-\infty}^{\infty} X_k e^{jk\Omega_p t}. \tag{3.10}$$

- The analysis equation:

$$X_k = \frac{1}{T_p} \int_0^{T_p} x(t) e^{-jk\Omega_p t} dt. \tag{3.11}$$

- Magnitude spectrum of $x(t)$: It is the plot[5] of $|X_k|$ vs k. That is, *it is the distribution of the amplitudes of the sinusoids as a function of the frequency.* Note that the index k *actually is an indicator of the frequency* $k\Omega_p$.
- Phase spectrum of $x(t)$: It is the plot of θ_k vs k *(i.e., distribution of phase vs frequency).*
- Power spectrum of $x(t)$: Since $|X_k|^2$ is the power of the kth harmonic,[6] the plot of $|X_k|^2$ vs k gives the *power spectrum (i.e., distribution of power vs frequency)* of $x(t)$.
- Power of $x(t)$: Since the complex sinusoids that constitute the signal $x(t)$ are at different frequencies (i.e., $k\Omega_p$, $k = 0, \pm 1, \pm 2, \ldots$), we get the power of $x(t)$ as[7]

$$p = \sum_{k=-\infty}^{\infty} |X_k|^2. \tag{3.12}$$

---

4. Dirichlet conditions for Fourier series:
   - $x(t)$ must be absolutely integrable over a period, and
   - $x(t)$ must have a finite number of exterma in any given time interval, i.e., there must be a finite number of maxima and minima in the interval.

5. The term *'spectrum'* is usually used to refer to the distribution of some quantity as a function of frequency.

6. Recall from Chapter 2 (see Section 2.2.2) that the power of a complex sinusoid is equal to the magnitude square of its amplitude.

7. Recall from Chapter 2 (see Section 2.2.3) that the power of a signal consisting of a sum of sinusoids is equal to the sum of the powers of the individual sinusoids, if the frequencies of the sinusoids are different.

Therefore, we get (using the definition of power in Eq. (2.5) of Chapter 2)

$$p = \frac{1}{T_p} \int_{-T_p/2}^{T_p/2} |x(t)|^2 dt = \sum_{k=-\infty}^{\infty} |X_k|^2. \tag{3.13}$$

Eq. (3.13) is known as *Parseval's Theorem* for periodic signals. That is, we can use either time-domain expression $\left(\frac{1}{T_p} \int_{-T_p/2}^{T_p/2} |x(t)|^2 dt\right)$ or frequency-domain expression $\left(\sum_{k=-\infty}^{\infty} |X_k|^2\right)$ for computing the power of a (periodic) signal, *depending upon which is easier to use.*

§ *Let $x(t)$ and $y(t)$ be periodic signals with periods T_x and T_y, respectively, and $z(t) = x(t) + y(t)$. Examine if $z(t)$ is periodic and if it is periodic, what is its period.*
§ *If $x(t)$ is periodic with period T_x, argue that the Fourier series representation given in Eq. (3.4) can't contain a complex sinusoid at frequency that is not an integer multiple of $1/T_x$.*

3.4 Properties of Fourier Series

We shall now briefly look at some of the important properties of Fourier series (FS). Knowing these properties will make it easy for us i) to evaluate FS of complicated signals, and ii) to predict the nature of signals based on their spectra and vice versa.

Let $x(t)$ and $y(t)$ be periodic signals with period T_p and FS coefficients X_k and Y_k, respectively. That is,

$$x(t) = \sum_{k=-\infty}^{\infty} X_k e^{jk\Omega_p t}, \qquad y(t) = \sum_{k=-\infty}^{\infty} Y_k e^{jk\Omega_p t}$$

$$X_k = \frac{1}{T_p} \int_0^{T_p} x(t) e^{-jk\Omega_p t} dt, \quad Y_k = \frac{1}{T_p} \int_0^{T_p} y(t) e^{-jk\Omega_p t} dt.$$

3.4.1 Linearity

Let us construct a signal $z(t)$ as linear combination of $x(t)$ and $y(t)$. That is,

$$z(t) = \alpha x(t) + \beta y(t), \tag{3.14}$$

where α and β are two scalars (real or complex). Clearly, $z(t)$ is also periodic with period T_p. Then, *how are the FS coefficients, Z_k, of $z(t)$ related to that of $x(t)$ and $y(t)$?*

By the definition of FS, we get

$$
\begin{aligned}
Z_k &= \frac{1}{T_p}\int_0^{T_p} z(t)e^{-jk\Omega_p t}dt \\
&= \frac{1}{T_p}\int_0^{T_p} \left[\alpha x(t) + \beta y(t)\right] e^{-jk\Omega_p t}dt \\
&= \frac{1}{T_p}\int_0^{T_p} \alpha x(t)e^{-jk\Omega_p t}dt + \frac{1}{T_p}\int_0^{T_p} \beta y(t)e^{-jk\Omega_p t}dt \\
&= \alpha\frac{1}{T_p}\int_0^{T_p} x(t)e^{-jk\Omega_p t}dt + \beta\frac{1}{T_p}\int_0^{T_p} y(t)e^{-jk\Omega_p t}dt \\
&= \alpha X_k + \beta Y_k. \qquad (3.15)
\end{aligned}
$$

Thus, *FS coefficients of linear combination of signals are given by the corresponding linear combination of the FS series coefficients of the individual signals.*

3.4.2 Time Shifting

Let $z(t) = x(t - t_0)$ where t_0 is an arbitrary time offset.[8] Then, as before, we get

$$
\begin{aligned}
Z_k &= \frac{1}{T_p}\int_0^{T_p} z(t)e^{-jk\Omega_p t}dt = \frac{1}{T_p}\int_0^{T_p} x(t-t_0)e^{-jk\Omega_p t}dt \\
&= \frac{1}{T_p}\int_{-t_0}^{T_p-t_0} x(u)e^{-jk\Omega_p (u+t_0)}du \qquad : u = t - t_0 \\
&= e^{-jk\Omega_p t_0}\frac{1}{T_p}\int_{-t_0}^{T_p-t_0} x(u)e^{-jk\Omega_p u}du \\
&= e^{-jk\Omega_p t_0}X_k \qquad (3.16)
\end{aligned}
$$

This implies that:

$$
|Z_k| = |e^{-jk\Omega_p t_0}X_k| = |e^{-jk\Omega_p t_0}|\cdot|X_k| = |X_k|, \qquad (3.17)
$$

$$
\{\text{Phase of } Z_k\} = \{\text{Phase of } X_k\} - k\Omega_p t_0. \qquad (3.18)
$$

Thus, *shifting the signal in time-domain results in a phase change in the frequency-domain, while the magnitude spectrum remains unaffected.*

---

8. If $t_0 > 0$, then $z(t)$ is obtained by shifting $x(t)$ to the right by t_0 on the time axis. This is also known as *delaying* $x(t)$ *by* t_0. If $t_0 < 0$, then $z(t)$ will be left-shifted (or, advanced in time) version of $x(t)$.

3.4.3 Frequency Shifting or Modulation

Let $z(t) = e^{jm\Omega_p t}x(t)$ where m is an arbitrary integer. Then, we get

$$\begin{aligned} Z_k &= \frac{1}{T_p}\int_0^{T_p}\left[e^{jm\Omega_p t}x(t)\right]e^{-jk\Omega_p t}dt \\ &= \frac{1}{T_p}\int_0^{T_p} x(t)e^{-j(k-m)\Omega_p t}dt \\ &= X_{k-m}. \end{aligned} \tag{3.19}$$

As an example, let $m = 10$ and $f_p = 10\ KHz$. Then, Eq. (3.19) implies that multiplying $x(t)$ by $e^{jm\Omega_p t}$ results in shifting the spectrum to the right by $10m$ KHz (kilo Hertz) = 0.1 MHz (mega Hertz). This principle is used in communication systems for shifting a signal from one frequency range to another for transmission purpose.

3.4.4 Time Reversal

Let $z(t) = x(-t)$, time-reversed $x(t)$. Then, we get[9]

$$\begin{aligned} Z_k &= \frac{1}{T_p}\int_0^{T_p} x(-t)e^{-jk\Omega_p t}dt \\ &= -\frac{1}{T_p}\int_0^{-T_p} x(u)e^{jk\Omega_p u}du \qquad : u = -t \\ &= \frac{1}{T_p}\int_{-T_p}^{0} x(u)e^{jk\Omega_p u}du = \frac{1}{T_p}\int_{-T_p}^{0} x(u)e^{-j(-k)\Omega_p u}du \\ &= X_{-k}. \end{aligned} \tag{3.20}$$

That is, *reversing a signal in time-domain results in reversing its spectrum in frequency-domain.*

3.4.5 Even-Symmetric Signal

Let $x(t)$ be an even-symmetric signal. That is, $x(-t) = x(t)$. Let us examine the Fourier series (FS) coefficients of $x(t)$.

9. Recall that $\int_a^b x(t)dt = -\int_b^a x(t)dt$ for any given function $x(t)$.

Consider X_{-k} given by (let $\tilde{T} = T_p/2$),

$$\begin{aligned}
X_{-k} &= \frac{1}{T_p}\int_{-\tilde{T}}^{\tilde{T}} x(t)e^{jk\Omega_p t}dt \\
&= \frac{-1}{T_p}\int_{\tilde{T}}^{-\tilde{T}} x(-u)e^{-jk\Omega_p u}du \qquad : u = -t \\
&= \frac{1}{T_p}\int_{-\tilde{T}}^{\tilde{T}} x(u)e^{-jk\Omega_p u}du \qquad : x(-u) = x(u) \\
&= X_k
\end{aligned} \tag{3.21}$$

This implies that:

$$|X_{-k}|e^{j\theta_{-k}} = |X_k|e^{j\theta_k}. \tag{3.22}$$

That is, *the phase and magnitude spectra of an even-symmetric signal are even-symmetric*. Thus, we have,

$$|X_{-k}| = |X_k|, \qquad \text{and} \qquad \theta_{-k} = \theta_k. \tag{3.23}$$

3.4.6 Odd-Symmetric Signal

Let $x(t)$ be an odd-symmetric signal. That is, $x(-t) = -x(t)$. Then, proceeding as above, we get,

$$X_{-k} = -X_k \tag{3.24}$$

This means that,

$$\begin{aligned}
&|X_{-k}|e^{j\theta_{-k}} = -|X_k|e^{j\theta_k} = |X_k|e^{j(\theta_k \pm \pi)} \\
&\Longrightarrow \ |X_{-k}| = |X_k|, \ \text{ and } \theta_{-k} = \theta_k \pm \pi.
\end{aligned} \tag{3.25}$$

That is, *the magnitude spectrum of an odd-symmetric signal is even-symmetric.*

3.4.7 Real Signal

Let $x(t)$ be a real signal. That is, $x^*(t) = x(t)$. Then, proceeding as above, we get

$$X_{-k} = X_k^*, \tag{3.26}$$

which implies

$$\begin{aligned}
&|X_{-k}|e^{j\theta_{-k}} = |X_k|e^{-j\theta_k} \\
&\Longrightarrow \ |X_{-k}| = |X_k|, \ \text{ and } \theta_{-k} = -\theta_k.
\end{aligned} \tag{3.27}$$

That is, *the magnitude spectrum of a real signal is even-symmetric and phase spectrum is odd-symmetric.*

From Eqns. (3.23), (3.25) and (3.27), we can conclude the following:

- *For a real signal, irrespective of any symmetry, the magnitude spectrum is even-symmetric and the phase spectrum is odd-symmetric.*
- *The phase spectrum of an even-symmetric real signal is 0 or $\pm\pi$ for all frequencies. That is, the FS coefficients X_k are purely real.*
  E.g., $x(t) = A_1\cos(100\pi t) + A_2\cos(300\pi t) + A_3\cos(1000\pi t)$ where A_1, A_2 and A_3 are real numbers.
- *The phase spectrum of an odd-symmetric real signal is odd-symmetric and the phase values are given by $\pm\pi/2$. That is, the FS coefficients X_k are purely imaginary.*
  E.g. $x(t) = A_1\sin(100\pi t) + A_2\sin(300\pi t) + A_3\sin(1000\pi t)$ where A_1, A_2 and A_3 are real numbers.

3.5 Real Fourier Series

Let $x(t)$ be a real periodic signal.[10] Then, using Eq. (3.4), write its Fourier series as,

$$\begin{aligned} x(t) = \sum_{k=-\infty}^{\infty} X_k e^{jk\Omega_p t} &= \cdots + X_{-2}e^{-j2\Omega_p t} + X_{-1}e^{-j\Omega_p t} \\ &+ X_0 + X_1 e^{j\Omega_p t} + X_2 e^{j2\Omega_p t} + \cdots \\ &= X_0 + \left[X_{-1}e^{-j\Omega_p t} + X_1 e^{j\Omega_p t}\right] \\ &+ \left[X_{-2}e^{-j2\Omega_p t} + X_2 e^{j2\Omega_p t}\right] + \cdots. \end{aligned} \tag{3.28}$$

Since $X_{-k} = X_k^*$ as $x(t)$ is real, we get

$$\begin{aligned} X_{-k}e^{-jk\Omega_p t} + X_k e^{jk\Omega_p t} &= X_k^* e^{-jk\Omega_p t} + X_k e^{jk\Omega_p t} \\ &= 2 \cdot \text{Real}\left[X_k e^{jk\Omega_p t}\right] \\ &= 2X_{k,r}\cos(k\Omega_p t) - 2X_{k,i}\sin(k\Omega_p t), \end{aligned} \tag{3.29}$$

10. The material in Section 3.5 helps us to understand how the Fourier series simplifies when we consider only real signals.

where $X_{k,r}$ and $X_{k,i}$ are the real and imaginary parts, respectively, of X_k (i.e., $X_k = X_{k,r} + jX_{k,i}$). Substituting Eq. (3.29) in Eq. (3.28), we get

$$\begin{aligned} x(t) &= X_0 + 2\left[X_{1,r}\cos(\Omega_p t) - X_{1,i}\sin(\Omega_p t)\right] \\ &\quad + 2\left[X_{2,r}\cos(2\Omega_p t) - X_{2,i}\sin(2\Omega_p t)\right] + \cdots \\ &= X_0 + 2\sum_{k=1}^{\infty} X_{k,r}\cos(k\Omega_p t) - 2\sum_{k=1}^{\infty} X_{k,i}\sin(k\Omega_p t) \\ &= a_0 + \sum_{k=1}^{\infty} a_k\cos(k\Omega_p t) + \sum_{k=1}^{\infty} b_k\sin(k\Omega_p t) \end{aligned} \tag{3.30}$$

where the coefficients a_0, a_k and b_k for $k = 1, 2, \ldots, \infty$ are given by (let $\tilde{T} = T_p/2$)

$$a_0 = X_0 = \frac{1}{T_p}\int_{-\tilde{T}}^{\tilde{T}} x(t)dt \tag{3.31}$$

$$\begin{aligned} a_k = 2X_{k,r} &= \frac{2}{T_p}\text{Real}\left[\int_{-\tilde{T}}^{\tilde{T}} x(t)e^{-jk\Omega_p t}dt\right] \\ &= \frac{2}{T_p}\int_{-\tilde{T}}^{\tilde{T}} x(t)\cos(k\Omega_p t)dt \end{aligned} \tag{3.32}$$

$$\begin{aligned} b_k = -2X_{k,i} &= \frac{-2}{T_p}\text{Imag}\left[\int_{-\tilde{T}}^{\tilde{T}} x(t)e^{-jk\Omega_p t}dt\right] \\ &= \frac{2}{T_p}\int_{-\tilde{T}}^{\tilde{T}} x(t)\sin(k\Omega_p t)dt. \end{aligned} \tag{3.33}$$

Thus, *Eqns. (3.30)–(3.33) shows that a real signal can be represented using a real FS where the basic signals are real sinusoids rather than complex sinusoids, and the FS coefficients* $\{a_0, a_k, b_k\}$ *are real scalars.*

Let us consider the following special cases:

- $x(t)$ is Real and Even-Symmetric: Then, it is easy to show that $b_k = 0$ for all k. Therefore, we get from Eq. (3.30)

$$x(t) = a_0 + \sum_{k=1}^{\infty} a_k\cos(k\Omega_p t). \tag{3.34}$$

- $x(t)$ is Real and Odd-Symmetric: Then, it is easy to show that $a_0 = 0$ and $a_k = 0$ for all k. Therefore, we get from Eq. (3.30)

$$x(t) = \sum_{k=1}^{\infty} b_k \sin(k\Omega_p t). \tag{3.35}$$

3.6 Examples of Fourier Series Evaluation

We shall now present a few examples to illustrate i) how to evaluate the FS coefficients, and ii) how to exploit the properties of FS while evaluating the FS.

Example 1: Consider an example of a periodic square wave signal (Fig. 3.3(a)) of period $T_p = 0.5$. One period of the signal is defined as,

$$x(t) = \begin{cases} 1 & \text{for } 0 \leq t < 0.25 \\ 0 & \text{for } 0.25 \leq t < 0.5. \end{cases} \tag{3.36}$$

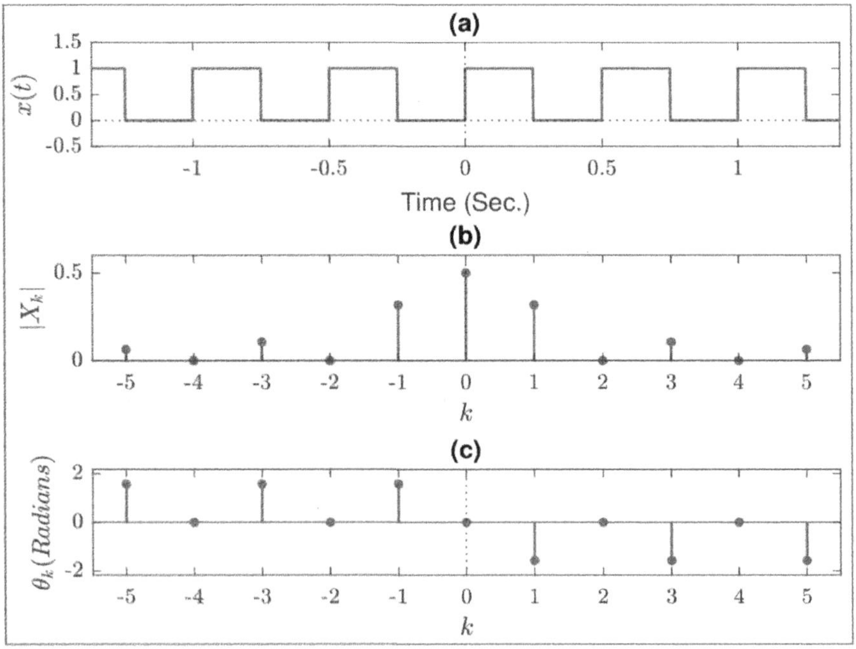

Figure 3.3. Periodic signal and its Fourier spectrum. (a) the periodic signal (real), (b) Magnitude spectrum, and (c) Phase spectrum.

Note that $T_p = 0.5$, $f_p = 1/T_p = 2$, $\Omega_p = 2\pi f_p = 4\pi$. Since $x(t)$ is zero during $0.25 \leq t < 0.5$, we get the complex FS coefficients as[11]

$$X_0 = \frac{1}{0.5}\int_0^{0.5} x(t)dt = \frac{1}{0.5}\int_0^{0.25} dt = 0.5.$$

$$\begin{aligned} X_k &= \frac{1}{T_p}\int_0^{T_p} x(t)e^{-jk\Omega_p t}dt = \frac{1}{0.5}\int_0^{0.5} x(t)e^{-jk\Omega_p t}dt \\ &= 2\int_0^{0.25} x(t)e^{-j4\pi kt}dt + 2\int_{0.25}^{0.5} x(t)e^{-j4\pi kt}dt \\ &= 2\int_0^{0.25} e^{-j4\pi kt}dt + 0 = 2\left.\frac{e^{-j4\pi kt}}{-j4\pi k}\right|_0^{0.25} = \frac{1-e^{-j\pi k}}{j2\pi k}. \end{aligned} \tag{3.37}$$

Therefore, we get (since $e^{j\pi} = -1 = e^{-j\pi}$)

$$\begin{aligned} &X_0 = 0.5, \\ &X_1 = \tfrac{1}{j\pi}, \quad X_{-1} = \tfrac{1}{-j\pi} \\ &X_2 = 0, \quad X_{-2} = 0 \\ &X_3 = \tfrac{1}{j3\pi}, \quad X_{-3} = \tfrac{1}{-j3\pi} \\ &X_4 = 0, \quad X_{-4} = 0 \\ &X_5 = \tfrac{1}{j5\pi}, \quad X_{-5} = \tfrac{1}{-j5\pi} \\ &\vdots \qquad\quad \vdots \end{aligned}$$

We see that $X_{-k} = X_k^*$, which should be so since the signal $x(t)$ is real. Figure 3.3 also shows the magnitude spectrum ($|X_k|$ vs k; Fig. 3.3(b)) and phase spectrum (θ_k vs k; Fig. 3.3(c)) of the given signal, $x(t)$. It can be seen that the magnitude spectrum is symmetric and the phase spectrum is anti-symmetric.

Example 2: Let us construct a signal $y(t)$ (Fig. 3.4(a)) using the above signal $x(t)$ as,

$$y(t) = x(t + 0.125).$$

Note that $y(t)$ is obtained by shifting $x(t)$ to the left by 0.125. It is easy to see that the resulting $y(t)$ is a real even-symmetric signal with period $T_p = 0.5$. One period

11. Since $X_k = \frac{1}{T_p}\int_{t_0}^{t_0+T_p} x(t)e^{-jk\Omega_p t}dt$, where t_0 can be anything, we should choose t_0 such that the resulting expression for $x(t)$ in the interval $[t_0, t_0 + T_p]$ is easy to integrate.

of $y(t)$ is given by

$$y(t) = \begin{cases} 1 & \text{for } -0.125 \leq t < +0.125 \\ 0 & \text{for } +0.125 \leq t < +0.375. \end{cases} \tag{3.38}$$

As before, we get the complex FS coefficients of $y(t)$ as

$$\begin{aligned} Y_0 &= \frac{1}{0.5} \int_{-0.125}^{0.375} x(t)dt = 2 \int_{-0.125}^{0.125} dt = 0.5 \\ Y_k &= \frac{1}{0.5} \int_{-0.125}^{0.375} x(t)e^{-jk\Omega_p t} dt = 2 \int_{-0.125}^{0.125} e^{-j4k\pi t} dt \\ &= 2 \left. \frac{e^{-j4\pi kt}}{-j4\pi k} \right|_{-0.125}^{0.125} = \frac{\sin(\pi k/2)}{\pi k}. \end{aligned} \tag{3.39}$$

Therefore, we get (since $\sin(k\pi) = 0$)

$$\begin{aligned} &Y_0 = 0.5, \\ &Y_1 = \tfrac{1}{\pi}, \quad Y_{-1} = \tfrac{1}{\pi} \\ &Y_2 = 0, \quad Y_{-2} = 0 \\ &Y_3 = \tfrac{-1}{3\pi}, \quad Y_{-3} = \tfrac{-1}{3\pi} \\ &Y_4 = 0, \quad Y_{-4} = 0 \\ &Y_5 = \tfrac{1}{5\pi}, \quad Y_{-5} = \tfrac{1}{5\pi} \\ &\vdots \qquad\quad \vdots \end{aligned}$$

We see that $Y_{-k} = Y_k^* = Y_k$, which should be so since the signal $y(t)$ is real and even-symmetric. The corresponding magnitude and phase spectrum are shown in Figs. 3.4(b) and 3.4(c), respectively. Here again, the magnitude spectrum is symmetric. The phase spectrum, unlike in Example 1, is symmetric as the signal $y(t)$ is real and symmetric.

Example 3: Construct another signal $z(t)$ (Fig. 3.5(a)) using the above $x(t)$ as,

$$z(t) = x(t) - 0.5.$$

Note that $z(t)$ is obtained by subtracting 0.5 from $x(t)$ for all values of t. This results in pulling down the waveform by 0.5. It is easy to see that the resulting $z(t)$ is a real and odd-symmetric signal with period $T_p = 0.5$. One period of $z(t)$ is

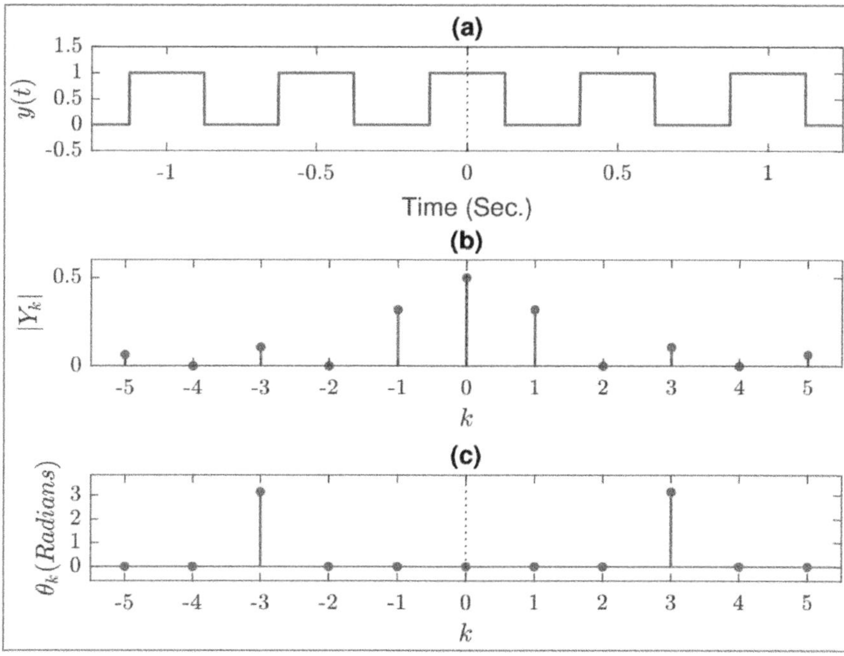

Figure 3.4. Periodic signal and its Fourier spectrum. (a) the periodic signal (real and even-symmetric), (b) Magnitude spectrum, and (c) Phase spectrum.

given by,

$$z(t) = \begin{cases} -0.5 & \text{for} \quad -0.25 \le t < 0 \\ +0.5 & \text{for} \quad 0 \le t < 0.25. \end{cases} \tag{3.40}$$

We get the complex FS coefficients of $z(t)$ as,

$$\begin{aligned} Z_0 &= \frac{1}{0.5}\int_{-0.25}^{0.25} x(t)dt \\ &= 2\int_{-0.25}^{0} (-0.5)dt + 2\int_{0}^{0.25} (0.5)dt = 0 \\ Z_k &= 2\int_{-0.25}^{0.25} x(t)e^{-jk\Omega_p t}dt \\ &= 2\int_{-0.25}^{0} (-0.5)e^{-j4k\pi t}dt + 2\int_{0}^{0.25} (0.5)e^{-j4k\pi t}dt \\ &= -\left.\frac{e^{-j4\pi kt}}{-j4\pi k}\right|_{-0.25}^{0} + \left.\frac{e^{-j4\pi kt}}{-j4\pi k}\right|_{0}^{0.25} = \frac{1-\cos(\pi k)}{j2k\pi}. \end{aligned} \tag{3.41}$$

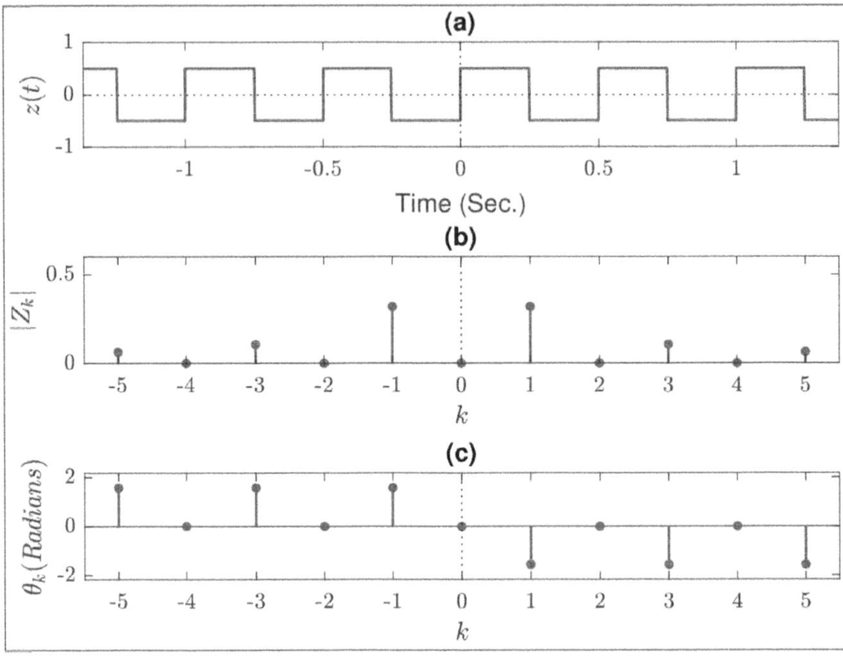

Figure 3.5. Periodic signal and its Fourier spectrum. (a) the periodic signal (real and odd-symmetric), (b) Magnitude spectrum, and (c) Phase spectrum.

Therefore, we get (since $\cos(\pi) = -1$ and $\cos(2\pi) = 1$)

$$\begin{aligned}
&Z_0 = 0 \\
&Z_1 = \frac{1}{j\pi}, \quad Z_{-1} = \frac{1}{-j\pi} \\
&Z_2 = 0, \quad Z_{-2} = 0 \\
&Z_3 = \frac{1}{j3\pi}, \quad Z_{-3} = \frac{1}{-j3\pi} \\
&Z_4 = 0, \quad Z_{-4} = 0 \\
&Z_5 = \frac{1}{j5\pi}, \quad Z_{-5} = \frac{1}{-j5\pi} \\
&\vdots \qquad\qquad \vdots
\end{aligned}$$

Thus, we see that $Z_{-k} = Z_k^* = -Z_k$, which should be so since the signal $z(t)$ is real and odd-symmetric. As can be seen from Fig. 3.5, the magnitude spectrum (Fig. 3.5(b)) is once again symmetric and the phase spectrum (Fig. 3.5(c)) is anti-symmetric. Also, the even indexed (i.e., $k = 0, \pm 2, \pm 4, \ldots$) Fourier coefficients are all zeros and the corresponding phases are also zeros.

§ *Use the properties of FS to verify the above results for Y_k and Z_k. Isn't this a much easier approach?*

3.7 Limitations of Fourier Series

We shall now point out some limitations of Fourier series.

3.7.1 Gibbs Phenomenon

Let us define a signal $\tilde{x}(t)$ as

$$\tilde{x}(t) = \sum_{k=-N}^{N} X_k e^{jk\Omega_p t}$$

where N is a positive integer. Based on the definition of FS in Eq. (3.4), we expect $\tilde{x}(t)$ to converge to $x(t)$ as $N \to \infty$. Indeed, this happens if the signal $x(t)$ is continuous, i.e., there are no discontinuities in the signal waveform (e.g., the triangular periodic signal defined in Section 1.2.2).

However, the situation is different when the signal contains discontinuities, as in a periodic square-wave signal as in Examples 1, 2 and 3). For such signals, as $N \to \infty$, $\tilde{x}(t)$ converges to $x(t)$ for all values of t except at points of discontinuities. "Overshoots" appear near the discontinuities. This is known as *Gibbs phenomenon.* As $N \to \infty$, the height of the overshoot remains constant, while its width approaches zero. Thus, as $N \to \infty$, the energy in the overshoots approaches zero, and hence the power of $\tilde{x}(t)$ approaches that of $x(t)$.

§ *Why does Gibbs phenomenon occur?*

3.7.2 Dirichlet Conditions

For a periodic signal $x(t)$ to be represented by Fourier series, it must satisfy the following conditions (for any t_0).

- It must be absolutely integrable over the interval $[t_0, t_0 + T_p]$. That is,

$$\int_{t_0}^{t_0+T_p} |x(t)|\, dt < \infty. \tag{3.42}$$

- It must have a finite number of maxima and minima over the interval $[t_0, t_0 + T_p]$.
- It must have a finite number of discontinuities, all of finite size, over the interval $[t_0, t_0 + T_p]$.

These are known as *Dirichlet conditions* for existence of FS.

So far, we have discussed the FS representation of periodic signals only. Almost all of the real-life signals are aperiodic in nature. How do we find the Fourier representation of such aperiodic signals? This is what is explained in the next chapter.

3.8 Concluding Remarks

We can make the following concluding remarks based on the material presented above.

- Fourier series gives a frequency-domain representation of periodic signals. It can be used to assess the effective bandwidth of periodic signals.
- Spectra of periodic signals are discrete-frequency in nature.
- The properties of Fourier series can be used to simplify the effort (integration, computations etc.) necessary in evaluating the Fourier series coefficients.
- For real signals, the complex Fourier series can be expressed as a real Fourier series where the basic/fundamental signals are real sinusoids instead of complex sinusoids.
- For real even-symmetric or odd-symmetric signals, the Fourier series coefficients also have certain well defined symmetry properties. These properties can be exploited to simplify the computations or to verify the solutions.

Exercises

3.1. The waveform shown in Fig. P3.1 is the periodic voltage of a signal generator. Find the Fourier coefficients and write down the terms of the series through the 6th harmonic.

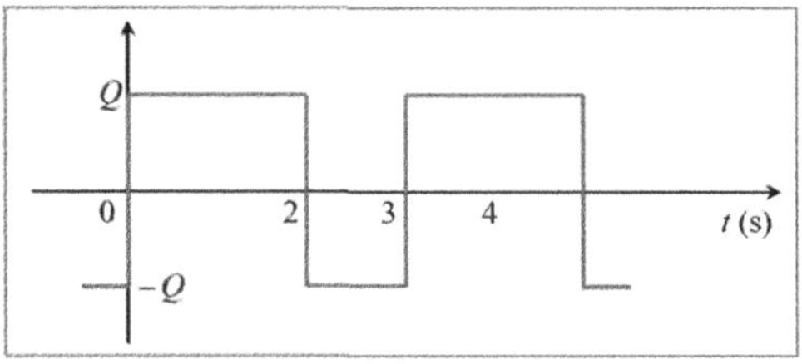

Figure P3.1.

3.2. The rectangular wave in Fig. P3.1 is modified as follows. The positive part of the wave now extends from 0 to 1 and the negative part from 1 to 3 s, and this pattern is repeated. In addition, the positive amplitude becomes $3Q/2$ and the negative amplitude $-Q/2$.

(a) Find the coefficients of the Fourier series.
(b) Write down the terms of the series through the third harmonic.

3.3. A square wave periodic signal is shown in Fig. P3.3 without an origin of time.

(a) Select a time origin to make the resulting function even, and determine the Fourier coefficients. Choose the one period integration interval in such a way that the evaluation of the integrals is simplified. Note how the harmonic amplitudes vary with the number.

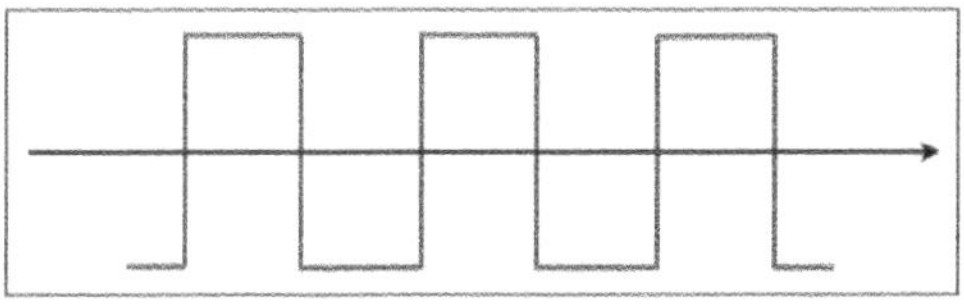

Figure P3.3.

(b) For the square wave in Fig. P3.3, choose the origin of time in order to make the function an odd function and hence derive its Fourier coefficients.

3.4. (a) Determine the coefficients of a series of sines representing the square wave in Fig. P3.3 with a proper choice of the time origin.
(b) Confirm your result, using the cosine series found in problem 3.3(a) with an appropriate shift of the time origin

3.5. The graph sketched in Fig. P3.4 represents half of a periodic function. Sketch the second half of the function so that

(a) The function is even
(b) The function is odd

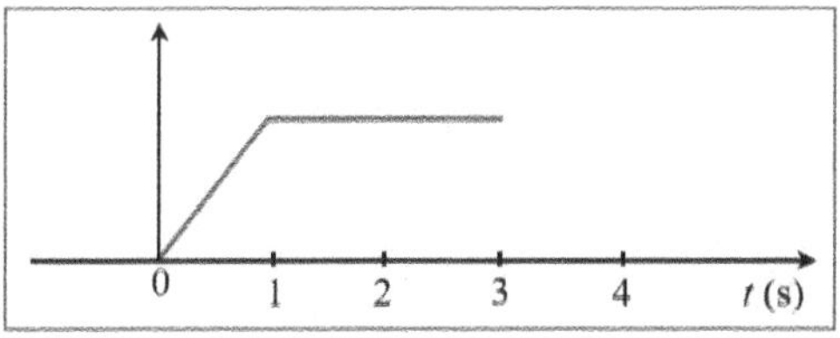

Figure P3.4.

(c) Carry out a development of the Fourier series similar to the one carried out in the case of an even function to show that the b_k coefficients in the Fourier series representation of an odd function can be determined by integrating over only the positive half-period and then multiplying by 2. You should obtain the following result:

$$a_k = 0$$
$$b_k = \frac{4}{T}\int_0^{T/2} f(t)\sin(k\Omega_p t)dt$$

3.6. Consider the full-wave rectified sinusoid signal ($x_a(t)$) in Fig. P3.5. Here, A is the maximum amplitude of $x_a(t)$.

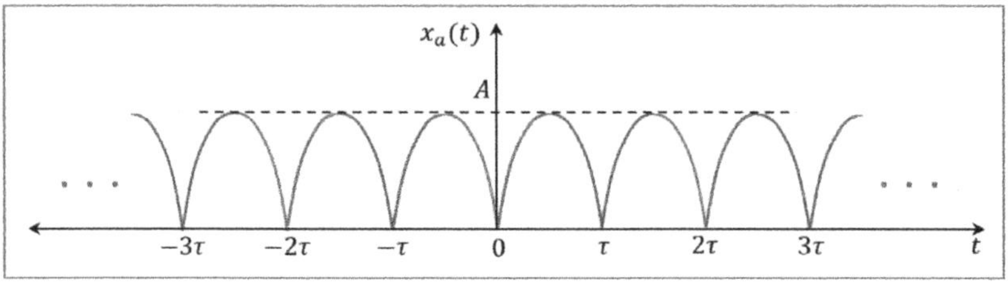

Figure P3.5.

(a) Derive an expression for the Fourier spectrum (X_k's for $k = 0, \pm1, \pm2, \pm3, \ldots$) of the signal, $x_a(t)$ (*Hint*: Using the parameters provided in the figure, find an expression for the full-wave rectified sinusoid signal $x_a(t)$ and then work on it).
(b) Determine the average power (p) of the signal, $x_a(t)$.
(c) Plot the power spectral density of $x_a(t)$ (i.e., the plot of $|X_k|^2$ as a function of $k = 0, \pm1, \pm2, \pm3, \ldots$). Show the plot for $k = 0, \pm1, \pm2$, and ±3.
(d) Prove the validity of Parseval's relation (i.e., the average power estimated directly from $x_a(t)$ is equal to the power estimated from the Fourier coefficients) for this signal.

3.7. Find the Fourier series for the following functions:

(a) $8\sin(7t)$,
(b) $\cos(2t) + \sin(4t)$,
(c) $\sin^2(t)$.

3.8. A periodic signal $x(t) = x(-t)$ is real, symmetric about $t = 0$. It has a period $T_p = \frac{2\pi}{\Omega_p}$. Show that its Fourier series, $x(t) = \ldots + X_{-1}e^{-1j\Omega_p t} + X_0 e^{0j\Omega_p t} + X_1 e^{1j\Omega_p t} + \ldots$, degenerates to become a cosine series of the form, $x(t) = a_0 + 2a_1 cos(\Omega_p t) + 2a_2\cos(2\Omega_p t) + \ldots$, where $a_0, a_1, \ldots$ are real.

3.9. A signal that is sometimes used in communications systems is the raised cosine pulse. Figure P3.9 shows a signal $x(t)$ that is a periodic sequence of these raised cosine pulses with equal spacing between them.

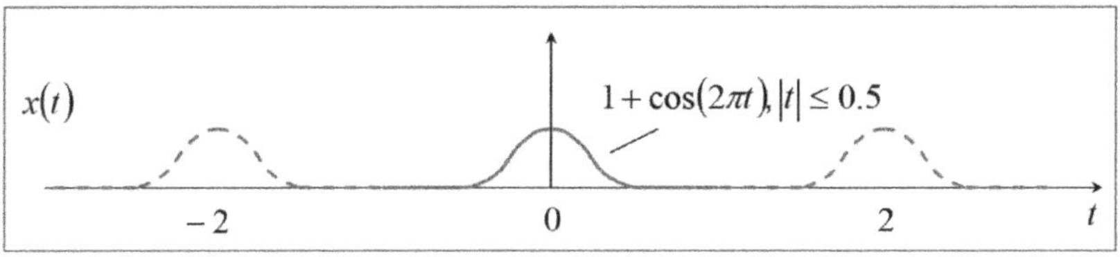

Figure P3.9.

(a) Show that the first three terms in the Fourier series expansion of x(t) are given by,

$$x(t) = \frac{1}{2} + \frac{8}{3\pi}\cos(\pi t) + \frac{1}{2}\cos(2\pi t) + \ldots$$

3.10. Show that the Fourier series coefficients for the signal $x(t)$ shown in Fig. P3.10 below are given by,

$$X_k = \frac{2}{k\pi}\left[1 - \cos\left(\frac{k\pi}{2}\right)\right]\sin\left(\frac{k\pi}{2}\right)$$

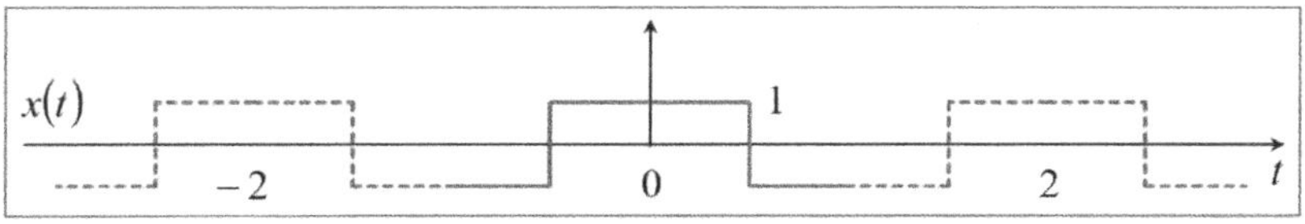

Figure P3.10.

(a) Hence, show that

$$X_k = \begin{cases} \frac{2}{k\pi} & \text{for } k = \ldots, 1, 5, 9, \ldots \\ -\frac{2}{k\pi} & \text{for } k = \ldots, 3, 7, 11, \ldots \\ 0 & \text{for } k = \ldots, 0, 2, 4, 6, 8, \ldots \end{cases}$$

(b) Then show that the Fourier series expansion for the signal can be written as,

$$x(t) = \frac{4}{\pi}\sum_{m=0}^{\infty}\frac{(-1)^m}{2m+1}\cos\left((2m+1)\pi t\right).$$

DOI: 10.1561/9781680839791.ch4

Chapter 4

Fourier Transform

In Chapter 3, we discussed Fourier series as a tool for frequency-domain representation of periodic signals. But, most of the signals in practical situations/applications are non-periodic (or, aperiodic) in nature. In this chapter, we discuss *the representation of non-periodic signals in terms of sinusoids*. The resulting frequency-domain representation is known as *Fourier transform*. Clearly, the Fourier transform approach has much more practical relevance and use compared to Fourier series. As a special case, we also consider periodic signals as well.

In this chapter, the following important points are discussed in detail.

- The role of sinusoids as basic building block for synthesizing signals.
- Representation of aperiodic signals using sinusoids.
- The *continuous-frequency nature of the spectra of aperiodic signals*, as compared to the discrete-frequency spectra of periodic signals.
- Various *properties of Fourier transform representation*, and their use in practice.
- Concept of *impulse function* and its use in signal processing.
- Fourier transform of periodic signals.
- Fourier transform as a fundamental tool for *spectral estimation.*

4.1 Introduction to Fourier Transform

Let $x(t)$ *be a periodic signal with period* T_p. Then, from our discussion of Fourier series in Chapter 3, we may recall the following points (with $f_p = \frac{1}{T_p}$ and $\Omega_p = 2\pi f_p$).

- Fourier series representation of $x(t)$ is given by,

$$x(t) = \sum_{k=-\infty}^{\infty} X_k e^{jk\Omega_p t}, \tag{4.1}$$

$$X_k = \frac{1}{T_p} \int_{-T_p/2}^{T/2} x(t) e^{-jk\Omega_p t} dt, \tag{4.2}$$

for $k = 0, \pm 1, \pm 2, \ldots$.
- Complex sinusoids of frequencies

$$\{\cdots, -3\Omega_p, -2\Omega_p, -\Omega_p, 0, \Omega_p, 2\Omega_p, 3\Omega_p, \cdots\}$$

and corresponding (complex) amplitudes

$$\{\cdots, X_{-3}, X_{-2}, X_{-1}, X_0, X_1, X_2, X_3, \cdots\}$$

are used as component signals to express the signal $x(t)$.
- The Fourier coefficient X_k gives the strength (i.e., magnitude and phase) of the kth harmonic $e^{jk\Omega_p t}$ in the given signal $x(t)$. In other words, *Fourier series gives a frequency-domain representation of periodic signals.*
- The magnitude spectrum (i.e., $|X_k|$ vs k), which is the distribution of the amplitudes of the sinusoids as a function of frequency, gives explicit information on the frequencies present in $x(t)$ (Note: k *indicates the frequency* $k\Omega_p$). The magnitude spectrum can be used to assess the *effective bandwidth* of the signal. Effective bandwidth refers to the maximum frequency present in $x(t)$ with non-negligible amplitude.

But, *Fourier series representation is applicable for periodic signals only.* Since most of the practical signals we deal with are non-periodic in nature, a natural question arises:

How to deal with non-periodic signals?

The answer to this question is *Fourier transform.*

§ *Fourier transform is the approach to obtain frequency-domain representation of non-periodic signals.*

4.2 Fourier Transform: Development and Interpretation

Let $x(t)$ be a non-periodic signal. Note that a non-periodic signal can be considered as a periodic signal where the period is infinity. We shall use this principle for deriving Fourier transform based on our knowledge of Fourier series.

Let us construct a periodic signal $x_p(t)$ (shown in Fig. 4.1(b)) with period T_p by periodically repeating the non-periodic signal $x(t)$ (shown in Fig. 4.1(a)) as[1]

$$\begin{aligned} x_p(t) &= x(t) \qquad \text{for} \quad -\tfrac{T_p}{2} < t < \tfrac{T_p}{2} \\ \text{and} \quad x_p(t+mT_p) &= x_p(t) \qquad \text{for any integer } m. \end{aligned} \tag{4.3}$$

Hence, we can write a Fourier series representation for $x_p(t)$ as

$$x_p(t) = \sum_{k=-\infty}^{\infty} X_k e^{jk\Omega_p t} \tag{4.4}$$

$$X_k = \frac{1}{T_p} \int_{-T_0/2}^{T_0/2} x_p(t) e^{-jk\Omega_p t} dt, \tag{4.5}$$

for $k = 0, \pm 1, \pm 2, \ldots$. Now, based on Eqs. (4.3), (4.4) and (4.5), we have the following remarks (with $f_p = \frac{1}{T_p}$).

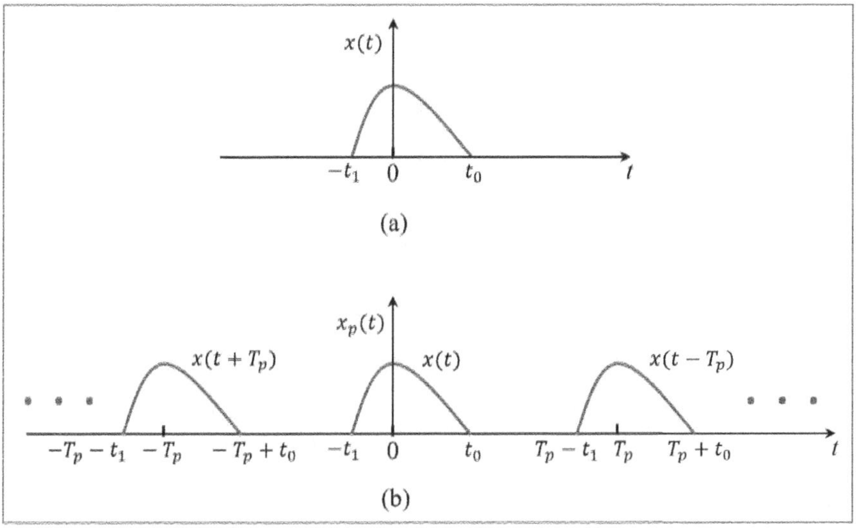

Figure 4.1. Construction of the periodic signal, $x_p(t)$ of period T_p shown in (b), using the non-periodic signal, $x(t)$ (a).

1. Without loss of generality, it is assumed that $x(t) = 0$ for $|t| > \frac{T_p}{2}$.

- As $T_p \to \infty$ (or, $f_p \to 0$), the periodic signal $x_p(t)$ will approach the non-periodic signal $x(t)$.
- Let $f_1 < f < f_2$ denote a small frequency range. The number of complex sinusoids (i.e., $e^{jk\Omega_p t}$) in $x_p(t)$ which lie in this range is given by $\frac{f_2 - f_1}{f_p}$. Then, consider the following.
  - As $T_p \to \infty$, the number of complex sinusoids that fall within the small frequency range tends to infinity.
  - The power of the signal is given by the sum of $|X_k|^2$ of all the sinusoids that make up the signal. Since the power of a given signal $x(t)$ is fixed,[2] it is easy to see that the values of $|X_k|^2$ must tend to zero as[3] $T_p \to \infty$.
  - Therefore, we cannot use X_k to denote the frequency-domain representation of $x_p(t)$ (or, $x(t)$) as $T_p \to \infty$.
- Let us define

$$X(kf_p) \triangleq T_p X_k = \frac{X_k}{f_p} \quad \Rightarrow \quad X_k = f_p X(kf_p). \tag{4.6}$$

Then, as $T_p \to \infty$, we have

$$\begin{aligned} x_p(t) &\to x(t), \qquad & f_p \to df, \qquad & kf_p \to f, \\ X_k &\to X(f)df, \qquad & X_k T_p \to X(f), & \end{aligned} \tag{4.7}$$

where df is an arbitrarily small frequency (almost zero) and f is the continuous-frequency variable.

§ *Examine the units of* X_k *and* $X(f)$.

- Using Eq. (4.6), we can rewrite Eqs. (4.4) and (4.5) as

$$x_p(t) = \sum_{k=-\infty}^{\infty} \left[f_p X(kf_p) \right] e^{j2\pi k f_p t} \tag{4.8}$$

$$T_p X_k = \int_{-T_p/2}^{T_p/2} x_p(t) e^{-j2\pi k f_p t} dt. \tag{4.9}$$

2. The power of a non-periodic finite-energy signal is zero.

3. For example, consider the frequency region from 100Hz to 110Hz, and the following values of f_p: 10Hz, 1Hz, 0.1Hz, 0.01Hz, 0.001Hz, 0.0001Hz, ... Then, as f_p keeps decreasing, the number of harmonics that fall between 100Hz and 110Hz keeps increasing.

With $T_p \to \infty$ and substituting Eq. (4.7) in Eqs. (4.8) and (4.9), we get,

$$x(t) = \int_{-\infty}^{\infty} [X(f)df]\, e^{j2\pi ft} = \int_{-\infty}^{\infty} X(f)e^{j2\pi ft} df \tag{4.10}$$

$$X(f) = \int_{-\infty}^{\infty} x(t)e^{-j2\pi ft} dt. \tag{4.11}$$

Eq. (4.11) is called the *forward Fourier transform* (the analysis equation). That is, getting the frequency-domain $X(f)$ from the time-domain $x(t)$. Equation (4.10) is called the *inverse Fourier transform* (the synthesis equation). That is, getting the time-domain $x(t)$ from the frequency-domain[4] $X(f)$.

We have the following remarks on the Fourier transform representation given in Eqs. (4.10) and (4.11).

- Complex sinusoids of all frequencies (i.e., $-\infty < f < \infty$; continuous-variable) are required for representing non-periodic signals. But the Fourier transform $X(f)$ will be different for each signal.
- The amplitudes of the complex sinusoids used in Fourier transform are given by $X(f)df$, compared to X_k in Fourier series (Note: df is arbitrarily close to zero).
- Because $X(f)$ is a complex number, we can express it as

$$X(f) = |X(f)|e^{j\theta(f)} \tag{4.14}$$

$$\text{where} \quad |X(f)| = \sqrt{X_r^2(f) + X_i^2(f)} \tag{4.15}$$

$$\text{and} \quad \theta(f) = \tan^{-1}\left(\frac{X_i(f)}{X_r(f)}\right) \tag{4.16}$$

Here, $X(f) = X_r(f) + jX_i(f)$ with $X_r(f)$ and $X_i(f)$ being the real and imaginary parts, respectively, of $X(f)$.

  – Magnitude Spectrum of $x(t)$: It is the plot of $|X(f)|$ *vs* f *(i.e., distribution of magnitude vs frequency).*

4. The inverse and forward Fourier transform equations can also be expressed in terms of the radian frequency variable $\Omega = 2\pi f$ as,

$$x(t) = \frac{1}{2\pi}\int_{-\infty}^{\infty} X\left(\frac{\Omega}{2\pi}\right) e^{j\Omega t} d\Omega \tag{4.12}$$

$$X\left(\frac{\Omega}{2\pi}\right) = \int_{-\infty}^{\infty} x(t)e^{-j\Omega t} dt. \tag{4.13}$$

- Phase Spectrum of $x(t)$: It is the plot of $\theta(f)$ *vs* f *(i.e., distribution of phase vs frequency).*

- The Fourier transform $X(f)$ gives the frequency composition of the given signal $x(t)$. That is,

  § *By observing the function $X(f)$, we can know which all frequencies (or, frequency regions) are required to construct the signal $x(t)$.*
  § *Fourier transform gives a frequency-domain representation of non-periodic signals.*

- Energy (density) Spectrum of $x(t)$: As shown earlier in Chapter 2, we get the energy of the signal $x(t)$ as (see Eq. (2.4)),

$$e = \int_{-\infty}^{\infty} |x(t)|^2 dt = \int_{-\infty}^{\infty} |X(f)|^2 df. \tag{4.17}$$

Eq. (4.17) is known as the *Parseval's Theorem for non-periodic signals*, which says that we can use either time-domain expression or frequency-domain expression for computing the energy of a signal, depending upon which is easier to use.

Therefore, the *energy density spectrum (i.e., distribution of energy vs frequency)* of $x(t)$ is given by $|X(f)|^2$. Consequently, the energy of $x(t)$ in a frequency range $f_3 < f < f_4$ can be obtained by,

$$\int_{f_3}^{f_4} |X(f)|^2 df. \tag{4.18}$$

Example: Consider an example of a rectangular pulse signal ('$x(t) = \mathsf{rect}(t)$', shown in Fig. 4.2(a)) defined as follows:

$$x(t) = \begin{cases} 1 & \text{for } |t| \leq 0.5 \\ 0 & \text{for } |t| > 0.5. \end{cases} \tag{4.19}$$

Its Fourier transform is calculated as,

$$\begin{aligned} X(f) &= \int_{-\infty}^{\infty} x(t) e^{-j2\pi f t} dt \\ &= \int_{-0.5}^{0.5} e^{-j\Omega t} dt = \left[\frac{e^{-j\Omega t}}{-j\Omega} \right]_{-0.5}^{0.5} \end{aligned}$$

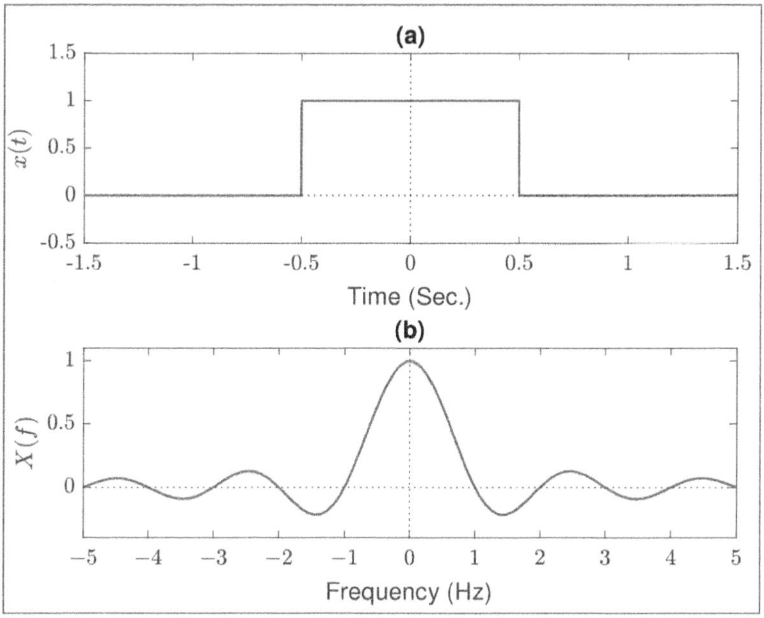

Figure 4.2. FT of rectangular pulse: (a) $x(t) = \text{rect}(t)$, and (b) Fourier Spectrum ($X(f) = \text{sinc}(f)$).

$$= \frac{e^{-j0.5\Omega} - e^{j0.5\Omega}}{-j\Omega} = \frac{-2j\sin(0.5\Omega)}{-j\Omega}$$

$$= \frac{\sin(0.5\Omega)}{0.5\Omega} = \frac{\sin(\pi f)}{\pi f} = \text{sinc}(f). \tag{4.20}$$

Figure 4.2(b) shows $\text{sinc}(f)$ as a function of frequency (f). It can be seen that the maximum of $\text{sinc}(f)$ is 1.0 and it occurs at $f = 0$. It can also be seen that the first zero crossings (on either side of the frequency axis) of $\text{sinc}(f)$ is at 1 and -1, respectively, which corresponds to the reciprocal of the width (T) of the rectangular pulse. In this case, $T = 1$. The subsequent zero crossings are at frequencies $f_k = k/T;\ k = \pm 2, \pm 3, \ldots$. The width of the main lobe of $\text{sinc}(f)$ is $\frac{2}{T}$. As T increases, the main lobe width decreases and vice versa. In the limiting case when $T \to \infty$, the sinc function becomes an impulse function.

4.3 Properties of Fourier Transform

The FT theorems and properties are tabulated and is available in Appendix B. We shall now briefly look at some of the important properties of Fourier transform (FT). Let $x(t)$ and $y(t)$ be two non-periodic signals with FTs $X(f)$ and $Y(f)$, respectively. That is,

$$x(t) = \int_{-\infty}^{\infty} X(f)e^{j2\pi ft}df, \qquad y(t) = \int_{-\infty}^{\infty} Y(f)e^{j2\pi ft}df$$

$$X(f) = \int_{-\infty}^{\infty} x(t)e^{-j2\pi ft}dt, \qquad Y(f) = \int_{-\infty}^{\infty} y(t)e^{-j2\pi ft}dt.$$

4.3.1 Linearity

Let us construct a new signal $z(t)$ as the linear combination of $x(t)$ and $y(t)$. That is,

$$z(t) = \alpha x(t) + \beta y(t), \tag{4.21}$$

where α and β are two scalars (real or complex). Then, we can obtain the FT of $z(t)$ as

$$\begin{aligned}
Z(f) &= \int_{-\infty}^{\infty} z(t)e^{-j2\pi ft}dt \\
&= \int_{-\infty}^{\infty} [\alpha x(t) + \beta y(t)]\, e^{-j2\pi ft}dt \\
&= \int_{-\infty}^{\infty} \alpha x(t)e^{-j2\pi ft}dt + \int_{-\infty}^{\infty} \beta y(t)e^{-j2\pi ft}dt \\
&= \alpha \int_{-\infty}^{\infty} x(t)e^{-j2\pi ft}dt + \beta \int_{-\infty}^{\infty} y(t)e^{-j2\pi ft}dt \\
&= \alpha X(f) + \beta Y(f).
\end{aligned} \tag{4.22}$$

Thus, *FT of a linear combination of signals is given by the corresponding linear combination of the FTs of the individual signals.*

4.3.2 Time Shifting

Let $z(t) = x(t - t_0)$ where t_0 is an arbitrary time offset. Then, the FT of $z(t)$ is given by

$$\begin{aligned}
Z(f) &= \int_{-\infty}^{\infty} z(t)e^{-j2\pi ft}dt = \int_{-\infty}^{\infty} x(t - t_0)e^{-j2\pi ft}dt \\
&= \int_{-\infty}^{\infty} x(u)e^{-j2\pi f(u+t_0)}du \qquad : u = t - t_0
\end{aligned}$$

$$= e^{-j2\pi f t_0} \int_{-\infty}^{\infty} x(u) e^{-j2\pi f u} du$$
$$= e^{-j2\pi f t_0} X(f). \quad (4.23)$$

Let $\theta_z(f)$ and $\theta_x(f)$ be the phase spectra of $z(t)$ and $x(t)$, respectively. Then, we can express Eq. (4.23) as

$$|Z(f)| e^{j\theta_z(f)} = e^{-j2\pi f t_0} |X(f)| e^{j\theta_x(f)}$$
$$= |X(f)| e^{j(\theta_x(f) - 2\pi f t_0)} \quad (4.24)$$
$$\Longrightarrow \quad |Z(f)| = |X(f)|, \text{ and} \quad (4.25)$$
$$\theta_z(f) = \theta_x(f) - 2\pi f t_0. \quad (4.26)$$

Thus, we see that *shifting a signal in time-domain results in a phase change in the frequency-domain, while the magnitude spectrum remains unaffected.* Note also that this phase change is a linear function of frequency.

4.3.3 Frequency Shifting or Modulation

Let $z(t) = e^{j2\pi f_0 t} x(t)$ where f_0 is a given non-zero frequency. Then, we get its FT as

$$Z(f) = \int_{-\infty}^{\infty} \left[e^{j2\pi f_0 t} x(t) \right] e^{-j2\pi f t} dt$$
$$= \int_{-\infty}^{\infty} x(t) e^{-j2\pi (f - f_0) t} dt$$
$$= X(f - f_0), \quad (4.27)$$

by the definition of FT.

The frequency shifting property of the FT is illustrated in Fig. 4.3. It can be seen that the spectrum of $z(t)$ is a frequency shifted (to the right in the frequency axis) version of $|X(f)|$. On the other hand, if the frequency shift is by $-f_0$, then the spectrum would shift to the left in the frequency axis. The corresponding spectrum would be, $Z(f) = X(f + f_0)$. As an example, let $f_0 = 1$ MHz. Then, Eq. (4.27) implies that multiplying $x(t)$ by $e^{j2\pi f_0 t}$ results in shifting the spectrum to the right by 1 MHz. This principle is used in communication systems for shifting a signal from one frequency range to another for transmission purpose.

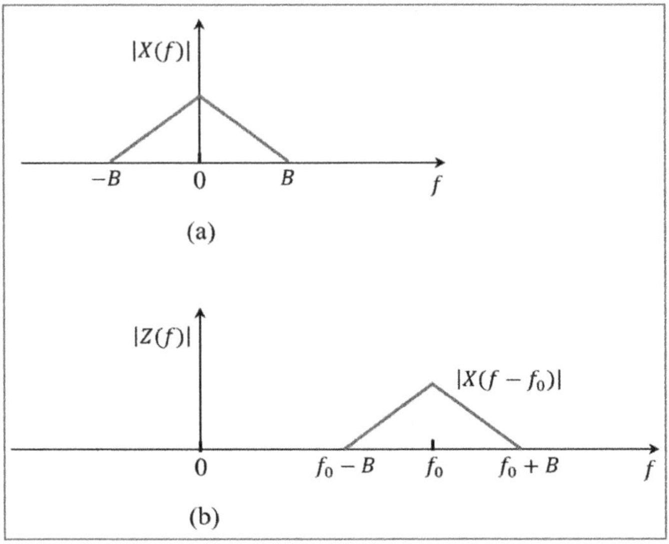

Figure 4.3. Illustration of 'frequency shifting' property of the Fourier transform. (a) spectra of $x(t)$, and (b) spectra of $z(t) = e^{j2\pi f_0 t}x(t)$.

4.3.4 Time Reversal

Let $z(t) = x(-t)$. Then, we get

$$\begin{aligned} Z(f) &= \int_{-\infty}^{\infty} x(-t)e^{-j2\pi f t}dt \\ &= -\int_{\infty}^{-\infty} x(u)e^{j2\pi f u}du \colon u = -t \\ &= \int_{-\infty}^{\infty} x(u)e^{j2\pi f u}du = \int_{-\infty}^{\infty} x(u)e^{-j2\pi(-f)u}du \\ &= X(-f). \end{aligned} \tag{4.28}$$

That is, *reversing a signal in time-domain results in reversing its spectrum in the frequency-domain.*

4.3.5 Time Scaling

Let $z(t) = x(t/a)$ where a is a real number. This is called time-scaling. This results in $z(t)$ being an expanded or compressed version of[5] $x(t)$.

5. To see this, define $t' = t/a$ and, for example, consider the interval $-10 < t < 10$. If $a = 2$, then this interval in t corresponds to the interval $-5 < t' < 5$ in t'. Similarly, if $a = 0.5$, it corresponds to the interval $-20 < t' < 20$ in t'. Thus, $|a| > 1$ results in expanding the signal $x(t)$ and $|a| < 1$ results in compressing the signal $x(t)$.

The FT of $z(t)$ is derived *for the case of* $a > 0$ and is given by,

$$Z(f) = aX(af). \tag{4.29}$$

Let us now consider the case when $a < 0$. Noting that $a = -|a|$ when $a < 0$, we get

$$\begin{aligned}
Z(f) &= \int_{-\infty}^{\infty} x(t/a)e^{-j2\pi ft}dt \\
&= a\int_{\infty}^{-\infty} x(u)e^{-j2\pi fau}du \qquad : u = t/a \\
&= -|a|\int_{\infty}^{-\infty} x(u)e^{-j2\pi(af)u}du \\
&= |a|\int_{-\infty}^{\infty} x(u)e^{-j2\pi(af)u}du \\
&= |a|X(af).
\end{aligned} \tag{4.30}$$

Comparing Eqs. (4.29) and (4.30), we see that we can take $|a|X(af)$ as the FT of the time-scaled signal $x(t/a)$ for any a (i.e., $a > 0$ or $a < 0$).

Note that $X(af)$ results in scaling in the frequency-domain. In particular, we note the following:

- Case 1: $|a| > 1$ - Results in expanding $x(t)$ in time-domain and compressing $X(f)$ in frequency-domain.
- Case 2: $|a| < 1$ - Results in compressing $x(t)$ in time-domain and expanding $X(f)$ in frequency-domain.

Example: Consider an example of a rectangular pulse ('$x(t) = \text{rect}(t)$') (Fig. 4.4(a)) and its time scaled pulse $z(t) = x\left(\frac{t}{1.5}\right)$ as shown in Fig. 4.4(b) below. The first rect signal has a width of $T = 1$ second and the second one has a width of $T = 1.5$ seconds. The corresponding spectra are shown in Figs. 4.4(c) and 4.4(d), respectively. As explained earlier, $z(t)$ is a time-expanded version of $x(t)$ and correspondingly, its FT, $Z(f)$ is a frequency compressed version of $X(f)$. This is clear from the two sinc functions in the figure. The maximum amplitude of the sinc function in Fig. 4.4(c) is 1.0 and the zero crossings of it occurs in frequencies $f_k = k/1.0;\; k = \pm1, \pm2, \ldots$. For the second sinc function in Fig. 4.4(d), which corresponds to the FT of time-expanded rect function ($z(t)$), the frequency compression is very clear. Its zero crossings are at frequencies, $f_k = k/1.5;\; k = \pm1, \pm2, \ldots$. The maximum amplitude of the sinc function is 1.5 (see Eq. (4.30)), which is the time-scaling factor.

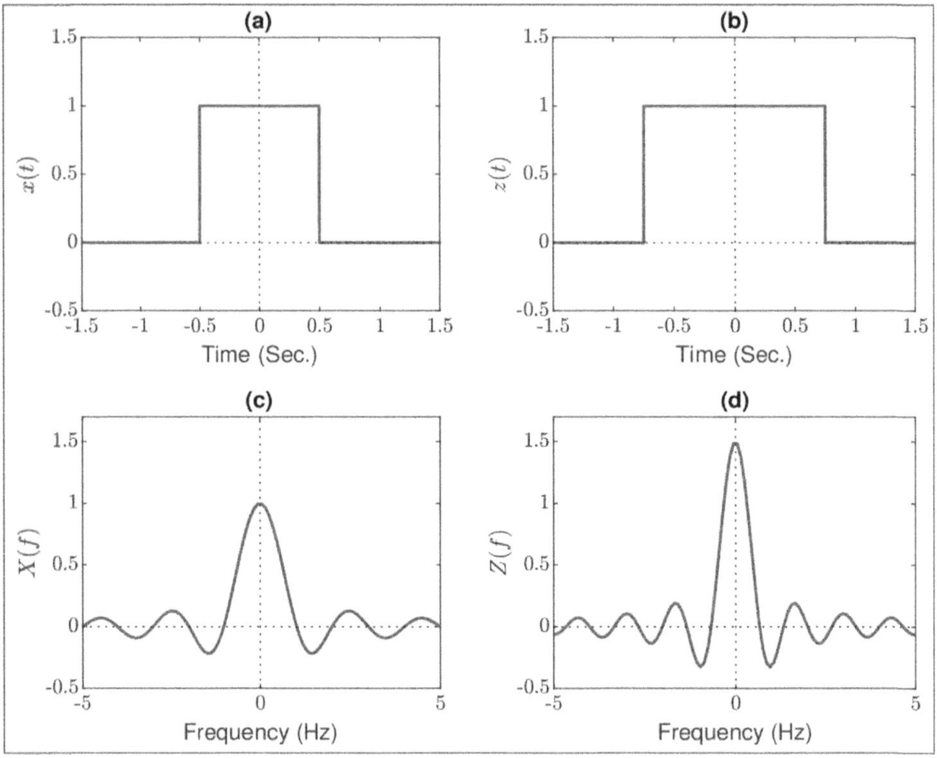

Figure 4.4. FT of rectangular pulses: (a) rect(t) (b) rect($\frac{t}{1.5}$), (c) Fourier spectrum of rect(t), and (d) Fourier spectrum of rect($\frac{t}{1.5}$).

4.3.6 Time Scaling and Delay

Let $z(t) = x\left(\frac{t-t_0}{a}\right)$ where t_0 and a are real numbers. Clearly, $z(t)$ is a time-scaled and time-shifted version of $x(t)$. The FT of $z(t)$ (assuming $a > 0$ for convenience) is given by

$$\begin{aligned} Z(f) &= \int_{-\infty}^{\infty} x\left(\frac{t-t_0}{a}\right) e^{-j2\pi f t} dt \\ &= a \int_{-\infty}^{\infty} x(u) e^{-j2\pi f (au+t_0)} du \qquad : u = (t-t_0)/a \\ &= a e^{-j2\pi f t_0} \int_{-\infty}^{\infty} x(u) e^{-j2\pi (af) u} du \\ &= a e^{-j2\pi f t_0} X(af). \end{aligned} \tag{4.31}$$

As before, if $a < 0$, then we would get,

$$Z(f) = |a| e^{-j2\pi f t_0} X(af). \tag{4.32}$$

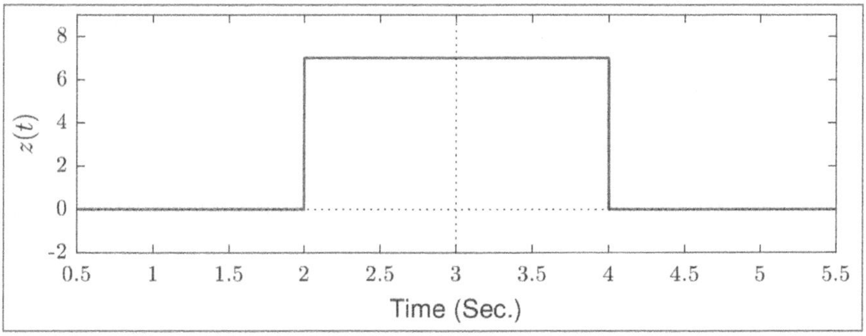

Figure 4.5. Amplitude scaled and shifted rectangular function.

Example: Let us *apply the time scaling and delaying to the problem of a rectangular pulse as given in the following expression*:

$$z(t) = \begin{cases} 7 & \text{for } 2 \leq t \leq 4 \\ 0 & \text{for } t < 2,\ t > 4. \end{cases} \tag{4.33}$$

Figure 4.5 illustrates the given function $z(t)$. Clearly, this signal is a time-scaled and time-shifted version of the standard $\text{rect}(t)$ function which is defined as

$$\text{rect}(t) = \begin{cases} 1 & \text{for } -0.5 \leq t \leq 0.5 \\ 0 & \text{for } t < -0.5,\ t > 0.5. \end{cases} \tag{4.34}$$

So, we can express $z(t)$ as

$$z(t) = 7\text{rect}\left(\frac{t - t_0}{a}\right). \tag{4.35}$$

Comparing the waveforms of $z(t)$ and $\text{rect}(t)$, we see that the points $t = -0.5$ and $t = 0.5$ in $\text{rect}(t)$ correspond to $t = 2$ and $t = 4$, respectively, in $z(t)$. That is,

$$(t - t_0)/a = -0.5 \qquad \text{when } t = 2 \tag{4.36}$$

$$(t - t_0)/a = +0.5 \qquad \text{when } t = 4. \tag{4.37}$$

So, we can determine t_0 and a by solving the equations:

$$2 = -0.5a + t_0 \qquad \text{and} \qquad 4 = 0.5a + t_0$$

which results in $t_0 = 3$ and $a = 2$. Therefore, we get

$$z(t) = 7\text{rect}\left(\frac{t - 3}{2}\right). \tag{4.38}$$

Since the FT of $\text{rect}(t)$ is $\text{sinc}(f)$, we can get the FT of $z(t)$ using Eq. (4.31) as

$$Z(f) = 7\left[2e^{-j2\pi f3}\text{sinc}(2f)\right] = 14e^{-j6\pi f}\text{sinc}(2f). \tag{4.39}$$

4.3.7 Time Differentiation

Let $z(t) = \frac{dx(t)}{dt}$. To determine the FT of $z(t)$, recall from Eq. (4.10) the inverse FT expression:

$$x(t) = \int_{-\infty}^{\infty} X(f)e^{j2\pi ft}df. \tag{4.40}$$

Differentiating both sides of this equation, we get

$$\begin{aligned}\frac{dx(t)}{dt} &= \int_{-\infty}^{\infty} X(f)\left[e^{j2\pi ft}(j2\pi f)\right]df \\ &= \int_{-\infty}^{\infty} [j2\pi f X(f)]\, e^{j2\pi ft}df.\end{aligned} \tag{4.41}$$

Eq. (4.41) is in the form of a standard inverse FT where the time-domain signal is $\frac{dx(t)}{dt}$ with its corresponding FT $j2\pi f X(f)$. Therefore, we get the FT of $z(t)$ as,

$$Z(f) = j2\pi f X(f). \tag{4.42}$$

From Eq. (4.42), we note that *differentiation in time-domain results in attenuating low-frequency part of the spectrum and amplifying high-frequency part of the spectrum in frequency-domain*. In particular,[6] note that $Z(0) = 0$ whether or not $X(0) = 0$ at $f = 0$.

4.3.8 Duality

If $X(f)$ is the FT of $x(t)$, then the FT of $X(t)$ is given by[7] $x(-f)$. Similarly, the FT of $X(-t)$ is given by $x(f)$. Using this, we can easily see that the FT of $\text{sinc}(t)$ is $\text{rect}(f)$, since we know that the FT of $\text{rect}(t)$ is $\text{sinc}(f)$.

Example: Consider an example of a rectangular pulse signal ('$X(f) = \text{rect}(f)$', in the frequency domain as shown in Fig. 4.6(a)) defined as follows:

$$X(f) = \begin{cases} 1 & \text{for } |f| \leq 0.5 \\ 0 & \text{for } |f| > 0.5. \end{cases} \tag{4.43}$$

6. Note that $X(0)$ is the total area under $x(t)$ (i.e., $\int_{-\infty}^{\infty} x(t)dt$).

7. Prove this by applying simple change of variables in Eq. (4.11).

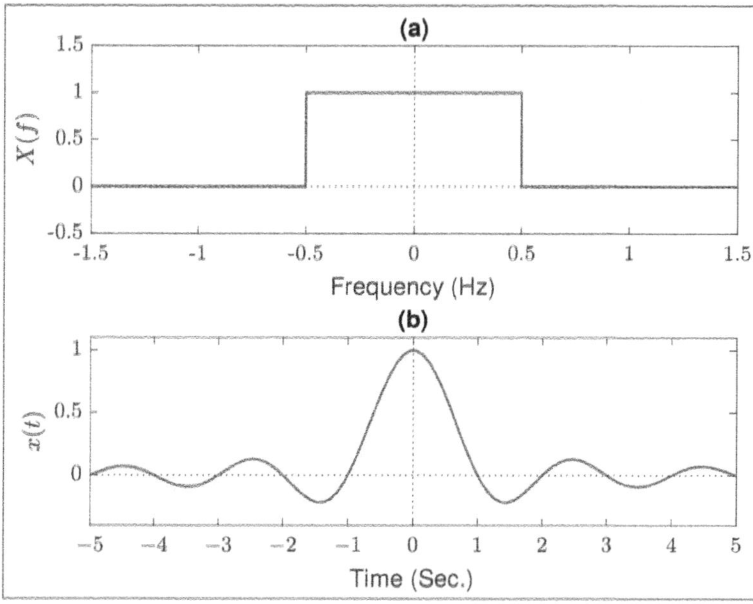

Figure 4.6. FT of rectangular pulse in the frequency domain: (a) rect(f), (b) Fourier spectrum (sinc(t)).

By taking the inverse Fourier transform, we get

$$
\begin{aligned}
x(t) &= \int_{-\infty}^{\infty} X(f) e^{j2\pi f t} df \\
&= \int_{-0.5}^{0.5} e^{j\omega t} df = \left[\frac{e^{j\omega t}}{j2\pi t} \right]_{-0.5}^{0.5} \\
&= \frac{e^{j2\pi 0.5t} - e^{j2\pi 0.5t}}{j2\pi t} = \frac{2j \sin(\pi t)}{j2\pi t} \\
&= \frac{\sin(\pi t)}{\pi t} = \text{sinc}(t).
\end{aligned} \tag{4.44}
$$

The resulting sinc function is plotted in Fig. 4.6(b). This explains the duality property of the FT.

4.3.9 Symmetric Signals

As in the case of Fourier series, the FT $X(f)$ also has some special features when $x(t)$ is real and/or symmetric. We list them below:

- Real $x(t)$: That is, $x^*(t) = x(t)$. Then, we will have $X(-f) = X^*(f)$. This results in *even-symmetric magnitude spectrum and odd-symmetric phase* *spectrum*:

$$|X(-f)| = |X(f)|, \qquad \theta(-f) = -\theta(f). \tag{4.45}$$

- Real and Even-Symmetric $x(t)$: That is, $x^*(t) = x(t)$ and $x(-t) = x(t)$. Then, we will have,

$$X(-f) = X^*(f) = X(f). \tag{4.46}$$

  That is, $X(f)$ is real and even-symmetric. The phase $\theta(f)$ is either 0 or $\pm\pi$.
- Real and Odd-Symmetric $x(t)$: That is, $x^*(t) = x(t)$ and $x(-t) = -x(t)$. Then, we will have,

$$X(-f) = -X(f) = X^*(f). \tag{4.47}$$

  That is, $X(f)$ is imaginary and odd-symmetric. The phase $\theta(f)$ is $\pm\pi/2$.
- Real $X(f)$: That is, $X^*(f) = X(f)$. Then, we will have,

$$x(-t) = x^*(t). \tag{4.48}$$

  That is, $x(t)$ is conjugate-symmetric.
- Real and Even-Symmetric $X(f)$: That is, $X^*(f) = X(f)$ and $X(-f) = X(f)$. Then, we will have,

$$x(-t) = x^*(t) = x(t). \tag{4.49}$$

  That is, $x(t)$ is real and even-symmetric.

A summary of the symmetry properties for the FT is shown in Fig. 4.7.

4.4 Fourier Transform of Periodic Signals

Recall that the main motivation for the development of FT is that Fourier series cannot be used to obtain frequency-domain representation of non-periodic signals. While FT is applicable for non-periodic signals, we could ask the following question:

Is Fourier transform also applicable for periodic signals?

The answer is 'yes' and we look into some details of this in this section.

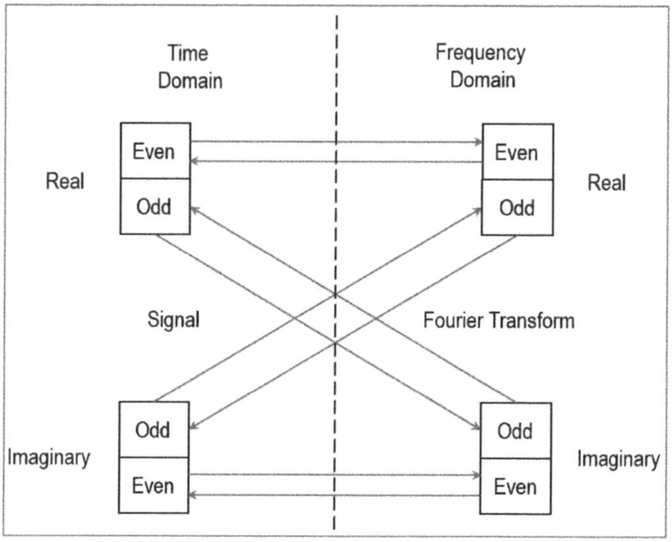

Figure 4.7. Summary of symmetry properties of FT.

4.4.1 Difficulty with Sinusoids

Consider a complex sinusoid given by,

$$x(t) = Ae^{j(2\pi f_0 t+\theta)} = \tilde{A}e^{j2\pi f_0 t}, \tag{4.50}$$

where A, f_0 and θ are the real-amplitude, frequency and phase, respectively, and $\tilde{A} = Ae^{j\theta}$. Then, it is easy to see that the Fourier series coefficients of $x(t)$ are given by,

$$X_k = \begin{cases} \tilde{A} & \text{for } k = 1 \\ 0 & \text{for } k \neq 1. \end{cases} \tag{4.51}$$

That is, *the frequency-domain representation of a complex sinusoid has only one non-zero value and this corresponds to the frequency of the sinusoid.*

We may also apply the standard FT approach for determining the FT of the complex sinusoid given above. Let $X(f)$ denote its FT. Based on our knowledge of the signal $\tilde{A}e^{j2\pi f_0 t}$ and its Fourier series, we know that $X(f)$ can be non-zero only at the frequency $f = f_0$. Further, we also know that its FT $X(f)$ should obey the following (see Eq. (4.17))

$$e = \int_{-\infty}^{\infty} |x(t)|^2 dt = \int_{-\infty}^{\infty} |X(f)|^2 df, \tag{4.52}$$

because of the Parseval's theorem. On the other hand, for the complex sinusoid $\tilde{A}e^{j2\pi f_0 t}$, the energy is clearly infinity. This implies that its FT $X(f)$ at $f = f_0$

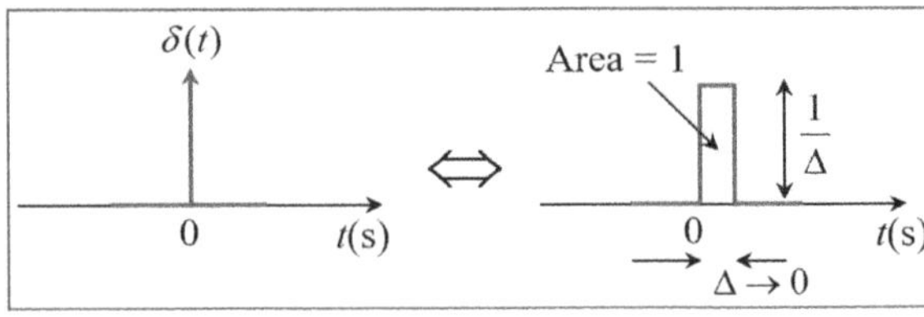

Figure 4.8. Impulse function.

should be also infinity! (since $X(f)$ in this case is non-zero only at $f = f_0$). In other words, we should have

$$\lim_{df \to 0} |X(f_0)|^2 df = \infty \tag{4.53}$$

where df is an arbitrarily small frequency interval around $f = f_0$. To handle this very special and difficult situation, we need to understand the concept behind the so-called *impulse function.*

4.4.2 Impulse Function

The impulse function (denoted by $\delta(t)$) and illustrated in Fig. 4.8 is defined as,

$$\delta(t) = \begin{cases} 1.0 & \text{for } \; t = 0 \\ 0 & \text{elsewhere} \end{cases} \tag{4.54}$$

We also note that

$$\int_{-\infty}^{\infty} \delta(t) dt = 1. \tag{4.55}$$

So, $\delta(t)$ is a function having infinite amplitude and zero width such that its area is equal to 1. An impulse sitting at $t = t_0$ is given by $\delta(t - t_0)$. We may also define the impulse function using a limiting form as[8]

$$\lim_{\Delta \to 0} \left\{ \delta(t) = \frac{1}{\Delta} \quad \text{for } \; 0 \leq t < \Delta \right\}. \tag{4.57}$$

8. The definitions given in Eqs. (4.54), (4.55), and (4.57) can be generalized to the case of the impulse $\delta(t - t_0)$ sitting at $t = t_0$ as

$$\delta(t - t_0) = 0 \quad \text{for } \; t \neq t_0 \tag{4.56a}$$

$$\int_{-\infty}^{\infty} \delta(t - t_0) dt = 1 \tag{4.56b}$$

$$\lim_{\Delta \to 0} \left\{ \delta(t - t_0) = \tfrac{1}{\Delta} \quad \text{for } \; t_0 \leq t < t_0 + \Delta \right\}. \tag{4.56c}$$

4.4.3 Properties of Impulse Function

Based on the above introduction to impulse function, we can list the following properties.

- Unit Area: As defined by Eq. (4.55), the area under an impulse function is equal to 1 (for any t_0):

$$\int_{-\infty}^{\infty} \delta(t - t_0)dt = 1. \tag{4.58}$$

- Sampling Property: By multiplying a signal with an impulse function, we can obtain the value of the signal at the location of the impulse. That is,

$$x(t)\delta(t - t_0) = x(t_0)\delta(t - t_0) = x(t_0). \tag{4.59}$$

Consequently, a discrete-time signal $x(nT_s)$ can be obtained from a continuous-time signal $x(t)$ by multiplying $x(t)$ with a train (or, sequence) of impulses spaced T_s apart. That is,

$$\begin{aligned} y(t) = x(t) \times \sum_{n=-\infty}^{\infty} \delta(t - nT_s) &= \sum_{n=-\infty}^{\infty} x(nT_s)\delta(t - nT_s) \\ &= \cdots + x(-2T_s)\delta(t + 2T_s) + x(-T_s)\delta(t + T_s) + x(0)\delta(t) \\ &\quad + x(T_s)\delta(t - T_s) + x(2T_s)\delta(t - 2T_s) + \cdots \end{aligned} \tag{4.60}$$

That is,

$$y(t) = \begin{cases} x(kT_s)\delta(t - kT_s) & \text{if } \; t = kT_s \\ 0 & \text{if } \; t \neq kT_s. \end{cases} \tag{4.61}$$

for $k = \ldots, -2, -1, 0, 1, 2, \ldots$ and T_s is the sampling interval.[9] Thus, $y(t)$ is a discrete-time signal, that is, non-zero only at the discrete-time instants $kT_s, k = 0, \pm 1, \pm 2, \ldots$.

9. There is yet another property known as *shifting property* of impulse function. We will study this later.

- <u>Flat Fourier Transform</u>: The FT of $\delta(t - t_0)$ is given by[10]

$$\begin{aligned}\mathcal{F}[\delta(t-t_0)] &= \int_{-\infty}^{\infty} \delta(t-t_0)e^{-j2\pi f t}dt \\ &= \int_{-\infty}^{\infty} \delta(t-t_0)e^{-j2\pi f t_0}dt \quad : Eq.\ (4.59) \\ &= e^{-j2\pi f t_0}\int_{-\infty}^{\infty} \delta(t-t_0)dt \\ &= e^{-j2\pi f t_0},\end{aligned} \tag{4.62}$$

because of Eq. (4.58). Thus, the magnitude of the FT of $\delta(t - t_0)$ is *flat over the complete frequency range* ($-\infty < f < \infty$) and its phase is given by $-2\pi f t_0$.

4.4.4 Fourier Transform of Sinusoids

Since the FT of a time-domain impulse $\delta(t - t_0)$ is $e^{-j2\pi f t_0}$, let us examine what is the inverse FT of an impulse $\delta(f - f_0)$ in frequency-domain, where f_0 is a frequency offset. Using Eq. (4.10), we obtain[11]

$$\begin{aligned}\mathcal{F}^{-1}[\delta(f-f_0)] &= \int_{-\infty}^{\infty} \delta(f-f_0)e^{j2\pi f t}df \\ &= \int_{-\infty}^{\infty} \delta(f-f_0)e^{j2\pi f_0 t}df \quad : Eq.\ (4.59) \\ &= e^{j2\pi f_0 t}\int_{-\infty}^{\infty} \delta(f-f_0)df \\ &= e^{j2\pi f_0 t},\end{aligned} \tag{4.63}$$

which is nothing but a complex sinusoid of frequency f_0. Thus,

> § *The FT of a complex sinusoid of frequency f_0 is an impulse $\delta(f - f_0)$ located at frequency f_0.*

Generalizing this to the complex sinusoid $x(t) = \tilde{A}e^{j2\pi f_0 t}$ given in Eq. (4.50), we get its FT as

$$\mathcal{F}\left[\tilde{A}e^{j2\pi f_0 t}\right] = \tilde{A}\delta(f - f_0). \tag{4.64}$$

10. The notation '$\mathcal{F}[x(t)]$' refers to the 'Fourier transform of $x(t)$'.
11. The notation '$\mathcal{F}^{-1}[X(f)]$' refers to the 'inverse FT of $X(f)$'.

Since the FT given in Eq. (4.64) is an impulse function (which has infinite amplitude), we see that this satisfies Eq. (4.53).

4.4.5 Fourier Transform of Periodic Signals

We shall now use the result in Eq. (4.64) to determine the FT of any periodic signal. Let $x(t)$ be a periodic signal with period $T_p = \frac{1}{f_p}$, where f_p is the frequency of the periodic signal, and Fourier series representation given by

$$x(t) = \sum_{k=-\infty}^{\infty} X_k e^{j2\pi k f_p t}. \tag{4.65}$$

Since $e^{j2\pi k f_p t}$ is a complex sinusoid of frequency kf_p, using the linearity property of FT, we can get the FT of $x(t)$ as

$$\begin{aligned} X(f) = \mathcal{F}[x(t)] &= \mathcal{F}\left[\sum_{k=-\infty}^{\infty} X_k e^{j2\pi k f_p t}\right] \\ &= \sum_{k=-\infty}^{\infty} \mathcal{F}\left[X_k e^{j2\pi k f_p t}\right] \\ &= \sum_{k=-\infty}^{\infty} X_k \mathcal{F}\left[e^{j2\pi k f_p t}\right] \\ &= \sum_{k=-\infty}^{\infty} X_k \delta(f - kf_p). \end{aligned} \tag{4.66}$$

That is,

$$X(f) = \begin{cases} X_k \delta(f - kf_p) & \text{if } f = kf_p \\ 0 & \text{if } f \neq kf_p. \end{cases} \tag{4.67}$$

for $k = \ldots, -2, -1, 0, 1, 2, \ldots$ and f_p is the fundamental frequency. Thus, we see that

> § *The FT of a periodic signal exists only at frequencies corresponding to its harmonics* kf_p. *For this reason, such spectra are also known as* <u>*line spectra*</u>.

In other words, *knowledge of the Fourier series coefficients is sufficient to determine the FT of periodic signals.*

4.5 Dirichlet Conditions

A signal $x(t)$ that satisfies the following conditions is guaranteed to have a Fourier transform:

- It must be absolutely integrable. That is,

$$\int_{-\infty}^{+\infty} |x(t)|\, dt < \infty. \tag{4.68}$$

- It must have only finite number of maxima and minima within any finite interval.
- It must have only finite number of discontinuities, all of finite size, within any finite interval.

These conditions are known as *Dirichlet conditions* for existence of FT.

4.6 Fourier Transform Summary

In Table 4.1, we summarize the key properties of Fourier transform for continuous-time signals.

In Table 4.2, we give the Fourier transforms of some of the most commonly used signals.

Recall that $u(t)$ is the unit step function which is defined as

$$u(t) = \begin{cases} 1 & \text{for } t \geq 0 \\ 0 & \text{for } t < 0 \end{cases} \tag{4.69}$$

Therefore, the meaning of the expression $e^{-at}u(t)$ is as follows:

$$e^{-at}u(t) = \begin{cases} e^{-at} & \text{for } t \geq 0 \\ 0 & \text{for } t < 0. \end{cases} \tag{4.70}$$

Thus, the one-line expression $e^{-at}u(t)$ is an easier way to express the 2-line equation given by Eq. (4.70).

Table 4.1. Properties of Fourier transform.

Property	Time domain	Frequency domain
Linearity	$\alpha x(t)+\beta y(t)$	$\alpha X(f)+\beta Y(f)$
Time shifting	$x(t-t_0)$	$e^{-j2\pi f t_0}X(f)$
Time Scaling	$x\,(t/a)$	$\lvert a\rvert X(af)$
Time shifting and Time Scaling	$x\left(\frac{t-t_0}{a}\right)$	$\lvert a\rvert e^{-j2\pi f t_0}X(af)$
Time reversal	$x(-t)$	$X(-f)$
Frequency shifting	$e^{j2\pi f_0 t}x(t)$	$X(f-f_0)$
Modulation	$x(t)\cos(2\pi f_0 t)$	$\dfrac{X(f-f_0)+X(f+f_0)}{2}$
Modulation	$x(t)\sin(2\pi f_0 t)$	$\dfrac{X(f-f_0)-X(f+f_0)}{2j}$
Differentiation	$\dfrac{dx(t)}{dt}$	$j2\pi f X(f)$
Real Signal	$x^*(t)=x(t)$	$X(-f)=X^*(f)$
Real and Even-Symmetric	$x^*(t)=x(t)=x(-t)$	$X(-f)=X^*(f)=X(f)$
Real and Odd-Symmetric	$x^*(t)=x(t)=-x(-t)$	$X(-f)=X^*(f)=-X(f)$

Table 4.2. Fourier transform of selected signals.

Time domain	Frequency domain
$x(t)$	$X(f)=\int_{-\infty}^{\infty}x(t)e^{-j2\pi f t}dt$
$\mathrm{rect}(t)$	$\mathrm{sinc}(f)=\dfrac{\sin(\pi f)}{\pi f}$
$\delta(t)$	1
$\delta(t-t_0)$	$e^{-j2\pi f t_0}$
$e^{j2\pi f_0 t}$	$\delta(f-f_0)$
$e^{-at}u(t)$	$\dfrac{1}{a+j2\pi f}$ if $a>0$

4.7 Concluding Remarks

- Fourier transform can be used to give frequency-domain representation of non-periodic as well as periodic signals. In contrast to this, Fourier series can be used only for periodic signals.
- Fourier transform of non-periodic signals are continuous in the frequency-domain, whereas that of periodic signals is discrete (line spectra).
- Fourier transform can be used to assess the effective bandwidth of signals.
- The various properties of Fourier transform can be used to simplify the effort (integration, computations etc.) necessary in evaluating the Fourier transform or to verify the solutions.

Exercises

4.1. Consider the continuous signals shown in Fig. P4.1.

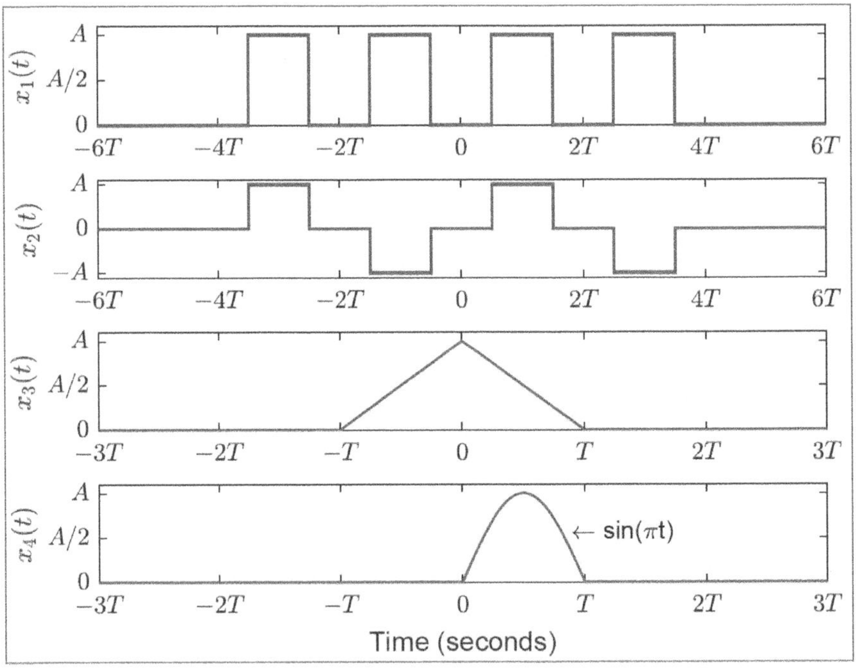

Figure P4.1.

(a) Find the spectrum of the signals in Fig. P4.1 using exclusively (only) the spectra of basic signals in combination with the appropriate properties of the Fourier transform.

(b) Sketch roughly the amplitude spectra of $x_1(t), x_2(t), x_3(t)$ and $x_4(t)$.

4.2. The analog signal in Fig. P4.2(a) represents the impulse response of a biological system. The amplitude of the corresponding transfer function ($H(f)$, the Fourier transform of the impulse response $h(t)$) is determined in two ways, and the results are depicted in Fig. P4.2(b) and Fig. P4.2(c), respectively.

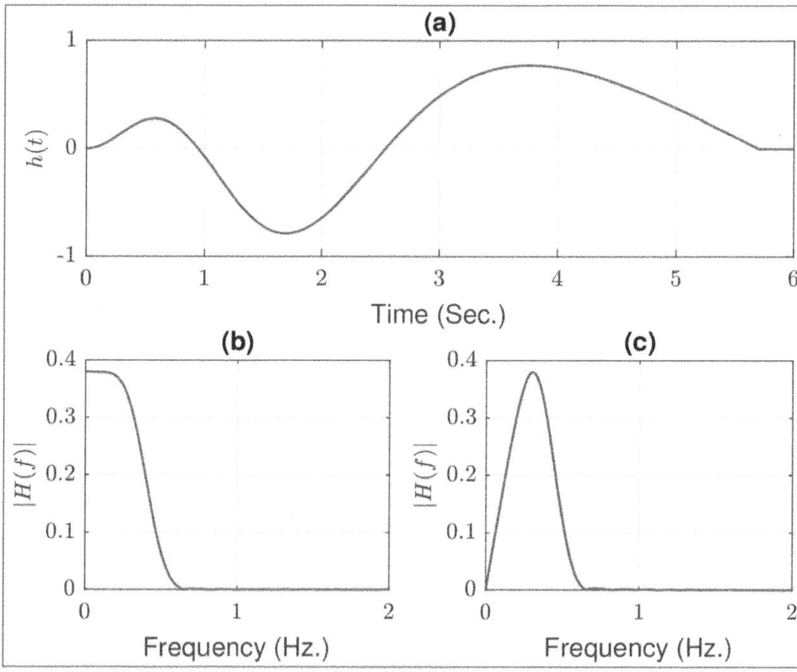

Figure P4.2.

Express $|H(f)|$ for $f = 0$ of $h(t)$. Using the result argue which of the results is incorrect.

4.3. An ultrasound system for investigation of the blood flow in the body uses a pulse given by:

$$p(t) = v(t)\sin(2\pi f_0 t).$$

Here, f_0 is the frequency of the pulse and $v(t)$ is the Hamming window given by,

$$v(t) = \begin{cases} 0.54 + 0.46\cos\left(\pi \frac{t}{T}\right) & \text{for } |t| < T \\ 0 & \text{otherwise} \end{cases}$$

The spectrum of $v(t)$ is

$$V(f) = \frac{1.08 - 0.64T^2 f^2}{2\pi f \left(1 - 4T^2 f^2\right)} \sin(2\pi f T).$$

The window ($v(t)$) and its amplitude spectra ($V(f)$) are shown in Fig. P4.3.

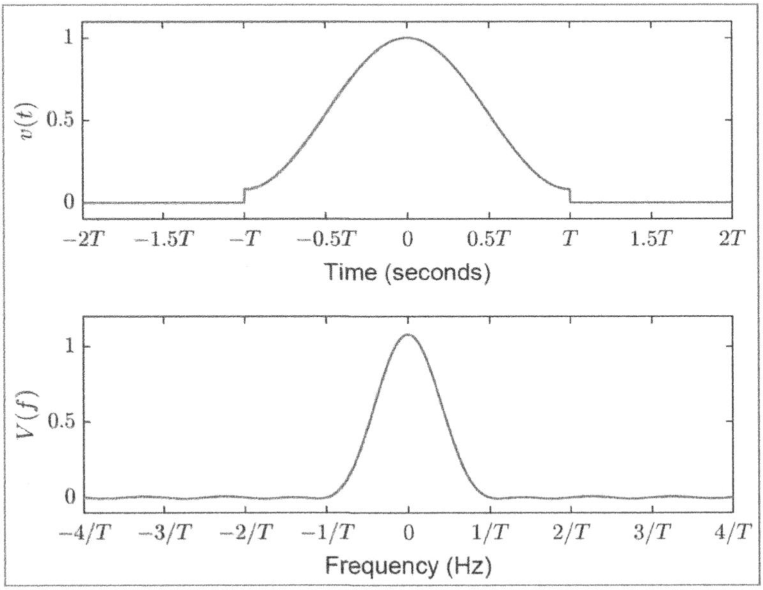

Figure P4.3.

Write down how one can find the spectrum $P(f)$ of $p(t)$ out from $V(f)$ by using appropriate properties of the Fourier transform. Show that $P(f)$ is concentrated around $f = \pm f_0$, when $T \gg \frac{1}{f_0}$.

4.4. To estimate the blood velocity, two identical pulses ($p(t)$) are transmitted one after the other separated by T_{prf} seconds. Two signals $y_1(t)$ and $y_2(t)$ are received, respectively. For the sake of simplicity, it is assumed that the zero time instance ($t = 0$) for the two signals $y_1(t)$ and $y_2(t)$ coincides with the time instance when the respective pulses have been transmitted. The signal reflected by the blood can be approximated with a single reflector with reflection coefficient a, such that the received signal from the first pulse transmission can be written as:

$$y_1(t) = ap(t - t_0) + g_v(t),$$

where a and t_0 are real positive constants. $g_v(t)$ is the signal received from the tissue surrounding the blood. At the next pulse transmission, the received signal is:

$$y_2(t) = ap(t - t_0 - t_s) + g_v(t),$$

where

$$t_s = \frac{2V_b}{c} T_{prf}.$$

V_b is the speed of the blood, c is the speed of propagation of ultrasound signal. Figure P4.4 shows the two received signals.

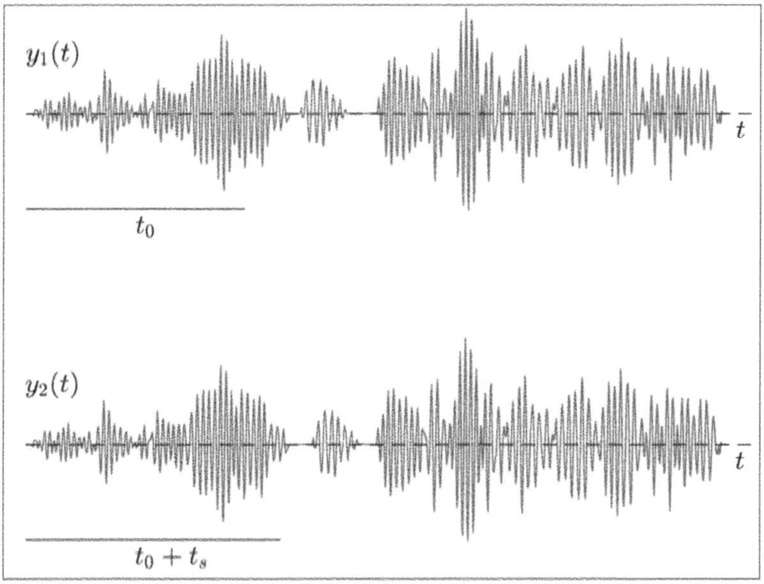

Figure P4.4.

(a) The signal from the tissue, $g_v(t)$ is much stronger than the signal from the blood. To remove its influence, the two received signals are subtracted from each other. The resulting signal is denoted as $d(t)$. Show that $g_v(t)$ is not present in the signal $d(t)$. Write down the spectrum $D(f)$ expressed through $P(f)$, a, t_0 and t_s.

(b) In the last question, we assume that $V_b = 0.75$m/s and $c = 1500$ m/s. For real applications, $T_{prf} = 200\mu$s, $f_0 = 5$Mhz, and $T = 2\mu$s. Explain why $d(t)$ has no spectral components for $|f| = f_0$, for the chosen parameters. Derive and give your reasons, that if one chooses other values for T and T_{prf}, then this situation can be avoided.

4.5. Determine the signals that have the following Fourier transform (FT):

(a)

$$X(\Omega) = \begin{cases} 0 & \text{for } 0 \le |\Omega| \le \Omega_p \\ 1 & \text{for } \Omega_p \le |\Omega| \le \pi. \end{cases}$$

(b)

$$X(\Omega) = \cos^2(\Omega).$$

(c)

$$X(\Omega) = \begin{cases} 1 & \text{for } \Omega_p - \frac{\delta\Omega}{2} \le |\Omega| \le \Omega_p + \frac{\delta\Omega}{2} \\ 0 & \text{otherwise.} \end{cases}$$

(d) For the spectrum given in Fig. P4.5 below:

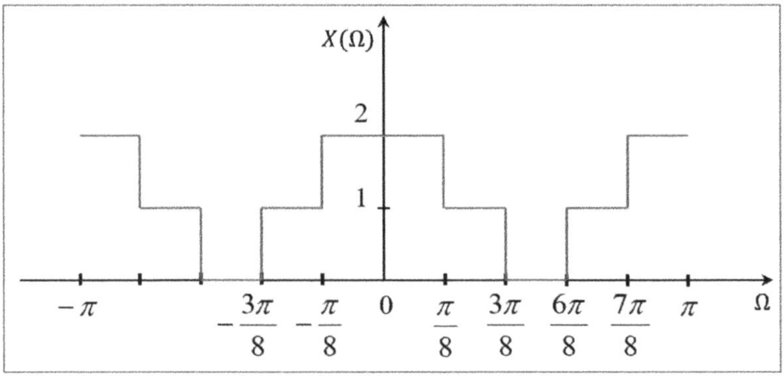

Figure P4.5.

4.6. A single pulse emitted by a medical device for imaging purpose is shown in Fig. P4.6. Suggest a mathematical expression for the pulse. Write down a formula from which its energy can be calculated. Suggest a reasonable upper bound for this energy. What is the pulse's power?

4.7. The pulse in Question 4.6 is found to have a one-sided magnitude spectrum as shown in Fig. P4.7. Suggest a value for f_0. Will the phase spectrum of the signal be 0? If the device is emitting a series of these pulses at the rate of one pulse per 107 seconds, sketch the magnitude spectrum.

4.8. Consider the block diagram representation of a linear time-invariant (LTI) system as shown in Fig. P4.8. Here $x(t)$ and $y(t)$ are the input and output, respectively of the system. $h(t)$ in the picture is the impulse response of the system. The system function $H(f)$ (Fourier transform of

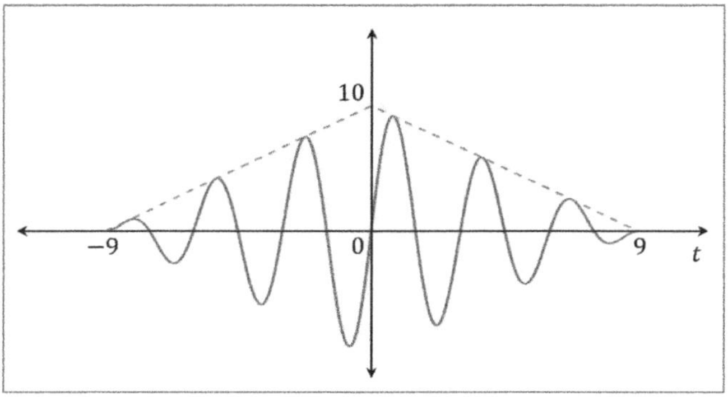

Figure P4.6.

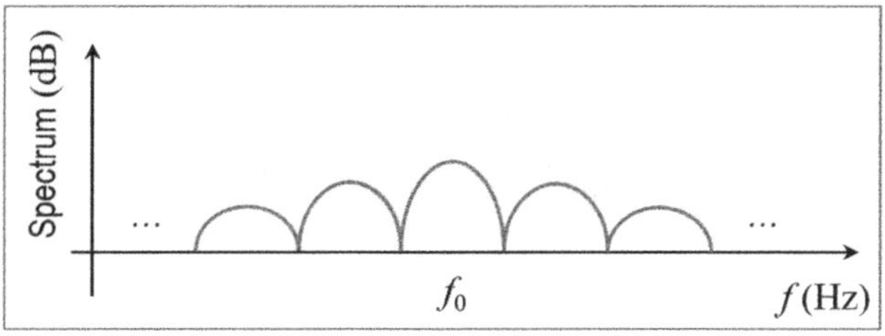

Figure P4.7.

$h(t)$) is given by,

$$H(f) = \cos\left(\frac{\pi}{2f_c} f\right) \text{rect}\left(\frac{f}{2f_c}\right).$$

Here, t is the time variable, f is the frequency variable and f_c is a given specific frequency in Hz.

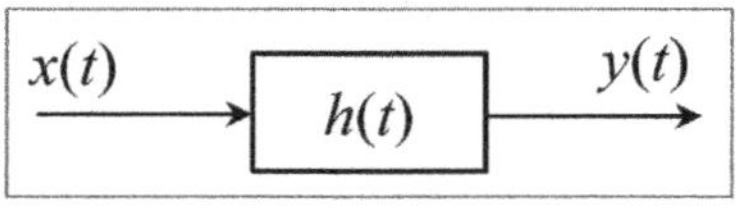

Figure P4.8.

(a) Determine the impulse response $h(t)$ of the system (*Hint*: $h(t) \leftrightarrow H(f)$ and $\mathcal{F}^{-1}[\text{rect}(f)] = \text{sinc}(t)$).

4.9. Two periods of a periodic signal, $x(t)$, are shown in Fig. P4.9.

(a) Write an expression for $x(t)$ and determine its Fourier transform, $X(f)$.

(b) Another signal, $y(t)$, is constructed as $y(t) = x(t) + x(t-1)$. Is $y(t)$ periodic? Justify your answer.

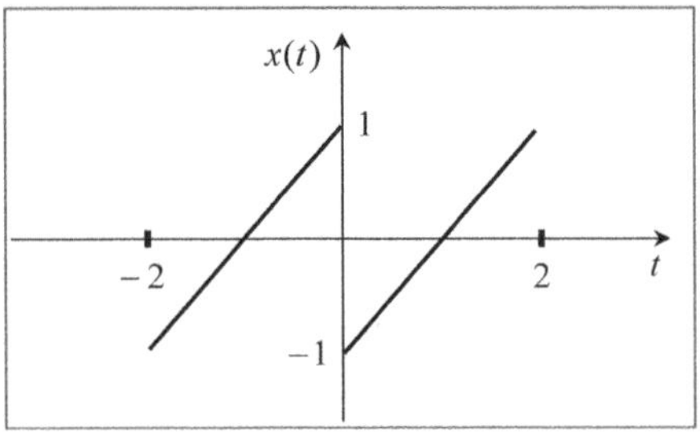

Figure P4.9.

(c) Using your solution to Question (a), determine the Fourier transform of the signal, $y(t)$, given in Question (b).
(d) Let $z(t) = x(t) + x(-t)$. Without doing any derivations, give the Fourier transform, $Z(f)$, of $z(t)$. Justify your answer.

4.10. Consider the pulse signal, $x(t)$, shown in Fig. P4.10.

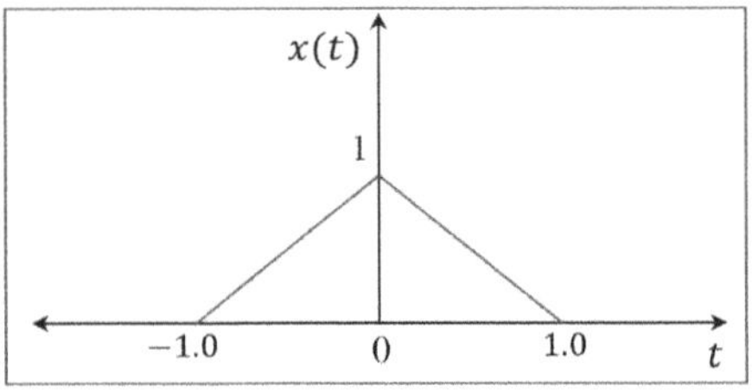

Figure P4.10. Caption

(a) Write down the expression for $x(t)$.

Based on the signal $x(t)$ shown in Fig. P4.10, a new signal $y(t)$ is defined as

$$\begin{aligned} y(t) &= x(t) \quad \text{for} \quad -1 < t \leq 0 \\ &= 0 \quad \text{otherwise} \end{aligned}$$

(b) Determine the Fourier transform, $Y(f)$, of $y(t)$ and using $Y(f)$, develop an approach to obtain the Fourier transform, $X(f)$ of $x(t)$ (*Hint*: Express $x(t)$ in terms of $y(t)$).
(c) Fourier transform of $\text{rect}(t)$ is $\text{sinc}(f)$. Use this information to obtain the Fourier transform, $X(f)$, of $x(t)$ shown in Fig. P4.10.

4.11. Part of a signal with biological origin consists of two full periods of a sinusoidal signal. This partial signal is denoted as $g(t)$ and can be seen in Fig. P4.11, where $f_0 = 3$kHz, and d is a positive constant.

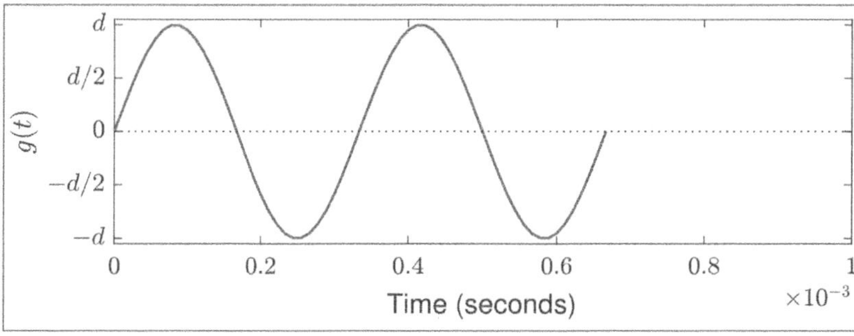

Figure P4.11.

(a) Show that the spectrum $G(f)$ of the signal can be written as

$$G(f) = \frac{d}{\pi f_0} \times \frac{\sin\left(2\pi \frac{f}{f_0}\right)}{1 - \left(\frac{f}{f_0}\right)^2} e^{jk_1},$$

where k_1 depends on f. Then, find k_1. (Do not calculate Fourier integrals.)

(b) The biological system sends the signal $g_p(t)$ consisting of $g(t)$ repeated with a period T, in other words:

$$g_p(t) = \sum_{m=-\infty}^{\infty} g(t - mT).$$

Sketch the numerical values of the spectrum $G_p(k)$[12] of $g_p(t)$, for frequencies below 8kHz. Give the frequencies of the spectral components, when $T = 1$ms. Explain qualitatively, how the magnitude of the spectrum ($|G_p(k)|$) changes when T assumes different values in the interval from 1ms to 2/3ms.

(c) The biological system can maintain the RMS $\{g_p(t)\}$[13] approximately constant in the said-interval of T, by changing d.

12. $G_p(k) = c_k$, where c_k are the coefficients of the Fourier series.

13. RMS means the 'root-mean-square' value and can be calculated as $\sqrt{\frac{1}{T}\int_T g_p^2(t)dt}$.

Find RMS$\{g_p(t)\}$ for $T = 1$ms. Derive how d must change as a function of T and f_0, so that RMS$\{g_p(t)\}$ is kept constant for $\frac{2}{3}$ms $\leq T \leq 1$ms.

4.12. A signal $x(t)$ is defined as

$$x(t) = e^{j2\pi t} \quad \text{for} \quad 0 \leq t \leq 1$$
$$= 0 \qquad \text{otherwise}$$

(a) Determine the Fourier transform, $X(f)$, of $x(t)$.
(b) Let $y(t) = w(t)e^{j2\pi t}$. Determine $w(t)$ such that $y(t)$ is same as $x(t)$ given in Problem 1. Determine the Fourier transform, $W(f)$, of $w(t)$. Express the Fourier transform, $Y(f)$, of $y(t)$ in terms of $W(f)$. Show that the resulting $Y(f)$ is same as $X(f)$ obtained from Question (a).
(c) Draw the magnitude and phase spectra of $x(t)$.

4.13. Find the Fourier transform of the half-cosine pulse defined as,

$$x(t) = \text{rect}\left(\frac{t}{T}\right) a \cos\left(\frac{\pi t}{T}\right).$$

4.14. A single raised cosine pulse is defined as,

$$\text{rect}(t)\,[1 + \cos(2\pi t)].$$

Find its Fourier spectrum.

4.15. The spectrum of the Gaussian pulse $x(t)$ shown in Fig. P4.12 is $X(f)$.

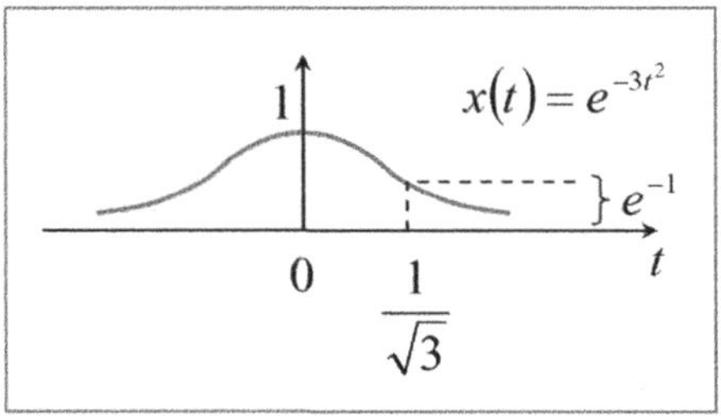

Figure P4.12.

(a) Express $\frac{dX(f)}{df}$ in terms of $X(f)$.
(b) Find the shape of $X(f)$ if $\frac{dX(f)}{df} = -\frac{2\pi^2}{3} f X(f)$.

DOI: 10.1561/9781680839791.ch5

Chapter 5

Complex Signals

A number of signal processing applications make use of the complex signals. Some examples include the characterization of the Fourier transform, blood velocity estimations, and modulation of signals in telecommunications. Furthermore, a number of signal processing concepts are easier to derive, explain and understand using complex notation. It is much easier, for example to add the phases of two complex exponentials such as $x(t) = e^{j\phi_1}e^{j\phi_2}$, than to manipulate trigonometric formula, such as $\cos(\phi_1)\cos(\phi_2)$ in the rectangular representation.

In this chapter, the following points are discussed in detail:

- Complex signals in general.
- Fourier spectrum of complex signals.
- Linear processing of complex signals.
- Hilbert transform and their properties.
- Analytic signals and their use.

5.1 Introduction to Complex Signals

A complex analog signal $x(t)$ is formed by the signal pair $\{x_R(t), x_I(t)\}$, where both $x_R(t)$ (real part) and $x_I(t)$ (imaginary part) are real signals. The relationship between these signals for a complex signal is given by,

$$x(t) = x_R(t) + jx_I(t), \tag{5.1}$$

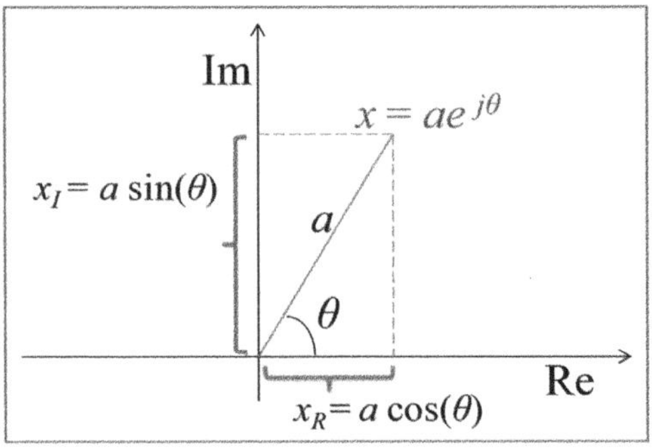

Figure 5.1. Illustration of the relationship between the real (x_R) and imaginary (x_I) parts of the complex number x and its magnitude (a) and phase (θ).

where $j = \sqrt{-1}$. A complex discrete [or digital] signal $x[n]$ is defined in a similar manner,

$$x[n] = x_R[n] + jx_I[n]. \tag{5.2}$$

A complex number x can be represented by its real (x_R) and imaginary (x_I) parts or by its magnitude ($a = |x|$) and phase ($\theta = \angle x$), respectively. The relationship between these parameters is illustrated in Fig. 5.1. The real part, x_R, is the projection of x onto the real axis and the imaginary part, x_I, is the projection of x onto the imaginary axis.

Complex signals (as a function of time) are defined in both continuous time (t) and discrete time (n) as shown below:

$$x(t) = a(t)e^{j\theta(t)} \quad \text{and} \quad x[n] = a[n]e^{j\theta[n]}. \tag{5.3}$$

Here, the various quantities are provided in the table below:

Table 5.1. Various quantities associated with complex signals.

Continuous time	Discrete time
$a(t) = \sqrt{x_R^2(t) + x_I^2(t)}$	$a[n] = \sqrt{x_R^2[n] + x_I^2[n]}$
$\theta(t) = \arctan\frac{x_I(t)}{x_R(t)}$	$\theta[n] = \arctan\frac{x_I[n]}{x_R[n]}$
$x_R(t) = a(t)\cos(\theta(t))$	$x_R[n] = a[n]\cos(\theta[n])$
$x_I(t) = a(t)\sin(\theta(t))$	$x_I[n] = a[n]\sin(\theta[n])$

The magnitudes $a(t)$ and $a[n]$ are also known as the *envelopes* of $x(t)$ and $x[n]$, respectively.

5.1.1 Some Useful Rules and Identities

Many applications require the conversion of the complex number between the rectangular (trigonometric) form and phasor (polar) form. The conversion is done by Euler's formula:

$$e^{j\theta} = \cos(\theta) + j\sin(\theta) \tag{5.4}$$

$$\cos(\theta) = \frac{e^{j\theta} + e^{-j\theta}}{2} \tag{5.5}$$

$$\sin(\theta) = \frac{e^{j\theta} - e^{-j\theta}}{2j} \tag{5.6}$$

Table 5.2 below shows some useful identities.

Table 5.2. Some useful identities.

a	θ	$ae^{j\theta}$
1	0	$e^{j0} = 1$
1	$\pm\pi$	$e^{\pm j\pi} = -1$
1	$\pm n\pi$	$e^{\pm jn\pi} = -1$; n odd integer
1	$\pm 2\pi$	$e^{\pm j2\pi} = 1$
1	$\pm 2n\pi$	$e^{\pm jn2\pi} = 1$; n integer
1	$\pm\frac{\pi}{2}$	$e^{\pm j\frac{\pi}{2}} = \pm j$
1	$\pm\frac{n\pi}{2}$	$e^{\pm j\frac{n\pi}{2}} = \pm j$; $n = 1, 5, 9, 13, \ldots$
1	$\pm\frac{n\pi}{2}$	$e^{\pm j\frac{n\pi}{2}} = \mp j$; $n = 3, 7, 11, 15, \ldots$

5.1.2 Phasors

The word *phasor* is often used by mathematicians to denote any complex number. In engineering, it is frequently used to denote a complex exponential function of constant modulus and linear phase, that is a function of pure harmonic behavior. Here is an example of such a phasor:

$$x(t) = Ae^{j2\pi f_0 t}, \tag{5.7}$$

which has a constant modulus A and a linearly varying phase. It is not uncommon that the modulus and phase are plotted separately. Different ways to depict the phasors are illustrated in Fig. 5.2.

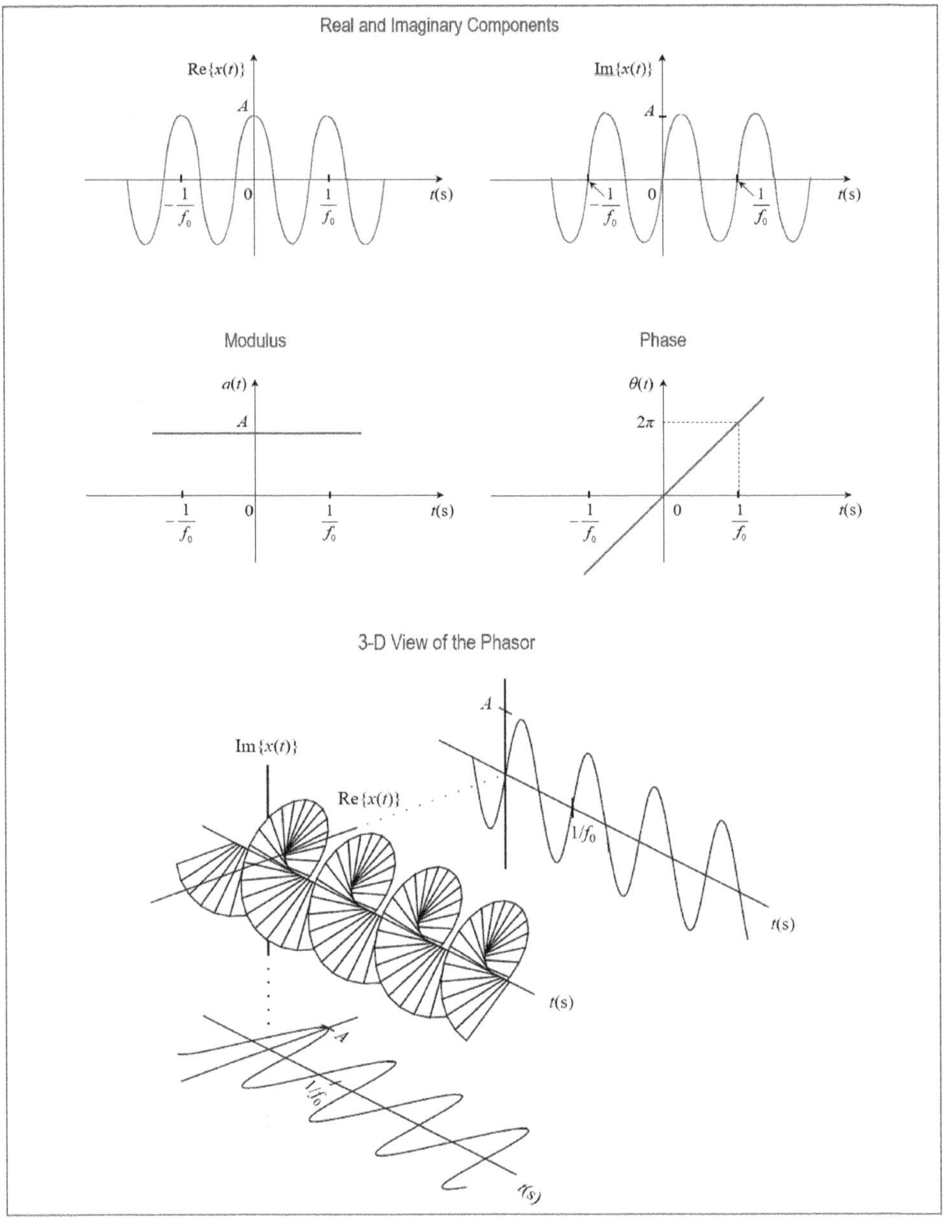

Figure 5.2. Different views of the phasor $Ae^{j2\pi f_0 t}$.

A phasor, whose real part is an even function and imaginary part is an odd function is said to be *Hermitian.* A phasor whose real and imaginary parts are odd, is said to be an *anti-Hermitian.*

5.2 Spectrum of Complex Signals

The spectrum of a complex signal can be found by the Fourier method. In the following, we will derive the spectrum $X(f)$ of the complex signal $x(t) = x_R(t) + jx_I(t)$ as a linear combination of the spectra $X_R(f)$ and $X_I(f)$ of the real valued signals $x_R(t)$ and $x_I(t)$, respectively.

One consequence of the fact that $x(t)$ or $x[n]$ is complex, is that the typical odd/even symmetry of the spectrum are lost. It is easy to demonstrate that the following expression is valid for complex signals:

$$x^*(t) \leftrightarrow X^*(-f) \text{ and } x^*[n] \leftrightarrow X^*(-f) \tag{5.8}$$

where $x^*(t)$ is the complex conjugate of $x(t)$, and ($\leftrightarrow$) represents a Fourier transform pair.

Let the complex signal $x(t)$ be expressed as,

$$x(t) = x_R(t) + jx_I(t). \tag{5.9}$$

Let the spectra of the real and imaginary parts be $X_R(f)$ and $X_I(f)$, respectively. i.e., $x_R(t) \leftrightarrow X_R(f)$ and $x_I(t) \leftrightarrow X_I(f)$. The real part of $x(t)$ can be expressed as[1]:

$$x_R(t) = \frac{1}{2}\left(x(t) + x^*(t)\right). \tag{5.10}$$

Using the linearity property of the Fourier transform, we get

$$X_R(f) = \frac{1}{2}\left(X(f) + X^*(-f)\right) \tag{5.11}$$

In a similar way, we get

$$x_I(t) = -j\frac{1}{2}\left(x(t) - x^*(t)\right) \tag{5.12}$$

and

$$X_I(f) = -j\frac{1}{2}\left(X(f) - X^*(-f)\right). \tag{5.13}$$

Similar expressions can be derived for discrete complex signals as well.

1. Remember that $(a + jb)^* = (a - jb)$.

Table 5.3. Properties of the FT for analog complex signals.

Let $x(t) \leftrightarrow X(f)$; $x_1(t) \leftrightarrow X_1(f)$; and $x_2(t) \leftrightarrow X_2(f)$

Property	Time and Frequency Domain
Linearity	$ax_1(t) + bx_2(t) \leftrightarrow aX_1(f) + bX_2(f)$
Symmetry	$X(t) \leftrightarrow x(-f)$
Scaling	$x(kt) \leftrightarrow \frac{1}{\|k\|} X\left(\frac{f}{k}\right)$
Time reversal	$x(-t) \leftrightarrow X(-f)$
Time shifting	$x(t + t_0) \leftrightarrow X(f)e^{j2\pi f t_0}$
Frequency shift	$x(t)e^{-2\pi f_0 t} \leftrightarrow X(f + f_0)$
Time differentiation	$\frac{d^p x(t)}{dt^p} \leftrightarrow (j2\pi t)^p X(f)$
Frequency differentiation	$(-j2\pi t)^p x(t) \leftrightarrow \frac{d^p X(f)}{df^p}$ p is a real number.
Convolution	$x_1(t) \otimes x_2(t) \leftrightarrow X_1(f)X_2(f)$ $x_1(t)x_2(t) \leftrightarrow X_1(f) \otimes X_2(f)$
Parseval's theorem	$\int_{-\infty}^{\infty} x_1(t)x_2^*(t)dt = \int_{-\infty}^{\infty} X_1(f)X_2^*(f)df$

Table 5.4. Properties of the FT for digital complex signals.

Let $x[n] \leftrightarrow X(f); x_1[n] \leftrightarrow X_1(f); x_2[n] \leftrightarrow X_2(f)$

Property	Time and Frequency Domain
Linearity	$ax_1[n] + bx_2[n] \leftrightarrow aX_1(f) + bX_2(f)$
Time reversal	$x[-n] \leftrightarrow X(-f)$
Time shifting	$x[n + n_0] \leftrightarrow X(f)e^{j2\pi f n_0 T_s}$; T_s is the sampling period.
Frequency shift	$x[n]e^{-2\pi f_0 n T_s} \leftrightarrow X(f + f_0)$
Differentiation	$(j2\pi n T_s)^p x[n] \leftrightarrow \frac{d^p X(f)}{df^p}$
Convolution	$x_1[n] \otimes x_2[n] \leftrightarrow X_1(f)X_2(f)$ $x_1[n]x_2[n] \leftrightarrow X_1(f) \otimes X_2(f)$
Parseval's theorem	$\sum_{-\infty}^{\infty} x_1[n]x_2^*[n] = \frac{1}{f_s} \int_{-f_s/2}^{f_s/2} X_1(f)X_2^*(f)df$

5.2.1 Properties of the Fourier Transform of Complex Signals

The basic set of properties of the Fourier transform for real signals is also valid for complex signals. Table 5.3 gives a short overview of the properties of the Fourier transform for analog complex signals. Table 5.4 gives the corresponding properties for the digital complex signals.

5.3 Linear Processing of Complex Signals

A complex signal consists of two real signals – one for the real part and the other for the imaginary part. The linear processing of a complex signal, such as filtration with a linear time-invariant (LTI) filter, corresponds to applying the processing to both the real and imaginary parts of the signal.

The filtering of the complex signal $x_R(t) + jx_I(t)$ with a filter whose impulse response $h(t)$ is real, is equivalent to two filtering operations – one for the real part and the other for the imaginary part as follows:

$$h(t) \otimes (x_R(t) + jx_I(t)) = h(t) \otimes x_R(t) + j\,(h(t) \otimes x_I(t)). \tag{5.14}$$

Here, $\otimes$ denotes the convolution operation. If the filter has a complex impulse response (i.e., $h(t) = h_R(t) + jh_I(t)$), then the operation corresponds to four real filtering operations as shown in Fig. 5.3.

Example of an often-used digital filter with complex impulse response is the one given by,

$$h_m[n] = \frac{1}{N} e^{jm\frac{2\pi}{N} nT_s}; \quad 0 \le n \le N-1, \tag{5.15}$$

where T_s is the sampling time and m is an integer. N in Eqn. (5.15) is the length of the filter. The filter coefficients are clearly complex. The transfer function of the filter is derived as,

$$H_m(f) = \frac{1}{N}\left(\frac{\sin\pi(fNT_s - m)}{\sin\pi(fT_s - m/N)}\right) e^{-j\pi(N-1)(fT_s - m/N)}. \tag{5.16}$$

From this equation, it is clear that the filter transfer function is a function of the parameter m.

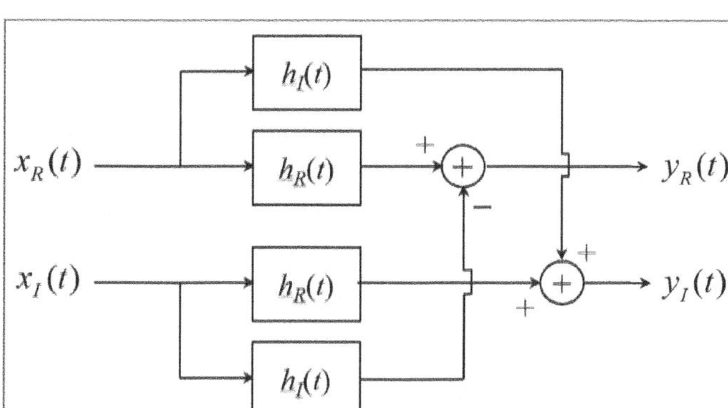

Figure 5.3. Filtering of complex signals.

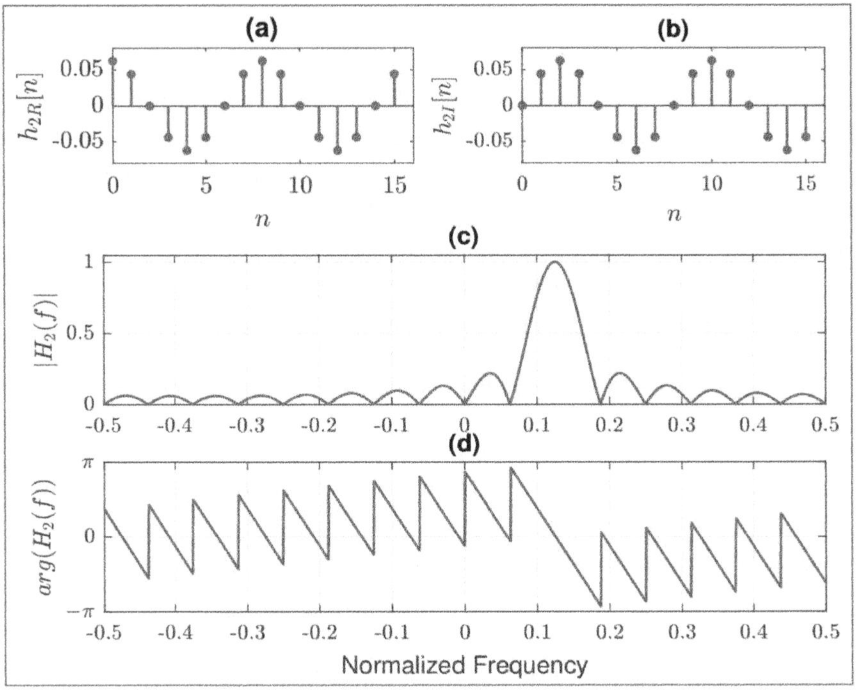

Figure 5.4. Impulse response and transfer function of the complex filter for parameter set $\{N, m\} = \{16, 2\}$; (a) Real part of the impulse response, (b) Imaginary part of the impulse response, (c) Magnitude response of the filter, and (d) Phase response of the filter.

Figure 5.4 illustrates both the impulse response and the transfer function of the filter for parameter set $\{N, m\} = \{16, 2\}$. The real ($h_{2R}[n]$) and imaginary ($h_{2I}[n]$) components of the impulse response of the filter is shown in Figs. 5.4(a) and (b), respectively. Its magnitude response is shown in Fig. 5.4(c) and the phase response in Fig. 5.4(d).

5.4 Hilbert Transform

Signal processing is a fast growing area today and a desired effectiveness in utilization of bandwidth and energy makes the progress even faster. Special signal processors have been developed to make it possible to implement the theoretical knowledge in an efficient way. Signal processors are nowadays frequently used in equipment for radio, transportation, medicine, production, etc.

In 1743 a famous Swiss mathematician named Leonard Euler (1707–1783) derived the formula,

$$e^{jx} = \cos(x) + j\sin(x). \tag{5.17}$$

Nearly 150 years later, the physicist Arthur E. Kennelly and the scientist Charles P. Steinmetz used this formula to introduce the complex notation of harmonic waveforms in electrical engineering, that is

$$e^{j\Omega t} = \cos(\Omega t) + j\sin(\Omega t). \tag{5.18}$$

Later on, in the beginning of the 20th century, the German scientist David Hilbert (1862–1943) finally showed that the function $\sin(\Omega t)$ in Eq. (5.18) is the Hilbert transform of $\cos(\Omega t)$ in the same equation. This gave us the $\pm\frac{\pi}{2}$ phase-shift operator, which is a basic property of the Hilbert transform.

A real function, $x(t)$, and its Hilbert transform, $x_H(t)$, are related to each other in such a way that they together create the so called *strong analytic signal.* The strong analytic signal can be written with an amplitude and a phase where the derivative of the phase can be identified as the instantaneous frequency. The Fourier transform of the strong analytic signal gives us a one-sided spectrum in the frequency domain. It is not hard to see that a function and its Hilbert transform also are orthogonal. This orthogonality is not always realized in applications because of truncation in numerical calculations. However, a function and its Hilbert transform has the same energy and therefore the energy can be used to measure the calculation accuracy of the approximated Hilbert transform.

Hilbert Transform – Definition: Signal $x_H(t)$ obtained from a real signal $x(t)$ by shifting the phase angles of all spectral components of $x(t)$ by ± 90 degrees is known as the Hilbert transform of the signal. That is,

$$x_H(t) = \frac{1}{\pi}\int_{-\infty}^{\infty}\frac{x(\tau)}{t-\tau}d\tau, \tag{5.19}$$

and the inverse Hilbert transform is given by,

$$x(t) = -\frac{1}{\pi}\int_{-\infty}^{\infty}\frac{x_H(\tau)}{t-\tau}d\tau. \tag{5.20}$$

5.4.1 Hilbert Transform as a Filter

The Hilbert transform, $x_H(t)$, defined in the time domain is the convolution between the Hilbert transformer $\frac{1}{\pi t}$ and the function $x(t)$. That is,

$$x_H(t) = x(t) \otimes \frac{1}{\pi t}, \tag{5.21}$$

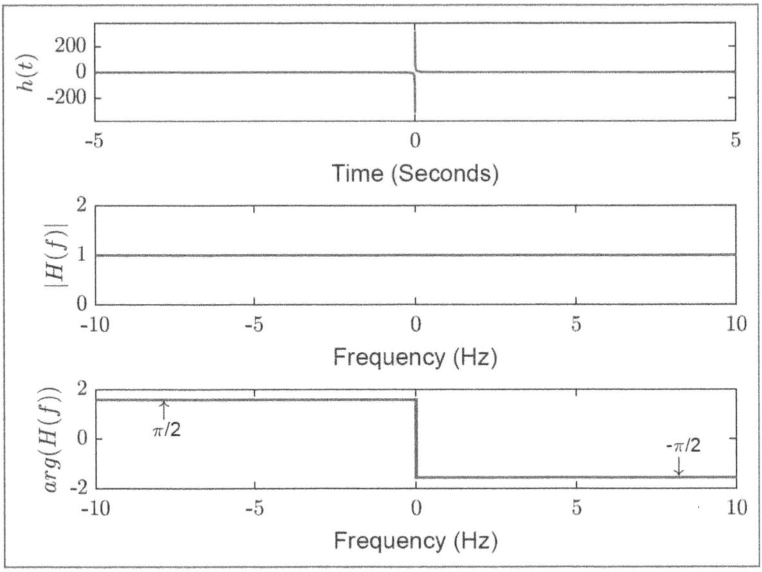

Figure 5.5. Impulse response (Top), magnitude response (Middle) and phase response (Bottom) of the Hilbert transformer.

where $\otimes$ denotes the convolution operation and $\frac{1}{\pi t}$ is the Hilbert transform operator. We have,

$$\frac{1}{\pi t} \leftrightarrow -j\text{sgn}(f), \tag{5.22}$$

Fourier transform pair, where

$$-j\text{sgn}(f) = \begin{cases} e^{-j\pi/2} & f > 0 \\ 0 & f = 0 \\ e^{j\pi/2} & f < 0 \end{cases} \tag{5.23}$$

The impulse response, $h(t) = \frac{1}{\pi t}$, of the Hilbert transform operator and the corresponding frequency response are shown in Fig. 5.5. It can be seen that the impulse response has a discontinuity at $t = 0$ and it is inversely proportional to t. Its magnitude response $|H(f)|$ is constant (= 1) throughout the frequencies. The phase response is $-\frac{\pi}{2}$ for positive values of the frequencies and $+\frac{\pi}{2}$ for negative values of the frequencies. Also, there is a discontinuity of the phase response at $f = 0$.

Example: The Hilbert transform of rectangular pulse train is shown in Fig. 5.6. As can be seen from the figure, the Hilbert transform becomes infinite at the discontinuities of the rectangular pulse train.

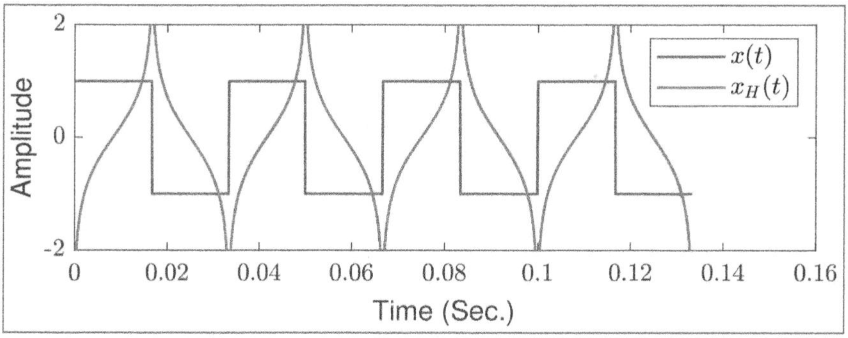

Figure 5.6. Hilbert transform of a rectangular pulse train.

5.4.2 Properties of Hilbert Transform

- The Hilbert transform of a constant is zero.
- The Hilbert transform of a Hilbert transform of a function is the negative of the original function.
- A function and its Hilbert transform are orthogonal over the infinite interval.
- The Hilbert transform of a real function is a real function.
- The Hilbert transform of a sine function is a cosine function, the Hilbert transform of a cosine function is the negative of the sine function.
- The Hilbert transform of an even function is an odd function and the Hilbert transform of an odd function is an even function.
- The Hilbert transform of the convolution of the two functions is the convolution of one with the Hilbert transform of the other.

- Linearity:

$$\mathcal{H}\{ax_1(t) + bx_2(t)\} = ax_{H1}(t) + bx_{H2}(t), \tag{5.24}$$

  where a and b are scalar constants and $\mathcal{H}$ is the Hilbert transform operator.
- Time shift:

$$\mathcal{H}\{x(t + t_0)\} = x_H(t + t_0). \tag{5.25}$$

- Applying two times the Hilbert transform gives the original signal:

$$\mathcal{H}\{\mathcal{H}\{x(t)\}\} = -x(t). \tag{5.26}$$

- The inverse Hilbert transform:

$$x(t) = \mathcal{H}^{-1}\{x_H(t)\} = -\int_{-\infty}^{\infty} \frac{x_H(t)}{\pi(t-u)} du = x_H(t) \otimes \frac{-1}{\pi t}. \tag{5.27}$$

- Even/Odd Property:

$$\begin{aligned} x(t) \ \text{even} &\Leftrightarrow x_H(t) \ \text{odd} \\ x(t) \ \text{odd} &\Leftrightarrow x_H(t) \ \text{even} \end{aligned} \tag{5.28}$$

- Conservation of energy:

$$\int_{-\infty}^{\infty} x^2(t)dt = \int_{-\infty}^{\infty} x_H^2(t)dt. \tag{5.29}$$

- Orthogonality:

$$\int_{-\infty}^{\infty} x(t)x_H(t)dt = 0. \tag{5.30}$$

- Time-derivative:

$$\mathcal{H}\left\{\frac{dx(t)}{dt}\right\} = \frac{d}{dt}\mathcal{H}\{x(t)\}. \tag{5.31}$$

- Convolution:

$$\mathcal{H}\{x(t) \otimes y(t)\} = x(t) \otimes y_H(t) = x_H(t) \otimes y(t). \tag{5.32}$$

5.5 Analytic Signals

An *analytic signal* (denoted as $z_x(t)$ of an analog signal $x(t)$) is a signal whose spectrum is *one sided*. For analog signals, this means that their spectrum is $\equiv 0$ for $f > 0$ or $f < 0$. Analytic discrete time signals have a spectrum which is $\equiv 0$ for $-\frac{f_s}{2} < f < 0$ and in the corresponding parts of the periodic spectrum, or $0 < f < \frac{f_s}{2}$ and the corresponding parts of the periodic spectrum. This condition for an analytic signal gives the connection between the real and imaginary parts of the complex signal.

5.5.1 Analytic Analog Signals

If a real signal $x(t)$ with frequency spectrum $X(f)$ is taken as a starting point, then the following relations will be valid for the respective analytic signal, $z_x(t)$, and its spectrum:

$$z_x(t) \leftrightarrow Z_x(f) = \begin{cases} 2X(f) & \text{for } f > 0 \\ X(f) & \text{for } f = 0 \\ 0 & \text{for } f < 0. \end{cases} \tag{5.33}$$

This relation can be expressed in a more compact form as[2]:

$$Z_x(f) = [1 + \mathsf{sgn}(f)]\, X(f). \tag{5.34}$$

Since $\mathsf{sgn}(f)$ is the Fourier spectrum of the function $j\frac{1}{\pi t}$ ($j\frac{1}{\pi t} \leftrightarrow \mathsf{sgn}(f)$), then the above equation is equivalent to:

$$z_x(t) = \left(\delta(t) + j\frac{1}{\pi t}\right) \otimes x(t). \tag{5.35}$$

Here, we introduce the signal $x_H(t)$, known as the Hilbert transform of $x(t)$ and given by,

$$x_H(t) = x(t) \otimes \frac{1}{\pi t}. \tag{5.36}$$

It can be seen that

$$z_x(t) = x(t) + j x_H(t). \tag{5.37}$$

Notice that the complex conjugate $z_x^*(t)$ is also analytic with spectrum given by,

$$Z_x^*(f) = \begin{cases} 0 & \text{for } f > 0 \\ X(f) & \text{for } f = 0 \\ 2X(f) & \text{for } f < 0. \end{cases} \tag{5.38}$$

and that consequently:

$$x(t) = \frac{1}{2}\left(z_x(t) + z_x^*(t)\right). \tag{5.39}$$

If $z_x(t)$ is written in the form,

$$z_x(t) = a_z(t) e^{j\theta_z(t)} \tag{5.40}$$

then

$$x(t) = a_z(t)\cos(\theta_z(t)) \;\text{ and }\; x_H(t) = a_z(t)\sin(\theta_z(t)). \tag{5.41}$$

If the analytic signal $z_x(t)$ is filtered using a filter with real impulse response, then the output signal $y(t)$ will be:

$$y(t) = h(t) \otimes z_x(t) = h(t) \otimes x(t) + j h(t) \otimes x_H(t). \tag{5.42}$$

2. $\mathsf{sgn}(t)$ returns the sign of the argument. It returns $+1$ if $t > 0$, -1 if $t < 0$, and 0 if $t = 0$.

If the Hilbert transform of $h(t)$ is denoted by $h_H(t)$, then one gets:

$$y(t) = x(t) \otimes (h(t) + jh_H(t)) = x(t) \otimes z_H(t), \tag{5.43}$$

where $z_H(t)$ is the Hilbert transform of $h(t)$.

This operation is sometimes useful when one wants to work with analytic signals, but has only a real signal to start with. Notice that $h_H(t)$ is usually *noncausal*.

Using the symmetry property of the Fourier transform, it can be shown that the real and imaginary parts of the spectrum of a real signal form a Hilbert pair, that is each can be obtained from the other using a Hilbert transform. This, and a number of other properties of the Hilbert transform can be found in the Section 5.4.2. The symbol $\mathcal{H}$ is used to denote the Hilbert transform, i.e., $x_H(t) = \mathcal{H}\{x(t)\}$.

5.5.2 Analytic Discrete-time Signals

If a real discrete-time signal $x[n] \leftrightarrow X(f)$ is used as a basis, then the corresponding analytic signal[3] $z_x[n]$ will be given as

$$z_x[n] \leftrightarrow Z_x(f) = \begin{cases} 2X(f) & \text{for } pf_s < f < (2p+1)\frac{f_s}{2} \\ X(f) & \text{for } f = pf_s \\ 0 & \text{otherwise} \end{cases} \tag{5.44}$$

where p is an integer and f_s is the sampling frequency. Let us consider the following spectrum:

$$z_x[n] \leftrightarrow Z_x(f) = \begin{cases} +1 & \text{for } pf_s < f < (2p+1)\frac{f_s}{2} \\ 0 & \text{for } f = pf_s \\ -1 & \text{for } (2p-1)\frac{f_s}{2} < f < pf_s. \end{cases} \tag{5.45}$$

The discrete-time signal that corresponds to this spectrum is

$$j\frac{2}{n\pi}\sin^2\left(n\frac{\pi}{2}\right). \tag{5.46}$$

It follows directly that

$$z_x[n] = \left(\delta[n] + j\frac{2}{n\pi}\sin^2\left(n\frac{\pi}{2}\right)\right) \otimes x[n]. \tag{5.47}$$

3. The use of the term analytic in relation to discrete-time signals leads to mathematical difficulties. Many of them can, however, be circumvented if one applies the fact, that a given digital signal corresponds to an equivalent analog signal.

By the analogy with the relation in Eqn. (5.36), we denote the signal

$$x_H[n] = x[n] \otimes j\frac{2}{n\pi}\sin^2\left(n\frac{\pi}{2}\right) \tag{5.48}$$

as the Hilbert transform of $x[n]$. The relation between $z_x[n]$, $x[n]$ and $x_H[n]$ is given by,

$$z_x[n] = x[n] + jx_H[n]. \tag{5.49}$$

Similar to Section 10.3.7, it can be shown that

$$h[n] \otimes z_h[n] = x[n] \otimes z_h[n], \tag{5.50}$$

where

$$z_h[n] = h[n] + jh_H[n] = h[n] + jh[n] \otimes \frac{2}{n\pi}\sin^2\left(n\frac{\pi}{2}\right). \tag{5.51}$$

5.6 Instantaneous Amplitude and Frequency

Let us consider the band-limited real signal $g(t)$ with a band width f_g. The amplitude spectrum of such a signal is shown in the left subplot of Fig. 5.7. For this particular case, an analytic signal can be created by frequency-shifting the spectrum of the signal with an offset f_0, where $|f_0| > f_g$. This operation is illustrated in the second sub-plot of Fig. 5.7. In other words, we create a signal $z_0(t) = g(t)e^{j2\pi f_0 t}$, which is analytic if $|f_0| > f_g$. It can be represented as real and imaginary parts as:

$$z_0(t) = g_0(t) + jg_{0H}(t), \tag{5.52}$$

where $g_{0H}(t)$ is the Hilbert transform of the real signal $g_0(t) = g(t)\cos(2\pi f_0 t)$. Using the Euler's relations, we find that

$$g_0(t) = g(t)\cos(2\pi f_0 t) \text{ and } g_{0H}(t) = g(t)\sin(2\pi f_0 t). \tag{5.53}$$

If we assume that $g(t)$ gets only non-negative values ($g(t) \geq 0$), then it follows immediately that:

$$a_z(t) = g(t) \text{ and } \theta_z(t) = 2\pi f_0 t, \tag{5.54}$$

where, $a_z(t)$ is the envelope of the analytic signal and $\theta_z(t)$ is the phase of the analytic signal. The *instantaneous* amplitude of $z_0(t)$ is then $g(t)$, and the value $\theta_z'(t)/2\pi = \frac{1}{2\pi}\frac{d\theta_z(t)}{dt}$ is said to be the *instantaneous* frequency of the signal.

An example of an analytic analog signal and its various parameters are shown in Fig. 5.8.

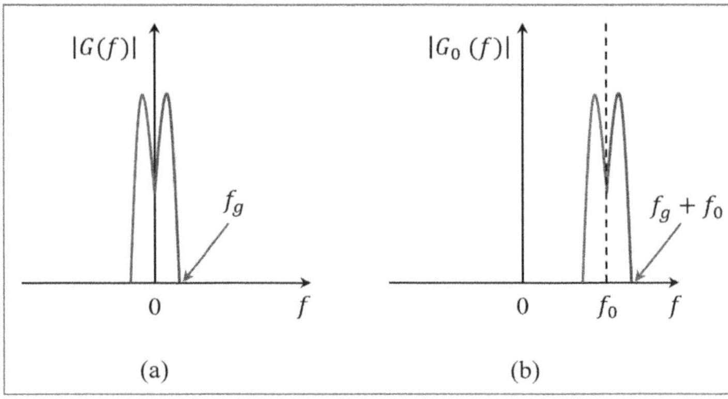

Figure 5.7. Construction of an analytic signal.

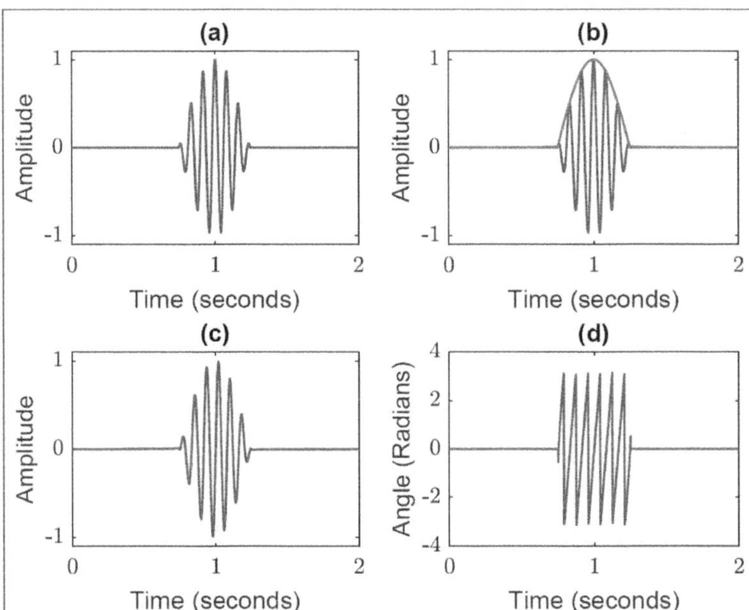

Figure 5.8. Instantaneous amplitude and phase for an analytic signal; (a) the analytic signal, $g_0(t)$, (b) Signal and its envelope (or instantaneous amplitude) ($a_z(t)$ in red), (c) Hilbert transform, $g_{0H}(t)$, and (d) Instantaneous phase, $\theta_z(t)$.

In the case that $g(t)$ can be both positive and negative, the amplitude $a_z(t)$ will be the absolute value of $g(t)$, i.e., $a_z(t) = |g(t)|$. The phase $\theta_z(t)$ of such analytic signal will jump from $\pm\pi$ in those time instances, when $g(t)$ changes its sign. The first derivative with respect to time, $\theta_z'(t)$, will still be proportional to the instantaneous frequency, except for the points of discontinuity. This is illustrated in Fig. 5.9.

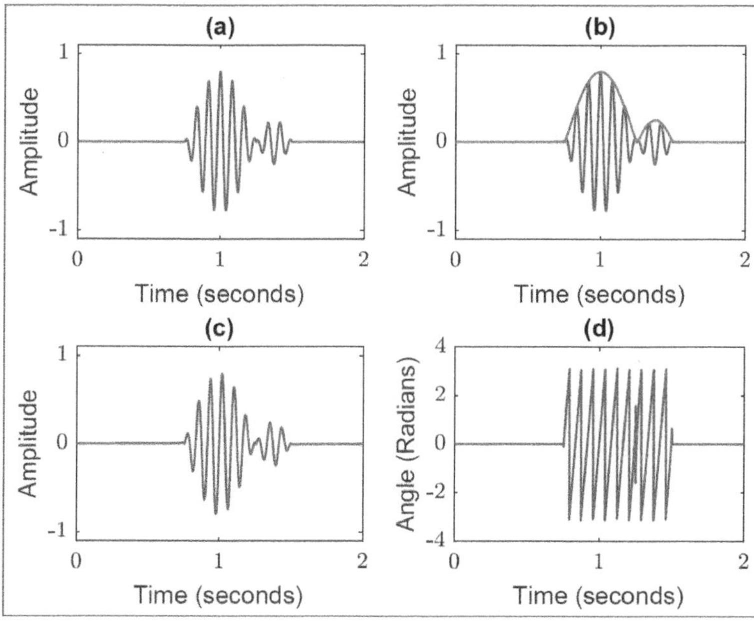

Figure 5.9. Instantaneous amplitude and phase for an analytic signal with a point of discontinuity; (a) the analytic signal, $g_0(t)$, (b) Signal and its envelope (or instantaneous amplitude) ($a_z(t)$ in red), (c) Hilbert transform, $g_{0H}(t)$, and (d) Instantaneous phase, $\theta_z(t)$.

This figure clearly shows the phase jump in the instantaneous phase plot (Fig. 5.9(d)) at the point of discontinuity in the analytic signal.

The concepts of instantaneous amplitude and frequency can be transferred formally to the typical case, in which the analytic signal $z(t)$ is given in the form $a_z(t)e^{j\theta_z(t)}$. Figure 5.10 shows an example in which $\theta_z(t)$ varies non-linearly in time.

These concepts are also applied to non-analytic signals, and the same can be defined for discrete-time signals as well. The phase of the signal is given by:

$$\theta_z(t) = \arctan\left(\frac{x_H(t)}{x(t)}\right). \tag{5.55}$$

The instantaneous frequency is proportional to the derivative of the phase with respect to time, and can be found from:

$$\frac{\theta_z'(t)}{2\pi} = \frac{1}{2\pi}\left(\frac{x(t)x_H'(t) - x_H(t)x'(t)}{x^2(t) + x_H^2(t)}\right). \tag{5.56}$$

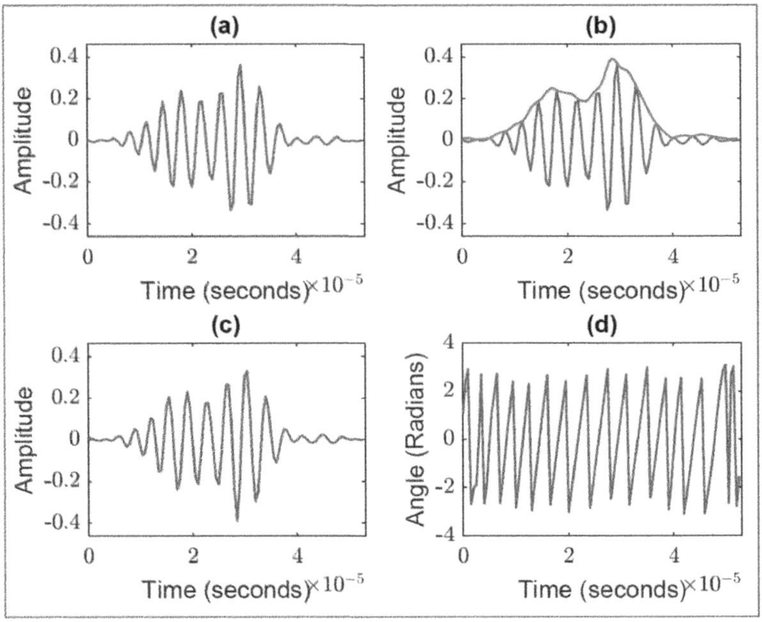

Figure 5.10. Instantaneous amplitude and phase for an analytic signal with non-linear variation of the phase ($\theta_z(t)$); (a) the analytic signal, $g_0(t)$, (b) Signal and its envelope (or instantaneous amplitude) ($a_z(t)$ in red), (c) Hilbert transform, $g_{0H}(t)$, and (d) Instantaneous phase, $\theta_z(t)$.

5.7 Concluding Remarks

- Introduced both the analog and discrete complex signals.
- Fourier transform of complex signals were discussed in detail.
- Theory of linear processing of complex signals introduced.
- Hilbert transform and its properties were discussed in detail.
- The concept of analytic signals and instantaneous amplitude and frequency were also introduced.

Exercises

5.1. Consider the real signal $x(t)$ with spectrum $X(f)$. Some properties of the Hilbert transform $y(t) = \mathcal{H}\{x(t)\} = \frac{1}{\pi t} \otimes x(t)$ are to be investigated. The impulse response of the Hilbert transformer is indicated by $h(t)$.

(a) Give $H(f) \leftrightarrow h(t)$ and sketch the magnitude and phase of $H(f)$. How does the Hilbert transform affect $|X(f)|$ and $\arg\{X(f)\}$?

(b) Is $y(t)$ real, imaginary or complex? What symmetry does the Hilbert transform of the even part $x_e(t)$ and the odd part $x_o(t)$ have?

(c) Compute the Hilbert transform of the Fourier series

$$x(t) = \frac{A_0}{2} + \sum_{k=1}^{\infty} [a_k \cos(f_0 kt) + b_k \sin(f_0 kt)], \quad f_0 > 0.$$

5.2. Given the signal $x(t) = \frac{f_g}{\pi} \times \frac{1-\cos(f_g t)}{f_g t}$ and its Hilbert transform $x_H(t) = \mathcal{H}\{x(t)\}$.

(a) Calculate and sketch $X(f) \leftrightarrow x(t)$. (*Note:* use the multiplication theorem of the Fourier transform.)

(b) Determine $X_H(f) \leftrightarrow x_H(t)$ graphically and calculate $x_H(t)$. (*Note:* consider the discussion on the Hilbert transform in time-domain and frequency-domain)

(c) Let $x_a(t) = x(t) + jx_H(t)$. Which special property exhibits $X_a(f) \leftrightarrow x_a(t)$?

(d) Calculate $X_a(f)$ formally in dependence on $X(f)$.

5.3. For $\Omega_0 > 0$, calculate the Hilbert transform in the frequency domain for the following functions:

(a) $e^j\Omega_0 t$,

(b) $sin\Omega_0 t$,

(c) $\cos \Omega_0 t$, and

(d) $\cos 2\Omega_0 t$.

5.4. What is the connection between the energy of a signal $x(t)$ and its Hilbert transformed signal $x_H(t)$? Assume that the signal has no DC component.

DOI: 10.1561/9781680839791.ch6

Chapter 6

Analog Systems

In Chapters 1 to 4, we studied about signals and representation of signals in time and frequency domains. The focus of this chapter is *analog systems* that work with these signals. We will find that Fourier transform (FT) is a very useful tool in working with and understanding systems.

In a general sense, *we may define a system as anything that performs a function.* That is, *it operates on something and produces something else.* We may also say that *a system is anything that responds (by producing an output) when excited or stimulated (by an input).* The schematic of a system with input signal $x(t)$ and output signal $y(t)$ is shown in Figure 6.1. A system can be an electrical system (e.g., electrical circuits), a mechanical system (e.g., automobile engines), a biological system (e.g., human vocal tract or heart or brain, etc.), a computer system, etc.

In this chapter, we will discuss the following topics in detail.

- Classification of systems.
- Characterization (time-domain and frequency-domain) of systems, leading to the concepts of *convolution, impulse response and frequency response.*
- Processing and analysis of signals using systems (filters).
- Determination of impulse response or frequency response of systems, and output of a system for given input, etc.
- Importance of sinusoids and FT in the study of systems.

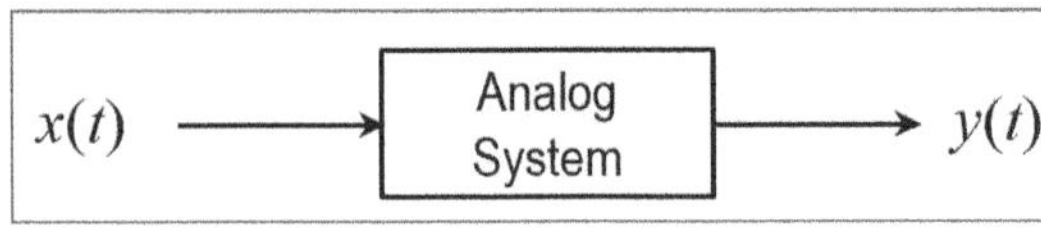

Figure 6.1. A system with input signal $x(t)$ and output signal $y(t)$.

6.1 Classification of Systems

We give below a few ways of classifying systems.

6.1.1 Continuous-Time or Discrete-Time Systems

Just as we have continuous-time and discrete-time signals, we can also define continuous-time and discrete-time systems. A *continuous-time system* operates on continuous-time inputs and produces continuous-time outputs (e.g., mobile-phone, communication channel, etc.). Similarly, a *discrete-time system* operates on discrete-time inputs and produces discrete-time outputs (e.g., computers, digital filters, etc.). In this chapter, we will only be considering continuous-time systems.

6.1.2 Time-Invariant Systems

Let $y(t)$ be the output of a system when the input is $x(t)$. Then, the system is said to be time-invariant if its output is $y(t - t_0)$ when the input is $x(t - t_0)$. That is, *in time-invariant systems, a time-shift in the input results in the same time-shift in its output, without changing the shape of the output.*

6.1.3 Causal Systems

A system is said to be causal if its output at any instant t_0 depends on its input $x(t)$ for time $t \leq t_0$ only. That is, *the output of a causal system depends only on the present and past values of its input.*

6.1.4 Linear or Non-linear Systems

Let $y_1(t)$ and $y_2(t)$ be the outputs of a system when the inputs are $x_1(t)$ and $x_2(t)$, respectively. Then, the system is said to be linear if its output is $\alpha y_1(t) + \beta y_2(t)$ (where α and β are scalars) when the input is $\alpha x_1(t)+\beta x_2(t)$. That is, *linear systems obey the principle of superposition.* Non-linear systems do not obey the principle of superposition.

6.2 Description of a System

Consider a linear system whose output is $y(t)$ when the input is $x(t)$. If we want to use this system in any particular application, we need to know beforehand what function this system is performing. In other words, *we need to have a description of the system.* There are different ways to describe a system, some of which are listed below.

6.2.1 Input-Output Equation

We may describe a system by stating how the output is related to its input. For example, we have

$$y(t) = 7x(t) + 8x(t-3). \tag{6.1}$$

6.2.2 Frequency Response

We may describe a system by specifying its frequency response. *Frequency response (denoted by $H(f)$) is given by the ratio of the Fourier transforms of the output to input.* For example, the frequency response of the system described in Eq. (6.1) is given by[1]

$$H(f) = \frac{Y(f)}{X(f)} = 7 + 8e^{-j3\Omega}. \tag{6.2}$$

6.2.3 Impulse Response

We may describe a system by specifying its impulse response. *Impulse response (denoted by $h(t)$) of a system is its response (or output) when the input is an impulse $\delta(t)$ at time $t = 0$.* For example, the impulse response of the system described in Eq. (6.1) is given by[2]

$$h(t) = y(t)|_{x(t)=\delta(t)} = 7\delta(t) + 8\delta(t-3). \tag{6.3}$$

Recall that *the frequency response $H(f)$ of a system is the FT of its impulse response $h(t)$.* Therefore, *the impulse response of a system can be obtained by determining the*

1. To determine the FT $Y(f)$ of $y(t)$ given by Eq. (6.1), apply the *linearity property of FT.* That is,

$$\begin{aligned} \mathcal{F}[y(t)] &= \mathcal{F}[7x(t) + 8x(t-3)] = 7\mathcal{F}[x(t)] + 8\mathcal{F}[x(t-3)] \\ \Rightarrow \quad Y(f) &= 7X(f) + 8e^{-j3\Omega}X(f) = \left[7 + 8e^{-j3\Omega}\right]X(f) \end{aligned}$$

since $\mathcal{F}[x(t-t_0)] = e^{-j\Omega t_0}X(f)$. Thus, we have $Y(f) = H(f)X(f)$ for linear systems.

2. The notation $y(t)|_{x(t)=\delta(t)}$ means "$y(t)$ when $x(t) = \delta(t)$".

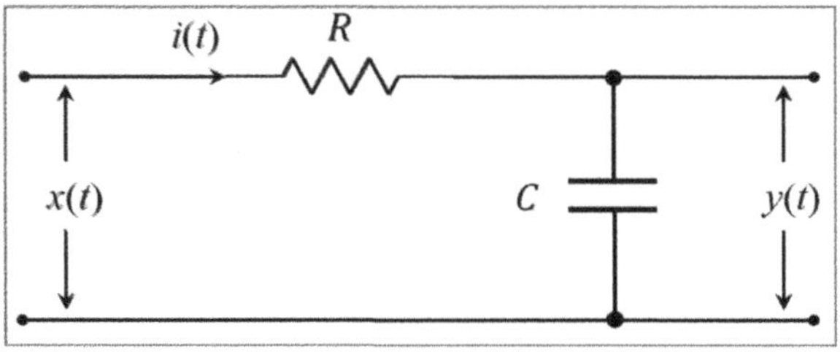

Figure 6.2. RC circuit.

inverse FT of its frequency response. Note that Eq. (6.2) is the FT of Eq. (6.3) since $\mathcal{F}[\delta(t-t_0)] = e^{-j\Omega t_0}$.

Note that from the linear system theory, *the output $y(t)$ of a system can be obtained by convolving the input $x(t)$ with its impulse response $h(t)$*. That is, we have[3]

$$y(t) = x(t) \otimes h(t) = \int_{-\infty}^{\infty} x(\tau)h(t-\tau)d\tau \tag{6.4a}$$

$$= \int_{-\infty}^{\infty} h(\tau)x(t-\tau)d\tau. \tag{6.4b}$$

§ *Determine the output $y(t)$ of the system specified by Eqs. (6.1)–(6.3) for the following inputs: $x(t) = 10$ for all t, $x(t) = 3e^{j\Omega_0 t}$, $x(t) = 10 + 3\cos(\Omega_0 t)$.*

§ *In the RC-circuit (Fig. 6.2), note that $y(t) = x(t) - Ri(t)$ where $i(t)$ is the current flowing through the resistor R. Noting that $i(t) = C\frac{dy(t)}{dt}$, determine the frequency response of the circuit.*

6.3 Why Study of Systems is Important?

In many practical applications, we will need to deal with systems and their inputs and outputs. Broadly speaking, in most of the practical applications, we will need to do one or more of the following three tasks.

6.3.1 System Identification

This deals with *acquiring the knowledge of the system*. For example, to properly design the transmitter and receiver in a communication system, we should have a very

3. For the sake of clarity, I am using the notation '⊗' to denote convolution, though some text books uses '∗' to denote convolution.

good understanding of the channel (or medium) through which the signal propagates. The channel may be described by means of its frequency response or impulse response (ignoring the noise). So, as a first step to designing the transmitter/receiver, we need to develop a mechanism to obtain the channel characteristics. This may be done, for example, by transmitting a known signal, $x(t)$ and measuring the corresponding output, $y(t)$. From $x(t)$ and $y(t)$, $H(f)$, the system function can be found as, $H(f) = \frac{Y(f)}{X(f)}$, thus identifying the system.

6.3.2 Output Evaluation

This deals with *evaluating the output* $(y(t))$ *of a system (for specific inputs,* $x(t)$*) if its frequency response* $(H(f))$ *or impulse response* $(h(t))$ *is known*. This can be done either using $Y(f) = X(f)H(f)$ or using the convolution operation as in Eq. (6.4). In the above example of a communication system, once we know the channel, we can use it to simulate the transmission of various signals and understand how the output will be for each of these signals. This will help us to design the receiver and signals.

6.3.3 Input Estimation

This deals with *estimating the input,* $x(t)$*, of a system if its output,* $y(t)$ *and the knowledge of the frequency response,* $H(f)$ *or impulse response,* $h(t)$ *of the system are given*. This may be done, for example, using $X(f) = \frac{Y(f)}{H(f)}$. We could use such an approach in the above example of a communication system to determine the transmitted signal.

The topics covered in this chapter equip us with the basic knowledge necessary to understand and perform the above tasks.

6.4 Frequency Response

Since $Y(f) = H(f)X(f)$, we can write

$$|Y(f)| = |H(f)| \cdot |X(f)|, \quad \theta_y(f) = \theta_x(f) + \theta_h(f) \tag{6.5}$$

where $\theta_y(f)$, $\theta_x(f)$ and $\theta_h(f)$ are the phase characteristics of $Y(f)$, $X(f)$ and $H(f)$, respectively. Equation (6.5) clearly shows how the system is affecting the input. In other words, *we can shape the spectrum of a signal* $x(t)$ *by passing this signal through an appropriately chosen filter*. For example, consider a signal $x(t)$ whose spectrum is as shown in Figure 6.3(a). Let us specify two filters $H_1(f)$ and $H_2(f)$

as[4] (see Figures 6.3(b) and (c))

$$|H_1(f)| = \begin{cases} 1 & \text{for } |f| \le 30 \\ 0 & \text{for } |f| > 30. \end{cases} \tag{6.6}$$

$$|H_2(f)| = \begin{cases} 1.75 & \text{for } 80 \le |f| \le 95 \\ 0 & \text{for } 80 > |f| > 95. \end{cases} \tag{6.7}$$

Let $y_1(t)$ and $y_2(t)$ be the outputs of these filters when $x(t)$ is applied to their inputs. Then, we get

$$Y_1(f) = H_1(f)X(f) \quad \text{and} \quad Y_2(f) = H_2(f)X(f). \tag{6.8}$$

With $H_1(f)$ and $H_2(f)$ given as above, we find that

$$Y_1(f) = \begin{cases} X(f) & \text{for } |f| \le 30 \\ 0 & \text{for } |f| > 30. \end{cases} \tag{6.9}$$

$$Y_2(f) = \begin{cases} 1.75X(F) & \text{for } 80 \le |f| \le 95 \\ 0 & \text{for } 80 > |f| > 95. \end{cases} \tag{6.10}$$

Thus, we find that *using the filters $H_1(f)$ and $H_2(f)$, we are able to extract different frequency regions in the spectrum of the input signal $x(t)$* (see Figures 6.3(d) and (e)).

The above principle can be generalized to *transform a given signal $x(t)$ into another signal $y(t)$ with any specific characteristics we desire.* What we need to do is to *select an appropriate filter $H(f)$ such that $Y(f) = H(f)X(f)$ satisfies our requirements.* To illustrate this, we consider an example below.

Example: In an experiment aimed at recording a particular electrical activity of the human body, the signal obtained from the bio-medical sensor is $x(t)$ given by

$$x(t) = d(t) + p(t) + n(t).$$

Here, $d(t)$ is the desired bio-medical signal and its spectrum lies in the frequency band $|f| \le 30$, $p(t)$ is the 50 Hz power-line interference signal whose spectrum consists of two lines at $f = \pm 50$ Hz, and $n(t)$ is a noise pick-up whose spectrum lies in the band $|f| \le 300$. *We want to use some filters to clean up the signal $x(t)$ (i.e., reject the interference and reduce the noise) so as to obtain a better estimate of $d(t)$.*

Solution: Because the spectrum of the interference $p(t)$ lies outside the spectrum of the desired signal $d(t)$, we can easily reject this interference by passing $x(t)$ through

4. The notation $80 \le |f| \le 95$ means $-95 \le f \le -80$ and $80 \le f \le 95$. Further, $|f| \le 30$ means $-30 \le f \le 30$, and $|f| > 30$ means $-30 > f > 30$.

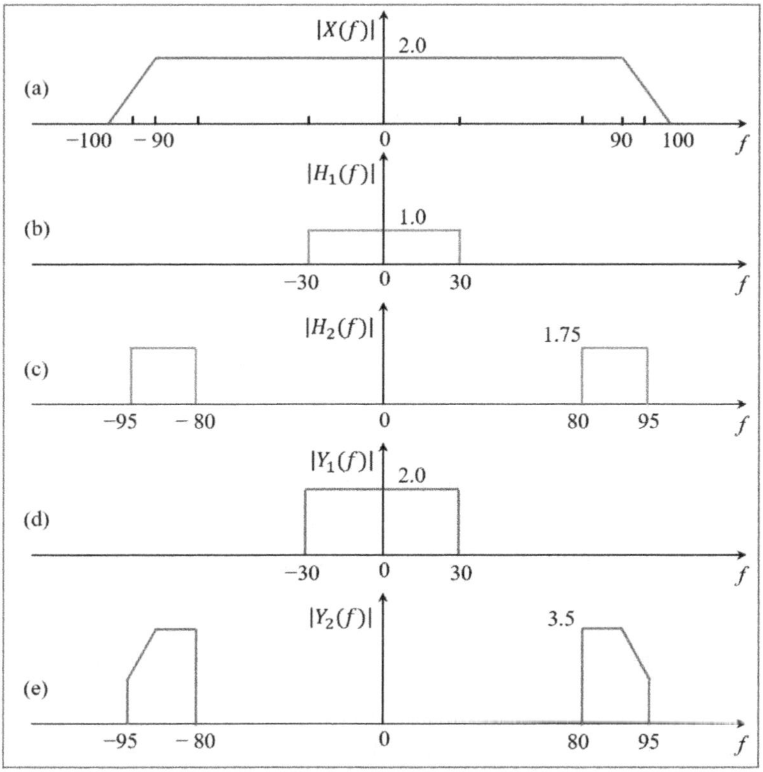

Figure 6.3. Example of a signal $x(t)$ being filtered by filters $H_1(f)$ and $H_2(f)$, resulting in outputs $y_1(t)$ and $y_2(t)$, respectively.

the filter $H_1(f)$ specified in Eq. (6.6). Let $y_1(t)$ be the resulting output. Then, $y_1(t)$ satisfies Eq. (6.9). Since $Y_1(f)$ is zero for $|f| > 30$, we find that filtering by $H_1(f)$ also helps to reject all the noise that lies in the frequency region $30 < |f| < 300$. Thus, we can write the output $y_1(t)$ as[5]

$$y_1(t) = d(t) + \tilde{n}(t),$$

where $\tilde{n}(t)$ is the result of filtering the noise $n(t)$ with $H_1(f)$. Further, the spectrum of $\tilde{n}(t)$ is now limited to $|f| \leq 30$, from $|f| \leq 300$. Thus, we find that *the output signal $y_1(t)$ is a much cleaner version compared to the original recorded signal $x(t)$ since the interference has been totally rejected and the noise has been significantly reduced.*

5. Note that $Y_1(f) = H_1(f)X(f) = H_1(f)[D(f) + P(f) + N(f)] = H_1(f)D(f) + H_1(f)P(f) + H_1(f)N(f) = D(f) + 0 + \tilde{N}(f)$ where $\tilde{N}(f) = H_1(f)N(f)$. Taking inverse FT, we get $y_1(t)$.

§ *Assume that a signal $x(t)$ is passed separately through two filters whose outputs are given by*

$$y(t) = x(t) + x(t - T) \tag{6.11a}$$

$$z(t) = x(t) - x(t - T) \tag{6.11b}$$

where T denotes a time-shift. Determine the expressions for the impulse responses and frequency responses of the two filters generating $y(t)$ and $z(t)$. Verify that the first filter attenuates the signal $x(t)$ around the frequency $1/(2T)$ Hz, and the second filter attenuates the signal $x(t)$ around the frequency[6] *0 Hz.*

6.5 Sinusoidal Input to Linear Systems

Let $x(t)$ and $y(t)$ be the input and output, respectively, of a linear system whose frequency response is given by $H(f)$. Let us choose the input to be a complex sinusoid given by

$$x(t) = A_0 e^{j(2\pi f_0 t + \theta_0)} \quad \Rightarrow \quad X(f) = A_0 e^{j\theta_0} \delta(f - f_0) \tag{6.12}$$

where A_0, f_0 and θ_0 are the amplitude, frequency and phase, respectively, of the sinusoid. Then, the filter output is given by

$$\begin{aligned} Y(f) &= X(f)H(f) = A_0 e^{j\theta_0} \delta(f - f_0) H(f) \\ &= A_0 e^{j\theta_0} H(f_0) \delta(f - f_0) && (6.13) \\ \Longrightarrow \quad y(t) &= A_0 e^{j\theta_0} H(f_0) e^{j2\pi f_0 t} = H(f_0) \left[A_0 e^{j(2\pi f_0 t + \theta_0)} \right] \\ &= H(f_0) x(t) && (6.14) \\ &= A_0 |H(f_0)| e^{j(2\pi f_0 t + \theta_0 + \theta_h(f_0))} && (6.15) \end{aligned}$$

where $\theta_h(f_0)$ is the phase of $H(f_0)$, i.e., $H(f_0) = |H(f_0)| e^{j\theta_h(f_0)}$. We note the following from Eqs. (6.13)–(6.15):

- When the input of a linear system is a sinusoid of frequency f_0, the resulting output is also a sinusoid at the same frequency f_0, with a different amplitude and phase.
- Amplitude and phase of the output sinusoid are $A_0|H(f_0)|$ and $\theta_0 + \theta_h(f_0)$, respectively. That is, the input amplitude and phase are changed by the

6. To determine the impulse responses, set the input to $\delta(t)$. Show that $h_1(t) = \delta(t) + \delta(t - T)$, $H_1(f) = 1 + e^{-j2\pi f T}$, $|H_1(f)|^2 = 4\cos^2(\pi f T)$, and $h_2(t) = \delta(t) - \delta(t - T)$, $H_2(f) = 1 - e^{-j2\pi f T}$, $|H_2(f)|^2 = 4\sin^2(\pi f T)$.

magnitude and phase, respectively, of the filter frequency response $H(f)$ at $f = f_0$.

6.5.1 Generalization to Multiple Sinusoids

We now generalize the above result to the case where the input $x(t)$ consists of a sum of M sinusoids rather than one sinusoid. That is,

$$x(t) = \sum_{i=1}^{M} A_i e^{j(2\pi f_i t + \theta_i)} \tag{6.16}$$

$$\Rightarrow \quad X(f) = \sum_{i=1}^{M} A_i e^{j\theta_i} \delta(f - f_i) \tag{6.17}$$

where A_i, f_i and θ_i are the amplitude, frequency and phase, respectively, of the ith sinusoid. The filter output is given by,

$$Y(f) = X(f)H(f) = \sum_{i=1}^{M} A_i e^{j\theta_i} \delta(f - f_i) H(f)$$

$$= \sum_{i=1}^{M} A_i e^{j\theta_i} H(f_i) \delta(f - f_i) \tag{6.18}$$

$$\Longrightarrow \quad y(t) = \sum_{i=1}^{M} A_i H(f_i) e^{j(2\pi f_i t + \theta_i)} \tag{6.19}$$

$$= \sum_{i=1}^{M} A_i |H(f_i)| e^{j(2\pi f_i t + \theta_i + \theta_h(f_i))} \tag{6.20}$$

where $\theta_h(f_i)$ is the phase of $H(f_i)$, i.e., $H(f_i) = |H(f_i)| e^{j\theta_h(f_i)}$. Based on these results, we can conclude the following.

§ *The output of a linear system cannot contain a frequency that is not present at its input.*

6.6 Partial Fraction Approach

It is often necessary for us to find the time-domain output $y(t)$ of a system from its FT, $Y(f)$ or the impulse response $h(t)$ of a system from its frequency response $H(f)$. This is a trivial task if the frequency-domain expression looks like the ones given in Tables 4.1 or 4.2 (see Section 17.6). However, often, the expressions are not as simple as these. In such cases, *the method of partial fraction is used to express a complicated expression in the form of a sum of several simple expressions whose inverse transforms can be found based on table look-up.*

Let $Y(f)$ be a complicated expression whose inverse FT we want to find. The method of partial fraction determines scalars $\alpha_1, \alpha_2, \cdots, \alpha_m$ such that $Y(f)$ can be written as

$$Y(f) = \alpha_1 Y_1(f) + \alpha_2 Y_2(f) + \cdots + \alpha_m Y_m(f) \tag{6.21}$$

where $Y_1(f), Y_2(f), \cdots, Y_m(f)$ are simple expressions whose inverse transforms are known. Then, we can get the inverse FT of $Y(f)$ as,

$$y(t) = \alpha_1 y_1(t) + \alpha_2 y_2(t) + \cdots + \alpha_m y_m(t) \tag{6.22}$$

where $y_i(t)$ is the inverse FT of $Y_i(f)$ for $i = 1, 2, \ldots, m$.

Example: As an example, let us consider the problem where we have

$$Y(f) = \frac{1}{(1 + j\Omega)(3 + j\Omega)}. \tag{6.23}$$

While we know the inverse Fourier transforms of $\frac{1}{1+j\Omega}$ and $\frac{1}{3+j\Omega}$, it is not easy to determine the inverse Fourier transform of their product. So, express $Y(f)$ as a sum of these easy factors as

$$Y(f) = \frac{1}{(1 + j\Omega)(3 + j\Omega)} = \frac{a}{1 + j\Omega} + \frac{b}{3 + j\Omega} \tag{6.24}$$

where the scalars a and b are to be determined. To determine a and b, combine the last two terms on the right-hand-side (RHS) of Eq. (6.24) to result in

$$\frac{1}{(1 + j\Omega)(3 + j\Omega)} = \frac{3a + b + j\Omega(a + b)}{(1 + j\Omega)(3 + j\Omega)}. \tag{6.25}$$

Since the denominators are the same on both sides of Eq. (6.25), *the numerators also must be the same on both sides.* Noting that the numerator on RHS is a polynomial in the variable Ω, *we can equate the coefficients of* Ω^0, Ω^1, Ω^2, $\cdots$ *on both sides.* Doing this, we get

$$3a + b = 1 \qquad \text{and} \qquad j(a + b) = 0. \tag{6.26}$$

Solving these two equations for a and b, we get $a = 1/2$ and $b = -1/2$. Therefore, we get from Eq. (6.24)

$$Y(f) = \frac{0.5}{1 + j\Omega} - \frac{0.5}{3 + j\Omega} \tag{6.27}$$

$$\implies y(t) = 0.5e^{-t}u(t) - 0.5e^{-3t}u(t), \tag{6.28}$$

since $e^{-ct}u(t)$ (with $c > 0$) is the inverse FT of $1/(c + j\Omega)$.

6.7 Convolution

The output of a linear time invariant (LTI) system for any input, $x(t)$, can be found by convolution operation between the input, $x(t)$ and the system impulse response (or system function), $h(t)$. Consider a linear system with frequency response $H(f) \leftrightarrow h(t)$. Then, we know that its output (in frequency-domain) is given by,

$$Y(F) = X(f)H(f), \tag{6.29}$$

where $X(f)$ is the FT of the input, $x(t)$. *The inverse FT of both sides of Eq. (6.29) is given by the convolution of the filter impulse response $h(t)$ with the input $x(t)$* as,

$$\begin{aligned} y(t) &= x(t) \otimes h(t) \\ &= \int_{-\infty}^{\infty} x(\tau)h(t-\tau)d\tau = \int_{-\infty}^{\infty} h(\tau)x(t-\tau)d\tau. \end{aligned} \tag{6.30}$$

It is also to be noted that convolution is commutative. This implies that $y(t) = h(t) \otimes x(t) = x(t) \otimes h(t)$. This means that the order in which these two functions convoluted does not influence the output.

6.7.1 Convolution Explained

The convolution operation is explained briefly in the following:

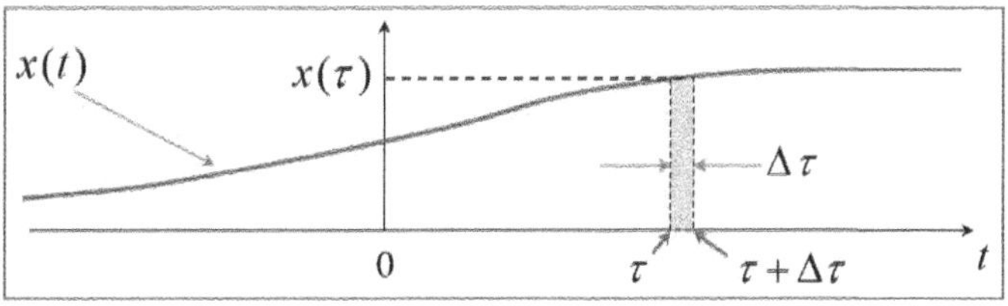

Figure 6.4. Convolution explained.

Let $x(t)$ in Fig. 6.5 be a continuous time function of time, which is input to a LTI system. The output $y(t)$ of the system is the convolution of $x(t)$ with the system function $h(t)$. This is explained here.

At time τ, consider the interval $[\tau, \tau + \triangle\tau]$. For sufficiently small $\triangle\tau$, the area under $x(t)$ inside the interval (shaded area in Fig. 6.5) can be approximated by an impulse function of strength $x(\tau)\triangle\tau$.

By definition, the response of a LTI system to an unit impulse at $t = 0$ (i.e., $\delta(t)$) is $h(t)$. Therefore, the response of the system to an impulse (delta) function of strength $x(\tau)\triangle\tau$ occurring at $t = \tau$ is,

$$\triangle y(t) = x(\tau)h(t - \tau)\triangle\tau. \tag{6.31}$$

In the limiting case as $\triangle\tau \to 0$, apply the principle of superposition to obtain the overall system response to the input $x(t)$ as,

$$y(t) = \int_{-\infty}^{\infty} dy(t) = \int_{-\infty}^{\infty} x(\tau)h(t - \tau)d\tau. \tag{6.32}$$

This is the convolution operation and is denoted as $y(t) = x(t) * h(t)$ or $y(t) = x(t) \otimes h(t)$.

We can get the second integral expression from the first integral in Eq. (6.30) by applying a change of variables of the form $u = t - \tau$. Note that *the integration is done with respect to the variable* τ. In other words, inside the integral, t acts like a constant. For example, the expression for computing $y(t)$ at instant $t = 2$ is

$$y(2) = \int_{-\infty}^{\infty} h(\tau)x(2 - \tau)d\tau. \tag{6.33}$$

Let us examine $x(t - \tau)$.

$$\text{Let} \quad z(\tau) = x(-\tau). \tag{6.34}$$

$$\text{Then,} \quad x(t - \tau) = x(-(\tau - t)) = z(\tau - t) \tag{6.35}$$

$$\Longrightarrow \quad y(t) = \int_{-\infty}^{\infty} z(\tau - t)h(\tau)d\tau. \tag{6.36}$$

Eqs. (6.34)–(6.36) implies the following approach to compute the filter output $y(t)$ for any value of t.

- First, construct $z(\tau)$ by time-reversing $x(\tau)$.
- If $t > 0$, then $x(t - \tau)$ is obtained by shifting $z(\tau)$ to the right by t. That is, shift $x(-\tau)$ to the right by t.
- If $t < 0$, then $x(t - \tau)$ is obtained by shifting $z(\tau)$ to the left by t. That is, shift $x(-\tau)$ to the left by t.

- Compute $y(t)$ by integrating the product $z(\tau - t)h(\tau)$ over the complete range of τ for which this product is non-zero.

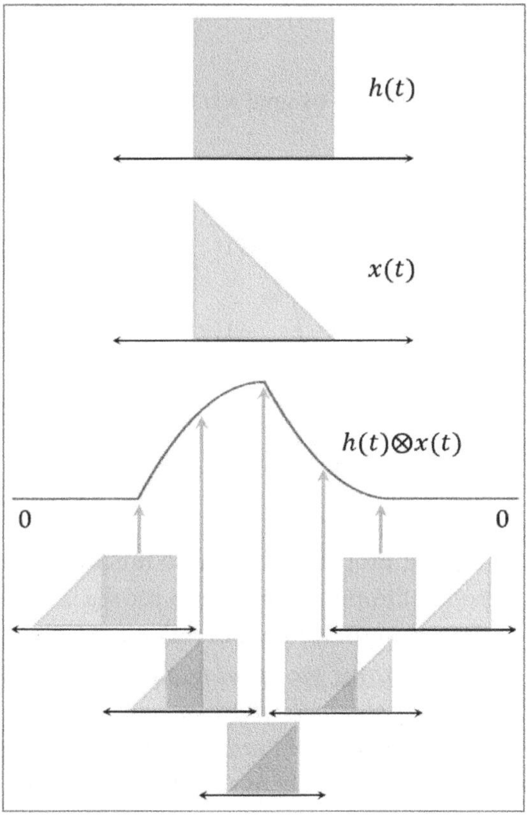

Figure 6.5. Illustration of the convolution operation of two functions in the time domain.

Eqs. (6.29) and (6.30) show that *multiplying two signals in frequency-domain is equivalent to convolving these signals in time-domain.*[7] Thus, we have two approaches for computing the filter output $y(t)$ given $x(t)$ and $h(t)$, and are given below.

- Use the convolution integral given by Eq. (6.30).
- Determine the Fourier transforms $X(f)$ and $H(f)$, and find the inverse transform of the product $X(f)H(f)$.

We can use either of these approaches, depending on which is easier to implement in the given application/problem.

7. Similarly, *multiplying two signals in time-domain is equivalent to convolving them in frequency-domain.*

A graphical illustration of the convolution operation of two functions ($h(t)$ and $x(t)$) is given in Fig. 6.5.

6.8 Shifting Property of Impulse Function

Let us consider the convolution of a signal $x(t)$ with an impulse $\delta(t - t_0)$ located at $t = t_0$. We obtain,

$$\begin{aligned}
x(t) \otimes \delta(t - t_0) &= \int_{-\infty}^{\infty} x(t - \tau)\delta(\tau - t_0)d\tau, \\
&= \int_{-\infty}^{\infty} x(t - t_0)\delta(\tau - t_0)d\tau, \\
&= x(t - t_0) \int_{-\infty}^{\infty} \delta(\tau - t_0)d\tau, \\
&= x(t - t_0),
\end{aligned} \tag{6.37}$$

where we used the sampling property of impulse function to get the second line and the unit-area property of impulse function to get the last line (see Eqs. (4.59)–(4.61) in Chapter 4). Equation (6.37) implies that *convolving a signal $x(t)$ with an impulse $\delta(t - t_0)$ results in shifting the signal to the location of the impulse.* This is known as the *shifting property of impulse function.*

Using this property, it is easy for us to examine what happens in frequency-domain when we do sampling in time-domain. Let us sample a continuous-time signal $x(t)$ at the rate of $f_s = 1/T_s$ samples/second. (see Eqs. (4.59)–(4.61) in Chapter 4). This is done by multiplying $x(t)$ with a periodic impulse train $i(t) = \sum_{n=-\infty}^{\infty} \delta(t - nT_s)$ of period T_s, resulting in,

$$y(t) = x(t) \times i(t) = x(t) \sum_{n=-\infty}^{\infty} \delta(t - nT_s). \tag{6.38}$$

FT of the impulse train is given by,

$$I(f) = \mathcal{F}[i(t)] = f_s \sum_{k=-\infty}^{\infty} \delta(f - kf_s). \tag{6.39}$$

FT of the sampled signal $y(t)$ can be obtained by convolving $X(f)$ with $I(f)$, resulting in

$$\begin{aligned} Y(f) &= X(f) \otimes I(f) = X(f) \otimes \left[f_s \sum_{k=-\infty}^{\infty} \delta(f - kf_s) \right] \\ &= f_s \sum_{k=-\infty}^{\infty} [X(f) \otimes \delta(f - kf_s)] \\ &= f_s \sum_{k=-\infty}^{\infty} X(f - kf_s) \\ &= f_s \left[\cdots + X(f + 2f_s) + X(f + f_s) + X(f) \right. \\ &\qquad \left. + X(f - f_s) + X(f - 2f_s) + \cdots \right]. \end{aligned} \tag{6.40}$$

Thus, $Y(f)$ *is periodic with period* f_s, *and is obtained by adding periodically shifted copies of* $X(f)$. See Fig. 6.6 for illustration of these spectra. It can be seen that the spectra of the sampled signal repeats (with a period, f_s) on either side of the frequency axis and are scaled by the sampling frequency, f_s.

§ *Based on Figure 6.6, suggest an approach to reconstruct the analog signal* $x(t)$ *from its samples. What is the minimum possible sampling frequency?*

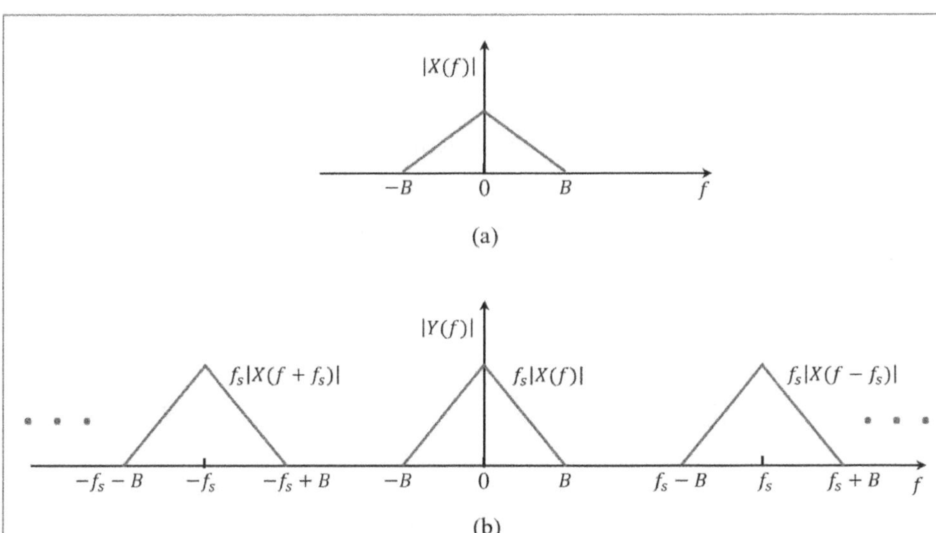

Figure 6.6. Spectra of (a) continuous-time signal $x(t)$, (b) sampled version of $x(t)$ with sampling frequency f_s.

6.9 Concluding Remarks

We can make the following concluding remarks based on the material presented above.

- Fourier transform (FT) is a very useful (and powerful) tool in the study of systems.
- Frequency response of a system specifies how the system is going to shape the input signal.
- The output of a linear system cannot contain a frequency that is not present in its input.
- The output of a linear system can be determined either using time-domain computations (i.e., convolution of input signal $x(t)$ and system impulse response $h(t)$) or using frequency-domain computations (i.e., multiplying the Fourier transforms of $x(t)$ and $h(t)$, and finding the inverse FT of the product).
- Partial fraction approach is very useful to determine the inverse Fourier transforms of complicated frequency-domain expressions.

Exercises

6.1. (a) An analog linear system has an impulse response $h(t) = e^{-at}u(t)$, where a is a real constant and $u(t)$ is a unit step function. Use convolution to find the response of the system ($y(t)$) to a unit step input (i.e., $x(t) = u(t)$, here $x(t)$ denotes the input to the system).
(b) Sketch the magnitude and phase response of the system. Determine the 3dB bandwidth of the system.

6.2. Determine the convolution of $x(t)$ and $u_a(t)$, where $x(t)$ is any signal. Derive the expression for the convolution of $x(t)$ with a time-shifted unit step function $u(t - t_0)$, where t_0 is the time delay.

6.3. The impulse response of an analog system is given by $h(t) = u(t) - u(t-2)$. Use convolution to find the response of this system to an input $x(t) = u(t) - u(t-3)$.

6.4. (a) An engineer is contemplating the use of a Gaussian pulse of the form $x(t) = e^{-\left(\frac{t}{c}\right)^2}$, where c is a suitable constant related to the pulse width, for transmitting digital information in a mobile communication system. The engineer finds that the spectrum of this pulse is $X(f) = be^{-(\pi cf)^2}$, where b is a constant and it depends on c. Sketch the pulse and its spectrum for $c = 5$. Using energy argument or

Parseval's theorem, derive a formula that can be used for calculating the constant b from c. Derive equations from which the -3 dB pulse width and the -3 dB bandwidth can be found.

(b) The Gaussian pulse in Question 6.4(a) is convolved with itself. Show that the resulting waveform is also Gaussian. Has the -3 dB pulse width increased? If yes, by how much?

6.5. An engineer designed a filter whose impulse response $h(t)$ is given by (αt), where α is a real positive scalar.

(a) When the engineer passed the signal $x(t) = \cos(200\pi t)$ through the designed filter, he observed that the filter output, $y(t)$, becomes zero (for all t) for some choice of α. For some other choice of α, the output becomes a scaled version of the input (i.e., $y(t) = \beta x(t)$, where β is a real scalar). Analyse this problem and determine the ranges of values of α such that the output will be a scaled version of the input. What is the expression for β?

(b) Assume that α is chosen from the range determined from Question (a). Determine the power and energy of the input and output of the filter. Plot the magnitude and phase spectra of the input and output of the filter.

6.6. In this problem, we consider a system for measuring the speed of objects. The operation is as follows. The system sends a radio-frequency (RF) pulse $p(t)$, like the one shown in Fig. P6.1 below:

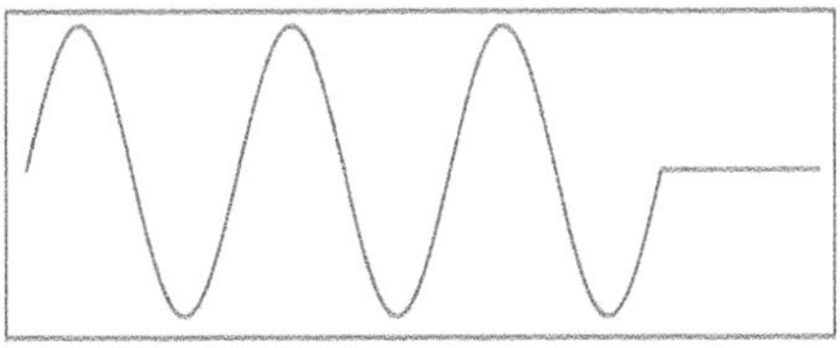

Figure P6.1. RF pulse ($p(t)$).

The signal is reflected by the object, whose velocity is being measured and the echo arrives at the receiver at time t_0. The received signal is given by:

$$g(t) = a \times p(\alpha t - t_0).$$

The shape of the signal is modified due to the motion of the object. The constant α is given by:

$$\alpha = 1 + \frac{2v}{c} = \frac{c + 2v}{c},$$

where v is the speed of the object and c is the speed of propagation of the ultrasound signal which is assumed to be much larger than the speed of the object which is being measured, i.e., $c \gg v$. a and t_0 are positive real constants. The RF pulse $p(t)$ is given by:

$$p(t) = \begin{cases} \sin(2\pi f_0 t) & \text{for} \quad 0 \leq t \leq \frac{M}{f_0} \\ 0 & \text{otherwise,} \end{cases}$$

where f_0 is a positive real number and M is a positive integer number. It can be assumed that $M \gg 1$ in the rest of the problem.

(a) Show how the spectrum $P(f)$ of $p(t)$ can be found through the use of appropriate theorems and properties of the Fourier transform. Express the spectrum as a function of f_0 and M. Sketch the amplitude density spectrum $|P(f)|$, and denote the placement of maxima and zeros (minima) in the spectrum.

(b) Find the spectrum $G(f)$ of $g(t)$, using $P(f)$ as a starting point and applying appropriate theorems and properties of the Fourier transform. Express the spectrum as a function of a, α, f_0, M and t_0. Show that the spectrum has maxima at $f = \pm\alpha f_0$. Sketch the amplitude density spectrum of the received signal for $\alpha = 1.1$, when the relevant parameters are fixed.

(c) The received signal is processed by the system shown in Fig. P6.2 below.

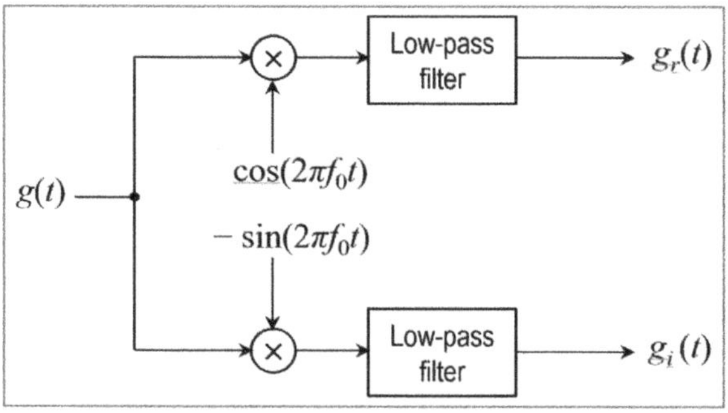

Figure P6.2. Detector scheme.

First the signal $g(t)$ is multiplied by the complex signal $g_d(t) = e^{-j2\pi f_0 t} = \cos(2\pi f_0 t) - j\sin(2\pi f_0 t)$. Then the resulting signals are passed through ideal low-pass filters. These low-pass filters remove all spectral components over the frequency f_0, and have an amplification of 1 in the pass-band. The result of this processing is the complex signal:

$$g_1(t) = g_r(t) + jg_i(t).$$

Sketch the amplitude spectrum of the complex signal $g_1(t)$ for $\upsilon = +0.05 \times c$ and for $\upsilon = -0.05 \times c$, by fixing the relevant parameters (choosing appropriate values). Show that the sign (positive or negative) of υ can be found from the maximum value in the amplitude spectrum of the signal $g_1(t)$.

DOI: 10.1561/9781680839791.ch7

Chapter 7

Sampling and Digital Signals

In practical applications of signals and systems, we often need to measure signals and analyse them to discover the characteristics of the signals and/or systems present in the particular application under consideration. For example, in the receiver of a communication system, we would depend a lot on the received signal to understand the nature of the channel through which the signal propagation takes place. Another example would be the non-invasively collected ECG signals from the chest sensors/electrodes of a subject/patient to process and diagnose the condition of the heart.

Most of the signals that we encounter in practice are analog in nature, i.e., continuous-time and continuous-amplitude signals. On the other hand, we use the power of digital computers and digital signal processors to do the analysis and/or processing of these signals, which require signals in the digital (i.e., discrete-time and discrete-amplitude signals) format. So, it becomes necessary to convert the analog signals into digital signals and this is done through the processes of sampling and quantization. The device which does sampling and quantization is known as the analog-to-digital converter (ADC).

It is very important to ensure that we do not lose any information through the process of sampling; otherwise, the inferences that we make about the analog signal based on our analysis of the sampled signal will not be useful. Since the process of quantization results in loss of accuracy in the signal amplitude, it is extremely important to design the quantizer to minimize this error.

The following topics are discussed in detail in this chapter.

- Theoretical concepts behind sampling of signals:
  - Relation between sampling frequency and signal bandwidth.
  - Concept of aliasing.
  - Reconstruction of analog signal from samples.
  - Nyquist's sampling theorem.
- Design aspects of quantizer.
- Definition and properties of Fourier transform (or spectrum) for discrete-time signals.
- Use of z-transform in the study of discrete-time signals.

7.1 Sampling of Analog Signals

Let $x_a(t)$ represent an analog signal and let $X_a(f)$ be its Fourier transform such that

$$|X_a(f)| = 0 \qquad \text{for} \quad |f| > B. \tag{7.1}$$

That is, $x_a(t)$ is a band-limited signal, and the maximum frequency component in[1] $x_a(t)$ is at $\pm B$ Hz.

Let us sample the continuous-time signal $x_a(t)$ at the rate of f_s samples per second, i.e., with a sampling interval of $T_s = 1/f_s$ seconds. This is shown in Fig. 7.1 below. An example of the signal $x_a(t)$ and its sampled versions are shown in Fig. 7.2. As can be seen, the sampled signals, $x[n]$ takes values of $x_a(t)$ at instances, $t = nT_s$, where $n = 0, \pm 1, \pm 2, \ldots$.

Recall from Chapter 6 that sampling can be done by multiplying $x_a(t)$ with a periodic train of impulses spaced at T_s seconds. This periodic impulse train and its

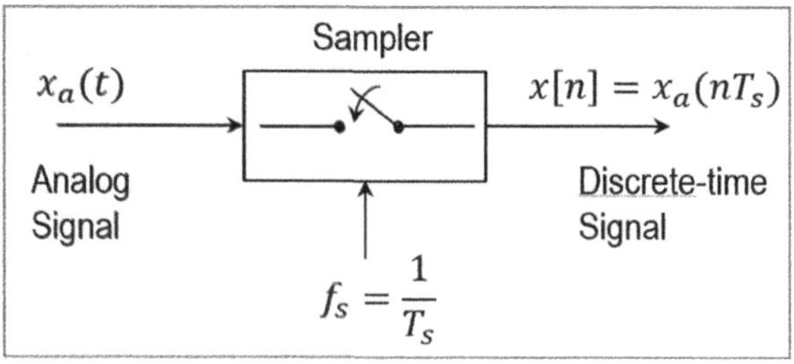

Figure 7.1. Schematic representation of periodic sampling of an analog signal.

1. For the signal $x_a(t)$ with its Fourier spectrum given by Eq. (7.1), we say that the 2-sided bandwidth is $2B$, or the one-sided bandwidth is B.

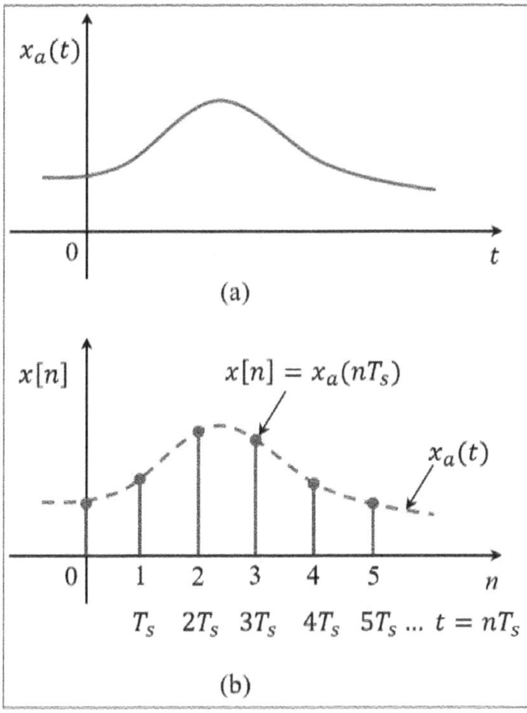

Figure 7.2. Example of an analog signal $x_a(t)$ and its sampled version, $x[n]$. (a) Analog signal and (b) Sampled (discrete-time) signal.

Fourier transform are given by,

$$i(t) = \sum_{n=-\infty}^{\infty} \delta(t - nT_s) \tag{7.2}$$

$$I(f) = f_s \sum_{k=-\infty}^{\infty} \delta(f - kf_s). \tag{7.3}$$

We get the sampled signal as,

$$\begin{aligned} x_s(t) = x_a(t) \sum_{n=-\infty}^{\infty} \delta(t - nT_s) &= \sum_{n=-\infty}^{\infty} x_a(nT_s)\delta(t - nT_s) \\ &= \sum_{n=-\infty}^{\infty} x[n]\delta(t - nT_s) \end{aligned} \tag{7.4}$$

$$= \cdots + x[-2]\delta(t + 2T_s) + x[-1]\delta(t + T_s) + x[0]\delta(t) + x[1]\delta(t - T_s) + x[2]\delta(t - 2T_s) + \cdots, \tag{7.5}$$

where we denote the samples of $x_a(t)$ by $x[n]$. That is,

$$x[n] \triangleq x_a(nT_s) = [x_a(t) \quad \text{at} \quad t = nT_s], \tag{7.6}$$

where n is an integer in the range $[-\infty, +\infty]$. At instants other than the sampling instants, i.e., $t \neq nT_s$, note that $x_s(t) = 0$.

Noting that Fourier transform of the product of two signals (in the time domain) is equal to the convolution of the Fourier transforms of the two signals (in the Fourier/frequency domain), the Fourier transform of the sampled signal can be obtained as

$$\begin{aligned} X(f) &= X_a(f) \otimes I(F) \\ &= X_a(f) \otimes \left[f_s \sum_{k=-\infty}^{\infty} \delta(f - kf_s) \right] \\ &= f_s \sum_{k=-\infty}^{\infty} [X_a(f) \otimes \delta(f - kf_s)] \end{aligned} \tag{7.7}$$

$$\begin{aligned} &= f_s \sum_{k=-\infty}^{\infty} X_a(f - kf_s) \\ &= f_s [\cdots + X_a(f + 2f_s) + X_a(f + f_s) + X_a(f) \\ &\qquad + X_a(f - f_s) + X_a(f - 2f_s) + \cdots]. \end{aligned} \tag{7.8}$$

Thus, $X(f)$ is obtained by adding periodically the shifted copies of $X_a(f)$. Figure 7.3 illustrates this for two choices of the sampling frequencies. Using Fig. 7.3, we can deduce several concepts underlying discrete-time signals.[2]

It can be seen from the above figure, that the spectra of the sampled signals overlap each other when the sampling frequency (f_2) is less than $2B$ (Fig. 7.3(c). On the other hand, if the sampling frequency is greater than $2B$, (Fig. 7.3(b)), the spectra does not overlap, making it easier to reconstruct the original signal from the sampled signal.

2. Since we have not quantized the values of the samples $x[n]$ into discrete-amplitude, strictly speaking, we should call $x[n]$ as a discrete-time signal (or sequence) and not digital signal.

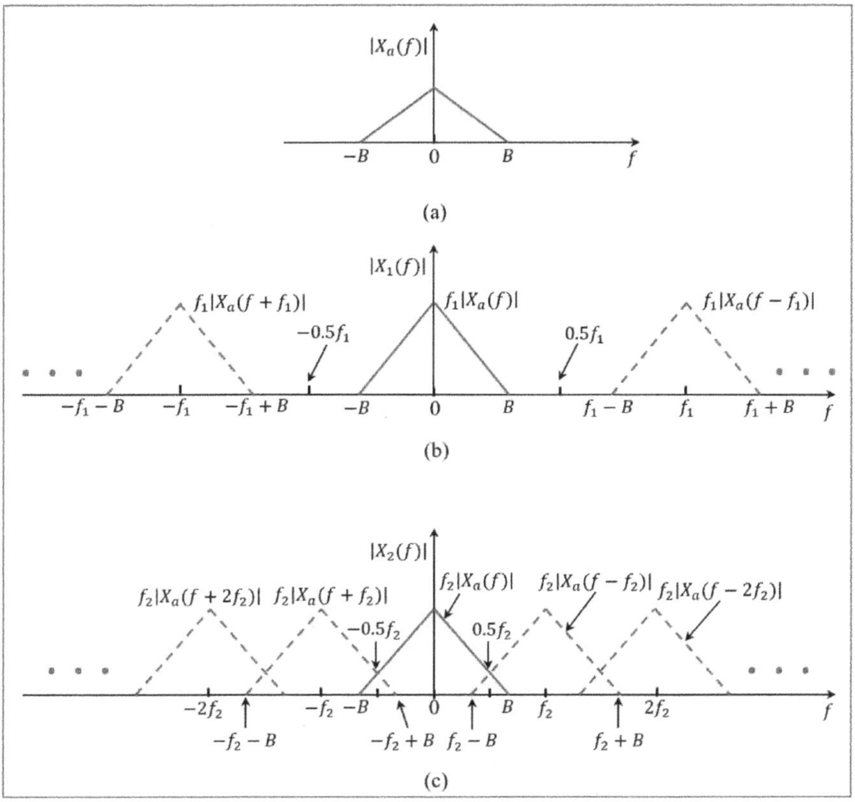

Figure 7.3. Spectra of signals before and after sampling (with different sampling frequencies): (a) Analog signal $x_a(t)$ with 2-sided bandwidth $2B$, (b) Sampled signal $x_1[n]$ with sampling frequency $f_1 > 2B$, and (c) Sampled signal $x_2[n]$ with sampling frequency $f_2 < 2B$.

7.2 Signal Reconstruction from Samples

To address the problem of reconstruction of a continuous-time signal from its samples, we ask the following questions.

> Is it possible to perfectly reconstruct the analog signal $x_a(t)$ from its samples $x[n]$? How can this be done? In other words, how can we obtain the spectrum $X_a(f)$ of the analog signal from the spectrum $X(f)$ of the sampled signal? What conditions must be satisfied for perfect reconstruction to be possible? Conversely, what are the conditions which make perfect reconstruction impossible?

Using Fig. 7.3, we shall answer these questions.

7.2.1 Perfect Reconstruction

Consider Fig. 7.3(b). Since the sampling frequency f_1 is chosen to be greater than $2B$, observe that the spectrum $f_1 X_a(f)$ in the frequency range $[-B, +B]$ is not affected by the higher frequency spectral images $f_1 X(f + kf_1)$ for $k = \pm 1, \pm 2, \ldots$.

Let us pass the samples $x[n]$ through an ideal low-pass filter (LPF) whose frequency response is given by,

$$H(f) = \begin{cases} \frac{1}{f_1} & \text{for } -0.5f_1 \leq f \leq 0.5f_1 \\ 0 & \text{otherwise.} \end{cases} \tag{7.9}$$

The impulse response of this LPF is given by,

$$h(t) = \mathcal{F}^{-1}[H(f)] = \mathsf{sinc}(f_1 t). \tag{7.10}$$

Therefore, we get the output of the LPF as,

$$\begin{aligned} \hat{x}_a(t) &= \{x[n]\} \otimes h(t) = x_s(t) \otimes h(t) \\ &= \sum_{n=-\infty}^{\infty} x[n]\mathsf{sinc}(f_1(t - nT_s)). \end{aligned} \tag{7.11}$$

It is easy to see from Eq. (7.8) and Fig. 7.3(b) that the spectrum of the LPF output will be $X_a(f)$. That is, the signal $\hat{x}_a(t)$ at the LPF output will be same as the original analog signal $x_a(t)$.

> § *Perfect reconstruction of the analog signal $x_a(t)$ from its samples taken at sampling rate f_1 is possible if the sampling rate is chosen to be greater than or equal to twice the maximum frequency B of $x_a(t)$.* That is,
>
> $$\text{Sampling rate} \geq 2B. \tag{7.12}$$
>
> This is known as *Nyquist's sampling theorem*. The frequencies $2B$ and B are known as Nyquist rate and Nyquist frequency, respectively, of the signal $x_a(t)$.

> § *If $f_1 \geq 2B$, then perfect reconstruction of $x_a(t)$ from its samples $x[n]$, taken at rate f_1, can be done by passing the samples through an ideal low-pass filter with cut-off frequencies $\pm 0.5f_1$ and gain $1/f_1$.*

7.2.2 Spectral Aliasing

Consider Fig. 7.3(c). In this case, the sampling rate f_2 does not obey Eq. (7.12). As a result, the higher frequency spectral images $f_2 X_a(f + f_2)$ and $f_2 X_a(f - f_2)$

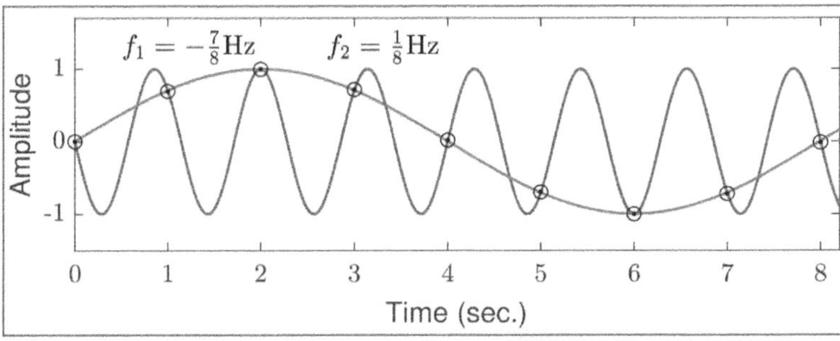

Figure 7.4. Illustration of aliasing.

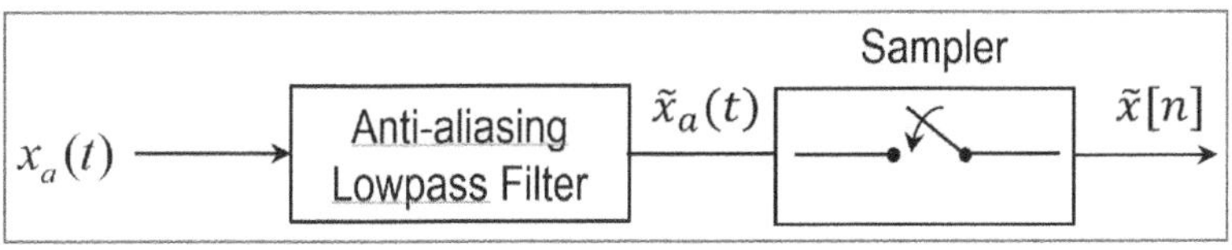

Figure 7.5. Use of anti-aliasing filter to avoid aliasing. If f_2 is the sampling rate, then the low-pass filter limits the bandwidth of $x_a(t)$ to $[-0.5f_2, 0.5f_2]$, resulting in $\tilde{x}_a(t)$.

overlap with the the spectrum $f_2 X_a(f)$ in the range $[-B, +B]$. This is called spectral aliasing. Because of aliasing, it is easy to see that we cannot recover the original spectrum $X_a(f)$ from $X(f)$.

§ *Perfect reconstruction of an analog signal $x_a(t)$ from its samples ($x[n]$) is impossible if the sampling rate is less than twice the maximum frequency B of the signal $x_a(t)$.* That is,

$$\text{Sampling rate} < 2B. \tag{7.13}$$

An example of aliasing is illustrated in Fig. 7.4, where two sinusoids with frequencies $f_1 = 1/8$ Hz and $f_2 = -7/8$ Hz yield identical samples when sampled at $f_s = 1$ Hz. This means that, when the signals are reconstructed from the sampled signal, it will never be able to obtain the original signal.

7.2.3 Anti-Aliasing Filtering

We can avoid aliasing by pre-filtering the signal $x_a(t)$ by a low-pass filter to limit the bandwidth of the signal to satisfy the condition given by Eq. (7.12). This filtering should be done before sampling, as shown in Fig. 7.5. Such a filter is known as *anti-aliasing filter or bandlimiting filter*. In Fig. 7.5, the spectrum of the samples $\tilde{x}[n]$ is free from aliasing with respect to the signal $\tilde{x}_a(t)$. That is, the signal $\tilde{x}_a(t)$ (not $x_a(t)$!) can be perfectly reconstructed from the samples $\tilde{x}[n]$.

From power and complexity considerations, it is good to use an anti-aliasing filter to band-limit the signal, so that the sampling rate can be reduced, thus reducing the complexity of the whole process. However, this may result in loss of information which can never be recovered from the samples.

7.2.4 Choice of Sampling Frequency

Selection of the sampling frequency involves several tradeoffs.

- Sampling rate should be as low as possible to minimize (i) memory required to store the data, (ii) bandwidth required to transmit the data, (iii) power required to process the data at high-speed, (iv) cost required to design and fabricate components/devices to operate at high-speed, etc.
- Sampling rate should be as high as possible (i) to avoid aliasing, (ii) to relax the sharp cut-off characteristics required of filters used for bandlimiting the signals, etc.

7.3 Quantization

As mentioned earlier, digitization of signals involves sampling and quantization, which are performed by a device known as the analog-to-digital converter (ADC). In this section, we shall briefly touch upon quantization.

The process of converting a discrete-time continuous amplitude signal into a digital signal by expressing each sample value as a finite number of digits is called *quantization*. The task of the quantizer is to convert the input sample $x[n]$, which is continuous-valued, into a discrete-valued format. A simple quantizer, known as the *uniform quantizer*, is described here.

Let the dynamic range of the input samples $x[n]$ be $[-x_{max}, x_{max}]$ with $x_{max} > 0$, being the maximum value of the signal. The quantizer divides this dynamic range into L equal intervals (also known as *quantization step-size or resolution*) of length $\Delta = \frac{2x_{max}}{L}$, and i^{th} interval is given by,

$$[d_{i-1}, d_i] \qquad \text{with} \quad d_i = -x_{max} + i\,\Delta, \tag{7.14}$$

for $i = 1, \ldots, L$. This is illustrated in Fig. 7.8(b). The discrete-valued output of the quantizer is given by the middle-point of each interval. Therefore, we get the quantizer output as

$$x_q[n] = \frac{d_{i-1} + d_i}{2} \qquad \text{if} \quad x[n] \in [d_{i-1}, d_i]. \tag{7.15}$$

If we have b-bit quantizer, then $L = 2^{b+1}$ and every sample $x_q[n]$ will be represented using a b-bit binary-word. Thus, with a sampling rate of f_s, the number of

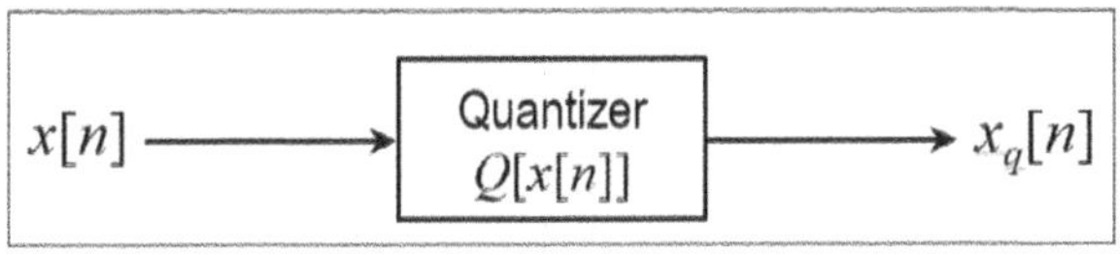

Figure 7.6. Schematic diagram of quantizer (actual system).

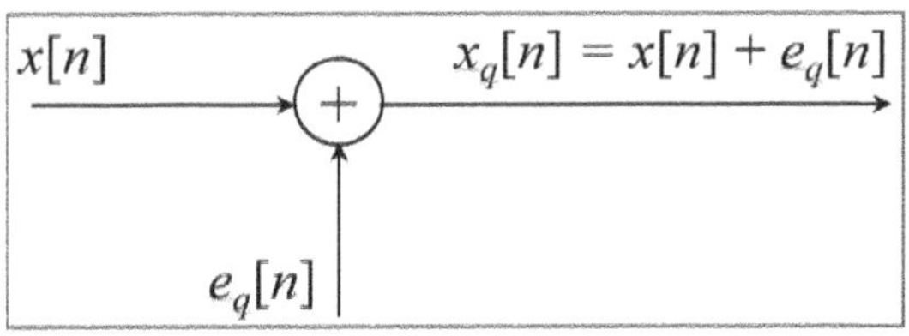

Figure 7.7. Mathematical model of quantization noise (mathematical model).

bits generated per second will be bf_s. Clearly, more the number of bits per sample used in the quantizer, less will be the error incurred in quantization, and more will be the storage space and/or speed required to handle the generated data.

7.3.1 Quantization of Continuous-Amplitude Signals

A digital signal is a sequence of numbers (samples) in which each number is represented by a finite number of digits (finite precision). The error introduced in representing the continuous-valued signal by a finite set of discrete-value levels is called *quantization error* or *quantization noise*. Figure 7.6 illustrates the block diagram of an actual quantizer and its mathematical model is given in Fig. 7.7.

We denote the quantizer operation on the samples $x[n]$ as $Q[x[n]]$ and the sequence of quantized samples as $x_q[n]$. Hence,

$$x_q[n] = Q[x[n]]. \tag{7.16}$$

The quantization error is a sequence $e_q[n]$ defined as

$$e_q[n] = x_q[n] - x[n]. \tag{7.17}$$

The quantization process is illustrated with an example below. Let us consider the discrete-time signal

$$x[n] = \begin{cases} 0.9^n & \text{for } n \geq 0 \\ 0 & \text{for } n < 0 \end{cases} \tag{7.18}$$

obtained by sampling the analog exponential signal $x_a(t) = 0.9^t$, $t \geq 0$ with a sampling frequency $f_s = 1Hz$ (see Fig. 7.8(a)). Observation of Table 7.1, which

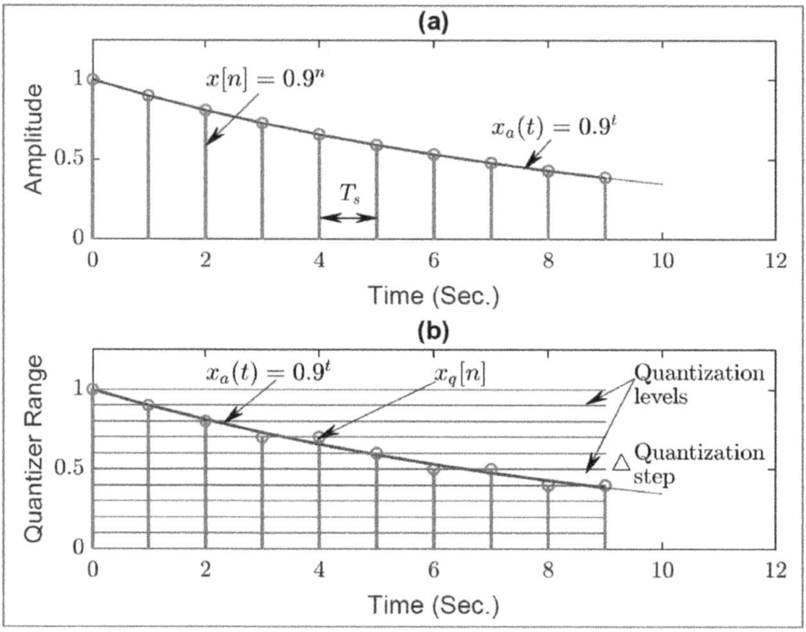

Figure 7.8. Illustration of quantization.

shows the values of the first 10 samples of $x[n]$, reveals that the description of the sample value $x[n]$ requires n significant digits.

Let us assume that we want to use only one significant digit. To eliminate the excess digits, we can either simply discard the extra digits (*truncation*) or discard them by rounding the resulting number (*rounding*). The resulting $x_q[n]$ corresponding to the truncation and rounding operations are given in Table 7.1. We discuss here the quantization by rounding.

The rounding process is graphically illustrated in Fig. 7.8(b). The values allowed in the digital signal are the *quantization levels*, whereas the distance $\triangle$ between two successive quantization levels is the *quantization step-size* or *resolution*. The rounding quantizer assigns each sample of $x[n]$ to the nearest quantization level. In contrast, a quantizer that performs truncation would have assigned each sample of $x[n]$ to the quantization level below it. The quantization error in rounding is limited to the range of $-\triangle/2$ to $\triangle/2$, that is

$$-\frac{\triangle}{2} \leq e_q[n] \leq \frac{\triangle}{2} \tag{7.19}$$

In other words, the instantaneous quantization error cannot exceed half of the quantization step-size (see Table 7.1).

Table 7.1. Numerical illustration of quantization with one significant digit using truncation or rounding.

n	$x[n]$ Discrete-time signal	$x_q[n]$ (Truncation)	$x_q[n]$ (Rounding)	$e_q[n]$ (Rounding)
0	1	1.0	1.0	0.0
1	0.9	0.9	0.9	0.0
2	0.81	0.8	0.8	−0.01
3	0.729	0.7	0.7	−0.029
4	0.6561	0.6	0.7	0.0439
5	0.59049	0.5	0.6	0.00951
6	0.531441	0.5	0.5	−0.031441
7	0.4782969	0.4	0.5	0.0217031
8	0.43046721	0.4	0.4	−0.03046721
9	0.387420489	0.3	0.4	0.012579511

If x_{min} and x_{max} represent the minimum and maximum values of $x[n]$ and L is the number of quantization levels, then

$$\triangle = \frac{x_{max} - x_{min}}{L - 1} = \frac{R}{L - 1} \tag{7.20}$$

The numerator of Eq. (7.20) is defined as the *dynamic range* (R) of the signal. In our example, $x_{max} = 1.0$, $x_{min} = 0$, and $L = 11$, which makes $\triangle = 0.1$. Note that if the dynamic range is fixed, increasing L results in a decrease of the quantization step-size. Thus the quantization error decreases and the accuracy of the quantizer increases. If the quantizer has $b + 1$ bits of accuracy and the quantizer covers the entire range R, the quantization step is

$$\triangle = \frac{R}{2^{b+1}}. \tag{7.21}$$

7.3.2 Analysis of Quantization Errors

To determine the effects of quantization on the performance of an ADC, we adopt a statistical approach because the quantization error is assumed to be random in nature. We model this error as noise that is added to the original (unquantized) signal (see Fig. 7.8(b)). If the input analog signal is within the range of the quantizer, the quantization error $e_q[n]$ is bounded in magnitude (i.e., $|e_q[n]| < \triangle/2$), and

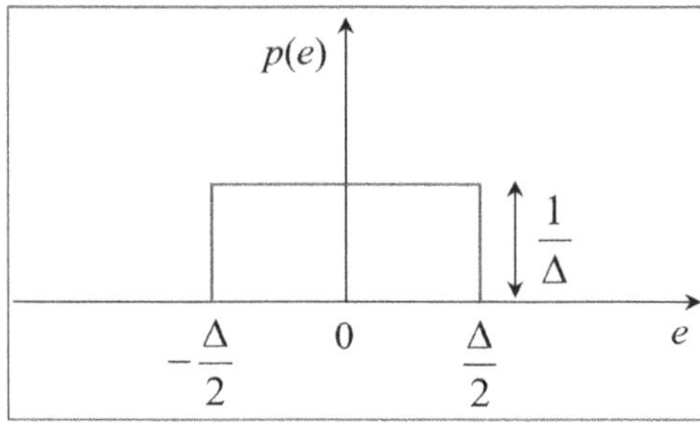

Figure 7.9. Probability density function for the quantization error.

the resulting error is called the *granular noise*. When the input falls outside the range of the quantizer (clipping), $e_q[n]$ becomes unbounded and results in *overload noise*. The following analysis is based on the assumption that there is no overload noise.

To carry out the analysis, we make the following assumptions about the statistical properties of $e_q[n]$:

- The error $e_q[n]$ is uniformly distributed over the range $-\frac{\triangle}{2} \leq e_q[n] \leq \frac{\triangle}{2}$.
- The error sequence $e_q[n]$ is a stationary white noise process.[3] In other words, the error $e_q[n]$ and error $e_q[m]$ for $m \neq n$ are uncorrelated.[4]
- The error sequence $e_q[n]$ is uncorrelated with the signal sequence $x[n]$.
- The signal sequence $x[n]$ is zero mean and stationary.

Under these assumptions, the effect of the additive noise $e_q[n]$ on the desired signal can be quantified by evaluating the signal-to-quantization-noise (power) ratio (SQNR), which can be expressed on a logarithmic scale (in dB) as

$$\text{SQNR} = 10\log_{10}\left(\frac{p_x}{p_n}\right) \tag{7.22}$$

where $p_x = \sigma_x^2 = E[x^2[n]]$ is the signal power and $p_n = \sigma_e^2 = E[e_q^2[n]]$ is the power of the quantization noise. Here, $E[.]$ is the expectation operation.

If the quantization error is uniformly distributed[5] in the range $(-\triangle/2,\ \triangle/2)$ as shown in Fig. 7.9, the mean value of the error is zero and the variance

3. This will be discussed in a later chapter.

4. This will be discussed in a later chapter.

5. Details of uniform PDF can be seen in Appendix B.

(the quantization noise power) is,

$$p_n = \sigma_e^2 = \int_{-\triangle/2}^{\triangle/2} e^2 p(e)de = \frac{1}{\triangle}\int_{-\triangle/2}^{\triangle/2} e^2 de = \frac{\triangle^2}{12}. \tag{7.23}$$

By combining Eq. (7.21) with Eq. (7.23) and substituting the result into Eq. (7.22), the expression for SQNR becomes

$$\text{SQNR} = 10\log_{10}\left(\frac{p_x}{p_n}\right) = 20\log_{10}\left(\frac{\sigma_x}{\sigma_e}\right), \tag{7.24}$$

$$\begin{aligned}
&= 20\log_{10}(\sigma_x) - 20\log_{10}(\sigma_e),\\
&= 20\log_{10}(\sigma_x) - 20\log_{10}(\triangle) + 10\log_{10}(12); \quad \because \sigma_e = \frac{\triangle}{\sqrt{12}},\\
&= 20\log_{10}(\sigma_x) - 20\log_{10}\left(\frac{R}{2^{b+1}}\right) + 10.79; \quad \because \triangle = \frac{R}{2^{b+1}},\\
&= 20\log_{10}(\sigma_x) - 20\log_{10}(R) + 6.02b + 16.81,\\
&= 6.02b + 16.81 - 20\log_{10}\left(\frac{R}{\sigma_x}\right)\text{dB}.
\end{aligned} \tag{7.25}$$

This equation shows that the SQNR depends on the range R of the ADC and the statistics of the input signal. Equation (7.25) shows that each additional bit (b) in the quantizer increases the SQNR by $\approx$ 6dB.

7.4 Concluding Remarks

- Theory behind sampling an analog signal is described.
- Spectral relationships between the analog and sampled signals described in detail with illustrations.
- The function of anti-aliasing filters in sampling is explained.
- Conditions for perfect reconstruction of signals from sampled signals are described.
- Quantization process and the analysis of errors associated with it has been discussed in detail.

Exercises

7.1. In this problem, consider a band-limited continuous-time signal $x_a(t)$ and its echo $x_a(t-\tau)$, where τ is a time-delay, arrive simultaneously at a TV receiver. The received analog signal is modelled as,

$$r_a(t) = x_a(t) + \alpha x_a(t-\tau), \quad |\alpha| < 1.$$

This signal is processed by the system shown in Fig. P7.1 to generate the analog signal $y_a(t)$ (In the figure, ADC: Analog to Digital Converter, DAC: Digital to Analog Converter, and f_s: Sampling frequency).

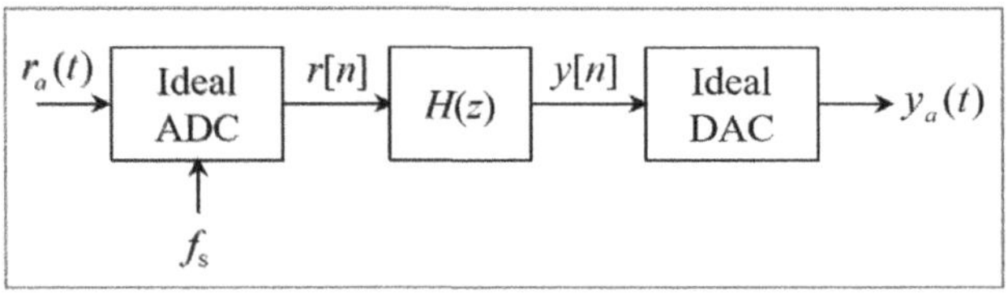

Figure P7.1.

(a) Is it possible to specify f_s and $H(z)$ so that $y_a(t) = x_a(t)$ (i.e., remove the "ghost" $x_a(t-\tau)$ of $x_a(t)$ from the received signal, $r_a(t)$)? If yes, discuss the choice of f_s and $H(z)$.

In the following, it is assumed that the signal $x_a(t)$ has the spectrum as shown in Fig. P7.2. Here, B is the one-sided bandwidth of the signal $x_a(t)$. This signal is passed through two systems as shown in Fig. P7.3(a) and P7.3(b), respectively (In Fig. P7.3(a) and P7.3(b), $(.)^2$ represents the squaring operation).

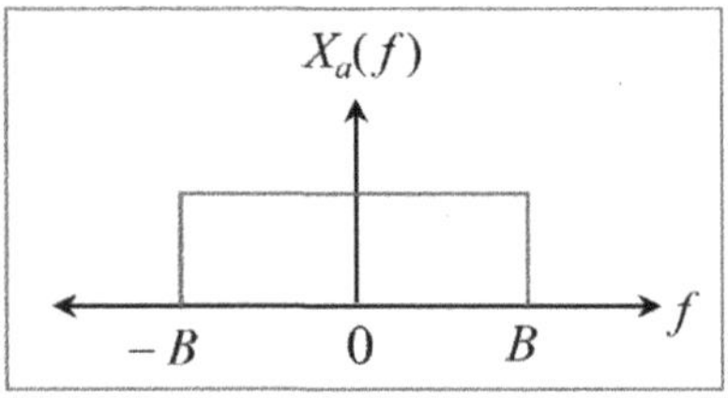

Figure P7.2.

(b) Sketch the spectra of $y[n]$ and $s[n]$ in Fig. P7.3(a) and Fig. P7.3(b), respectively if the sampling frequency $f_s = 2B$ is chosen.

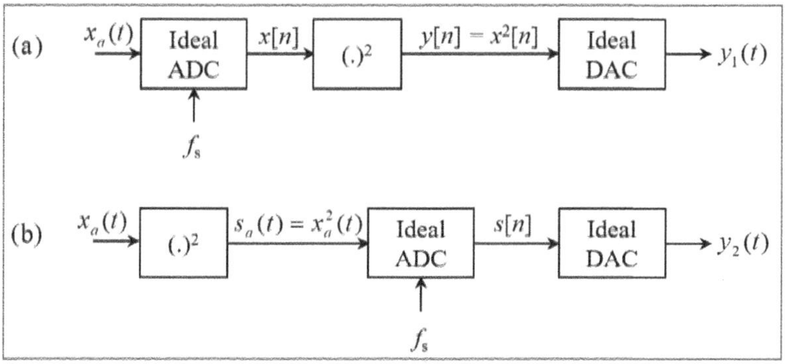

Figure P7.3.

(c) Determine $y_1(t)$ and $y_2(t)$ if $x_a(t) = \cos 2\pi f_0 t$, $f_0 = 20Hz$ and $f_s = 50Hz$. Are there any relationship between $y_1(t)$ and $y_2(t)$? (*Hint*: $\cos^2(\theta) = \frac{1}{2}[1 + \cos(2\theta)]$)

7.2. Consider the spectrum (base-band) $X_a(f)$ of an analog signal $x_a(t)$ as shown in Fig. P7.4. Using a scaled impulse train, $T_s \sum_{n=-\infty}^{\infty} \delta(t - nT_s)$, the analog signal $x_a(t)$ is sampled to produce the discrete-time signal $x[n]$. Here, T_s is the periodicity of the impulse train. f_c in the figure is the one-sided bandwidth of the signal $x_a(t)$.

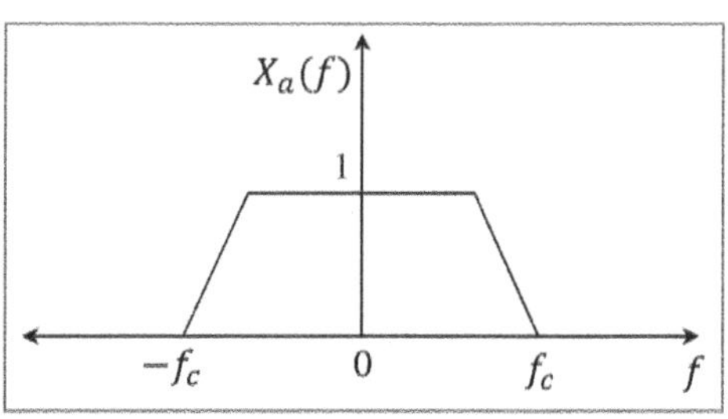

Figure P7.4.

(a) For this signal $x_a(t)$, what is the minimum sampling time T_s that results in alias-free spectrum ($X(f)$) of the sampled signal, $x[n]$? Plot the corresponding spectrum $X(f)$.

In the following, it is assumed that $f_c = 10^4$Hz and the signal $x_a(t)$ is modulated using a complex exponential $e^{j\Omega_0 t}$, where $\Omega_0 = \pi \times 10^5$ rad/sec (the arrangement shown in Fig. P7.5) to produce the modulated signal $y(t)$.

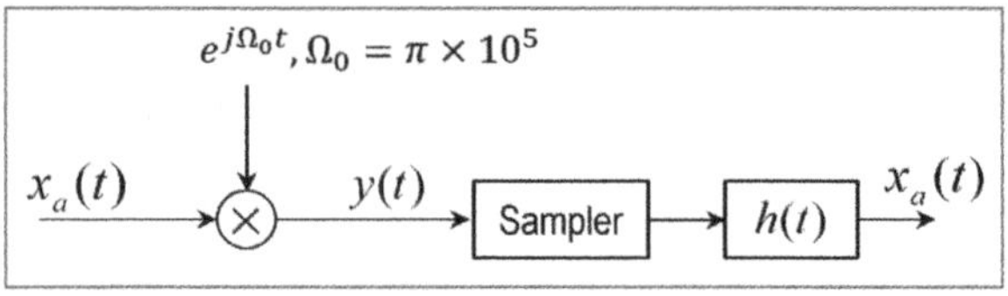

Figure P7.5.

(b) Sketch the Fourier spectrum $Y(f)$ of the modulated signal $y(t)$.

(c) The modulated signal $y(t)$ is then sampled and passed through a reconstruction filter (with filter impulse response $h(t)$) that demodulates $y(t)$ so that $x_a(t)$ can be retrieved at the output. What are the possible sampling frequencies? Sketch the ideal $|H(f)|$ for a suitable reconstruction filter, $h(t)$. $|.|$ here represents the magnitude (absolute value).

7.3. Normal and over-sampling ADC (analog to digital converter) with and without anti-aliasing filter: The block diagram in Fig. P7.6 shows different steps in sampling and reconstructing an analog signal, $x_a(t)$.

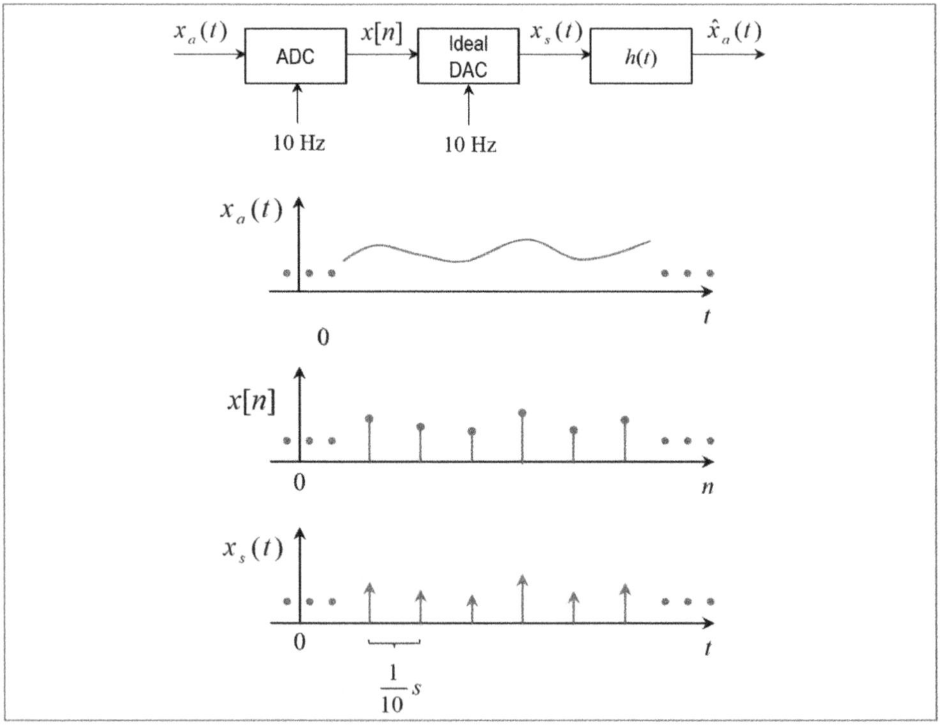

Figure P7.6.

The signal $x_a(t)$ has a bandwidth $|f| < 5$ Hz. It is digitized at 10 Hz to give $x[n]$. To re-construct $x_a(t)$ (the reconstructed signal is is denoted as $\hat{x}_a(t)$), $x[n]$ is passed into an ideal DAC (digital to analog converter) as shown. The output $x_s(t) = \ldots + x(0)\delta(t) + x(1)\delta\left(t - \frac{1}{10}\right) + \ldots$ (continuous-time representation of $x[n]$) of the ideal DAC gives a pulse which approximates an impulse for every $x[n]$. Various signals ($x_a(t)$, $x[n]$, and $x_s(t)$) are depicted in Fig. P7.6. Sketch the spectrum after DAC. Determine a suitable filter ($h(t)$) that can recover $x_a(t)$ from the DAC output (i.e., from $x_s(t)$).

7.4. Ideal and zeroth-order hold DAC: An ideal DAC converts $x[n]$ to an impulse train $x_s(t)$. A practical DAC on the other hand converts $x[n]$ into a step waveform $x_{DA}(t)$ as shown in Fig. P7.7 (i.e., a practical DAC gives a pulse of width $1/f_s$ seconds instead of an impulse for every $x[n]$; f_s is the sampling frequency). See that $x_{DA}(t) = x_s(t) \otimes \text{rect}(10t)$. Relate $X_{DA}(f)$ to $X(f)$. How can this zeroth order hold effect be compensated?

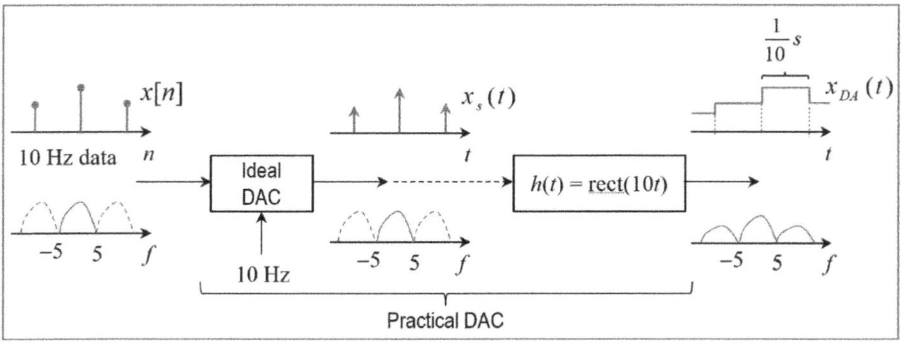

Figure P7.7.

7.5. The amplitude spectrum of one analog signal is shown in Fig. P7.8. The signal is sampled with a sampling interval $T_s = 140$ms (milliseconds).

Sketch roughly the amplitude spectrum of the sampled signal in the frequency range below 10Hz. Comment on the choice of sampling interval.

7.6. (a) Sketch $x[n] = \ldots, 0, 0, e^{-0.3}, e^{-0.6}, e^{-0.9}, e^{-1.2}, \ldots$, with $x[0] = e^{-0.3}$. Write down and sketch the continuous-time representation of $x[n]$, assuming that a sampling frequency of 0.2 Hz has been used. Determine the z-transform and spectrum of $x[n]$.

(b) The signal, $x[n]$, in Question (a) is obtained from sampling an analog signal, $x_a(t)$, directly. Using the exponential and unit step functions, suggest how $x_a(t)$ may look like. Give two possible answers. Can $x[n]$

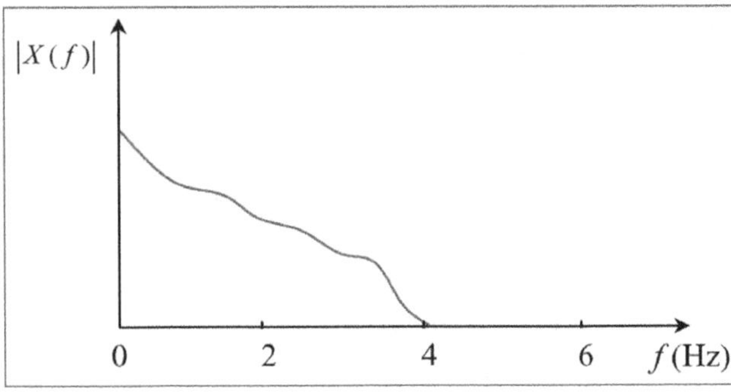

Figure P7.8.

be used to recover $x_a(t)$ perfectly? Why? By using a perfect D/A converter and an anti-aliasing filter, an analog signal, $x_b(t)$, is generated from $x[n]$. Write down a formula from which $x_b(t)$ may be calculated.

DOI: 10.1561/9781680839791.ch8

Chapter 8

z-Transform of Discrete Time Signals

Laplace transform is the concept of transforming a function of time into a function of frequency. For the analysis of continuous time LTI systems, Laplace transform has been used extensively. The variable used in the Laplace transform is $s = \sigma + j\Omega$, which represents a complex frequency, i.e., it is frequency with a real and imaginary part. The general expression for the two-sided Laplace transform of a continuous-time signal $x(t)$ is given by,

$$X(s) = \int_{-\infty}^{\infty} x(t)e^{-st}dt, \tag{8.1}$$

where the integral is taken as a line integral along a suitable closed contour in the complex $s-$plane. To learn more about Laplace transform, one may refer [8].

z-transform is the discrete-time version of the Laplace transform. Here, we use the variable $z = re^{j\theta}$, which is complex (with magnitude $|z| = r$, and $\angle z = \theta$) instead of s. By applying the z-transform to a sequence of data points, we create an expression that allows us to perform frequency-domain analysis of discrete-time signals.

With the z-transform, we can create transfer functions for digital systems, and we can find (and plot) poles and zeros on a complex plane for stability analysis. It can be used for finding the output of a system to an input, estimating the input signals based on the outputs and the transfer function of the system. The inverse

z-transform allows us to convert a z-domain transfer function into a *difference equation* that can be implemented in a computer or in a digital signal processor.

8.1 z-Transform

The z-transform of a discrete-time signal $x[n]$ is defined as,

$$X(z) = \sum_{n=-\infty}^{\infty} x[n]z^{-n} \tag{8.2}$$

where $z = re^{j\theta}$ is a complex variable, $r = |z|$ and $\theta = \angle z$. Then $X(z)$ can be expressed as,

$$X(z)|_{z=re^{j\theta}} = \sum_{n=-\infty}^{\infty} x[n]r^{-n}e^{-j\theta n} \tag{8.3}$$

Associated with each $X(z)$ is its *region of convergence* (ROC) which is the set of values of z for which $X(z)$ attains a finite value in the complex z-plane. In the ROC of $X(z)$, $|X(z)| < \infty$. But

$$\begin{aligned} |X(z)| &= \left| \sum_{n=-\infty}^{\infty} x[n]r^{-n}e^{-j\theta n} \right| \\ &= \sum_{n=-\infty}^{\infty} \left| x[n]r^{-n}e^{-j\theta n} \right| = \sum_{n=-\infty}^{\infty} \left| x[n]r^{-n} \right|. \end{aligned} \tag{8.4}$$

The problem of finding the ROC for $X(z)$ is equivalent to determining the range of values of r for which the sequence $x[n]r^{-n}$ is absolutely summable. To elaborate, let us express Eq. (8.4) as,

$$\begin{aligned} |X(z)| &= \sum_{n=-\infty}^{-1} \left| x[n]r^{-n} \right| + \sum_{n=0}^{\infty} \left| \frac{x[n]}{r^n} \right|, \\ &= \sum_{n=1}^{\infty} \left| x[-n]r^{n} \right| + \sum_{n=0}^{\infty} \left| \frac{x[n]}{r^n} \right|. \end{aligned} \tag{8.5}$$

If $X(z)$ converges in some region of the complex plane, both summations in Eq. (8.5) must be finite in that region. If the first sum in Eq. (8.5) converges, there must exist values of r small enough such that the product sequence $x[-n]r^n$, for

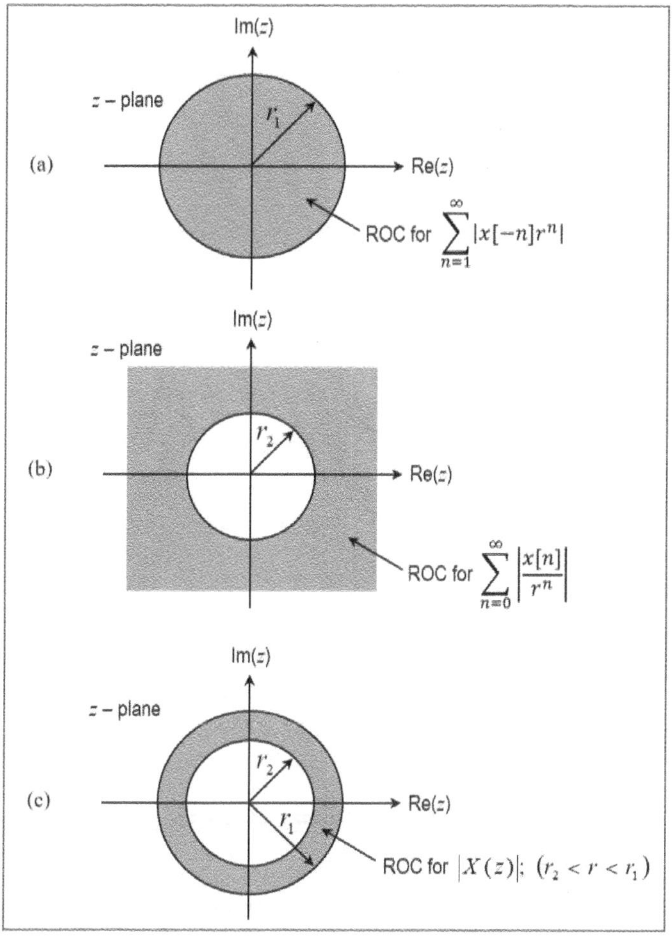

Figure 8.1. ROC for $X(z)$ and its corresponding causal and anti-causal components.

$1 \le n \le \infty$, is absolutely summable. Therefore, the ROC for the first sum consists of all points in a circle of radius r_1, where $r_1 < \infty$, as illustrated in Fig. 8.1(a). On the other hand, if the second sum in Eq. (8.5) converges, there must exist values of r large enough such that the sequence $x[n]/r^n$, for $0 \le n \le \infty$, is absolutely summable. Hence the ROC for the second sum in Eq. (8.5) consists of all points outside a circle of radius $r > r_2$, as illustrated in Fig. 8.1(b).

Since the convergence of $X(z)$ requires that both sums in Eq. (8.5) be finite, it follows that the ROC of $X(z)$ is generally specified as the annular region in the z-plane, $r_2 < r < r_1$, which is the common region where both sums are finite (see Fig. 8.1(c)). On the other hand, if $r_2 > r_1$, there is no common ROC for the two sums and hence $X(z)$ does not exist.

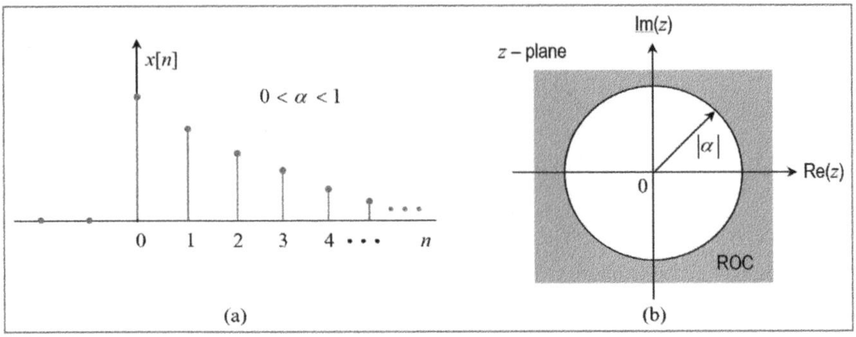

Figure 8.2. (a) The right-sided exponential signal, $x[n] = \alpha^n u[n]$, and (b) the ROC of $X(z)$.

8.2 Discrete-time Signal Types and z-Transforms

Below, some common discrete-time signal types and their corresponding z-transforms are explained.

8.2.1 Right-Sided Sequences

The sequence $x[n]$ is said to be right sided if $x[n] = 0$ for n less than a finite value. For such sequences, *the ROC is the exterior of a circle.*

$$E.g., \quad x[n] = \alpha^n u[n] \implies X(z) = \frac{1}{1 - \alpha z^{-1}}. \tag{8.6}$$

Here, $0 < \alpha < 1$. The ROC (Fig. 8.2(b)) corresponds to $\left|\alpha z^{-1}\right| < 1$ or $|z| > |\alpha|$.

8.2.2 Left-Sided Sequences

The sequence $x[n]$ is said to be left sided if $x[n] = 0$ for n greater than a finite value. For such sequences, the *ROC is the interior of a circle.*

$$E.g., \quad x[n] = -\alpha^n u[-n-1] \implies X(z) = -\frac{\alpha^{-1} z}{1 - \alpha^{-1} z}. \tag{8.7}$$

Here again, $0 < \alpha < 1$ and the ROC corresponds to $|\alpha^{-1} z| < 1$ or $|z| < |\alpha|$. This is shown in Fig. 8.3(b).

8.2.3 Two-Sided Sequences

For a two-sided sequence (i.e., $x[n]$ for $n = -\infty, \ldots, \infty$), the ROC is a ring formed by the intersection of the ROCs corresponding to the right-sided and

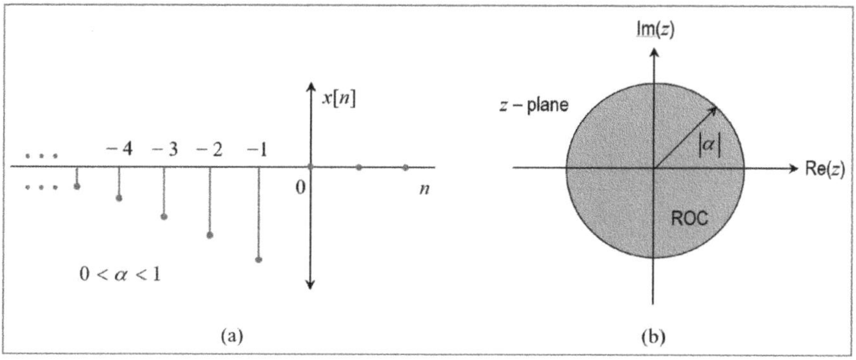

Figure 8.3. (a) The left-sided exponential signal, $x[n] = -\alpha^n u[-n-1]$, and (b) the ROC of $X(z)$.

left-sided parts of the sequence.

$$
\begin{aligned}
&E.g., \quad x[n] = \alpha^n u[n] + b^n u[-n-1] \\
&\Rightarrow \quad X(z) = \frac{1}{1-\alpha z^{-1}} + \frac{b^{-1}z}{1-b^{-1}z} = \frac{z}{z-\alpha} + \frac{z}{b-z}.
\end{aligned} \tag{8.8}
$$

In determining the convergence of $X(z)$, we consider two cases:

- $|b| < |\alpha|$: In this case, the two ROCs do not overlap, as shown in Fig. 8.4(a). Consequently, we can not find values of z for which both power series converge simultaneously. Clearly, in this case, $X(z)$ does not exist.
- $|b| > |\alpha|$: Clearly, in this case there is a ring in the z-plane where both power series converge simultaneously, as shown in Fig. 8.4(b). The ROC corresponds to $|\alpha| < |z| < |b|$.

8.2.4 Finite Duration Sequences

For a finite duration sequence (i.e., $x[n]$ for $n = n_1, \ldots, n_2$ where $-\infty < n_1 < n_2 < \infty$), the ROC is the entire z-plane, except possibly the points $z = 0$ and/or $z = \infty$.

§ *Are FIR filters stable? Why?*

8.3 Properties of ROC

There are several important properties for the ROCs, which are listed below:

- Bounded by circles (dependent only on $|z|$)
- Connected and bounded by poles or by ∞. (ROC cannot contain a pole.)

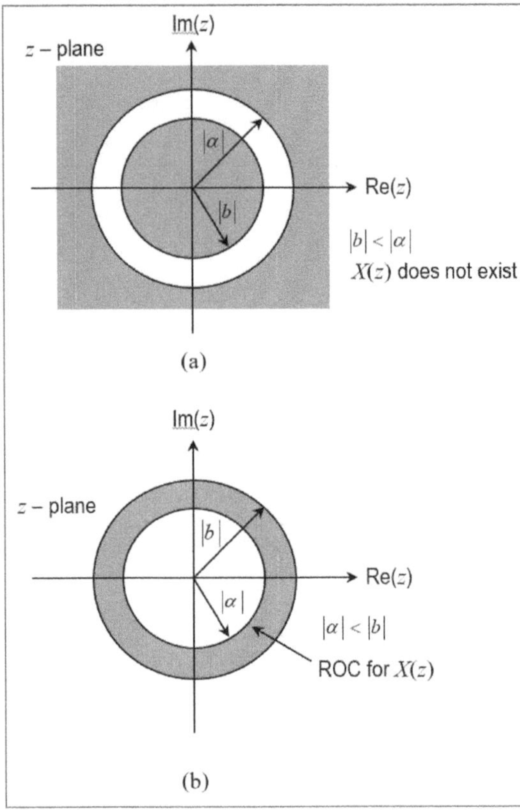

Figure 8.4. The ROC for the z-transform of both-sided exponential signal $x[n] = \alpha^n u[n] + b^n u[-n-1]$.

- If $x[n]$ is absolutely summable, then the ROC contains the unit circle, the system has a DTFT and is said to be "stable".
- A stable and causal sequence (system) has all its poles inside the unit circle.
- Right-sided infinite sequence: ROC lies outside the outermost pole. If also causal, then the ROC includes $|z| = \infty$.
- Left-sided infinite sequence: ROC lies inside the innermost pole. If also anti-causal, then the ROC includes $|z| = 0$.
- Two-sided infinite sequence: ROC lies between the two poles. The ROC does not include $|z| = 0$ or $|z| = \infty$.
- Finite-length sequence: Always converges. ROC always includes $0 < |z| < \infty$. ROC includes 0 if sequence is anti-causal, or includes ∞ if causal.

The properties of z-transform and z-transform pairs are provided as tables in Appendix B.

8.4 Relationship to Fourier Transform

With $z = re^{j\omega}$ where $r = |z|$ and $\omega = \tan^{-1}[\Im(z)/\Re(z)]$, we get

$$X\left(re^{j\omega}\right) = \sum_{n=-\infty}^{\infty} \left[x[n]r^{-n}\right] e^{-j\omega n}, \tag{8.9}$$

where, $\Im(z)$ is the imaginary part of z and $\Re(z)$ is the real part of z. We can infer the following from Eq. (8.9):

- z-Transform of a sequence $x[n]$ is equal to the Fourier transform of the sequence $x[n]r^{-n}$. Therefore, z-transform can exist even when Fourier transform may not exist.
- Fourier transform of a sequence $x[n]$ is equal to the z-transform evaluated on the unit circle in the complex plane (i.e., $z = e^{j\omega}$ when $r = |z| = 1$).

8.4.1 Common z-Transforms and Properties of z-Transform

The z-transforms of some of the elementary signals are given in Table 8.1. Table 8.2 lists some of the properties of z-transforms that will be useful in the study of signals. In Table 8.2, $X(z)$, $X_1(z)$ and $X_2(z)$ are the z-transforms of $x[n]$, $x_1[n]$ and $x_2[n]$, respectively. The contour '$c1$' indicated in the table is a closed contour

Table 8.1. Some common z-transform pairs.

Signal	z-Transform	ROC
$\delta[n]$	1	all z
$u[n]$	$\dfrac{1}{1-z^{-1}}$	$\|z\| > 1$
$a^n u[n]$	$\dfrac{1}{1-az^{-1}}$	$\|z\| > \|a\|$
$a^n u[-n]$	$\dfrac{1}{1-a^{-1}z}$	$\|z\| < \|a\|$
$-a^n u[-n-1]$	$\dfrac{1}{1-az^{-1}}$	$\|z\| < \|a\|$
$\cos(\omega_0 n)u[n]$	$\dfrac{1-z^{-1}\cos(\omega_0)}{1-2z^{-1}\cos(\omega_0)+z^{-2}}$	$\|z\| > 1$
$\sin(\omega_0 n)u[n]$	$\dfrac{z^{-1}\sin(\omega_0)}{1-2z^{-1}\cos(\omega_0)+z^{-2}}$	$\|z\| > 1$
$nu[n]$	$\dfrac{z^{-1}}{(1-z^{-1})^2}$	$\|z\| > 1$

Table 8.2. Properties of the z-transform for discrete-time signals.

Property	Time domain	z domain
Linearity	$a_1x_1[n] + a_2x_2[n]$	$a_1X_1(z) + a_2X_2(z)$
Time shifting	$x[n-k]$	$z^{-k}X(z)$
Time reversal	$x[-n]$	$X\left(z^{-1}\right)$
Conjugation	$x^*[n]$	$X^*\left(z^*\right)$
Scaling in time domain	$a^n x[n]$	$X\left(a^{-1}z\right)$
Differentiation in frequency domain	$nx[n]$	$-z\frac{dX(z)}{dz}$
Convolution	$x_1[n] \otimes x_2[n]$	$X_1(z)X_2(z)$
Correlation	$x_1[n] \otimes x_2^*[-n]$	$X_1(z)X_2^*\left(1/z^*\right)$
Multiplication (complex convolution)	$x_1[n]x_2[n]$	$\frac{1}{2\pi j}\int_{c1} X_1(v)X_2(z/v)\,v^{-1}dv$

in the ROC of $X_1(v)X_2(z/v)$. If we choose the contour of integration as the unit circle $z = e^{j\omega}$, we get (since $z^* = e^{-j\omega} = z^{-1}$)

$$x[n] = x_1[n] \otimes x_2^*[-n] \;\Rightarrow\; X(z) = X_1(z)X_2^*(z). \tag{8.10}$$

$$x[n] = x_1[n] \otimes x_1^*[-n] \;\Rightarrow\; X(z) = |X_1(z)|^2. \tag{8.11}$$

8.5 Inverse z-Transform

The inverse z-transform is formally defined by (using *Cauchy's integral theorem* [12])

$$x[n] = \frac{1}{2\pi j}\int_c X(z)z^{n-1}dz \tag{8.12}$$

where 'c' is a closed counter-clockwise contour in the ROC of $X(z)$. Three popular approaches for obtaining the inverse z-transform are:

- Direct evaluation of Eq. (8.12) by contour integration,
- Partial fraction expansion and table look-up,
- Power series expansion method.

8.5.1 Inverse z-Transform by Contour Integration

For rational z-transforms, the contour integral can be evaluated using *Cauchy's residue theorem* as

$$\begin{aligned}x[n] = &\text{ sum of the residues of } X(z)z^{n-1} \text{ at the poles}\\ &\text{of } X(z)z^{n-1} \text{ inside the contour } c.\end{aligned}$$

In general, if p_0 is a pole of $X(z)z^{n-1}$ (inside contour c) repeated m times, then the residue of $X(z)z^{n-1}$ at p_0 is given by

$$Res(z = p_0) = \frac{1}{(m-1)!}\left[\frac{d^{m-1}\psi(z)}{dz^{m-1}}\right]_{z=p_0} \tag{8.13}$$

where $\psi(z) = (z - p_0)^m X(z)z^{n-1}$.

Example: Consider $H(z) = \frac{1}{1-az^{-1}}$ with $|z| > |a|$. Then,

$$H(z)z^{n-1} = \frac{z^{n-1}}{1 - az^{-1}} = \frac{z^n}{z - a}. \tag{8.14}$$

We take the contour c as a circle at radius greater than $|a|$. For $n \geq 0$, the only pole inside c is $z = a$. Therefore $\psi(z) = (z - a)H(z)z^{n-1} = z^n$ and $m = 1$. Hence, we get

$$h[n] = Res(z = a) = z^n\Big|_{z=a} = a^n \quad \text{for } n \geq 0. \tag{8.15}$$

For $n < 0$, inside c, there is a single pole at $z = a$ and $|n|$-times repeated pole at $z = 0$. Let $n = -1$. Then, $\psi(z) = \frac{1}{z}$ for the pole $z = a$, and $\psi(z) = \frac{1}{z-a}$ for the pole at $z = 0$. Therefore, with $n = -1$ and $m = 1$, we get

$$h[-1] = Res(z = a) + Res(z = 0) = \frac{1}{z}\Big|_{z=a} + \frac{1}{z-a}\Big|_{z=0} = 0. \tag{8.16}$$

Now, let $n = -2$. Then, $\psi(z) = \frac{1}{z^2}$ for the pole $z = a$, and $\psi(z) = \frac{1}{z-a}$ for the twice-repeated pole at $z = 0$. Therefore, with $n = -2$, we get

$$\begin{aligned}h[-2] &= Res(z = a) + Res(z = 0)\\ &= \frac{1}{z^2}\Big|_{z=a} + \frac{d}{dz}\left(\frac{1}{z-a}\right)\Big|_{z=0} = 0.\end{aligned} \tag{8.17}$$

Similarly, it can be shown that $h[n] = 0$ for all $n < 0$. Thus, we get $h[n] = a^n u[n]$ as the inverse z-transform of the given $H(z)$.

8.5.2 Inverse z-Transform by Partial Fraction Expansion

In this approach, the given $X(z)$ is expressed as a linear combination of simpler terms as below.

$$X(z) = a_1 X_1(z) + a_2 X_2(z) + \cdots + a_K X_K(z) \tag{8.18}$$

where $X_i(z), i = 1, 2, \ldots, K$, are simpler z-transform expressions whose inverse transforms (i.e., $x_i[n], i = 1, 2, \ldots, K$) are easily available or known in the form of a table look-up, and $a_i, i = 1, 2, \ldots, K$, are scalars. Then, we get,

$$x[n] = a_1 x_1[n] + a_2 x_2[n] + \cdots + a_K x_K[n]. \tag{8.19}$$

This approach is particularly useful when $X(z)$ is a rational function. Let $X(z)$ be a *proper rational function*[1] with $z = p_1$ being a simple pole (i.e., non-repeated) and $z = p_2$ being a pole repeated m times. That is,

$$X(z) = \frac{1}{\left(1 - p_1 z^{-1}\right)\left(1 - p_2 z^{-1}\right)^m} = \frac{z^{m+1}}{(z - p_1)(z - p_2)^m}. \tag{8.20}$$

Clearly, $\frac{X(z)}{z}$ is a proper rational function. Its partial fraction expansion is given by,

$$\frac{X(z)}{z} = \frac{a_1}{z - p_1} + \sum_{k=1}^{m} \frac{a_{2k}}{(z - p_2)^k}. \tag{8.21}$$

Its residues, a_1 and a_{2k} are found as follows:

$$a_1 = (z - p_1)\frac{X(z)}{z}\bigg|_{z=p_1}, \tag{8.22}$$

$$a_{2k} = \frac{1}{(m-k)!} \times \frac{d^{m-k}}{dz^{m-k}}\left[(z - p_2)^m \frac{X(z)}{z}\right]\bigg|_{z=p_2} \tag{8.23}$$

for $k = 1, 2, \ldots, m$.

---

1. In proper rational functions, the degree of the numerator polynomial is less than that of the denominator. If the given rational function is not proper, then it can be written as the sum of a polynomial of finite length and a proper rational function. For example, $X(z) = \frac{1+3z^{-1}+(11/6)z^{-2}+(1/3)z^{-3}}{1+(5/6)z^{-1}+(1/6)z^{-2}} = 1 + 2z^{-1} + \frac{(1/6)z^{-1}}{1+(5/6)z^{-1}+(1/6)z^{-2}}$.

Example: Consider $X(z) = \frac{1}{1-1.5z^{-1}+0.5z^{-2}}$ with $|z| > 1$. Then,

$$X(z) = \frac{z^2}{z^2 - 1.5z + 0.5} = \frac{z^2}{(z-0.5)(z-1)} \tag{8.24}$$

$$\Rightarrow \quad \frac{X(z)}{z} = \frac{z}{(z-0.5)(z-1)} = \frac{a_1}{z-0.5} + \frac{a_2}{z-1}$$

$$a_1 = (z-0.5)\frac{X(z)}{z}\bigg|_{z=0.5} = \frac{z}{z-1}\bigg|_{z=0.5} = -1$$

$$a_2 = (z-1)\frac{X(z)}{z}\bigg|_{z=1} = \frac{z}{z-0.5}\bigg|_{z=1} = 2$$

$$\Rightarrow \quad X(z) = \frac{-z}{z-0.5} + \frac{2z}{z-1} = \frac{-1}{1-0.5z^{-1}} + \frac{2}{1-z^{-1}}$$

$$\Rightarrow \quad x[n] = -(0.5)^n u[n] + 2u[n]. \tag{8.25}$$

8.5.3 Inverse z-Transform by Power Series Expansion

Given a z-transform with its corresponding ROC, we expand $X(z)$ into a power series of the form

$$X(z) = \sum_{n=-\infty}^{\infty} c_n z^{-n}, \tag{8.26}$$

which converges in the given ROC. Then, it follows from the uniqueness of the z-transform that $x[n] = c_n$ for all n. For rational $X(z)$, this expansion may be performed by long division. To simplify this process, we may convert the given rational $X(z)$ into simpler functions using the method of partial fractions and then apply the long division approach.

Example: Consider $X(z) = \frac{1}{1-1.5z^{-1}+0.5z^{-2}}$ with two cases of ROCs given by (a) $|z| > 1$ and (b) $|z| < 0.5$.

In *Case (a)*, the ROC is the exterior of the circle $|z| = 1$. Therefore, the underlying sequence $x[n]$ must be a *right-sided one*. Therefore, we should seek for a power series expansion in the negative powers of z. By doing the polynomial division of the numerator of $X(z)$ by its denominator, we get,

$$X(z) = 1 + \frac{3}{2}z^{-1} + \frac{7}{4}z^{-2} + \frac{15}{8}z^{-3} + \frac{31}{16}z^{-4} + \cdots \tag{8.27}$$

$$\Rightarrow \quad x[n] = \begin{cases} \left\{1, \frac{3}{2}, \frac{7}{4}, \frac{15}{8}, \frac{31}{16}, \cdots\right\} & \text{for } n = 0, 1, 2, \ldots \\ 0 & \text{for } n < 0. \end{cases}$$

In *Case (b)*, the ROC is the interior of the circle $|z| = 0.5$. Therefore, the sequence $x[n]$ must be a *left-sided one*. Therefore, we should seek for a power series expansion in the positive powers of z. For this, the long division is done as,

$$\begin{aligned} X(z) &= \frac{1}{0.5z^{-2} - 1.5z^{-1} + 1} \\ &= 2z^2 + 6z^3 + 14z^4 + 30z^5 + 62z^6 + \cdots \end{aligned} \tag{8.28}$$

$$\Rightarrow \quad x[n] = \begin{cases} \{2, 6, 14, 30, 62, \cdots\} & \text{for } n = -2, -3, -4, -5, \ldots \\ 0 & \text{for } n \geq -1. \end{cases} \tag{8.29}$$

8.6 Concluding Remarks

- The z-transform, the associated ROCs, and its properties explained in detail.
- Methods for finding the inverse z-transform discussed.
- z-transform and its relationship to Fourier transform explained.

Exercises

8.1. Spectra of finite length sequences.

(a) Find the z-transform and the spectrum of the finite length geometric sequence:

$$x[n] = \ldots, 0, \alpha^0, \alpha^1, \ldots, \alpha^7, 0, 0, \ldots$$

(b) Find the z-transform and the spectrum of the finite length sinusoid:

$$x[n] = \ldots, 0, e^{0j2\pi 0.2}, e^{1j2\pi 0.2}, \ldots, e^{10j2\pi 0.2}, 0, 0, \ldots$$

(c) Find the z-transform and the spectrum of the finite length reciprocally decaying sequence:

$$x[n] = \ldots, 0, 1, \frac{1}{2}, \frac{1}{3}, \frac{1}{4} \cdots$$

(d) Spectrum of conjugated and reversed signal: $x[n]$ has a spectrum of $X(f) = e^{\sqrt{-f}}$. Find the spectra of $x^*[n]$ and $x[-n]$.

(e) Signal with conjugated spectrum: $x[n]$ has a spectrum of $X(f) = \sqrt{\cos(f)}$. Find the signal that has a spectrum of $X^*(f)$.

8.2. Inverse z-transform from partial fraction and long division.

(a) Find the signal $x[n]$ with z-transform $X(z) = \frac{z}{z^2+3z+2}$.

(b) Use long division to express $X(z) = \frac{z^2}{(z-1)^2}$ as a function of z^{-1}, and find $x[n]$.

8.3. Consider the system (mixer of two signals) shown in Fig. P8.3 below: The signals $x_1(t)$ and $x_2(t)$ are given by:

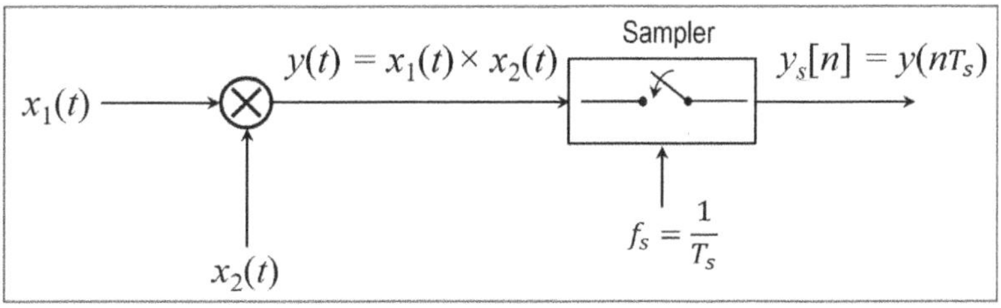

Figure P8.3. Mixer.

$$x_1(t) = 2\sin(2\pi f_1 t)u(t)$$
$$x_2(t) = 3\sin(2\pi f_2 t)u(t),$$

where $u(t)$ is the unit step function defined by:

$$u(t) = \begin{cases} 0 & \text{when } t < 0 \\ 1 & \text{when } t \geq 0. \end{cases}$$

The frequencies f_1 and f_2 are 1kHz and 2kHz, respectively. The sampling frequency f_s is 8kHz.

(a) Find the z-transform of $y_s[n]$.

(b) Sketch the zero-pole diagram of the output of the system. Label the zeros and the poles.

(c) The analog signal $y(t)$ is a linear combination of two oscillations with frequencies $f_3 = f_2 - f_1$ and $f_4 = f_2 + f_1$. A *digital* filter can be designed from the zero-pole diagram by placing zeros and poles at appropriate places. Suggest a digital filter that removes the signal oscillating at the frequency f_3 from the discrete-time signal $y_s[n]$. Write down the z-transform of the filter $H(z)$.

(d) Write down the expression for the transfer function $H(f)$ of the filter. Provide a labeled sketch of $|H(f)|$.

8.4. Inverse z-transform from partial fraction and long division.

(a) Find the signal $x[n]$ with z-transform $X(z) = \frac{z}{z^2+3z+2}$.

(b) Use long division to express $X(z) = \frac{z^2}{(z-1)^2}$ as a function of z^{-1}, and find $x[n]$.

DOI: 10.1561/9781680839791.ch9

Chapter 9

Fourier Spectra of Discrete-Time Signals

The Discrete Time Fourier Transform (DTFT), a member of the Fourier transform family, is often used to analyse samples of analog aperiodic signals. From discrete samples, DTFT produces a function of frequency that is a periodic summation of the continuous Fourier transform of the original analog signals. As we will see, DTFT is a continuous function of frequency, and a discrete version of it is the discrete Fourier transform (DFT), which will be discussed in detail in a later chapter.

9.1 Spectra of Discrete-Time Signals

The spectrum of a discrete-time signal indicates how the signal may be thought of as being composed of discrete-time complex exponentials. It describes the frequencies, amplitudes, and phases of these exponentials that combine to create the discrete-time signal. The individual complex exponential functions that sum up to give the signal are called the complex exponential components.

Observe from Figure 7.3 that *the spectrum $X(f)$ of a sampled signal is periodic with period equal to the sampling frequency f_s*. This can also be easily verified using Eq. (7.8) by showing that

$$X(f + mf_s) = X(f) \quad \text{for any integer } m. \tag{9.1}$$

Thus, sampling a signal in time-domain results in periodicity in the frequency-domain. Note that this periodicity of the spectrum is irrespective of whether or not the original analog signal $x_a(t)$ is periodic.

9.1.1 Discrete-Time Fourier Transform (DTFT)

Noting from Eq. (7.4) that the sampled signal $x[n]$ has a continuous-time representation given by,

$$x_s(t) = \sum_{n=-\infty}^{\infty} x[n]\delta(t - nT_s), \tag{9.2}$$

where $T_s = \frac{1}{f_s}$ is the sampling time and f_s is the sampling frequency. We can determine its Fourier transform as,

$$\begin{aligned} X(f) &\triangleq \mathcal{F}[x[n]] = \mathcal{F}[x_s(t)], \\ &= \int_{-\infty}^{\infty} \left[\sum_{n=-\infty}^{\infty} x[n]\delta(t - nT_s) \right] e^{-j2\pi f t} dt, \\ &= \sum_{n=-\infty}^{\infty} x[n] \int_{-\infty}^{\infty} \delta(t - nT_s) e^{-j2\pi f t} dt, \\ &= \sum_{n=-\infty}^{\infty} x[n] e^{-j2\pi f n T_s}, \end{aligned} \tag{9.3}$$

since $\int_{-\infty}^{\infty} \delta(t - nT_s)e^{-j2\pi f t} dt = e^{-j2\pi f n T_s}$. Equation (9.3) gives the expression[1] for determining the Fourier transform of any discrete-time signal. This is also known as discrete-time Fourier transform (DTFT), as opposed to the continuous-time Fourier transform (CTFT) we studied in Chapter 4. In DTFT, note that time is discrete and frequency is continuous. The Fourier series we studied in Chapter 3 is also known as continuous-time Fourier series (CTFS). For discrete-time periodic signals, we can define the so-called discrete-time Fourier series (DTFS).

As in Chapter 4, the plot of $|X(f)|$ vs f gives the magnitude spectrum of $x[n]$, and the plot of the phase of $X(f)$ vs f gives the phase spectrum of $x[n]$.

1. Using Eq. (9.3), it is also easy to see that $X(f + mf_s) = X(f)$ for any integer m, since $e^{j2\pi m} = 1$ and $f_s T_s = 1$.

9.1.2 Inverse DTFT

Given the Fourier transform $X(f)$, we can obtain the time-domain signal $x[n]$ as,[2]

$$x[n] = T_s \int_{-f_s/2}^{f_s/2} X(f)e^{j2\pi f n T_s} df. \tag{9.4}$$

Equation (9.4) gives the expression for determining the inverse Fourier transform in the case of discrete-time signals. Because the spectrum of a discrete-time signal is periodic, the evaluation of inverse DTFT involves integration over one period (i.e., $[-0.5f_s, 0.5f_s]$) of the spectrum only. This is unlike in the case of continuous-time signals where the evaluation of inverse Fourier transform requires integration of the spectrum over the complete frequency range $[-\infty, \infty]$.

9.2 Normalized Frequency Representation

Because the spectrum $X(f)$ is periodic with period $f_s = \frac{1}{T_s}$, it is enough to describe $X(f)$ over one period given by $[-0.5f_s, 0.5f_s]$ (or $[0, f_s]$). Therefore, for the sake of convenience of representation, it is common to normalize the frequency variable f by the sampling frequency f_s (i.e., $\frac{f}{f_s}$) and it is denoted as, $\tilde{f}$. In this normalized notation, one period of the spectrum spans the frequency range $[-0.5, +0.5]$. Correspondingly, *the normalized sampling frequency and normalized sampling period are given as* 1 *Hz and* 1 *second, respectively.*

In the normalized domain, i.e., with $f_s = 1$ and $T_s = 1$, we can express the forward and inverse Fourier transforms as,

$$X(\tilde{f}) = \sum_{n=-\infty}^{\infty} x[n]e^{-j2\pi \tilde{f} n} \tag{9.5}$$

$$x[n] = \int_{-0.5}^{0.5} X(\tilde{f})e^{j2\pi \tilde{f} n} d\tilde{f}. \tag{9.6}$$

2. Equation (9.4) for inverse DTFT can be obtained as follows:

$$\int_{-f_s/2}^{f_s/2} X(f)e^{j2\pi f m T_s} df = \int_{-f_s/2}^{f_s/2} \left[\sum_{n=-\infty}^{\infty} x[n]e^{-j2\pi f n T_s} \right] e^{j2\pi f m T_s} df$$

$$= \sum_{n=-\infty}^{\infty} x[n] \int_{-f_s/2}^{f_s/2} e^{j2\pi f [m-n] T_s} df = f_s x[m]$$

since the quantity $\int_{-f_s/2}^{f_s/2} e^{j2\pi f[m-n]T_s} df$ is equal to f_s if $m = n$, and is equal to 0 if $m \neq n$.

Note that the $X(\tilde{f})$ in Eq. (9.5) is periodic with period 1.0, because of the normalization. To obtain the DTFT expressions with unnormalized frequency (i.e., Eqs. (9.3) and (9.4)) from Eqs. (9.5) and (9.6), scale the frequency axis by the actual sampling frequency.[3]

9.3 Effect of Sampling Frequencies on the Spectrum

Consider a continuous-time signal $x_a(t)$ that is band-limited to $[-50, 50]$Hz as shown in Fig. 9.1(a). Let $x_1[n]$ denote the samples of $x_a(t)$ at sampling rate 200 Hz, i.e., sampling period $T_1 = \frac{1}{200}$ seconds. The spectrum $X_1(f)$ of $x_1[n]$ is shown in Fig. 9.1(b). Let us also create another sample-sequence $x_2[n]$ from $x_1[n]$ by artificially increasing the spacing between the samples of $x_1[n]$ to $T_2 = 2T_1 = \frac{1}{100}$. Then, we obtain

$$X_1(f) = \sum_{n=-\infty}^{\infty} x_1[n]e^{-j2\pi f n T_1} \tag{9.7}$$

$$\begin{aligned} X_2(f) &= \sum_{n=-\infty}^{\infty} x_2[n]e^{-j2\pi f n T_2} \\ &= \sum_{n=-\infty}^{\infty} x_1[n]e^{-j2\pi (2f) n T_1} \\ &= X_1(2f). \end{aligned} \tag{9.8}$$

Thus, the spectrum $X_2(f)$ of $x_2[n]$ is a compressed version of the spectrum $X_1(f)$ of $x_1[n]$, as illustrated in Fig. 9.1(c).

§ *Changing the sampling rate of a discrete-time signal results in scaling of the frequency axis: increase in sampling frequency results in expanding the frequency axis and decrease in sampling frequency results in compressing the frequency axis.*

On the other hand, if we had directly sampled the analog signal $x_a(t)$ at a rate of 100 Hz to produce the sequence $x_3[n]$, the resulting spectrum ($X_3(f)$) would be as shown in Fig. 9.1(d). It is interesting to compare the spectra shown in Figs. 9.1(c) and (d).

3. That is, replace the normalized frequency variable $\tilde{f}$ in Eqs. (9.5) and (9.6) by fT_s where f is the unnormalized frequency.

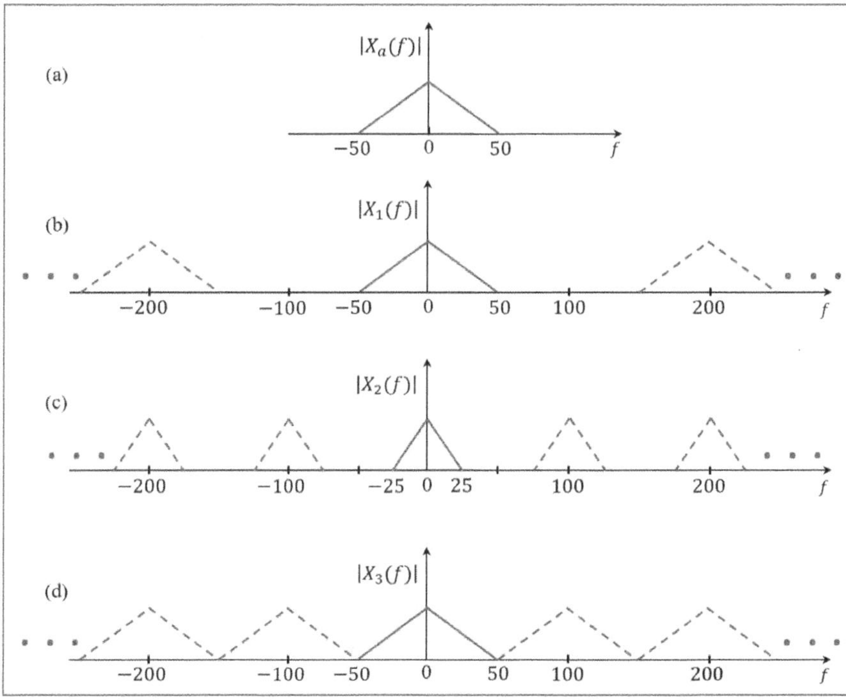

Figure 9.1. Illustration of the effect of different sampling frequencies: (a) spectrum of the analog signal $x_a(t)$, (b) spectrum of $x_1[n]$, obtained by sampling $x_a(t)$ at 200 Hz, (c) spectrum of $x_2[n]$, which is obtained from $x_1[n]$ by taking the sampling frequency as 100 Hz, and (d) spectrum of $x_3[n]$, which is obtained by sampling $x_a(t)$ at 100 Hz.

9.4 Frequency-Domain and Time-Domain Signal Properties

So far, we have introduced several methods for the frequency analysis of signals. Several methods are necessary to accommodate different types of signals. To summarize, the following frequency analysis tools have been introduced:

- The Fourier series for continuous-time periodic signals (FS),
- The Fourier transform for continuous-time aperiodic signals (FT),
- The Fourier series for discrete-time periodic signals (DTFS), and
- The Fourier transform for discrete-time aperiodic signals (DTFT).

Figure 9.2 summarizes the analysis and synthesis formulae for these types of signals.

As we can see, there are two time-domain characteristics that determine the type of signal spectrum we obtain. These are whether the time variable is continuous or

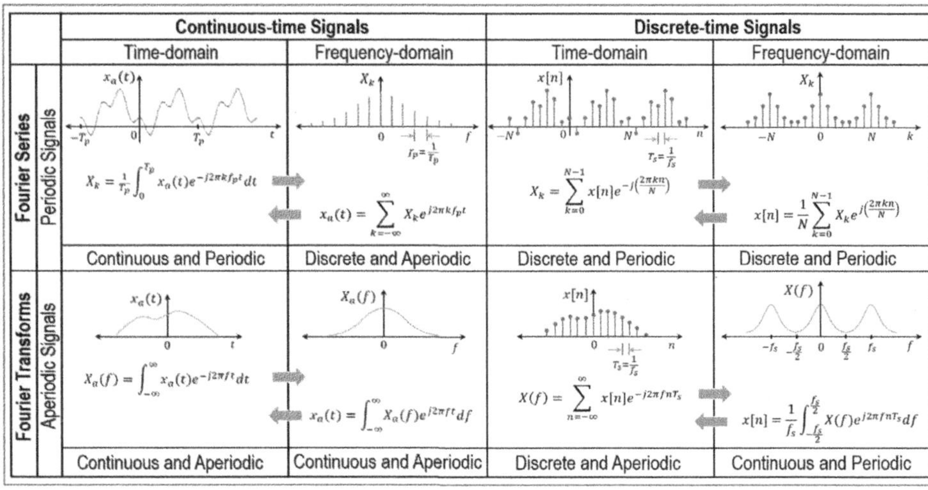

Figure 9.2. Summary of the analysis and synthesis formulae for different types of signals.

discrete, and whether the signal is periodic or aperiodic. The summary is,

- Continuous-time signals have aperiodic spectra.
- Discrete-time signals have periodic spectra.
- Periodic signals have discrete spectra.
- Aperiodic finite energy signals have continuous spectra.

9.5 Analysis of Discrete-Time Signals

In this section, we shall briefly introduce some tools (including DTFT) for the analysis of discrete-time signals.

9.5.1 Common Discrete-Time Signals

A number of basic signals appear often in the study of discrete-time signals and systems. They are listed below.

- *Unit Impulse Signal*: It is denoted by $\delta[n]$ and defined as

$$\delta[n] = \begin{cases} 1 & \text{for } n = 0 \\ 0 & \text{for } n \neq 0. \end{cases} \tag{9.9}$$

  This is also known as *Kronecker delta function*.[4]

4. The delta function (or, impulse function) $\delta(t)$ we studied in Chapter 4, where t denotes continuous-time, is known as the *Dirac delta function*.

- *Unit Step Signal*: It is denoted by $u[n]$ and defined as

$$u[n] = \begin{cases} 1 & \text{for } n \geq 0 \\ 0 & \text{for } n < 0. \end{cases} \tag{9.10}$$

- *Exponential Signal*: It is defined as

$$x[n] = a^n \quad \text{for all } n \tag{9.11}$$

where a is a scalar.

- *Sinusoidal Signal*: It is given by

$$x[n] = A\cos(2\pi \tilde{f} n + \theta) \quad \text{for all } n \tag{9.12}$$

where A is the amplitude, $\tilde{f} \in [-0.5, 0.5]$ is the normalized frequency, and $\theta \in [0, 2\pi]$ is the phase.

- *Periodic Signals*: A discrete-time signal $x[n]$ is said to be periodic with period N (N is a positive integer) if,

$$x[n] = x[n + kN], \qquad \text{for any integer } k. \tag{9.13}$$

9.5.2 Elementary Manipulations

Let $x[n]$ be a discrete-time signal. The following shifting operations are often required while working with signals.

- *Time Delayed Version*: The signal $y[n] = x[n-2]$ is obtained by delaying $x[n]$ by 2 samples in time. That is, shift $x[n]$ to the right on the time axis by 2 samples.
- *Time Advanced Version*: The signal $y[n] = x[n+2]$ is obtained by advancing $x[n]$ by 2 samples in time. That is, shift $x[n]$ to the left on the time axis by 2 samples.
- *Time Reversed Version*: The signal $y[n] = x[-n]$ is obtained by reflecting $x[n]$ on the time axis with respect to the origin $n = 0$.

The signal $y[n] = x[-n+2]$ is obtained by time-reversing $x[n]$ and shifting to the right by 2 samples (Note: $y[n] = z[n-2]$ where $z[n] = x[-n]$).

The signal $y[n] = x[-n-2]$ is obtained by time-reversing $x[n]$ and shifting to the left by 2 samples (Note: $y[n] = z[n+2]$ where $z[n] = x[-n]$).

9.5.3 Laplace Transform and z-Transform

The Laplace transform is the generalization of Fourier transform and for a signal $x(t)$, the Laplace transform is given by,

$$X(s) = \int_{-\infty}^{\infty} x(t)e^{-st}dt \tag{9.14}$$

where $s = \sigma + j\Omega$. The inverse Laplace transform is,

$$x(t) = \frac{1}{2\pi j}\int_{\sigma-\infty}^{\sigma+\infty} X(s)e^{st}ds \tag{9.15}$$

Now, from Eq. (9.14), we have,

$$X(s) = \int_{-\infty}^{\infty} x(t)e^{-st}dt \tag{9.16}$$

$$= \lim_{T_s \to 0} \sum_{n=-\infty}^{\infty} x(nT_s)\left(e^{sT_s}\right)^{-n} \tag{9.17}$$

where T_s is the sampling time. Equation (9.17), is the z-transform and is given by

$$X(z) = \sum_{n=-\infty}^{\infty} x[n]z^{-n} \tag{9.18}$$

where

$$x[n] = x(nT_s) \tag{9.19}$$

$$z = e^{sT_s} = e^{(\sigma+j\Omega)T_s}, \tag{9.20}$$

$$= e^{\sigma T_s} \times e^{j\Omega T_s} = re^{j\omega}, \tag{9.21}$$

where $r = e^{\sigma T_s}$ and $\omega = \Omega T_s$ (or $\tilde{f} = fT_s$). Here ω is the angular frequency of the discrete-time signal. When $T_s = 1$, then $r = e^{\sigma}$ and $\omega = \Omega$ (or $\tilde{f} = f$).

The relation between s-plane and z-plane is shown in Fig. 9.3. The $j\Omega$ (frequency) axis in the s-plane becomes a unit circle in the z-plane. Unit circle becomes the frequency (ω) axis in the z-plane. The entire left half of the s-plane is mapped *into* the unit circle and the right half of the s-plane is mapped *into* the exterior of the unit circle.

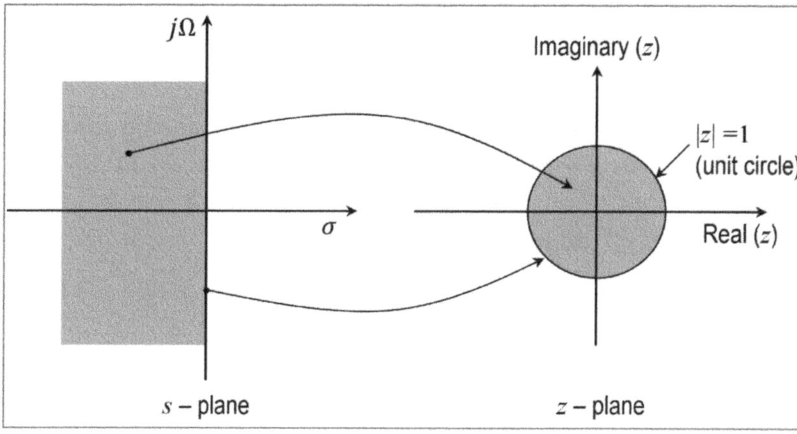

Figure 9.3. Relation between s-plane and z-plane.

9.5.4 z-Transform and DTFT

Recall from Chapter 4 that certain conditions must be satisfied by a signal if its Fourier transform should exist. Similarly, some conditions must be satisfied for the existence of DTFT also. On the other hand, the z-transform (ZT) is applicable to a much broader category of discrete-time signals and systems compared to DTFT. In simple words, the broader scope of applicability of ZT compared to DTFT follows from the following fact:

> § *DTFT is based on representation of signals (and systems) using complex sinusoids* $e^{j2\pi\tilde{f}n}$, *whereas ZT is based on representation of signals (and systems) using complex exponentials* z^n, *where* z *is a complex scalar.*

Therefore, the ZT of a discrete-time signal $x[n]$ is defined as

$$X(z) = \sum_{n=-\infty}^{\infty} x[n]z^{-n} = \sum_{n=-\infty}^{\infty} \left[x[n]r^{-n}\right] e^{-j2\pi\tilde{f}n} \tag{9.22}$$

where $z = re^{j2\pi\tilde{f}}$ is a complex variable[5] with $r = |z|$ and $2\pi\tilde{f}$ being the phase of z. Setting $|z| = r = 1$ in Eq. (9.22), we get,

$$X(z)|_{|z|=r=1} = \sum_{n=-\infty}^{\infty} x[n]e^{-j2\pi\tilde{f}n} = X(\tilde{f}). \tag{9.23}$$

5. Strictly speaking, z should be defined as $re^{j2\pi fT_s}$ where T_s denotes the sampling period. However, we assume that we are working with normalized frequency $(\tilde{f})$ and hence we use $z = re^{j2\pi\tilde{f}}$.

In the complex z-plane, $|z| = 1$ corresponds to a circle of radius 1 with centre at the origin. Therefore, we can state the following.[6]

> § *The DTFT $X(\tilde{f})$ of a discrete-time signal $x[n]$ can be obtained by evaluating its ZT $X(z)$ on the unit circle in the z-plane (i.e., by setting $z = e^{j2\pi \tilde{f}}$ in $X(z)$).*

9.5.4.1 Region of Convergence (ROC) of ZT

The ZT $X(z)$ exists when the infinite sum in Eq. (9.22) converges to a finite value. For this to happen, we must have

$$\sum_{n=-\infty}^{\infty} \left| x[n]r^{-n} \right| < \infty. \tag{9.24}$$

The range of r for which Eq. (9.24) is satisfied is known as the ROC of the ZT of $x[n]$. Therefore, when we give the ZT of a signal, we should also specify its corresponding ROC. In view of Eq. (9.23), we can say that the DTFT of $x[n]$ will exist only if the ROC of the ZT of $x[n]$ contains the unit circle $|z| = 1$.

9.5.4.2 Properties of ZT and DTFT

We shall now look at some of the properties of ZT and DTFT. Let $x[n]$ and $y[n]$ be two discrete-time signals with ZTs $X(z)$ and $Y(z)$, respectively, and DTFTs $X(\tilde{f})$ and $Y(\tilde{f})$, respectively.

- Linearity: Let $p[n] = \alpha x[n] + \beta y[n]$ where α and β are two scalars. Then, we can easily show that

$$P(z) = \alpha X(z) + \beta Y(z) \Longrightarrow P(\tilde{f}) = \alpha X(\tilde{f}) + \beta Y(\tilde{f}). \tag{9.25}$$

  Thus, *ZT and DTFT are linear transforms*.
- Time Shifting: Let $p[n] = x[n-k]$ where k is an arbitrary integer. Then, we get (with $m = n - k$)

$$P(z) = \sum_{n=-\infty}^{\infty} x[n-k]z^{-n} = \sum_{m=-\infty}^{\infty} x[m]z^{-m-k}$$

6. In view of the notation '$X(z)$' used for representing ZT, a better notation for DTFT would have been '$X\left(e^{j2\pi \tilde{f}}\right)$'. However, in this book, we stick to the notation $X(\tilde{f})$ for DTFT.

$$= z^{-k} \sum_{m=-\infty}^{\infty} x[m] z^{-m} = z^{-k} X(z).$$

$$\Rightarrow \quad P(z) = z^{-k} X(z), \quad P(\tilde{f}) = e^{-j2\pi \tilde{f} k} X(\tilde{f}). \tag{9.26}$$

In view of Eq. (9.26), z^{-1} is known as 1-sample delay operator. As in Chapter 4, the only effect of shifting a signal in time-domain is a linear phase change in the frequency-domain (no change in magnitude spectrum).

- Time Reversal: Let $p[n] = x[-n]$. Then, we get (with $m = -n$)

$$P(z) = \sum_{n=-\infty}^{\infty} x[-n] z^{-n} = \sum_{m=\infty}^{-\infty} x[m] z^{m}$$

$$= \sum_{m=-\infty}^{\infty} x[m] \left(z^{-1}\right)^{-m} = X(z^{-1}).$$

$$\Rightarrow \quad P(z) = X(z^{-1}), \quad P(\tilde{f}) = X(-\tilde{f}). \tag{9.27}$$

- Frequency Shifting: Let $p[n] = e^{j2\pi \tilde{f}_1 n} x[n]$ where $\tilde{f}_1$ is a given frequency. Then, we get

$$P(\tilde{f}) = \sum_{-\infty}^{\infty} \left[e^{j2\pi \tilde{f}_1 n} x[n] \right] e^{-j2\pi \tilde{f} n} = \sum_{-\infty}^{\infty} x[n] e^{-j2\pi (\tilde{f} - \tilde{f}_1) n}$$

$$= X(\tilde{f} - \tilde{f}_1). \tag{9.28}$$

- Symmetric Signals: For symmetric signals (even or odd), DTFT has same features as the CTFT in Chapter 4.
- Scaling in Time-Domain: Let $p[n] = a^n x[n]$ where a is a scalar. Then, we get (with $m = -n$)

$$P(z) = \sum_{n=-\infty}^{\infty} a^n x[n] z^{-n} = \sum_{n=-\infty}^{\infty} x[n] \left(a^{-1} z\right)^{-n}$$

$$= X(a^{-1} z). \tag{9.29}$$

Example 1: $x[n] = a^n u[n]$ (a is a scalar). This is called a *right-sided sequence*, because,

$$x[n] = \begin{cases} a^n & \text{for} \quad n \geq 0 \\ 0 & \text{for} \quad n < 0. \end{cases} \tag{9.30}$$

Then, we can obtain its ZT as,

$$X(z) = \sum_{n=-\infty}^{\infty} a^n u[n] z^{-n} = \sum_{n=0}^{\infty} a^n z^{-n} = \sum_{n=0}^{\infty} \left(az^{-1}\right)^n$$
$$= 1 + \left(az^{-1}\right)^1 + \left(az^{-1}\right)^2 + \left(az^{-1}\right)^3 + \cdots$$
$$= \frac{1}{1 - az^{-1}} \quad \text{if} \quad \left|az^{-1}\right| < 1. \tag{9.31}$$

The ROC of this $X(z)$ is given by $|az^{-1}| < 1$, i.e., $|z| > |a|$. In other words, $X(z)$ given by Eq. (9.31) will be finite only for the region specified by $|z| > |a|$. To obtain the corresponding DTFT, substitute $z = e^{j2\pi\tilde{f}}$ in $X(z)$, resulting in

$$X(\tilde{f}) = \frac{1}{1 - ae^{-j2\pi\tilde{f}}} \quad \text{with} \quad |a| < 1. \tag{9.32}$$

Thus, we see that the DTFT of the signal $a^n u[n]$ will exist only if $|a| < 1$. This implies that the DTFT does not exist for the unit step $u[n]$ since this corresponds to $|a| = 1$. However, the ZT exists even if $|a| = 1$ and is given by $\frac{1}{1-z^{-1}}$ with the ROC $|z| > 1$.

Example 2: $y[n] = -a^n u[-n-1]$ (a is a scalar). This is called a *strictly left-sided sequence*, because

$$y[n] = \begin{cases} -a^n & \text{for } n \le -1 \\ 0 & \text{for } n \ge 0. \end{cases} \tag{9.33}$$

Following the similar steps as in Example 1, we can obtain the ZT and DTFT of $y[n]$ as

$$Y(z) = -\frac{a^{-1}z}{1 - a^{-1}z} \quad \text{with} \quad |z| < |a| \tag{9.34}$$
$$= \frac{1}{1 - az^{-1}} \quad \text{with} \quad |z| < |a| \tag{9.35}$$
$$Y(\tilde{f}) = \frac{1}{1 - ae^{-j2\pi\tilde{f}}} \quad \text{with} \quad |a| > 1. \tag{9.36}$$

Comparing Eqs. (9.31) and (9.34)–(9.36), we see how important is the specification of ROC when working with ZTs. Two signals may have the same ZTs, but could be different in their respective ROCs. Again, we see that the DTFT does not exist for $u[-n-1]$ since this corresponds to $|a| = 1$. However, the ZT exists with $|a| = 1$ and is given by $\frac{1}{1-z^{-1}}$ with ROC $|z| < 1$.

§ *Show that the ZT of* $p[n] = a^n u[-n]$ *(a is a scalar) is given by* $\frac{1}{1-a^{-1}z}$ *with ROC* $|z| < |a|$. *Examine the DTFT in this case.*

9.6 Concluding Remarks

- A bandlimited continuous-time signal of 2-sided bandwidth $2B$ can be perfectly reconstructed from its samples taken at rate not less than $2B$ samples/second. An ideal low-pass filter of 2-sided bandwidth $2B$ can be used to reconstruct the continuous-time signal from its samples.
- The sampling process leads to spectral aliasing if the sampling rate is less than $2B$, and it is impossible to reconstruct the original analog signal from these samples. Aliasing can be prevented by using an anti-aliasing filter to bandlimit the signal prior to sampling.
- Choice of the sampling frequency and quantizer resolution involves several trade-offs.
- Fourier transform of a discrete-time signal (i.e., DTFT) is periodic with period equal to the sampling frequency.
- The DTFT can also be obtained by evaluating the z-transform (ZT) on the unit circle.
- It is convenient to use normalized frequency representation while dealing with DTFT and ZT of discrete-time signals.
- It is very important to specify the ROC while specifying the ZT of signals.

Exercises

9.1. Sketch the numerical values of the spectra for the signals (Fig. P9.1) in the frequency domain below 5kHz. Depict both the frequencies and the relative amplitude of the respective frequency components.

(a) The signal $g_0(t)$ in Fig. P9.1 is given by,

$$g_0(t) = a\cos(2\pi f_0 t),$$

where $f_0 = 200\text{Hz}$ and $a = 1$ is a real constant.

(b) The signal $g_s(t)$ in Fig. P9.1 is given by,

$$g_s(t) = \sum_{n=-\infty}^{\infty} g_0(t)\delta(t - nT_s),$$

where $T_s = 800\mu\text{s}$.

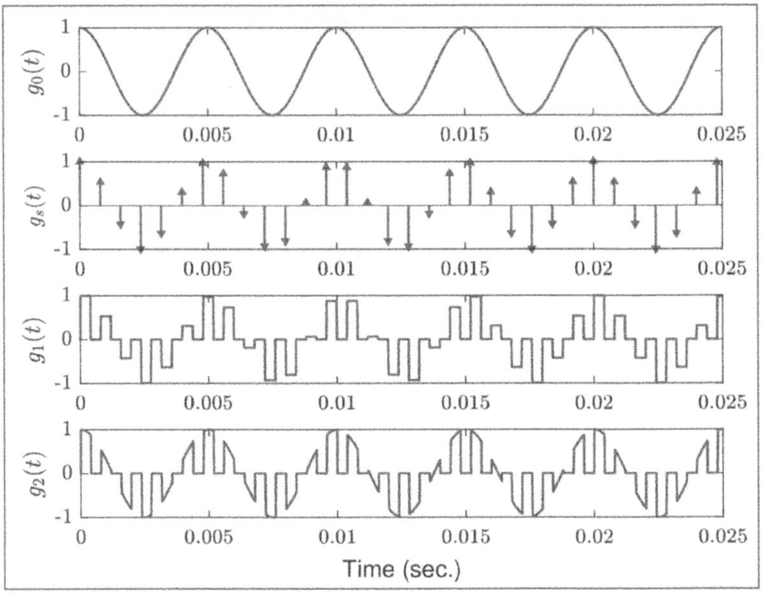

Figure P9.1.

(c) The signal $g_1(t)$ in Fig. P9.1 is given by,

$$g_1(t) = \begin{cases} g_0(nT_s) & \text{for } nT_s < t < (n + \frac{1}{2})T_s \\ 0 & \text{otherwise,} \end{cases}$$

where $T_s = 800\mu s$.

(d) The signal $g_2(t)$ in Fig. P9.1 is given by,

$$g_2(t) = \begin{cases} g_0(t) & \text{for } nT_s < t < (n + \frac{1}{2})T_s \\ 0 & \text{otherwise,} \end{cases}$$

where $T_s = 800\mu s$.

9.2. Spectra of finite length sequences.

(a) Find the z-transform and the spectrum of the finite length geometric sequence:

$$x[n] = \ldots, 0, \alpha^0, \alpha^1, \ldots, \alpha^7, 0, 0, \ldots$$

(b) Find the z-transform and the spectrum of the finite length sinusoid:

$$x[n] = \ldots, 0, e^{0j2\pi 0.2}, e^{1j2\pi 0.2}, \ldots, e^{10j2\pi 0.2}, 0, 0, \ldots$$

(c) Find the z-transform and the spectrum of the finite length reciprocally decaying sequence:

$$x[n] = \ldots, 0, 1, \frac{1}{2}, \frac{1}{3}, \frac{1}{4} \ldots$$

(d) Spectrum of conjugated and reversed signal: $x[n]$ has a spectrum of $X(\tilde{f}) = e^{\sqrt{-\tilde{f}}}$. Find the spectra of $x^*[n]$ and $x[-n]$.

(e) Signal with conjugated spectrum: $x[n]$ has a spectrum of $X(\tilde{f}) = \sqrt{\cos(\tilde{f})}$. Find the signal that has a spectrum of $X^*(f)$.

9.3. Sampling of rectangular, unit step and exponential waveforms:

(a) The signal $x_a(t)$ (shown in Fig. P9.3(a)) is sampled at a rate of 5 samples per second (i.e., $f_s = 5$Hz). Write down and sketch the continuous and discrete-time representation of the sampled signal. Write down their spectra.

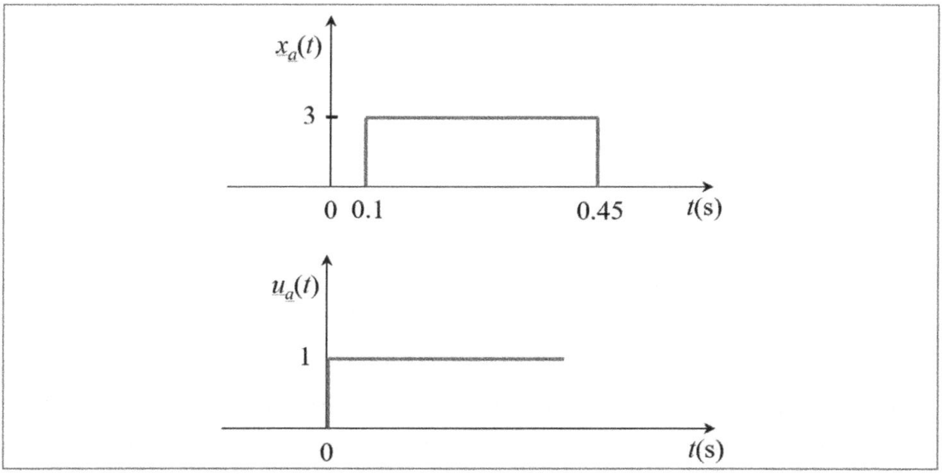

Figure P9.3.

(b) A sampling rate of 0.2Hz is used to sample the unit step signal $u_a(t)$ shown in Fig. P9.3(b). Write down and sketch the continuous and discrete-time representations of the sampled signal. Write down their spectra.

(c) An analog signal $x_a(t) = e^{-2t}u_a(t)$ (where $u_a(t)$ is the unit step function) is sampled at a rate of f_s Hz to generate the corresponding discrete-time signal, $x[n]$. Estimate the spectra of both $x_a(t)$ and $x[n]$. Are these different? Why? What should f_s be for these to have roughly the same shape?

9.4. Figure P9.4 shows the spectrum $X_a(f)$ of an analog signal $x_a(t)$. As you can see, all the frequency components lie in the low frequency region (i.e., low pass spectrum). Sketch the digital spectrum if the signal $x_a(t)$ is sampled at 50 Hz.

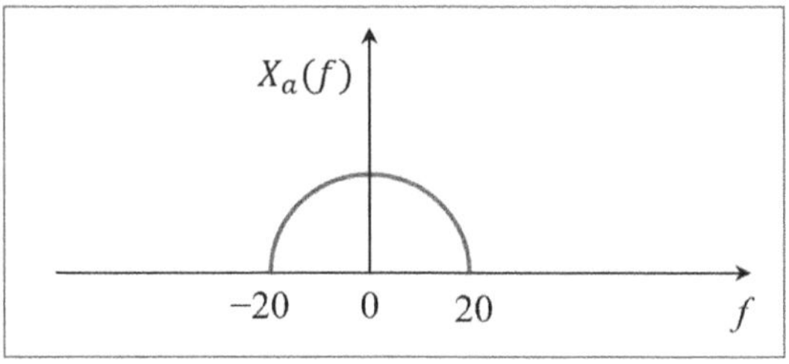

Figure P9.4.

9.5. The spectrum ($X_a(f)$) given in Fig. P9.5 is that of a real and band pass signal $x_a(t)$. It has a bandwidth of 40 Hz, but its highest frequency component is 120 Hz. The signal is sampled at 100 Hz. Sketch the digital spectrum and comment.

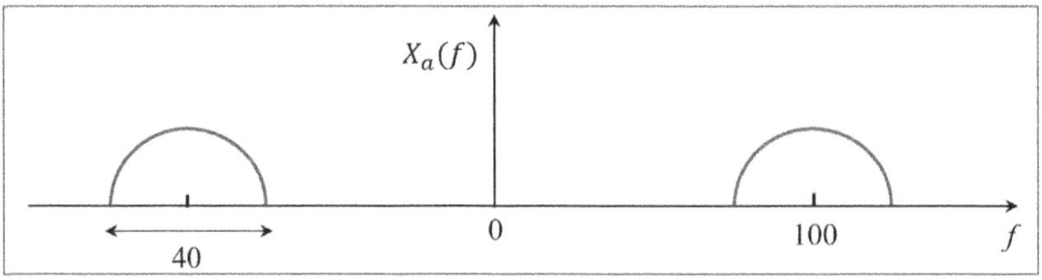

Figure P9.5.

9.6. An analog signal $x_a(t)$ is corrupted by an additive noise $v_a(t)$. Their power spectra ($S_{xx}(f)$ and $\Gamma_{vv}(f)$, respectively) are given in Fig. P9.6. Design an optimum digitization system (sampling system) for the signal which gives the best SNR (signal to noise ratio).

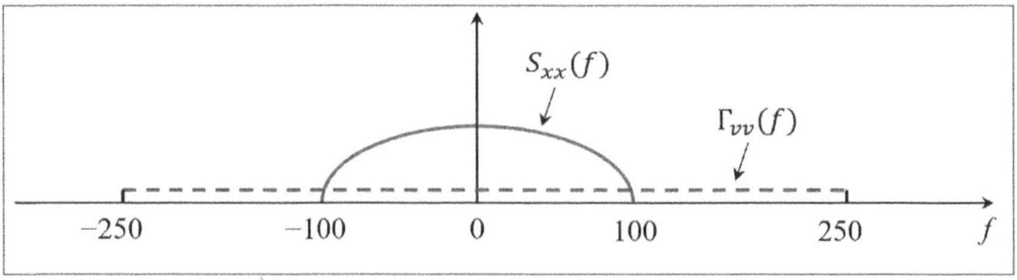

Figure P9.6.

9.7. The signal $x_a(t)$ has a spectrum $(X_a(f))$ as shown in Fig. P9.7. Only the part within $|f| < 20$ is important. Design a digitization system to sample $x_a(t)$ without using any anti-aliasing LPF (low pass filter) and also with no aliasing effects. Comment on the advantages and disadvantages.

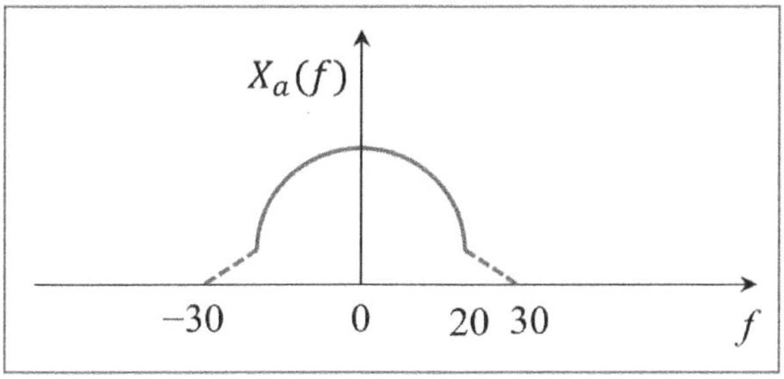

Figure P9.7.

(a) Design a digitization system to sample $x_a(t)$ in Fig. P9.7 at the lowest possible sampling rate.
(b) If we sample $x_a(t)$ in Fig. P9.7 directly at 50 Hz, will the important part of $x_a(t)$ be preserved?

9.8. Consider the analog signal:

$$x_a(t) = te^{-t}u(t),$$

where $u(t)$ is the unit step function, also known as the Heaviside function defined as:

$$u(t) = \begin{cases} 0 & \text{for } t < 0 \\ 1 & \text{for } t \geq 0. \end{cases}$$

The spectrum of the analog signal is given as:

$$te^{-t}u(t) \leftrightarrow \frac{1}{(1 + j2\pi f)^2}.$$

The function and its spectrum are shown in Fig. P9.8.

(a) Determine the spectrum of the sampled signal $x[n] = x_a(nT_s)$, where T_s is the sampling period.
(b) Plot the signals $x[n] = x_a(nT_s)$ for $T_s = 1/3$ sec and $T_s = 1$ sec, and their spectra.
(c) Plot the continuous-time signal $\hat{x}_a(t)$ after reconstruction with *ideal* band-limited interpolation.

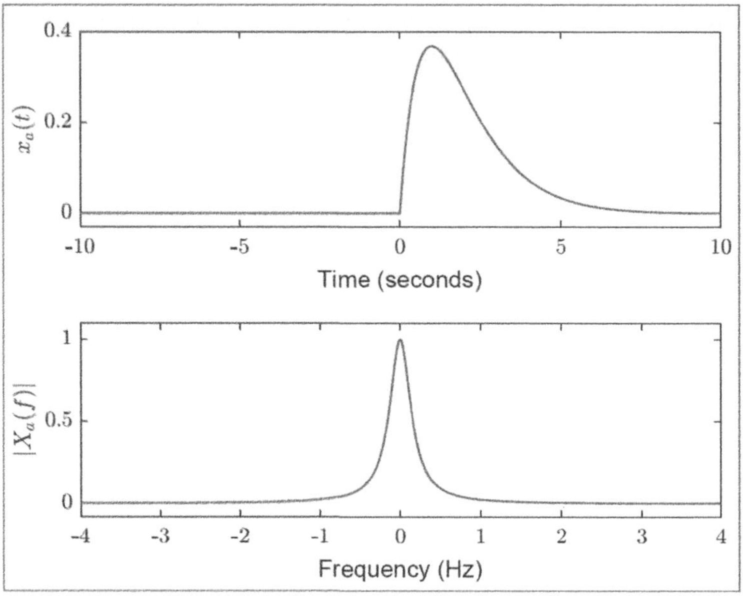

Figure P9.8. Plot of $x_a(t)$ (top) and $|X_a(f)|$ (bottom)

9.9. The real signal $g_m(t)$ has a real spectrum $G_m(f)$, which is shown in Fig. P9.9, where the median frequency f_0 is a constant and the bandwidth $B = \frac{f_0}{8}$.

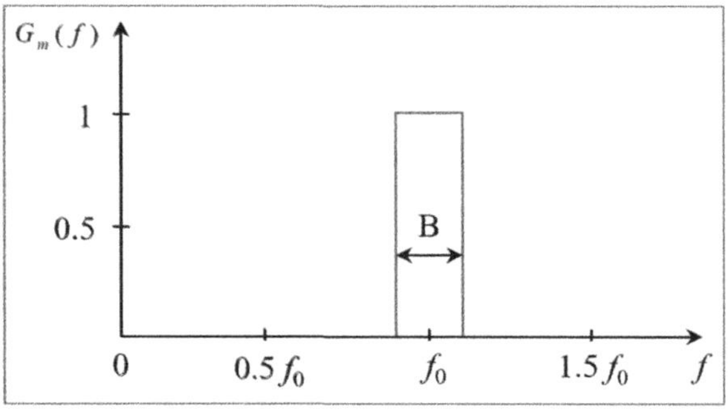

Figure P9.9.

(a) Sketch $g_m(t)$ roughly (groft).

(b) The signal $g_m(t)$ is converted to a digital signal $g_m[n]$ through a sampling operation. The sampling interval is $T_s = \frac{3}{4f_0}$. Does the chosen sampling frequency obey the sampling theorem? If not, suggest a sampling interval/frequency, that fulfils the sampling criterion. Sketch the spectrum of $g_m[n]$ in the frequency range below $2f_0$, when $T_s = \frac{3}{4f_0}$.

DOI: 10.1561/9781680839791.ch10

Chapter 10

Digital Systems

The contents of this chapter can be considered as the digital (strictly speaking, discrete-time) counter-part of the *analog systems* we studied in Chapter 6. So, we will find that the fundamentals behind the topics discussed in this chapter are same or significantly similar to that in Chapter 6.

Since the scope of our study is limited to systems whose inputs and outputs are signals, we may define *system as a device that is designed to process its input signals in a specified manner and generate output signals that meet well defined objectives.* Therefore, we may consider these systems as generalization of *filters*. For example, we may design filters (or, systems)

- to extract information about specific quantities of interest from noisy data (e.g., enhancing a desired signal buried in noise, recovering the transmitted data bits from received noisy signal etc.),
- to aid in tracking dynamic physical processes (e.g., missile guidance, jamming the radar of an enemy vehicle etc.),
- to aid in medical diagnosis by analyzing the measured biomedical signals (e.g., ECG, EEG, etc.), and
- to find efficient representation of signals (e.g., compression of speech and image data, etc.).

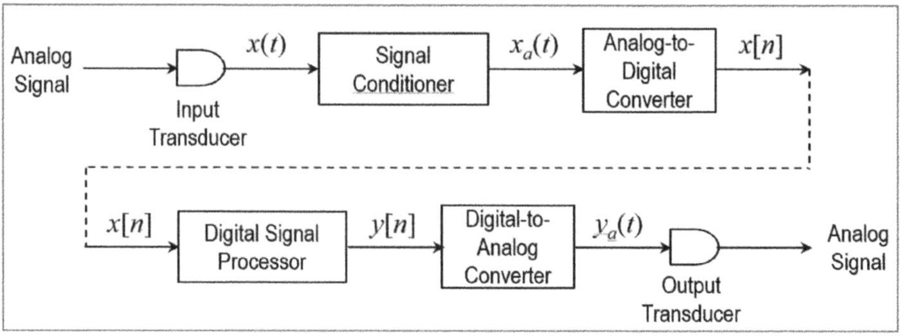

Figure 10.1. Block diagram of a practical signal processing system.

To introduce the relevance of the topics discussed in this chapter, we start with Fig. 10.1 which shows the block diagram of a practical signal processing system, highlighting the main sub-systems. The objective of any system is to process measured signals for the purpose of either extracting some important information or to transform the signal to another suitable form. Due to the many advantages of doing this processing in digital domain, it is now very common to use a digital signal processor (DSP) as the main processing unit in practical systems, as illustrated in Fig. 10.1. Let us take the processing of speech signals as an example.

- Input Transducer: In the case of speech signal, the input transducer is a microphone that converts the acoustic (sound) waves into an electrical signal, $x(t)$.
- Signal Conditioner: This is an analog circuit that prepares the signal for subsequent processing. The signal conditioning tasks involve band-limiting the measured signal to reject the noise outside the signal bandwidth (e.g., $[-5, +5]$ kHz) of interest, amplify the signal to match the dynamic range required by the circuits that follow this unit etc. The band-limiting operation also serves to prepare the signal for sampling without aliasing. This is what we have discussed earlier as the anti-aliasing filtering operation/process.
- Analog-to-Digital Converter (ADC): As discussed in Chapter 7, the ADC performs the tasks of sampling the analog signal $x_a(t)$ and digitizing the resulting samples using a quantizer. The ADC output samples, $x[n]$, are in the form of binary data-words (e.g., 5 or 6 bits per sample).
- Digital Signal Processor (DSP): This is the heart of the system, and its purpose is to process the digitized signal, $x[n]$, from the ADC. The DSP may be designed, for example, to identify what word/sentence was spoken, to identify who is the speaker, to find a compressed representation for the recorded

signal, to convert the signal into a form suitable for transmission, to extract some parameters from the signal, to improve the signal-to-noise ratio of the signal, to detect and suppress certain interferences, *etc.*

  - The DSP may be in the form of a software programme or hardware digital circuit.
  - The DSP unit may reside inside a digital computer or it may be a stand-alone unit. Binary (i.e., {0,1}) representation is used for the data (or signal samples) within the DSP.
  - While the DSP unit is designed to perform a specified task, the main constraints that must be taken into account during its design are hardware complexity, computational complexity, required speed of operation, required accuracy in the results, *etc.*
  - The main components of the digital circuits (or software programme) within the DSP are *adders, multipliers*, and *delay elements*. Using these *three basic building blocks*, all types of operations can be implemented.

- Digital-to-Analog Converter (DAC): This is the unit that converts the processed signal, $y[n]$, from the digital domain to the analog domain. It first converts the samples $y[n]$, which are in the form of binary data-words, into real numbers that represent voltage levels. The resulting sequence of samples is passed through a reconstruction filter (e.g., low-pass filter) to generate an analog speech signal $y_a(t)$. The function of DAC is thus the reverse of the function of ADC.
- Output Transducer: The output transducer, for example, a loud-speaker, converts the signal $y_a(t)$ into a sound wave that we can listen.

Our focus in this chapter is the study of the various fundamental aspects underlying the DSP unit in Fig. 10.1. We use 'digital system' to refer to any sub-system (or, operation) that is part of the DSP unit. By studying this chapter, we should be able to receive clarity and understanding on the following.

- Classification of digital systems:
  - linear or non-linear,
  - time-invariant or time-variant,
  - causal or non-causal,
  - stable or unstable
- Characterizing digital systems in time-domain and frequency-domain.
- Determining the output of a system for a given input, or determining the input of a system for a given output.
- Use of z-transform (ZT) in the study of digital signals and systems.

10.1 Input-Output Description of Systems

Consider, Fig. 10.2 which shows a digital system with input signal[1] $x[n]$ and output signal $y[n]$. For the sake of convenience, we use the representation '$\mathcal{D}[x[n]]$' to denote the 'output of the digital system when the input is $x[n]$'. Therefore, we can write

$$y[n] \triangleq \mathcal{D}[x[n]].$$

For example, consider the input-output relationship for an accumulator[2] as shown below:

$$y[n] = \sum_{k=-\infty}^{n} x[k] = \sum_{k=-\infty}^{n-1} x[k] + x[n] \tag{10.1}$$

$$= y[n-1] + x[n]. \tag{10.2}$$

We see that from Eq. (10.1), the output at time $n = n_0$ depends not only on the input at time $n = n_0$, but also on $x[n]$ at times $n = n_0 - 1, n_0 - 2$, and so on. From Eq. (10.2), we find that the system computes the current value of the output by adding (accumulating) the current value of the input to the past value of the output.

Let us have a close look at this simple system. Suppose that we are given the input signal $x[n]$ for $n \geq n_0$, and we wish to determine the output $y[n]$ for $n \geq n_0$. For $n = n_0, n_0 + 1, \ldots,$ Eq. (10.2) gives,

$$y[n_0] = y[n_0 - 1] + x[n_0], \tag{10.3}$$

$$y[n_0 + 1] = y[n_0] + x[n_0 + 1], \tag{10.4}$$

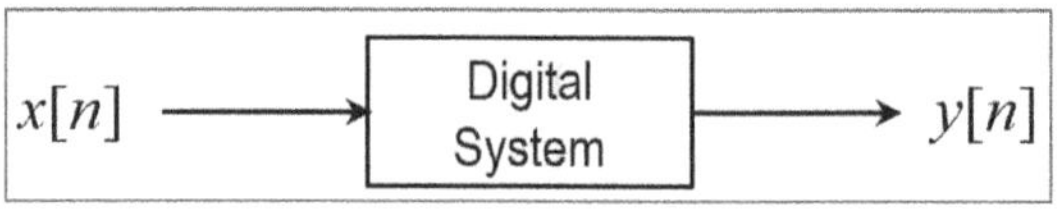

Figure 10.2. A digital system with input signal $x[n]$ and output signal $y[n]$.

1. We assume the normalized frequency representation throughout this chapter. So, the actual sampling frequency and sampling period are suppressed in the representations that we use here.

2. An accumulator basically computes the running sum of all the past input values up to the present time, $y[n] = \sum_{k=-\infty}^{n} x[k] = x[n] + x[n-1] + \ldots$.

and so on. Note that we have a problem in computing $y[n_0]$, as it depends on $y[n_0 - 1]$. However,

$$y[n_0 - 1] = \sum_{k=-\infty}^{n_0-1} x[k]. \tag{10.5}$$

That is, $y[n_0 - 1]$ "summarizes" the effect on the system from all the inputs which had been applied to the system before time n_0. Thus the response of the system for $n \geq n_0$ to the input $x[n]$ that is applied at time n_0 is the combined result of this input and all inputs that had been applied previously to the system. Consequently, $y[n], n \geq n_0$ is not uniquely determined by the input $x[n]$ for $n \geq n_0$.

The additional information required to determine $y[n]$ for $n \geq n_0$ is the *initial condition* $y[n_0 - 1]$. This value summarizes the effect of all previous inputs to the system. Thus, this initial condition together with the input sequence $x[n]$ for $n \geq n_0$ uniquely determine the output sequence $y[n]$ for $n \geq n_0$. If the initial condition is zero, we say that the system is *initially relaxed.*

10.2 Block Diagram Representation of Discrete-Time Systems

To build the block-diagram representation of discrete-time systems, we need to define the following basic building blocks:

- *An Adder*: Figure 10.3 illustrates a system (adder) that performs the addition of two signal sequences to form (the sum) another sequence. Here, the inputs to the adder is $x_1[n]$ and $x_2[n]$ and its output is $y[n] = x_1[n] + x_2[n]$. Note that it is not necessary to store the sequences in order to perform the operation. In other words, the addition operation is *memoryless.*
- *A Constant Multiplier*: This operation is depicted in Fig. 10.4, and simply represents applying a scale factor on the input $x[n]$. The output of this system is, $y[n] = ax[n]$, where a is a scalar. This operation is also *memoryless.*

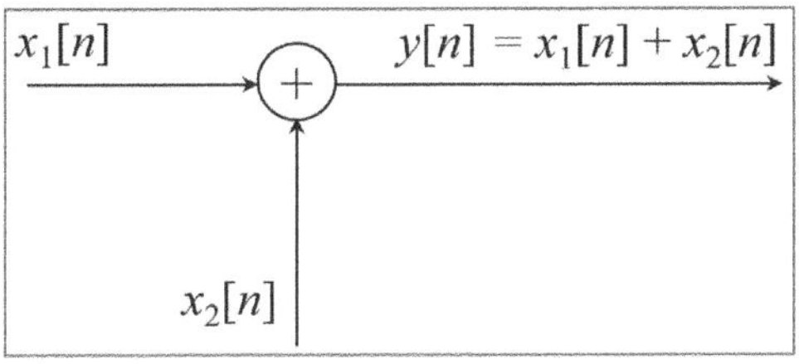

Figure 10.3. Graphical representation of an adder.

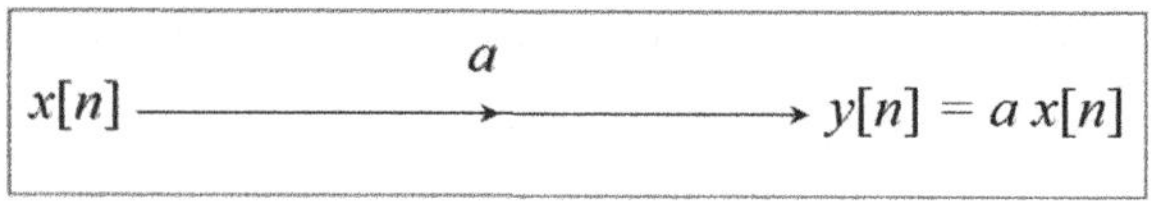

Figure 10.4. Graphical representation of a scalar multiplier.

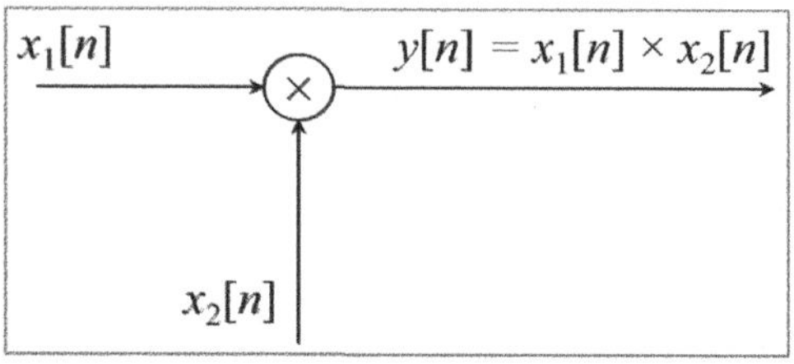

Figure 10.5. Graphical representation of a signal multiplier.

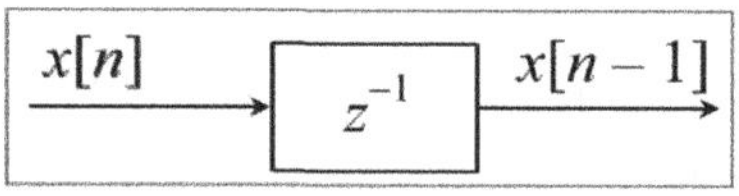

Figure 10.6. Graphical representation of the unit delay element.

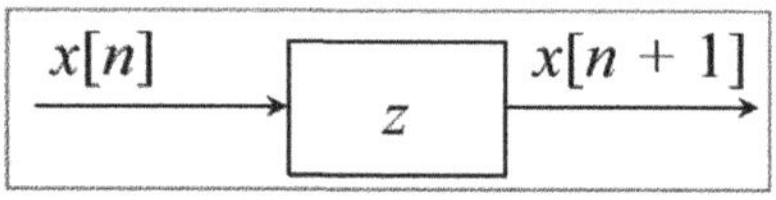

Figure 10.7. Graphical representation of the unit advance element.

- *A Signal Multiplier*: Figure 10.5 illustrates the multiplication of two signal sequences ($x_1[n]$ and $x_2[n]$) to form another sequence, $y[n] = x_1[n] \times x_2[n]$. This again is a *memoryless* operation.
- *A Unit Delay Element*: It is a special system that simply delays the input by one sample. Figure 10.6 illustrates such a system. We can see that the output of the system is stored in memory at time $n-1$ and is recalled from memory at time n to form $y[n] = x[n-1]$. This requires a *memory* and the use of the symbol[3] z^{-1} denote the unit delay.
- *A Unit Advance Element*: This unit advances the input $x[n]$ ahead by one sample (Fig. 10.7). It may be noted that such a system is physically impossible in real-time, since it involves looking into the future of the signal.

10.3 Classification of Digital Systems

There are several ways to classify digital systems, as we did for analog systems. We briefly describe below the different classes.

3. The use of this symbol z^{-1} is apparent from the discussion on z−transform.

10.3.1 Static and Dynamic Systems

A system $\mathcal{D}[\cdot]$ is said to be *static* or memoryless if its output at any instant n depends at most on the input sample at the same time, but not on past or future samples of the input. That is,

$$y[n] = \mathcal{D}[x[n]]. \tag{10.6}$$

Some examples are,

$$y_1[n] = ax[n], \tag{10.7}$$
$$y_2[n] = nx[n] + bx^3[n]. \tag{10.8}$$

A system is said to be *dynamic* if it has memory. If the output of a system at time n is completely determined by the input samples in the interval from $n - N$ to n $(N \geq 0)$, the system is said to have a memory of length N (finite memory). If $N = 0$, the system is static. If $N = \infty$, the system is said to have infinite memory. Some examples are shown below:

$$y_1[n] = x[n] + 3x[n-1], \tag{10.9}$$
$$y_2[n] = \sum_{k=0}^{N} x[n-k], \tag{10.10}$$
$$y_3[n] = \sum_{k=0}^{\infty} x[n-k]. \tag{10.11}$$

Here $y_1[n]$ is a dynamic system with finite memory of $N = 1$, $y_2[n]$ is again a finite memory (N) dynamic system, and $y_3[n]$ is a dynamic system with infinite memory.

10.3.2 Linear Systems

A system $\mathcal{D}[\cdot]$ is said to be *linear* if it satisfies the *principle of superposition.* That is, the system response to a linear combination of inputs should be equal to the linear combination of the system response to each of the input. Let $y_1[n]$ and $y_2[n]$ be the outputs of the system corresponding to the input signals $x_1[n]$ and $x_2[n]$, respectively. That is, $y_1[n] = \mathcal{D}[x_1[n]]$ and $y_2[n] = \mathcal{D}[x_2[n]]$. Then, the system is linear if,

$$\begin{aligned}\mathcal{D}\left[\alpha x_1[n] + \beta x_2[n]\right] &= \alpha\mathcal{D}[x_1[n]] + \beta\mathcal{D}[x_2[n]], \\ &= \alpha y_1[n] + \beta y_2[n],\end{aligned} \tag{10.12}$$

for any arbitrary input signals $x_1[n]$ and $x_2[n]$, and any arbitrary scalars α and β.

10.3.3 Time-Invariant Systems

A system $\mathcal{D}[\cdot]$ is said to be *time-invariant* if its input-output characteristics do not change with time. That is,

$$\mathcal{D}[x[n]] = y[n] \quad \Longrightarrow \quad \mathcal{D}[x[n-k]] = y[n-k] \tag{10.13}$$

for any input signal $x[n]$ and any time-shift k. Time-invariant digital systems are also known as shift-invariant systems.

10.3.4 Causal Systems

A system $\mathcal{D}[\cdot]$ is said to be *causal* if its output ($y[n]$) at any given time depends only on the inputs at that time and past (i.e., $x[n], x[n-1], x[n-2], \cdots$). If a system does not satisfy this condition, it is called *non-causal.*

10.3.5 Stable Systems

A system $\mathcal{D}[\cdot]$ is said to be *stable* in the sense of *bounded-input bounded-output* (BIBO), if its output is bounded in magnitude when it is excited by an input that is bounded in magnitude.

The condition that the input sequence $x[n]$ and the output sequence $y[n]$ are bounded is translated mathematically to mean that there exist some finite numbers M_x and M_y, such that

$$|x[n]| \leq M_x < \infty, \quad |y[n]| \leq M_y < \infty \tag{10.14}$$

for all n. If for some bounded input sequence $x[n]$, the output is unbounded (infinite), the system is classified as unstable.

10.3.6 Physically Realizable Systems

There are two properties of systems that are closely related to the physical realizability of systems. They are the *causality* and *stability*. Stable and causal systems are all physically realizable.

10.3.7 Linear Time-Invariant (LTI) Systems

The class of LTI systems is very important in many applications. A system $\mathcal{D}[\cdot]$ is said to be *linear time-invariant* (LTI) if it is linear and time-invariant. That is, for the signals we considered above, we should have

$$\mathcal{D}[\alpha x_1[n-l] + \beta x_2[n-k]] = \alpha y_1[n-l] + \beta y_2[n-k] \tag{10.15}$$

for any arbitrary input signals $x_1[n]$ and $x_2[n]$, any arbitrary scalars α and β, and any arbitrary integers l and k.

10.4 Outputs of LTI Systems (Convolution)

We shall now derive an expression for the output of LTI systems. Let us assume that the system $\mathcal{D}[\cdot]$ given in Fig. 10.2 is a LTI system. Further, let $h[n]$ denote the impulse response of this system, which is obtained by setting the input, $x[n]$, to a unit impulse at $n = 0$, i.e., $\delta[n]$. That is,[4]

$$h[n] = y[n]|_{x[n]=\delta[n]} = \mathcal{D}[\delta[n]]. \tag{10.16}$$

Let us consider an arbitrary input signal, as shown in Fig. 10.8. Observe from the figure that *the sample $x[k]$ at instant $n = k$ can be expressed in the form of a scaled version of a shifted unit impulse as $x[k]\delta[n-k]$* since

$$x[k]\delta[n-k] = \begin{cases} x[k] & \text{if } n = k \\ 0 & \text{if } n \neq k. \end{cases} \tag{10.17}$$

Therefore, we can express any sample of the signal as[5]

$$\begin{aligned} x[n] = \cdots &+ x[-1]\delta[n+1] + x[0]\delta[n] \\ &+ x[1]\delta[n-1] + x[2]\delta[n-2] + \cdots \end{aligned} \tag{10.18}$$

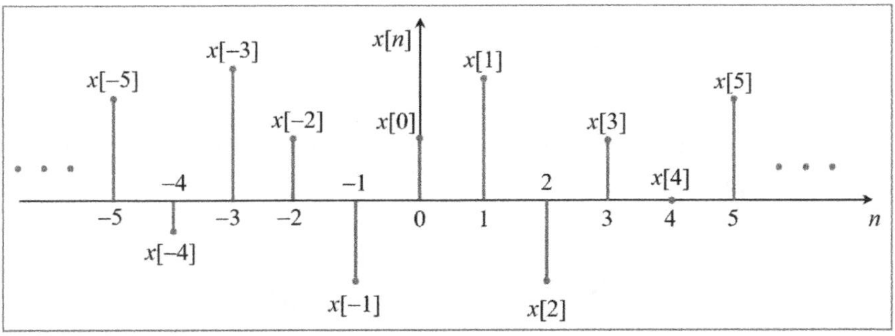

Figure 10.8. Any signal $x[n]$ can be considered as the sum of scaled unit impulses sitting at each value of n. For example, the sample $x[2]$ at $n = 2$ can be expressed as a signal $x_2[n] = x[2]\delta[n-2]$, which is zero for all $n \neq 2$. Similarly, the sample $x[-3]$ at $n = -3$ can be expressed as a signal $x_{-3}[n] = x[-3]\delta(n+3)$, which is zero for all $n \neq -3$.

4. The definition of impulse response given by Eq. (10.16) is applicable for any general system. This definition is not restricted to LTI systems.

5. Verify the validity of Eq. (10.18) by evaluating the left and right sides for any value of n. For example, if we put $n = 1$, then only the term $x[1]\delta[n-1]$ will remain on the right-side.

Observe that Eq. (10.18) expresses $x[n]$ as the sum of scaled and shifted unit impulses, the scale factors being the sample values $x[k]$ for each k. Since the system is LTI, we can obtain its output for the input $x[k]\delta[n-k]$ as

$$\begin{aligned}\mathcal{D}[x[k]\delta[n-k]] &= x[k]\mathcal{D}[\delta[n-k]] \\ &= x[k]h[n-k]. \end{aligned} \tag{10.19}$$

Therefore, we can obtain the output for the input, $x[n]$, as

$$\begin{aligned} y[n] = \mathcal{D}[x[n]] &= \mathcal{D}[\cdots + x[-1]\delta[n+1] + x[0]\delta[n] \\ &\quad + x[1]\delta[n-1] + x[2]\delta[n-2] + \cdots] \\ &= \cdots + \mathcal{D}[x[-1]\delta[n+1]] + \mathcal{D}[x[0]\delta[n]] \\ &\quad + \mathcal{D}[x[1]\delta[n-1]] + \mathcal{D}[x[2]\delta[n-2]] + \cdots \\ &= \cdots + x[-1]h[n+1] + x[0]h[n] \\ &\quad + x[1]h[n-1] + x[2]h[n-2] + \cdots \end{aligned} \tag{10.20}$$

which can be expressed in compact form as

$$y[n] = \sum_{m=-\infty}^{\infty} x[m]h[n-m] \triangleq x[n] \otimes h[n]. \tag{10.21}$$

Eq. (10.21) is known as the *discrete-time convolution* of sequences $\{x[n]\}$ and $\{h[n]\}$. It can also be expressed as

$$y[n] = \sum_{m=-\infty}^{\infty} h[m]x[n-m] \triangleq h[n] \otimes x[n]. \tag{10.22}$$

§ *The output of a LTI system is given by the convolution of its impulse response,* $h[n]$*, with the input,* $x[n]$*.*

Note from Eq. (10.22) that $y[n]$ will depend upon the future sample $x[n+1]$ if $h[-1] \neq 0$. Generalizing this, we can say that *a LTI system is causal if and only if its impulse response,* $h[n]$*, is zero for* $n < 0$. All physical systems are causal systems since they are unable to look into the future.

It can be shown that a LTI system will be BIBO (Bounded Input Bounded Output) stable if its impulse response satisfies the absolutely summable condition[6]

$$\sum_{n=-\infty}^{\infty} |h[n]| < \infty. \tag{10.23}$$

Note that

$$|H(\tilde{f})| = \left| \sum_{n=-\infty}^{\infty} h[n]e^{-j2\pi \tilde{f} n} \right| < \sum_{n=-\infty}^{\infty} |h[n]|. \tag{10.24}$$

In view of Eqs. (10.23) and (10.24), we can say the following:

A LTI system is BIBO stable if and only if the Fourier transform (DTFT), $H(\tilde{f})$, of its impulse response exists (i.e., $|H(\tilde{f})| < \infty$) for all $\tilde{f}$. That is, the region of convergence (ROC) of the $z-$transform, $H(z)$, of $h[n]$ must include the unit circle.

For physical realizability, a LTI system must be both causal and stable. That is, its impulse response must satisfy

$$h[n] = 0 \quad \text{for} \quad n < 0 \tag{10.25}$$

and

$$\sum_{n=0}^{\infty} |h[n]| < \infty. \tag{10.26}$$

10.5 Interconnection of Digital Systems

Digital systems can be interconnected to build larger systems. There are basically two ways by which they can be interconnected: in cascade (series) or in parallel.

In a cascade interconnection (Fig. 10.9), the output of the first system is

$$y_1[n] = \mathcal{D}_1[x[n]], \tag{10.27}$$

6. Let $x[n]$ be a bounded input, with $|x[n]| < \beta < \infty$. Then, we get

$$|y[n]| = \left| \sum_{m=-\infty}^{\infty} h[m]x[n-m] \right| < \sum_{m=-\infty}^{\infty} |h[m]x[n-m]|$$
$$= \sum_{m=-\infty}^{\infty} |h[m]| \cdot |x[n-m]| < \beta \sum_{m=-\infty}^{\infty} |h[m]|.$$

Therefore, for $y[n]$ to remain bounded, i.e., $|y[n]| < \infty$, we should have Eq. (10.23) satisfied.

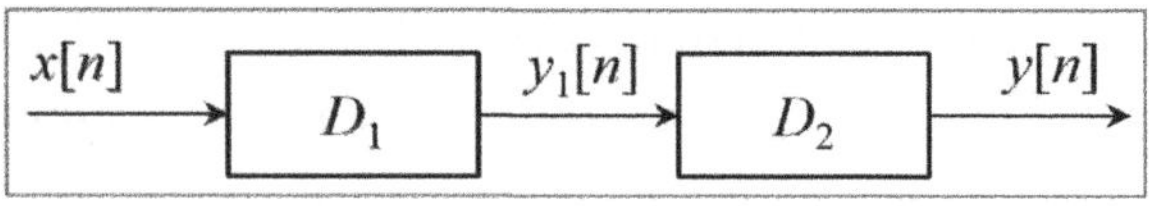

Figure 10.9. Cascading of two discrete-time linear systems.

and the output of the second system is,

$$\begin{aligned} y[n] &= \mathcal{D}_2[y_1[n]], \\ &= \mathcal{D}_2\{\mathcal{D}_1[x[n]]\}. \end{aligned} \tag{10.28}$$

We observe that the two systems $\mathcal{D}_1$ and $\mathcal{D}_2$ can be combined or consolidated to a single system with overall system function

$$\mathcal{D}_c \equiv \mathcal{D}_1\mathcal{D}_2. \tag{10.29}$$

Consequently, the output of the combined system can be expressed as

$$y[n] = \mathcal{D}_c[x[n]]. \tag{10.30}$$

In general, the order in which the operations $\mathcal{D}_1$ and $\mathcal{D}_2$ are performed is important. That is,

$$\mathcal{D}_1\mathcal{D}_2 \neq \mathcal{D}_2\mathcal{D}_1 \tag{10.31}$$

for arbitrary systems. However, if the systems are LTI, then $\mathcal{D}_c$ is time invariant and $\mathcal{D}_1\mathcal{D}_2 = \mathcal{D}_2\mathcal{D}_1$. That is the order in which the systems process the signal is not important.

In the parallel interconnection (Fig. 10.10), the output of the system $\mathcal{D}_1$ is $y_1[n]$ and that of the system $\mathcal{D}_2$ is $y_2[n]$. Hence the total output of the (parallel) interconnected system is,

$$\begin{aligned} y[n] &= y_1[n] + y_2[n], \\ &= \mathcal{D}_1[x[n]] + \mathcal{D}_2[x[n]], \\ &= (\mathcal{D}_1 + \mathcal{D}_2)[x[n]], \\ &= \mathcal{D}_p[x[n]], \end{aligned} \tag{10.32}$$

where $\mathcal{D}_p = \mathcal{D}_1 + \mathcal{D}_2$.

Just like we construct larger systems by interconnecting smaller ones, a larger system can be decomposed into smaller subsystems for the purpose of analysis and easy implementation (we shall see this later).

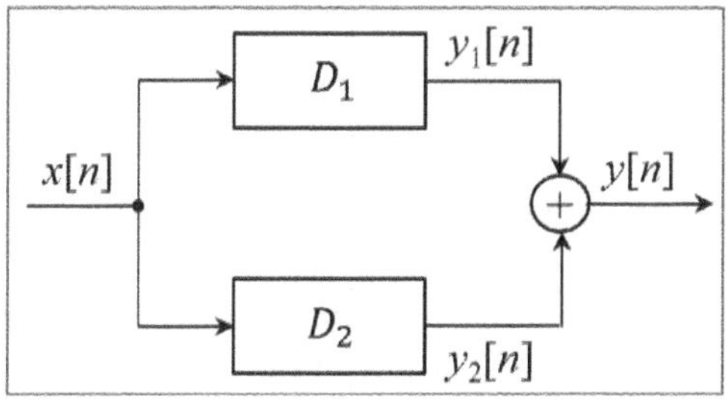

Figure 10.10. Parallel interconnection of two discrete-time linear systems.

10.6 Properties of Convolution and Interconnection of LTI Systems

Here, we investigate some important properties of convolution and interpret them in terms of interconnecting LTI systems. We have,

$$y[n] = x[n] \otimes h[n] = \sum_{k=-\infty}^{\infty} x[k]h[n-k] \tag{10.33}$$

where $\otimes$ represents the convolution operation. In Eq. (10.33), the impulse response $h[n]$ is folded and shifted. The input to the system is $x[n]$. We also know that,

$$y[n] = h[n] \otimes x[n] = \sum_{k=-\infty}^{\infty} h[k]x[n-k]. \tag{10.34}$$

Here, it is the input signal which is folded and shifted. The property embodies in Eq. (10.33) and Eq. (10.34) is called the *commutative law*. That is

$$x[n] \otimes h[n] = h[n] \otimes x[n]. \tag{10.35}$$

Alternatively, we may interpret this form of the convolution formula as a result of interchanging the roles of $x[n]$ and $h[n]$. In other words, we may regard $x[n]$ as the impulse response of the system and $h[n]$ as the input signal.

10.6.1 Identity and Shifting Properties

We also note that the unit sample response $\delta[n]$ is the identity element for convolution, that is

$$y[n] = x[n] \otimes \delta[n] = x[n]. \tag{10.36}$$

If we shift $\delta[n]$ by k, the convolution sequence is also shifted by k, that is

$$x[n] \otimes \delta[n-k] = y[n-k] = x[n-k]. \tag{10.37}$$

10.6.2 Associative Law

This law states that,

$$[x[n] \otimes h_1[n]] \otimes h_2[n] = x[n] \otimes [h_1[n] \otimes h_2[n]]\,. \tag{10.38}$$

From a physical point of view, we can interpret $x[n]$ as the input to a LTI system with impulse response $h_1[n]$. The output of the system, $y_1[n]$, becomes the input to a second LTI system with impulse response $h_2[n]$. Then the output is,

$$\begin{aligned} y[n] &= y_1[n] \otimes h_2[n], \\ &= [x[n] \otimes h_1[n]] \otimes h_2[n], \end{aligned} \tag{10.39}$$

which is precisely the left-hand side (LHS) of Eq. (10.38). Thus the LHS of Eq. (10.38) corresponds to having two LTI systems in cascade. Now, the right-hand side (RHS) of Eq. (10.38) indicates that the input $x[n]$ is applied to an equivalent system having an impulse response, say $h[n]$, which is equal to the convolution of $h_1[n]$ and $h_2[n]$. That is

$$h[n] = h_1[n] \otimes h_2[n], \tag{10.40}$$

and

$$y[n] = x[n] \otimes h[n]. \tag{10.41}$$

Furthermore, since the convolution operation satisfies the commutative property, one can interchange the order of the two systems with responses $h_1[n]$ and $h_2[n]$ without altering the overall input-output relationship. Associative law can be extended for more than two systems easily. Thus, if we have L, LTI systems with impulse responses $h_1[n], h_2[n], \ldots, h_L[n]$ in cascade, there is an equivalent LTI system having an impulse response, $h[n] = h_1[n] \otimes h_2[n] \otimes \ldots \otimes h_L[n]$.

Conversely, any LTI system can be decomposed into a cascade interconnection of subsystems.

10.6.3 Distributive Law

According to this law, we have,

$$x[n] \otimes [h_1[n] + h_2[n]] = x[n] \otimes h_1[n] + x[n] \otimes h_2[n]. \tag{10.42}$$

Interpreted physically, this law implies that if we have two LTI systems with impulse responses $h_1[n]$ and $h_2[n]$ which are excited by the same input signal $x[n]$ (or in a sense these two systems are connected in parallel), the sum of the two responses is identical to the response of the overall system with impulse response

$$h[n] = h_1[n] + h_2[n]. \tag{10.43}$$

Thus the overall system is viewed as a parallel combination of the two individual LTI systems. This can be generalized to the parallel combination of L, LTI systems. The overall system will have an impulse response,

$$h[n] = \sum_{j=1}^{L} h_j[n]. \tag{10.44}$$

Conversely, any LTI system can be decomposed into parallel interconnection of subsystems.

10.7 Characterizing Digital Systems

There are several ways to characterize (i.e., describe or represent) digital systems. Our discussion in this section is limited to only LTI systems.

10.7.1 Difference Equations and Block Diagram

We can describe a system by specifying how its output, $y[n]$, is related to its input, $x[n]$, which, in general, has the form,

$$y[n] = \mathcal{D}\left[y[n-1], \ldots, y(n-N), x[n], \ldots, x[n-M]\right] \tag{10.45}$$

where $\mathcal{D}[.]$ denotes some function of the quantities in brackets. Specifically, for a LTI system, the general form of the input-output relationship is,

$$y[n] = -\sum_{k=1}^{N} a_k y[n-k] + \sum_{k=0}^{M} b_k x[n-k] \tag{10.46}$$

where $\{a_k\}$ and $\{b_k\}$ are constant parameters that specifies the system and are independent of $x[n]$ and $y[n]$. The relationship in Eq. (10.46) is called a difference equation and is one of the ways to characterize the behaviour of a discrete-time LTI system.

For example, we have a system described by the following equations:

$$y[n] = 5v[n] + 6v[n-1] + 7v[n-2] \tag{10.47a}$$

$$v[n] = x[n] - 3v[n-1] - 4v[n-2]. \tag{10.47b}$$

Eqs. (10.47a) and (10.47b) are known as the *difference equations* that represent the underlying system. If we are given the input samples $x[n]$, it is very straightforward to compute the samples of the output, $y[n]$, using the above equations.

Observe that the implementation of Eq. (10.47) consists of the three basic units: *adders, multipliers, and delays*. Recall that z^{-1} represents the 1-sample delay operator.

The *complexity* of a digital system refers to the number of multiplications and additions it must perform during each sample interval. Since the input samples, $x[n]$, are coming at some rate (e.g., 1000 samples/second), it is necessary that the system is able to finish all the computations required to determine the output at time instant n before the next sample, $x[n+1]$, arrives. For obvious reasons, multiplication is much more costly to perform compared to addition. Sometimes, the number of delay elements is also considered while specifying the complexity.

10.7.2 Recursive and Non-recursive Digital Systems

Systems whose output depends on several past outputs as well as the present and several past inputs are called *recursive* systems. Equation (10.45) represents a causal and practically realizable recursive system. If $y[n]$ depends only on the present and past inputs, then

$$y[n] = \mathcal{D}\left[x[n], x[n-1], \ldots, x[n-M]\right]. \tag{10.48}$$

Such systems are called *non-recursive.*

10.7.3 Transfer Function

Transfer function (denoted as $H(z)$) of a LTI system is given by the ratio of the $z-$transforms of its output to input. That is

$$H(z) = \frac{Y(z)}{X(z)} \tag{10.49}$$

where $X(z)$ is the z-transform (ZT) of the input $x[n]$ and $Y(z)$ is the[7] ZT of the output $y[n]$.

7. Using Eq. (10.21), we can determine the ZT of the output, $y[n]$, of a LTI system with impulse response, $h[n]$, as (with $p = n - m$)

$$\begin{aligned} Y(z) &= \sum_{n=-\infty}^{\infty} y[n]z^{-n} = \sum_{n=-\infty}^{\infty}\sum_{m=-\infty}^{\infty} x[m]h[n-m]z^{-n} \\ &= \sum_{p=-\infty}^{\infty}\sum_{m=-\infty}^{\infty} x[m]h[p]z^{-(p+m)} = \sum_{p=-\infty}^{\infty} h[p]z^{-p} \sum_{m=-\infty}^{\infty} x[m]z^{-m} \\ &= H(z)X(z). \end{aligned} \tag{10.50}$$

Thus, we get the transfer function, $H(z)$, as defined in Eq. (10.49).

Recall from Chapter 8 that[8] $\mathcal{Z}[x[n-k]] = z^{-k}X(z)$. Using this and the linearity property of ZT, we can obtain the transfer function of the system given by Eq. (10.47) as follows:

$$\begin{aligned} Y(z) &= \mathcal{Z}\left[5v[n] + 6v[n-1] + 7v[n-2]\right] \\ &= 5\mathcal{Z}[v[n]] + 6\mathcal{Z}[v[n-1]] + 7\mathcal{Z}[v[n-2]] \\ &= 5V(z) + 6z^{-1}V(z) + 7z^{-2}V(z) \\ &= \left[5 + 6z^{-1} + 7z^{-2}\right]V(z) \end{aligned} \tag{10.51a}$$

$$\begin{aligned} V(z) &= \mathcal{Z}\left[x[n] - 3v[n-1] - 4v[n-2]\right] \\ &= X(z) - 3z^{-1}V(z) - 4z^{-2}V(z) \\ \Rightarrow \quad X(z) &= \left[1 + 3z^{-1} + 4z^{-2}\right]V(z) \end{aligned} \tag{10.51b}$$

$$V(z) = \frac{X(z)}{1 + 3z^{-1} + 4z^{-2}} \tag{10.51c}$$

$$\Rightarrow \quad H(z) = \frac{Y(z)}{X(z)} = \frac{5 + 6z^{-1} + 7z^{-2}}{1 + 3z^{-1} + 4z^{-2}}. \tag{10.51d}$$

Using Eq. (10.51d), we can obtain another difference equation for the output as follows. Using Eq. (10.51d), we can write

$$\begin{aligned} \left[1 + 3z^{-1} + 4z^{-2}\right]Y(z) &= \left[5 + 6z^{-1} + 7z^{-2}\right]X(z) \\ \Rightarrow \quad Y(z) + 3z^{-1}Y(z) + 4z^{-2}Y(z) &= 5X(z) + 6z^{-1}X(z) \\ &\quad + 7z^{-2}X(z). \end{aligned} \tag{10.52a}$$

Taking inverse ZT on both sides of Eq. (10.52a), we get

$$\begin{aligned} y[n] + 3y[n-1] + 4y[n-2] &= 5x[n] + 6x[n-1] + 7x[n-2] \\ \Longrightarrow \quad y[n] &= 5x[n] + 6x[n-1] + 7x[n-2] \\ &\quad - 3y[n-1] - 4y[n-2]. \end{aligned} \tag{10.52b}$$

The expression for the output, $y[n]$, given by Eq. (10.52b) is *recursive* in nature, as $y[n]$ depends on its own past values.

---

8. The notation $\mathcal{Z}[x[n]]$ implies 'the z-transform (ZT) of $x[n]$.'

Eq. (10.51d) shows that, in general, we can express the transfer function of a digital system as the ratio of two polynomials in z as

$$H(z) = \frac{N(z)}{D(z)} = \frac{\sum_{l=0}^{M} b_l z^{-l}}{\sum_{k=0}^{N} a_k z^{-k}}, \tag{10.53}$$

where b_l for $l = 0, 1, \ldots, M$, and a_k for $k = 0, 1, \ldots, N$, are the coefficients of the polynomials $N(z)$ and $D(z)$, respectively. Note that these coefficients correspond to the multiplication factors in the block-diagram[9] of the system.

10.7.4 Poles and Zeros

The values of z at which $H(z)$ becomes infinity are known as the poles of $H(z)$. Similarly, the values of z at which $H(z)$ becomes zero are known as the zeros of $H(z)$. Therefore, the poles and zeros of $H(z)$ can be obtained by determining the roots of its denominator and numerator polynomials, respectively. Since $H(z)$ becomes ∞ at its poles, it is clear that all the poles of $H(z)$ must lie outside its ROC.

For the $H(z)$ given in Eq. (10.51d), we get the poles as $\{(-1.5 + j1.323), (-1.5 - j1.323)\}$ and zeros as $\{(-0.6 + j1.02), (-0.6 - j1.02)\}$. It can be seen that both the numerator and denominator polynomials are of 2nd degree and hence there exits two roots each. Note that the poles and zeros occur in complex conjugate pairs since the polynomial coefficients are real.

A plot of the poles and zeros is known as the *pole-zero diagram*. The pole-zero diagram for the system in Eq. (10.51d) is illustrated in Fig. 10.11. It can be seen that both the poles and zeros lie outside the unit circle.

> § *From the pole-zero diagram of a transfer function, is it possible to obtain the expression for the transfer function?*

Poles of Causal and Stable LTI Systems: Let us consider a causal LTI system with impulse response given by $h[n] = a^n u[n]$, where a is a scalar. Recall from the example on the right-sided sequence in Section 8.1, that this results in $H(z) = \frac{1}{1-az^{-1}}$, which has one pole at $z = a$ and ROC is given by $|z| > |a|$. Thus, *the ROC does not include the pole*. On the other hand, recall that the ROC of a stable system must

9. Will be discussed in detail in the next chapter.

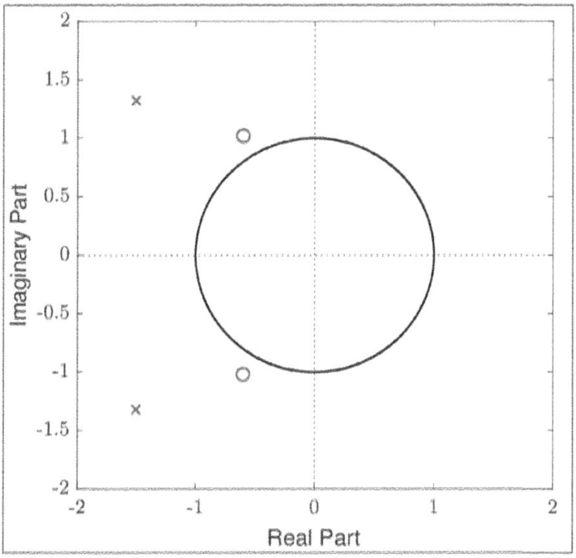

Figure 10.11. Pole-zero diagram. Here, the poles are denoted by xs and zeros are denoted by os.

include the unit circle. This implies that we should have $|a| < 1$ for this system to be stable. In other words, *the poles of a causal and stable system must lie inside the unit circle.*

10.7.5 Frequency Response

Frequency response (denoted $H(\tilde{f})$) of a LTI system is given by the Fourier transform of its impulse response, $h[n]$. It can also be obtained by evaluating the transfer function, $H(z)$, on the unit circle. That is,

$$H(\tilde{f}) = H(z)|_{z=e^{j2\pi\tilde{f}}} = \left.\frac{Y(z)}{X(z)}\right|_{z=e^{j2\pi\tilde{f}}} = \frac{Y(\tilde{f})}{X(\tilde{f})} \tag{10.54a}$$

$$= \sum_n h[n]e^{-j2\pi\tilde{f}n}. \tag{10.54b}$$

Thus, frequency response is equal to the ratio of the discrete-time Fourier transforms (DTFT) of the output and input. It is easy to see that $H(\tilde{f})$ is periodic with period equal to the normalized sampling frequency 1 (Note: $X(\tilde{f})$ and $Y(\tilde{f})$ are periodic).

We say that a digital system is real if the coefficients of the numerator and denominator polynomials ($N(z)$, $D(z)$) in its transfer function are real. This is equivalent to saying that the impulse response, $h[n]$, is real, i.e., $h^*[n] = h[n]$.

Noting that $\left(e^{j2\pi \tilde{f}}\right)^* = e^{-j2\pi \tilde{f}}$, we can write using Eq. (10.54b)

$$H^*(\tilde{f}) = \left[\sum_n h[n]e^{-j2\pi \tilde{f} n}\right]^* = \sum_n h^*[n]\left(e^{-j2\pi \tilde{f} n}\right)^*$$
$$= \sum_n h[n]e^{j2\pi \tilde{f} n} = H(-\tilde{f}). \tag{10.55}$$

Thus, for real systems, the magnitude of $H(\tilde{f})$ is even-symmetric and phase of $H(\tilde{f})$ is odd-symmetric. That is,

$$|H(\tilde{f})| = |H(-\tilde{f})| \quad \text{and} \quad \theta(-\tilde{f}) = -\theta(\tilde{f}), \tag{10.56}$$

where $\theta(\tilde{f})$ is the phase of $H(\tilde{f})$. As before, $|H(\tilde{f})|$ and $\theta(\tilde{f})$ are known as the magnitude spectrum and phase spectrum, respectively, of the system specified by the transfer function, $H(z)$.

10.7.6 Impulse Response

By definition, we get the impulse response as the system output (i.e., $y[n] = h[n]$), when we set the input as $x[n] = \delta[n]$. That is,

$$X(z) = 1 \quad \text{and} \quad X(\tilde{f}) = 1. \tag{10.57}$$

Using this in Eqs. (10.49) and (10.54), we get

$$Y(z) = H(z) \quad \text{and} \quad Y(\tilde{f}) = H(\tilde{f}). \tag{10.58}$$

Thus, the impulse response (denoted as $h[n]$) of a system can be obtained by finding

- either the inverse DTFT of its frequency response, $H(\tilde{f})$,
- or the inverse ZT of its transfer function, $H(z)$.

That is,

$$h[n] = \mathcal{F}^{-1}[H(\tilde{f})] = \int_{-0.5}^{0.5} H(\tilde{f})e^{j2\pi \tilde{f}}d\tilde{f}. \tag{10.59}$$

$$h[n] = \mathcal{Z}^{-1}[H(z)]. \tag{10.60}$$

10.7.7 Systems with Finite-Duration and Infinite-Duration Impulse Response

It is possible to subdivide the class of LTI systems into two types: Finite-duration impulse response (FIR) and Infinite-duration impulse response (IIR). Thus an FIR system has an impulse response which is zero outside some finite interval. Without loss of generality, we focus our attention on causal FIR systems, so that

$$h[n] = 0, \quad n < 0 \text{ and } n \geq M. \tag{10.61}$$

The convolution formula for such a system reduces to

$$y[n] = \sum_{k=0}^{M-1} h[k]x[n-k]. \tag{10.62}$$

In contrast, an IIR LTI system has an infinite duration impulse response and the output (convolution formula) is,

$$y[n] = \sum_{k=0}^{\infty} h[k]x[n-k]. \tag{10.63}$$

10.8 Determination of System Output

Consider a LTI system with impulse response $h[n]$ and transfer function $H(z)$. It becomes often necessary to determine the output, $y[n]$, of the system for a given input, $x[n]$. We can do this by any of the following methods.

- Transfer Function Approach: From $H(z)$ (the transfer function) and $X(z)$ (z-transform of the input, $x[n]$), the z-transform of the output, $Y(z)$ is obtained as $Y(z) = H(z) \cdot X(z)$. The corresponding output $y[n]$ is obtained by taking the inverse ZT of $Y(z)$.
- Frequency Response Approach: Given $H(\tilde{f})$, we can determine $y[n]$ by finding the inverse DTFT of $H(\tilde{f})X(\tilde{f})$.
- Convolution Approach: Given $h[n]$ and $x[n]$, the output is calculated as, $y[n] = h[n] \otimes x[n]$.
- Difference Equation Approach: Given $H(z)$, we can form the difference equations and determine the output, $y[n]$, as illustrated by Eq. (10.52).

10.8.1 Transfer Function Approach

Since $Y(z) = H(z)X(z)$, if we are given the system transfer function ($H(z)$), we can obtain the output, $y[n]$, by determining the inverse ZT of $H(z)X(z)$. Often, this is not an easy task, since the expression $H(z)X(z)$ may not be simple enough to write its inverse ZT by inspection. This occurs when $Y(z)$ has two or more poles (i.e., denominator is of degree 2 or more). In such cases, we resort to the method *partial fractions* as described in Chapter 8 (see Section 8.5.2).

Let $Y(z)$ be a complicated expression whose inverse ZT we want to find. The method of partial fraction determines scalars $\alpha_1, \alpha_2, \cdots, \alpha_m$ such that $Y(z)$ can be written as,

$$Y(z) = \alpha_1 Y_1(z) + \alpha_2 Y_2(z) + \cdots + \alpha_m Y_m(z), \tag{10.64}$$

where $Y_1(z), Y_2(z), \cdots, Y_m(z)$ are simple expressions whose inverse transforms are known. Then, we can get the inverse ZT of $Y(z)$ as,

$$y[n] = \alpha_1 y_1[n] + \alpha_2 y_2[n] + \cdots + \alpha_m y_m[n] \tag{10.65}$$

where $y_i[n]$ is the inverse ZT of[10] $Y_i(z)$ for $i = 1, 2, \ldots, m$. The scalars $\alpha_1, \alpha_2, \cdots, \alpha_m$ can be determined by following the method we outlined in Chapter 8.

Determination of Impulse Response, $h[n]$: In some applications, we may want to determine the impulse response, $h[n]$, of the system. As Eq. (10.60) shows, we can obtain $h[n]$ by finding the inverse ZT of $H(z)$. If $H(z)$ is not a simple function (i.e., it has more than one pole), we can use the method of partial fraction to obtain $h[n]$.

10.8.2 Convolution Approach

Given $h[n]$, we can determine $y[n]$ by convolving $h[n]$ with $x[n]$ as (see Eqs. (10.21) and (10.22))

$$y[n] = \sum_{m=-\infty}^{\infty} x[m]h[n-m] = \sum_{m=-\infty}^{\infty} h[m]x[n-m]. \tag{10.66}$$

We have also seen that (see Eq. (10.50)) *convolution in time-domain is equivalent to multiplication in frequency-domain*, as given by

$$Y(z) = H(z)X(z) \qquad \text{or} \qquad Y(\tilde{f}) = H(\tilde{f})X(\tilde{f}). \tag{10.67}$$

These relations are very useful while working with signals and systems.

10. While applying the partial fraction approach, we must ensure that the expression for $Y(z)$ is a proper rational function. That is, the degree of the numerator polynomial is less than that of the denominator.

The dual of the relations in Eqs. (10.66)–(10.67) arise when we consider multiplying two signals in time-domain. Recall that the *multiplication in time-domain results in convolution in frequency-domain.* That is, if $y[n]$ is obtained as $y[n] = x[n]h[n]$, then we have

$$Y(\tilde{f}) = \mathcal{F}[x[n]h[n]] = X(\tilde{f}) \otimes H(\tilde{f}) \tag{10.68}$$

$$= \int_{-0.5}^{0.5} X(u)H(\tilde{f} - u)du \tag{10.69}$$

$$= \int_{-0.5}^{0.5} H(u)X(\tilde{f} - u)du. \tag{10.70}$$

Since $X(\tilde{f})$ and $H(\tilde{f})$ are periodic, the convolution depicted by Eqs. (10.68)–(10.70) are known as *circular convolution.* This is because, the $H(\tilde{f} - u)$ and $X(\tilde{f} - u)$ are equivalent to circularly shifting one period of these spectra. As a result, for any value of $\tilde{f}$ in $Y(\tilde{f})$, the integration involves one complete period each from both $X(\tilde{f})$ and $H(\tilde{f})$.

10.8.3 Difference Between Linear and Circular Convolutions

Let $a[n]$ and $b[n]$ be N-point data sequences which are zero for n outside $\{0, 1, \ldots, N-1\}$. The linear convolution between $a[n]$ and $b[n]$ is given by

$$p_1[n] = a[n] \otimes b[n] = \sum_{l=0}^{N-1} a(l)b(n-l) \tag{10.71}$$

where $a[n] = 0$ and $b[n] = 0$ if $n \notin \{0, 1, \ldots, N-1\}$. The circular convolution between $a[n]$ and $b[n]$ is given by,

$$p_2[n] = a[n] \odot b[n] = \sum_{l=0}^{N-1} \tilde{a}(l)\tilde{b}(n-l) \tag{10.72}$$

where $\tilde{a}[n]$ and $\tilde{b}[n]$ are obtained by periodically extending $a[n]$ and $b[n]$, respectively.

10.9 Concluding Remarks

We conclude this chapter with the following remarks.

- Digital signal processing plays a very important role in many practical applications involving signals and systems.

- Linear time-invariant (LTI) systems can be characterized by impulse response, frequency response, transfer function, etc.
- Output of a LTI system can be obtained by convolving its impulse response with the input.
- Convolution in time-domain is equivalent to multiplication in frequency-domain, and multiplication in time-domain is equivalent to convolution in frequency-domain.
- A LTI system is stable if and only if the unit circle is inside the region of convergence of its transfer function. This is equivalent to saying that its frequency response must exist (be finite) for all frequencies.
- A causal LTI system is stable if and only if all the poles of its transfer function lie inside the unit circle.
- Physical realizability requires stability and causality.

Exercises

10.1. The output $y[n]$ of a digital system is related to its input $x[n]$ by,

$$y[n] = 0.98y[n-1] + 0.02x[n].$$

Find the system transfer function $H(z)$ and the impulse response $h[n]$.

10.2. The transfer function of a digital system is given by,

$$H(z) = \frac{0.03}{1 - 0.97z^{-1}}.$$

Sketch the magnitude response for $|\tilde{f}| < 1$ Hz. Is $H(z)$ a low-pass system?

10.3. In a biomedical engineering system, the power ($p[n]$) of the real signal $x[n]$ is determined as

$$p[n] = 0.02\left[x^2[n] + 0.98x^2[n-1] + 0.98^2x^2[n-2] + \ldots\right].$$

Show that this is equivalent to squaring $x[n]$ and passing this $x^2[n]$ into a filter with transfer function $H(z)$ given by,

$$H(z) = \frac{0.02}{1 - 0.98z^{-1}}.$$

10.4. The output $y[n]$ of a digital system is related to its input $x[n]$ by the following difference equation:

$$y[n] = 0.98y[n-1] + 0.02x^2[n].$$

If $x[n]$ is a real white noise sequence with power ($p[n]$) 3, find the mean value of $y[n]$ in the steady state.

10.5. The transfer function ($H(z)$) of an all-pass system is given in the equation below.

$$H(z) = \frac{1 - 0.8z}{z - 0.8}$$

(a) Plot the pole-zero diagram and the magnitude response of the system.

(b) Determine the phase response of $H(z)$ in Question (a) and check if it is negative over the positive main band.

10.6. The impulse response $h[n]$ of a digital system is as shown below:

$$h[n] = \begin{cases} 1 & \text{for } 2 \leq n \leq 4 \\ 0 & \text{otherwise,} \end{cases}$$

Find the output $y[n]$ of the system for the following input $x[n]$:

$$x[n] = \begin{cases} 1 & \text{for } 0 \leq n \leq 2 \\ 0 & \text{otherwise,} \end{cases}$$

10.7. For a digital system, the impulse response $h[n] = \delta[n-2]$. Find the output $y[n]$ of the system for the given input $x[n]$:

$$x[n] = \begin{cases} 1 & \text{for } n = 0, 2 \\ 2 & \text{for } n = 1 \\ 0 & \text{otherwise,} \end{cases}$$

10.8. The response $y[n]$ of a digital system to an impulse input (i.e., $x[n] = \delta[n]$) is given by,

$$y[n] = \begin{cases} 1 & \text{for } n = 0 \\ 2 & \text{for } n = 1 \\ 0 & \text{otherwise.} \end{cases}$$

Find the impulse response ($h[n]$) and frequency response ($H(\tilde{f})$) of the system. Using convolution, find the output of the system if the input is $x[n] = e^{j2\pi f_1 n}$.

10.9. We would like to implement an ideal low-pass filter with frequency response $H(\tilde{f})$ as shown in Fig. P10.9 below. Find the impulse response $h[n]$ of the filter. Express $y[n]$ in terms of $h[n]$ and the input $x[n]$. Is this filter implementable?

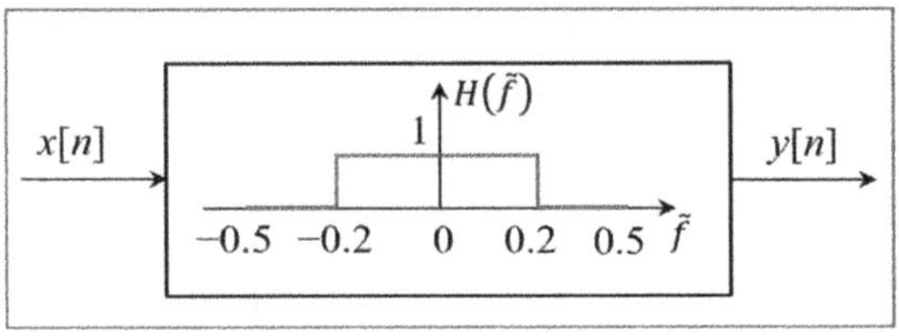

Figure P10.9. Frequency response of ideal low-pass filter.

10.10. Two digital linear systems with system impulse responses $h_1[n]$ and $h_2[n]$ are connected in parallel as shown in Fig. P10.10. Here $x[n]$ is the input to the system and $y[n]$ is the overall output of the system. $y_1[n]$ is the output of the first system to the input $x[n]$ and $y_2[n]$ is the corresponding output from the second system. Find the overall impulse and frequency responses ($h[n]$ and $H(\tilde{f})$, respectively) of the system.

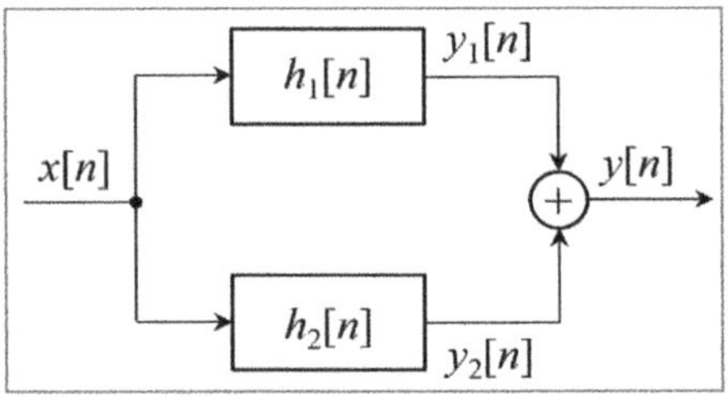

Figure P10.10. Parallel connection of two digital linear systems.

10.11. Figure P10.11 shows the cascading of two linear digital systems with impulse responses $h_1[n]$ and $h_2[n]$, respectively. Here $x[n]$ is the input to the system and $y[n]$ is the overall output of the system. $y_1[n]$ is the output of the first system to the input $x[n]$ and $y[n]$ is the output from the second system whose input is the output $y_1[n]$ from the first system. Find the overall impulse and frequency responses ($h[n]$ and $H(\tilde{f})$, respectively) of the system.

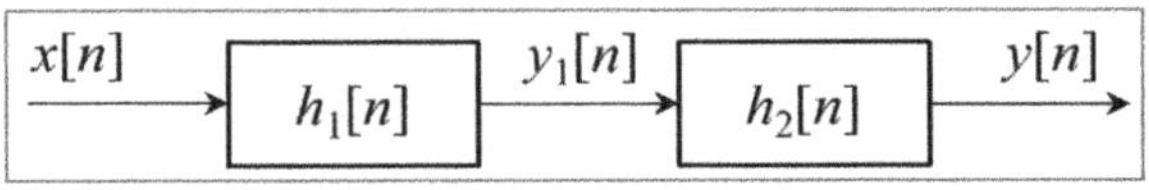

Figure P10.11. Cascading of two digital linear systems.

10.12. A FIR (finite impulse response) system has an impulse response $h[n]$ (given below) that is finite in length and symmetrical about the centre. Find its magnitude response ($|H(\tilde{f})|$). Show that it is a linear phase system or $\arg[H(\tilde{f})] \propto \tilde{f}$, where $\tilde{f}$ is the frequency.

$$h[n] = \begin{cases} 5 & \text{for } n = 0, 3 \\ 7 & \text{for } n = 1, 2 \\ 0 & \text{otherwise.} \end{cases}$$

10.13. A causal system is one whose output $y[n]$ depends on the present input $x[n]$ and its past values $x[n-1], x[n-2], \ldots$. A non-causal system has output that depends on future inputs and is not implementable. Is $H(z) = 4z + 1 + 2z^{-1} + 3z^{-3}$ causal?

10.14. A system has input $x[n]$, output $y[n]$, and $H(z) = 5z^2 + 1 + 2z^{-2} + 3z^{-3}$. The system is not causal as $H(z)$'s expansion has $5z^2$ term. Can the system be made causal and implementable by adding delays?

10.15. The decimator (with a compression ratio of 2) shown in Fig. P10.15 is used to reduce data rate and the sampling frequency from 8 to 4 Hz.

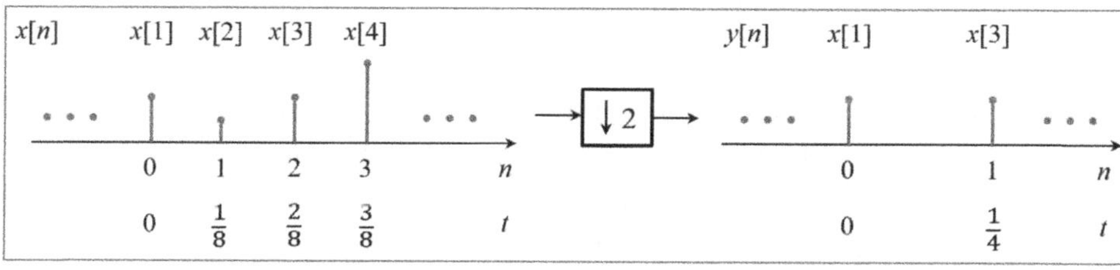

Figure P10.15. Decimator (compression ratio 2).

(a) Show that $Y(\tilde{f}) = 0.5\left[X(\tilde{f}) + X(\tilde{f} + 4)\right]$.

(b) From $Y(\tilde{f}) = 0.5\left[X(\tilde{f}) + X(\tilde{f} + 4)\right]$ above, draw the input and output spectra.

10.16. The operation of a digital filter with impulse response $h_N[n]$ and a transfer function $H_N(f)$ is described by the following difference equation:

$$y[n] = x[n] - \frac{1}{N}\sum_{k=1}^{N} x[n-k],$$

where $x[n]$ is the input signal to the filter, $y[n]$ is the output and N is a positive constant.

(a) Find and sketch the impulse response of the filter, $h_N[n]$, when $N = 1$ (i.e., $h_1[n]$) and $N = 2$ (i.e., $h_2[n]$).

(b) Write down $H_N(f)$ and sketch $|H_1(f)|$ in the frequency range $0 < f < f_s/2$, where f_s is the sampling frequency.

(c) The operation of the filter can be verbally described as: "Subtraction of the moving average of the last N samples from the newest sample of $x[n]$". Describe qualitatively how $|H_N(f)|$ changes with increasing N.

10.17. In this exercise, we will consider a device for recording the electrical signals from the heart (electrocardiogram, ECG). Simplified periodic signals are used in the development of the device. These signals are given by:

$$g(t) = g_1(t) + g_2(t) + g_3(t),$$

where

$$g_1(t) = \begin{cases} a\sin(2\pi f_1 t) & \text{for} \quad T_1 \le t \le 2.5T_1 \\ 0 & \text{otherwise} \end{cases}$$

$$g_2(t) = \begin{cases} b\sin(2\pi f_1 t) & \text{for} \quad 1.5T_1 \le t \le 2T_1 \\ 0 & \text{otherwise} \end{cases}$$

$$g_3(t) = \begin{cases} c\sin(2\pi f_2 t) & \text{for} \quad 4T_1 \le t \le 4T_1 + 0.5T_2 \\ 0 & \text{otherwise,} \end{cases}$$

where the signals $g_1(t)$, $g_2(t)$, $g_3(t)$, and $g(t)$ are all periodic with a period $T = 1$s. Furthermore, we have the following periods: $T_1 = 1/f_1$ and $T_2 = 1/f_2$, where $f_1 = 8$Hz and $f_2 = 2$Hz. A single period of the signal $g(t)$ is shown in Fig. P10.17 (top) and the corresponding amplitude spectrum is shown in Fig. P10.17 (bottom). It is assumed that the spectral component above 90Hz are negligibly small.

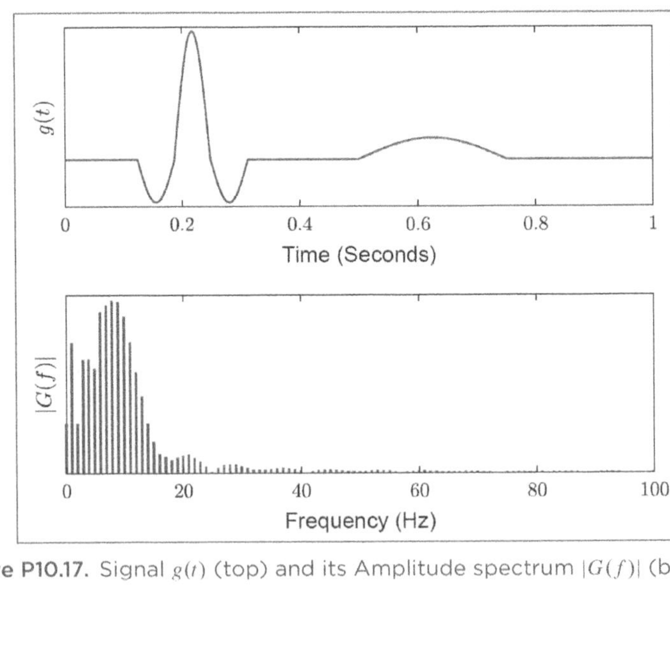

Figure P10.17. Signal $g(t)$ (top) and its Amplitude spectrum $|G(f)|$ (bottom).

(a) Express the power of the signal $g(t)$ as a function of a, b, and c, and suggest a sampling frequency such that there is an equivalence between the analog and the discrete-time periodic signals.

(b) We assume in the following, that the noise generated by the measurement process has negligibly little power, and is without any significance. The sampling frequency is chosen to be 200Hz, so that it is suitable for the consequent signal processing steps. A 50Hz interference signal is also present in the measured signal. It comes from the electric power net, and it should be removed by means of digital signal processing techniques. The following digital filter is used for the purpose:

$$h[n] = \delta[n] - h_1[n],$$

where

$$h_1[n] = \begin{cases} \frac{1}{N}\cos(2\pi f_3 n T_s) & \text{for } 0 \leq n \leq 2M-1 \\ 0 & \text{otherwise,} \end{cases}$$

where $f_3 = 50\text{Hz}$, T_s is the sampling period and M is a positive integer number. Write down the transfer function of the filter, and show that it removes the spectral component at 50Hz. Explain the influence of the filter's length $2M$ on its transfer function.

10.18. An analog signal of biological origin can be described by the following expression:

$$g_0(t) = g(t) \times \cos(2\pi f_0 t),$$

where f_0 is a constant, and $g(t)$ is a real signal, that can be seen in Fig. P10.18(a). $g(t)$ has a complex spectrum $G(f)$, and $|G(f)|$ is shown in Fig. P10.18(b) for positive frequencies (i.e., $f > 0$). Only the portion of the spectrum, in which the spectral components $|G(f)|$ have significant value (> 0) is depicted. After doing some measurements, it can be concluded that $|G(0)| = A$, $|G(50)| = A/5$, and $T_0 = 100$ ms.

(a) Give an expression of how to find A from $g(t)$, and give reasons why $A \neq 0$. Provide a labeled sketch of the spectrum of $g_0(t)$ by labelling the appropriate amplitude and frequency values, when $f_0 = 1750\text{Hz}$.

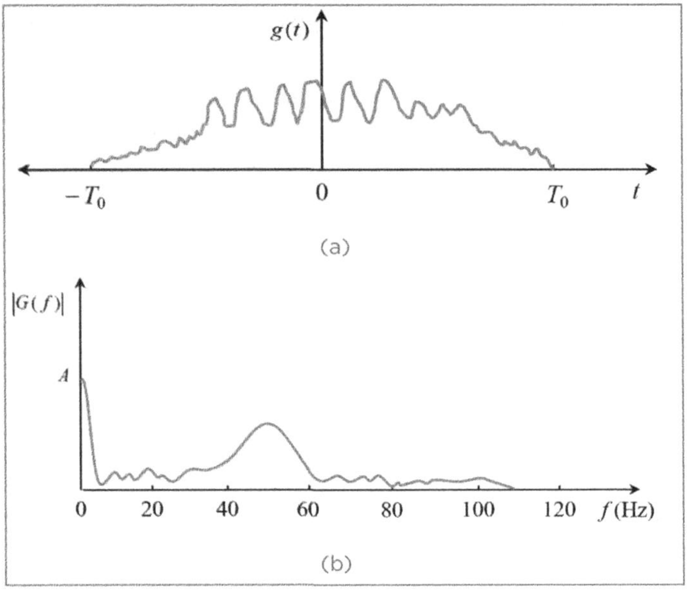

Figure P10.18. (a) Signal $g(t)$ (b) Complex Spectrum.

(b) A mathematical model of the signal $g(t)$ is needed for experiments. The modelled signal (denoted as $g_a(t)$) is chosen to be:

$$g_a(t) = g_1(t) + g_2(t),$$

where

$$g_1(t) = \begin{cases} a\sin(2\pi f_1 t) & \text{for } |t| < T_0 \\ 0 & \text{otherwise,} \end{cases}$$

and

$$g_2(t) = \begin{cases} b & \text{for } |t| < T_0 \\ 0 & \text{otherwise.} \end{cases}$$

In these expressions, a, b, and f_1 are constants. The magnitude (the value) of $2T_0 f_1$ is a suitable integer number. The spectrum of $g_a(t)$ is denoted as $G_a(f)$. Show that the real part of $G_a(f)$ is dependent only on $g_2(t)$, and that the imaginary part of $G_a(f)$ is dependent only on $g_1(t)$. Give an expression for calculating the energy of $g_a(t)$, and show that its energy is the sum of energies of $g_1(t)$ and $g_2(t)$.

(c) Argue that 50Hz will be an appropriate choice for f_1. Next, calculate the ratio between a and b for this value of f_1, so that

$|G_a(f)|$ becomes a reasonable approximation of $|G(f)|$ for frequencies $f = 0$ and $f = 50\text{Hz}$.

(d) $g(t)$ is converted to the discrete-time signal $g_d[n]$, which has a complex spectrum $G_d(f)$. The sampling interval is $T_s = 4\text{ms}$. Find the sampling frequency and sketch $|G_d(f)|$ in the frequency range below 500Hz (i.e., $0 \leq f \leq 500$). Discuss whether the conditions in the sampling theorem are fulfilled. Discuss if other values for T_s can be chosen, such that there is equivalence between $g(t)$ and $g_d[n]$.

10.19. In this exercise you will create a robust scheme to additive noise edge detector. In many applications, the edge of the pulse must be detected. Other applications require the removal of the low frequencies present in the signal. In both cases this can be done by a system with impulse response $h_1[n]$ which performs differentiation on the incoming data:

$$y[n] = x[n] - x[n-1],$$

where $x[n]$ is the input and $y[n]$ is the output of the system. A slightly more sophisticated version is the system with an impulse response $h_2[n]$ which processes the incoming signal as follows:

$$y[n] = x[n+1] - 2x[n] + x[n-1].$$

(a) Provide labelled sketches of the impulse responses $h_1[n]$ and $h_2[n]$. Sketch also their transfer functions $H_1(\tilde{f})$ and $H_2(\tilde{f})$ in the frequency range $\tilde{f} \in \{-\frac{1}{2}, \frac{1}{2}\}$, where $\tilde{f}$ is the normalized digital frequency, i.e., $\tilde{f} = \frac{f}{f_s}$ (f_s is the sampling frequency).

(b) Show that $h_2[n] = h_1[n] \otimes h_1[n+1]$, where $\otimes$ denotes the convolution. [*Hint*: You may want to use the equivalence between convolution in time domain and multiplication in frequency domain combined with the rule for time-shifting.]

(c) The input $x[n]$ to a system is often corrupted by additive noise as illustrated in Fig. P10.19(a). In this figure, the output of the system is $y[n] = x[n] \otimes h_2[n]$. The input is $x[n] = p[n] + d[n]$, where $p[n]$ is the signal of interest and $d[n]$ is the additive noise.

Assume that the input to the system, $p[n]$ is as shown in Fig. P10.19(b) and that there is no noise, i.e., $d[n] = 0$. Provide a labeled sketch of $y[n]$, the output of the system.

(d) Assume that the input to the system $p[n]$ is the same as in Question 10.19(c). Let there be noise this time, $d[n] = -\delta[n+1]$. Sketch the output of the system for $x[n] = p[n] + d[n]$.

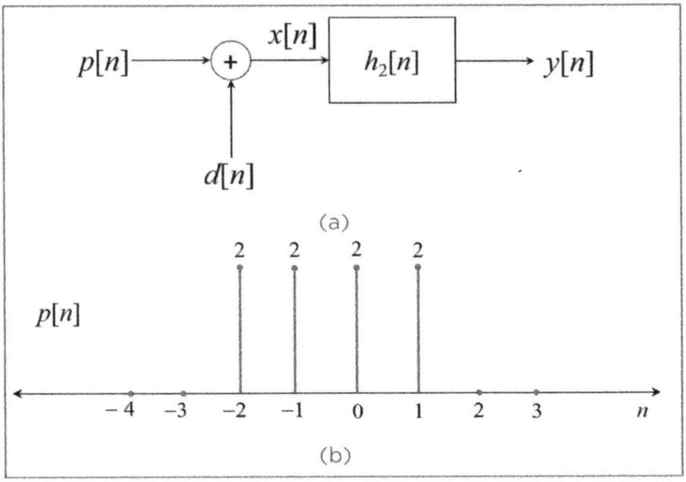

Figure P10.19.

(e) To remove noise, a filter with impulse response $h_s[n]$ is placed before $h_2[n]$ as shown in Fig. P10.20(a). The impulse response $h_s[n]$ is shown in Fig. P10.20(b).

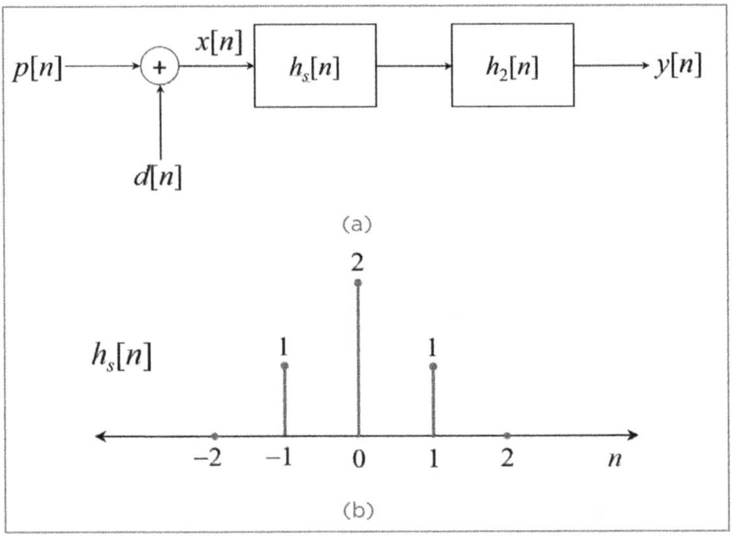

Figure P10.20.

Sketch the output of the combined system for the same input signal as in Question 10.19(c).

10.20. A MP3 player implements a simple tone control function by passing an audio signal, $x[n]$, into a linear filter with filter function $H_1(z) = \frac{1}{(1-0.9z^{-1})}$ to produce $y_1[n]$. Subsequently, this $y_1[n]$ is passed through a second linear filter, $H_2(z) = (1 - z^{-1})$, to produce $y_2[n]$. By sketching their magnitude responses, determine whether the two outputs will enhance the bass or treble of the input.

10.21. Consider the five pole-zero diagrams in Fig. P10.21. Each plot corresponds to a discrete-time linear time-invariant (LTI) system function, whose unit sample response is real. Each plot is drawn to scale. (*Note that*

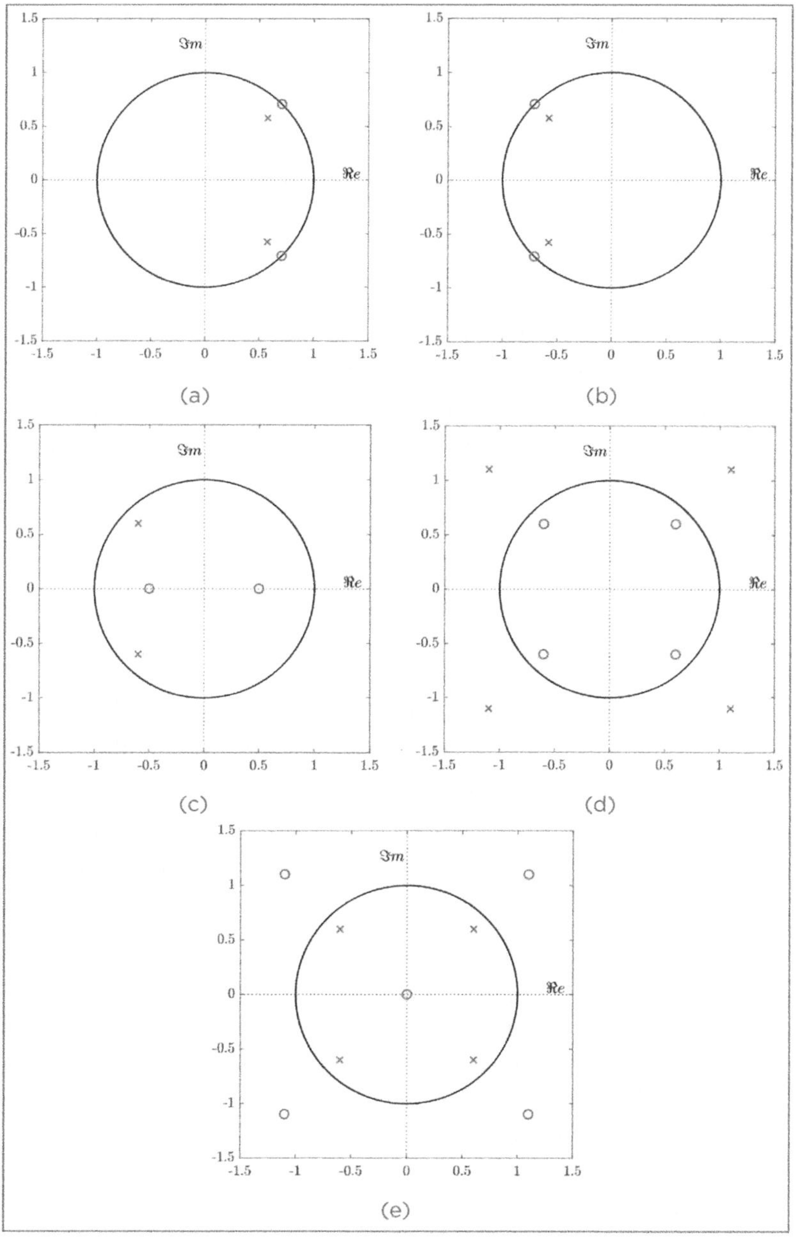

Figure P10.21. Pole-zero diagrams.

you have all the information to solve the equations in this problem, although some of the poles and zeros are not labelled).

(a) Which plot(s) can have a region of convergence (ROC) so that it corresponds to a causal and stable system? Explain.

(b) Consider the block diagram shown in Fig. P10.22. $H(z)$ is described by one or more of the pole-zero plots $A - E$ from Fig. P10.21. $G(z)$, which does not correspond to any of the pole-zero plots $A - E$ is a system such that $w[n] = x[n]$.

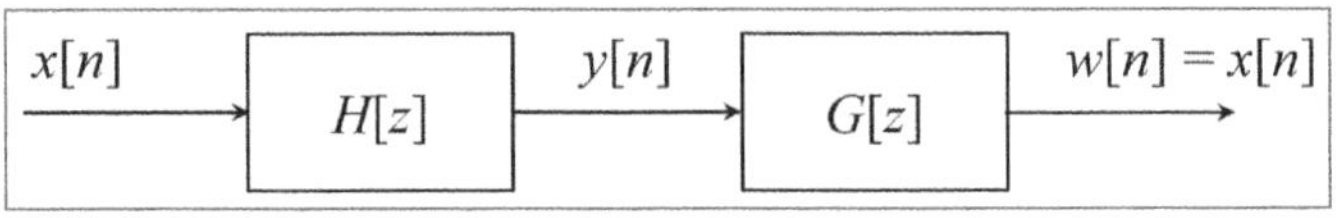

Figure P10.22. Cascaded systems.

(c) Which plot(s) correspond to $H(z)$ such that both $H(z)$ and $G(z)$ are causal and stable? Explain.

10.22. The block schematic of a LTI digital system is shown in Figure P10.23. Here $x[n]$ and $y[n]$ are the input and output, respectively of the system and $h[n]$ is the system impulse response. The system is described by the difference equation, $y[n] = x[n] - ax[n-1]$. Here, a is a real scalar variable.

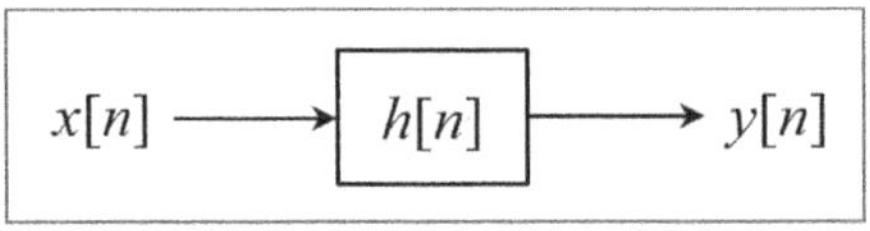

Figure P10.23.

(a) Determine the system function, $H(z) = \frac{Y(z)}{X(z)}$ and hence derive its impulse response, $h[n]$. Here, $X(z)$ and $Y(z)$ are the $z-$transforms of $x[n]$ and $y[n]$, respectively.

10.23. Figure P10.24 below shows the block diagram (arranged in parallel structure) of a causal LTI digital system with one input, $x[n]$, and one output, $y[n]$. In the figure, z^{-1} represents a unit sample delay.

(a) Derive the transfer function, $H(z) = \frac{Y(z)}{X(z)}$, for the given system. Find *also* its impulse response, $h[n]$ (**Hint**: Use the rule in finding the overall system function, $H(z)$ of systems connected in parallel).

(b) Draw the transposed direct form II structure of the system and verify that both the original and the transposed system have the same transfer function.

(c) Draw the pole-zero diagram of the system and comment on its stability.

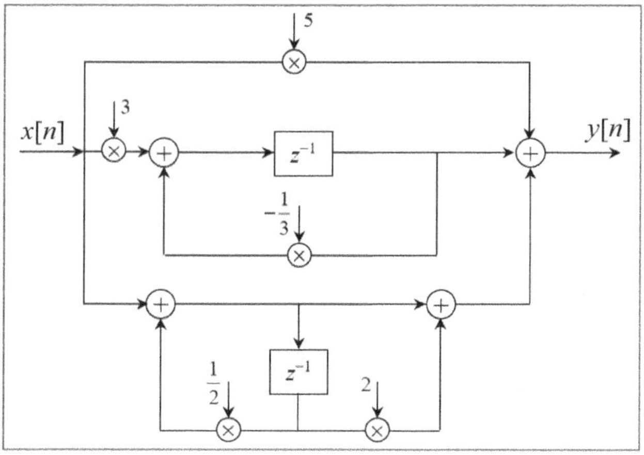

Figure P10.24.

10.24. The z−plane pole-zero diagram for a certain digital system is shown in Fig. P10.25. The system has unity gain at DC (i.e., $H(z) = 1$ at $z = 1$).

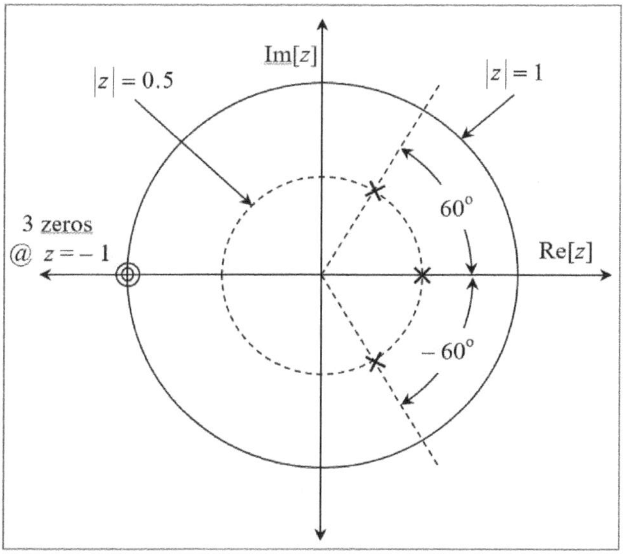

Figure P10.25.

(a) Determine the system function, $H(z)$ in the form

$$H(z) = A\left[\frac{\left(1 + a_1 z^{-1}\right)\left(1 + b_1 z^{-1} + b_2 z^{-2}\right)}{\left(1 + c_1 z^{-1}\right)\left(1 + d_1 z^{-1} + d_2 z^{-2}\right)}\right],$$

giving numerical values for the parameters $A, a_1, b_1, b_2, c_1, d_1$, and d_2 (Please note that the parameter, A in $H(z)$ is to set the DC gain of $H(z)$ to unity).

(b) Draw the direct form II (canonic form) structure of the complete system.

(c) Draw the cascade form structure of the system. Make each section canonic, with real coefficients. Also, make sure that the maximum order of the sections in cascade form to be two (i.e., 2nd order systems) (**Hint:** Consider $H(z)$ in Question 3.1 as, $H(z) = AH_1(z)H_2(z)$, with the numerical values of the parameters substituted. Here, $H_1(z)$ is a first order recursive system and $H_2(z)$ is a second order recursive system).

10.25. Consider the implementation of a causal linear time-invariant (LTI) recursive system as shown in Fig. P10.26. Here $x[n]$ and $y[n]$ are the input and output, respectively of the system. In the figure, the symbol z^{-1} denotes a unit sample delay and Q is a positive number ≤ 1 (i.e., $0 < Q \leq 1$). The system is assumed to be "at rest", i.e., $y[n] = 0$ for $n < 0$.

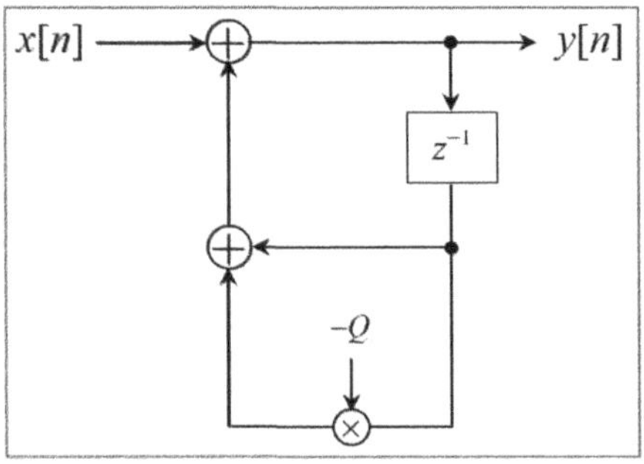

Figure P10.26.

(a) Show that the output $y[n] = D/Q$ for large values of the time index n when the input $x[n] = D$, where D is a positive constant. (**Hint**: Find $y[n]$ for $n = 0, 1, 2, \ldots$, etc., and develop a general series expression for $y[n]$, in terms of D, Q, and n. The requested solution can be derived from this.)

(b) Find the impulse response, $h[n] = IZT\{H(z)\}$, of the system and prove that the system in Figure P1 is stable if Q is in the range $0 < Q \leq 1$. IZT here refers to the inverse z−transform and $H(z)$ is the system transfer function.

(c) Determine the frequency response, $H(f)$, of the system in Figure P1 and show that its DC (0 Hz) gain, $H(0)$, is in the range $1 \leq H(0) < \infty$ for a given value of Q.

10.26. Figure P10.27 shows the block diagram of a causal LTI digital system with one input, $x[n]$, and one output, $y[n]$. In the figure, z^{-1} represents a unit sample delay.

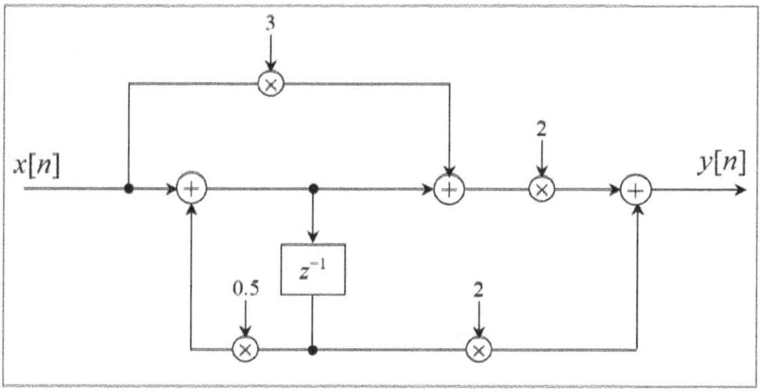

Figure P10.27.

(a) Derive the transfer function, $H(z)$, for the given system, sketch the pole-zero distribution in the z-domain and comment on its stability.
(b) Draw the direct form-I implementation structure of the given system.
(c) Reconfigure the system implementation structure in the standard canonic (direct form-II) form and draw it.
(d) Determine the system response, $y[n]$, to an input $x[n] = \left(\frac{3}{4}\right)^n u[n]$, where $u[n]$ is the discrete-time unit step function defined as,

$$u[n] = \begin{cases} 1 & \text{for } n \geq 0 \\ 0 & \text{for } n < 0. \end{cases}$$

10.27. Figure P10.28 shows the complete block diagram representation of a LTI system. Here X(z) and Y(z) are the $z-$transforms of the input ($x[n]$) and output ($y[n]$), respectively of the system. In the figure, z^{-1} represents a unit sample delay and K, α, β, and γ are scalar constants. W_1, W_2 and W_3 in the figure are temporary (intermediate) variables.

In the following, the value of γ is chosen to be -1.

(a) Determine the system function $H(z) = \frac{Y(z)}{X(z)}$.
(b) Is the structure provided in Fig. P10.28 canonic? If not, draw the canonic structure of the given LTI system.

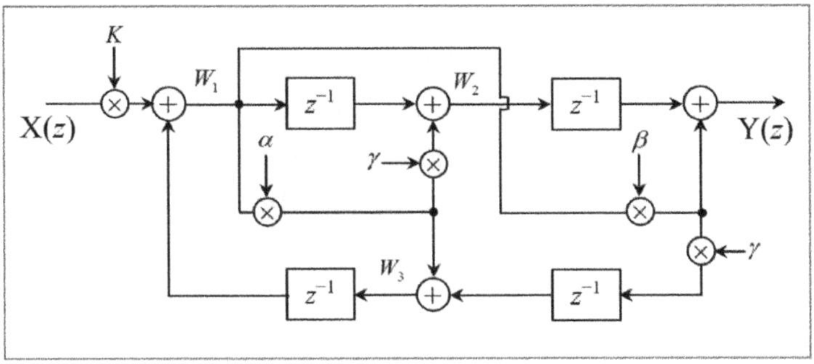

Figure P10.28.

(c) What should be the value of the multiplier coefficient K so that $H(z)$ has unity gain (i.e., all-pass) in all the frequencies.

(d) If the values of $\alpha = 1$, $\beta = 0.5$, $\gamma = -1$ and K is chosen to be the one obtained from Question (c), sketch the pole-zero distribution in the z-domain and comment on the stability of the system.

10.28. A researcher uses the derivative operator (filter) specified as $w[n] = x[n] - x[n-1]$, where $x[n]$ is the input and $w[n]$ is the output of the filter. The result is then passed through the moving-average (MA) filter given by $y[n] = \frac{1}{3}[w[n] + w[n-1] + w[n-2]]$, where $y[n]$ is the final output desired.

(a) Derive the transfer functions (in the $z-$domain) of the two filters individually as well as that of the combination.

(b) Derive the impulse response of each filter and that of the combination. Plot the three signals.

(c) Does it matter which of the two filters is placed first? Why (not)?

(d) The signal described by the samples $x[n] = \{\ldots, 0, 0, 0, 6, 6, 6, 6, 6, 6, 6, 6, 0, 0, \ldots\}$ is applied to the system. Here the first non-zero input corresponds to $n = 0$. Derive the values of the final output signal.

DOI: 10.1561/9781680839791.ch11

Chapter 11

Implementation of Digital Systems

This chapter focuses on the realization of LTI discrete-time systems either in software or hardware. Various structures for the realizations of FIR and IIR discrete-time systems are discussed in detail. Some of the important structures are the *cascade, parallel,* and *lattice structures*. A computationally efficient frequency-sampling realization of FIR system is also included in the discussion.

11.1 Structures for the Realization of Discrete-Time Systems

The important class of LTI discrete-time systems characterized by the general constant coefficient difference equation is considered here.

$$y[n] = -\sum_{k=1}^{N} a_k y[n-k] + \sum_{k=0}^{M-1} b_k x[n-k]. \tag{11.1}$$

By taking the $z-$transform of Eq. (11.1), the LTI system can also be characterized by the rational system function as,

$$H(z) = \frac{\sum_{k=0}^{M-1} b_k z^{-k}}{1 + \sum_{k=1}^{N} a_k z^{-k}}, \tag{11.2}$$

which is a ratio of two polynomials in z^{-1}. From Eq. (11.2), we obtain the *zeros* and *poles* of the system function, which depend on the choice of the system parameters $\{a_k\}$ and $\{b_k\}$. These parameters also determine the frequency response characteristics of the system. In this chapter, we shall show that Eq. (11.1) and Eq. (11.2) can be implemented in different ways depending on the form in which these two characterizations are arranged. The major factors that influence the choice of specific realizations are the *computational complexity*, *memory requirements*, and *finite-word-length effects* in computations.

The implementation of digital systems (both FIR and IIR types) require three basic operations: (i) multiplication, (ii) addition, and (iii) signal delaying.

The structures for FIR systems are described first followed by the structures for IIR systems. Alternative structures for both FIR and IIR systems are discussed towards the end of this chapter.

11.2 Structures for FIR Systems

The FIR system is described by the difference equation

$$y[n] = \sum_{k=0}^{M-1} b_k x[n-k], \tag{11.3}$$

or, equivalently the system function

$$H(z) = \sum_{k=0}^{M-1} b_k z^{-k}. \tag{11.4}$$

The unit sample response (impulse response) of the FIR system is identical to the coefficients $\{b_k\}$, that is,

$$h[n] = \begin{cases} b_n, & \text{for } 0 \le n \le M-1, \\ 0 & \text{Otherwise} \end{cases} \tag{11.5}$$

The implementation/realization schemes[1] discussed in the following sections are:

- Direct form realization,
- Cascade form realization,

1. The FIR system can be realized by the discrete Fourier transform (DFT) method as well.

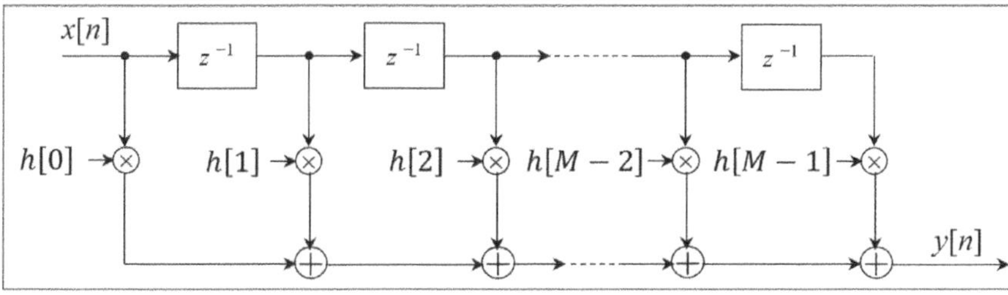

Figure 11.1. Direct-form realization of FIR system.

- Frequency-sampling realization, and
- Lattice realization

11.2.1 Direct-Form Structure

This structure (illustrated in Fig. 11.1) follows immediately from the non-recursive difference equation (Eq. (11.3)) or, equivalently by the convolution summation

$$y[n] = \sum_{k=0}^{M-1} h[k]x[n-k]. \tag{11.6}$$

It may be noted that the direst form structure requires $M-1$ memory locations, M multiplications and $M-1$ additions per each output point. It is also known as *transversal* or *tapped-delay-line* (TDL) structure.

When the FIR system is linear-phase, then $h[n] = \pm h[M-1-n]$. If $h[n] = +h[M-1-n]$, it is called an *even symmetric linear phase FIR system* and if $h[n] = -h[M-1-n]$, it is called an *odd symmetric linear phase FIR system.*

Expanding Eq. (11.6), we get the difference equation relating the input $x[n]$ and output $y[n]$ of the FIR system,

$$y[n] = h[0]x[n] + h[1]x[n-1] + \ldots + h[M-1]x[n-M-1]. \tag{11.7}$$

For an even symmetric linear phase FIR system, the summation in Eq. (11.7) can be written as,

$$y[n] = h[0]\,(x[n] + x[n-M-1]) + h[1]\,(x[n-1] + x[n-M-2]) + \ldots. \tag{11.8}$$

The direct form realization/implementation of the system in Eq. (11.8) is shown in Fig. 11.2, where the order of the filter M is odd. It can be seen that the number of multiplications is reduced to $M/2$ when M even and $(M-1)/2$ when M odd. The number of delay (memory) units remain the same in both the structures.

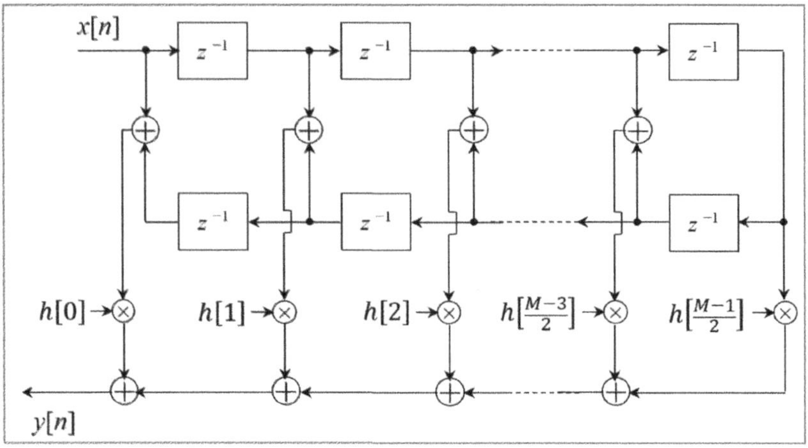

Figure 11.2. Direct-form realization of linear-phase FIR system when M Odd.

11.2.2 Cascade-Form Structure

This realization follows from the system function in Eq. (11.2). The system function is arranged in such a way that it becomes the product of several 2nd order FIR systems as,

$$H(z) = \prod_{k=1}^{K} H_k(z), \tag{11.9}$$

where

$$H_k(z) = b_{k0} + b_{k1}z^{-1} + b_{k2}z^{-2}, \quad k = 0, 1 \ldots, K \tag{11.10}$$

and K is the integer part of $(M+1)/2$. The filter parameter b_0 may be equally distributed among the K filter sections, such that $b_0 = b_{10}b_{20}\ldots b_{K0}$ or it may be assigned to a single filter section. The zeros of $H(z)$ are grouped in pairs to produce the second order FIR systems of the form in Eq. (11.10). It is always desirable to form pairs of complex-conjugate roots so that the coefficients $\{b_{ki}\}$ are real valued. If the roots are real-valued, they can be grouped in any manner. The block diagram representation of the cascade form realization (Fig. 11.3(a)) along with the basic second-order section (Fig. 11.3(b)) are shown in Fig. 11.3. The second-order section is the implementation of Eq. (11.10).

In the case of linear-phase FIR filters, the symmetry in $h[n]$ implies that the zeros of $H(z)$ also exhibit a form of symmetry. In particular, if z_k and z_k^* are a pair of complex-conjugate zeros, then $\frac{1}{z_k}$ and $\frac{1}{z_k^*}$ are also a pair of complex conjugate zeros (see Fig. 11.4). Consequently, we gain some simplifications by forming fourth

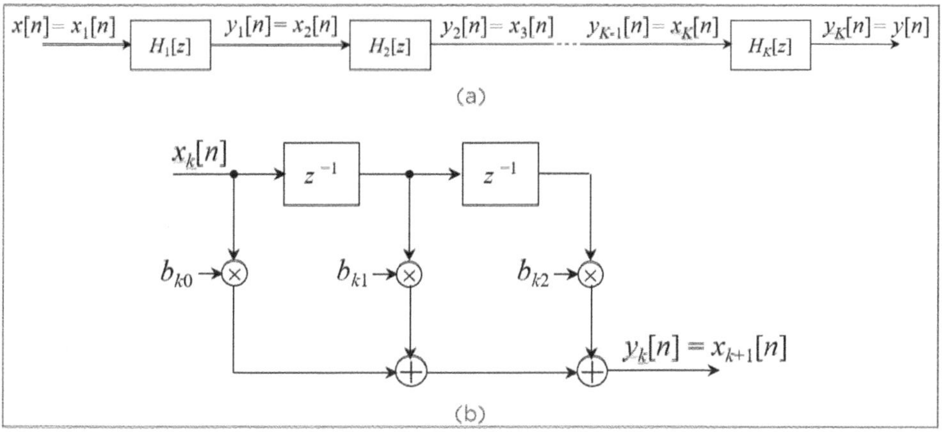

Figure 11.3. Cascade-form realization of a FIR system (a) and realization of each second order section (b).

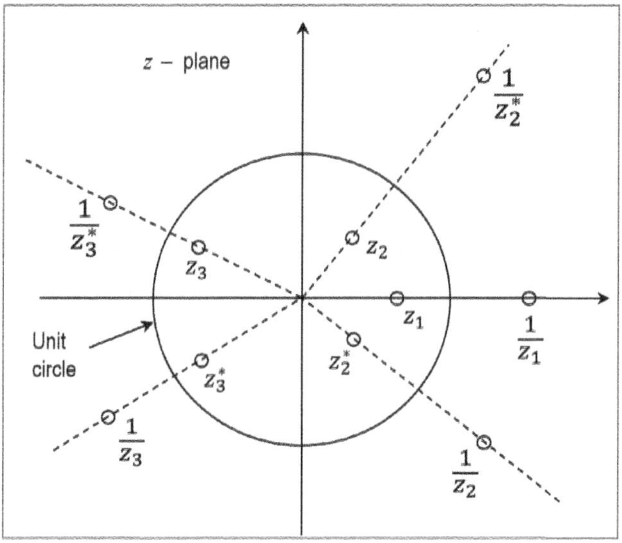

Figure 11.4. Symmetry of zero locations for a linear-phase FIR filter.

order sections of the FIR system

$$
\begin{aligned}
H_k(z) &= c_{k0}(1 - z_k z^{-1})(1 - z_k^* z^{-1})\left(1 - \frac{1}{z_k} z^{-1}\right)\left(1 - \frac{1}{z_k^*} z^{-1}\right), \\
&= c_{k0} + c_{k1} z^{-1} + c_{k2} z^{-2} + c_{k1} z^{-3} + c_{k0} z^{-4}, \\
&= c_{k0}(1 + z^{-4}) + c_{k1}(z^{-1} + z^{-3}) + c_{k2} z^{-2}, \qquad (11.11)
\end{aligned}
$$

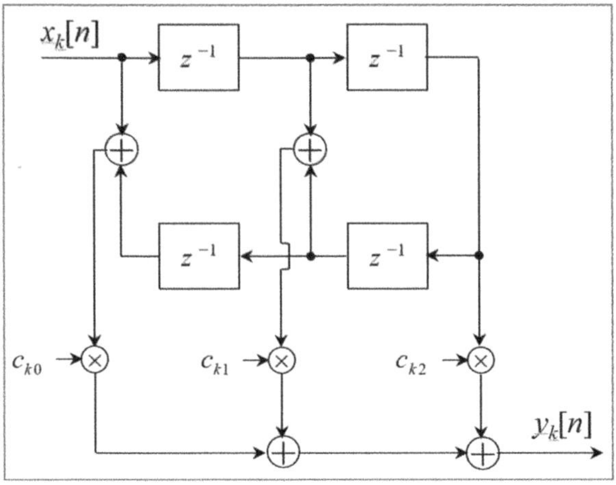

Figure 11.5. 4th-order section in cascade-form realization of an FIR system.

where the coefficients $\{c_{k1}\}$ and $\{c_{k2}\}$ are functions of z_k. Thus, by combining two pairs of zeros to form a fourth order filter section, we have reduced the number of multiplications from 6 (for two second-order systems) to 3. Figure 11.5 illustrates the basic fourth-order FIR filter structure (implementation of Eq. (11.11)).

11.2.3 Frequency-Sampling Structure

This is an alternative structure for a FIR filter in which the parameters that characterize the filter are the values of the desired frequency response instead of the impulse response $h[n]$. To derive the frequency sampling structure, we specify the desired frequency response at a set of equally spaced frequencies, namely,

$$\omega_k = \frac{2\pi k}{M}, \begin{cases} k = 0, 1, \ldots, \frac{M-1}{2}, & M \text{ odd}, \\ k = 0, 1, \ldots, \frac{M}{2} - 1, & M \text{ even} \end{cases} \tag{11.12}$$

and solve for the impulse response $h[n]$ of the FIR filter. Here M is the number of frequency points on the unit circle. Figure 11.6 illustrates the frequency sampling procedure, where $M = 12$.

Once $h[n]$ is obtained, we can write the frequency response as

$$H(\omega) = \sum_{n=0}^{M-1} h[n]e^{-j\omega n} \tag{11.13}$$

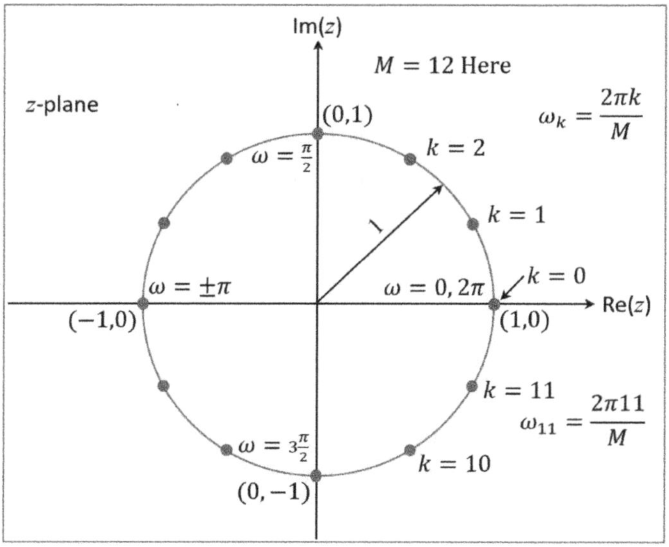

Figure 11.6. Illustration of frequency sampling.

and the values of $H(\omega)$ at frequencies $\omega_k = \frac{2\pi k}{M}$ are simply (for $k = 0, 1, \ldots, M-1$),

$$H[k] = H\left(\frac{2\pi k}{M}\right) = \sum_{n=0}^{M-1} h[n] e^{-j\frac{2\pi k}{M}n}. \tag{11.14}$$

The set of values $\{H[k]\}$ are called the frequency samples of $H(\omega)$ and it corresponds to the $M-$point DFT of $\{h[n]\}$.

By inverting Eq. (11.14), $h[n]$ can be expressed in terms of the frequency samples as

$$h[n] = \frac{1}{M} \sum_{k=0}^{M-1} H[k] e^{j\frac{2\pi k}{M}n}, \quad n = 0, 1, \ldots, M-1. \tag{11.15}$$

As can be seen from Eq. (11.15), it is the IDFT of the sequence $\{H[k]\}$. Now, we use this $h[n]$ to find its $z-$transform and we have,

$$H(z) = \sum_{n=0}^{M-1} h[n] z^{-n} = \sum_{n=0}^{M-1} \left[\frac{1}{M} \sum_{k=0}^{M-1} H[k] e^{j\frac{2\pi k}{M}n}\right] z^{-n}. \tag{11.16}$$

By interchanging the order of the two summations in Eq. (11.16) and performing the summation over the index n, we get

$$H(z) = \sum_{k=0}^{M-1} H[k] \left[\frac{1}{M} \sum_{n=0}^{M-1} \left(e^{j\frac{2\pi k}{M}} z^{-1} \right)^n \right]$$

$$= \frac{1 - z^{-M}}{M} \sum_{k=0}^{M-1} \frac{H[k]}{1 - e^{j\frac{2\pi k}{M}} z^{-1}}. \tag{11.17}$$

Thus, the system function $H(z)$ is characterized by the set of frequency samples $\{H[k]\}$ instead of $\{h[n]\}$.

We can view this FIR filter realization as a cascade of two filters. That is, $H(z) = H_1(z)H_2(z)$. Here, $H_1(z)$ is an all-zero filter with the system function

$$H_1(z) = \frac{1 - z^{-M}}{M}. \tag{11.18}$$

Its zeros are located at equally spaced points on the unit circle at $z_k = e^{j\frac{2\pi k}{M}}$, $k = 0, 1, \ldots, M-1$.

The second filter $H_2(z)$ with the system function

$$H_2(z) = \sum_{k=0}^{M-1} \frac{H[k]}{1 - e^{j\frac{2\pi k}{M}} z^{-1}} \tag{11.19}$$

consists of a parallel bank of M single-pole filters with resonant frequencies (pole locations), $p_k = e^{j\frac{2\pi k}{M}}$, $k = 0, 1, \ldots, M-1$.

Note that the pole locations and zero locations are same on the unit circle and they occur at $\omega_k = \frac{2\pi k}{M}$, which are the frequencies at which the desired frequency response is specified. This cascade realization is illustrated in Fig. 11.7. The implementation of the two sub-systems ($H_1(z)$ and $H_2(z)$) are clearly marked in the figure.

11.2.4 Lattice Structure for FIR Systems

This structure is used extensively in digital speech processing and in the implementation of the adaptive filters because of its modular structure. Let us begin the development of lattice structure by considering a sequence of FIR filters with system functions

$$H_m(z) = A_m(z), \quad m = 0, 1, \ldots, M-1 \tag{11.20}$$

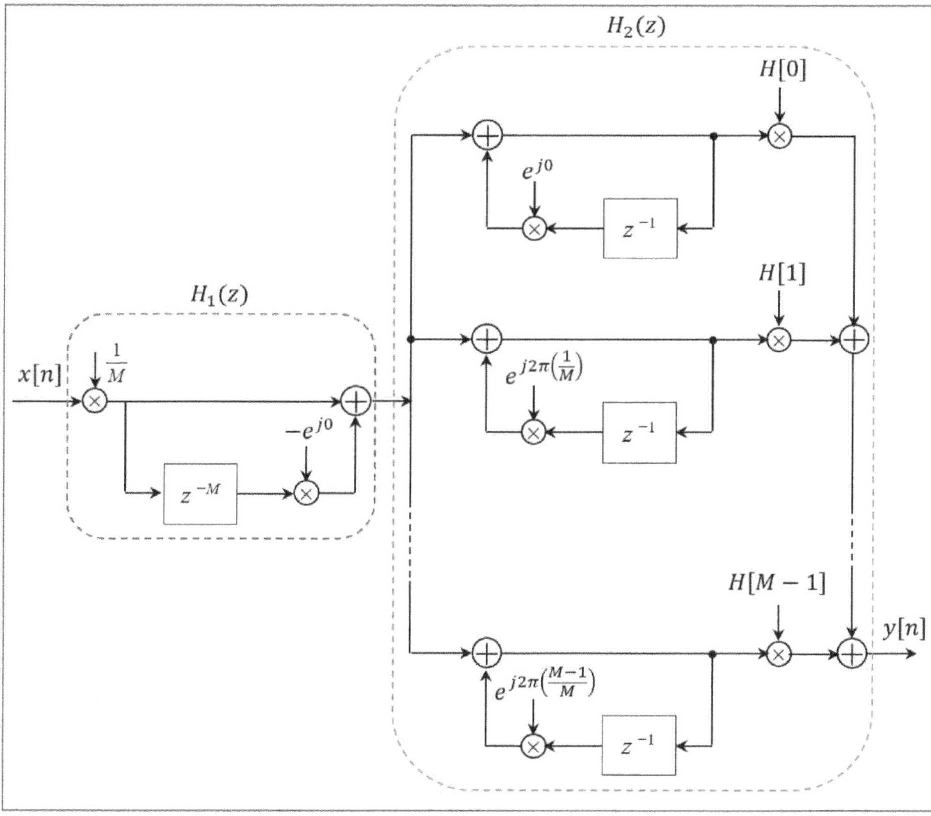

Figure 11.7. Frequency-sampling realization of an FIR system.

where, by definition $A_m(z)$ is a polynomial of the form,

$$A_m(z) = 1 + \sum_{k=1}^{m} \alpha_m[k] z^{-k}, \quad m \geq 1, \tag{11.21}$$

and $A_0(z) = 1$. The unit sample response of the mth filter is $h_m[0] = 1$ and $h_m[k] = \alpha_m[k], \ \ k = 1, 2, \ldots, m$. The subscript m on the polynomial $A_m(z)$ denotes the degree of the polynomial. For mathematical convenience, we define $\alpha_m[0] = 1$.

If $x[n]$ is the input sequence to the filter $A_m(z)$ and $y[n]$ is the output sequence, then

$$y[n] = x[n] + \sum_{k=1}^{m} \alpha_m[k] x[n-k]. \tag{11.22}$$

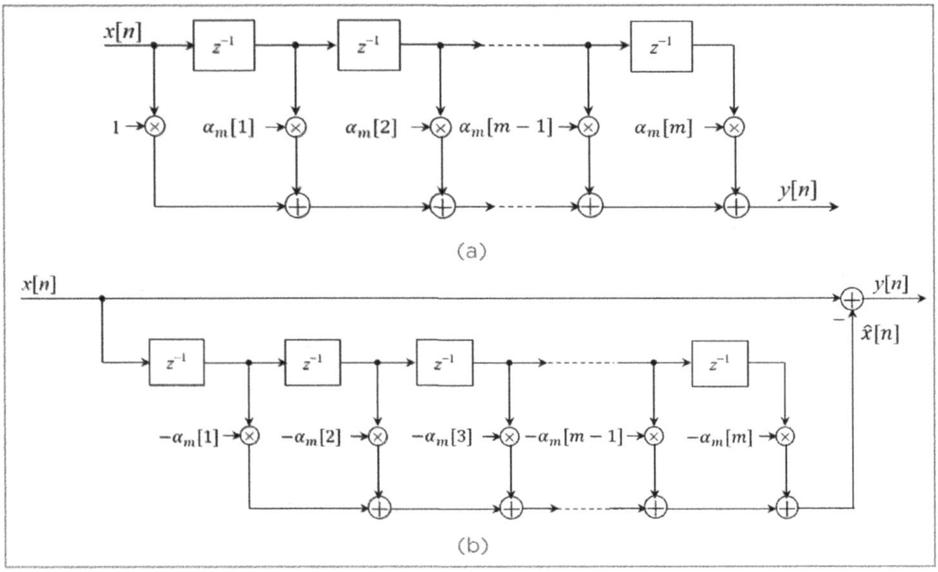

Figure 11.8. Two direct-form structures of an FIR system (prediction filter).

Two direct-form realizations of this FIR filter are shown in Fig. 11.8. The structures shown in Fig. 11.8 are related to the topic of linear prediction,[2] where

$$\hat{x}[n] = -\sum_{k=1}^{m} \alpha_m[k]x[n-k] \tag{11.23}$$

is the one-step forward predicted value of $x[n]$, using the m past inputs, and $y[n] = x[n] - \hat{x}[n]$, given by Eq. (11.22), represents the prediction error sequence. In this context, the top filter structure (Fig. 11.8(a)) is called the *prediction error filter.*

Now, suppose that we have a filter of order $m = 1$. The output of such a filter is

$$y[n] = x[n] + \alpha_1[1]x[n-1]. \tag{11.24}$$

This output can also be obtained from the first order or single-stage lattice filter, illustrated in Fig. 11.9, by exciting both of the inputs by $x[n]$ and selecting the output from the top branch. Thus, the output is exactly Eq. (11.24), if we select $K_1 = \alpha_1[1]$. The parameter K_1 in the lattice filter is called the *reflection coefficient.*

The different parameters of the above scheme are:

$$\begin{aligned} f_0[n] &= g_0[n] = x[n] \\ f_1[n] &= f_0[n] + K_1 g_0[n-1] = x[n] + K_1 x[n-1] \\ g_1[n] &= K_1 f_0[n] + g_0[n-1] = K_1 x[n] + x[n-1] \end{aligned} \tag{11.25}$$

2. J. Makhoul, "Linear prediction: A tutorial review", *Proceedings of the IEEE,* vol.63, No.5, pp.561–580, 1975.

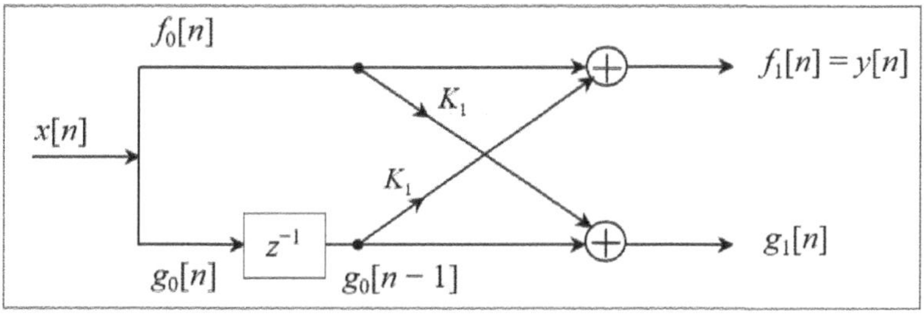

Figure 11.9. Single-stage lattice filter.

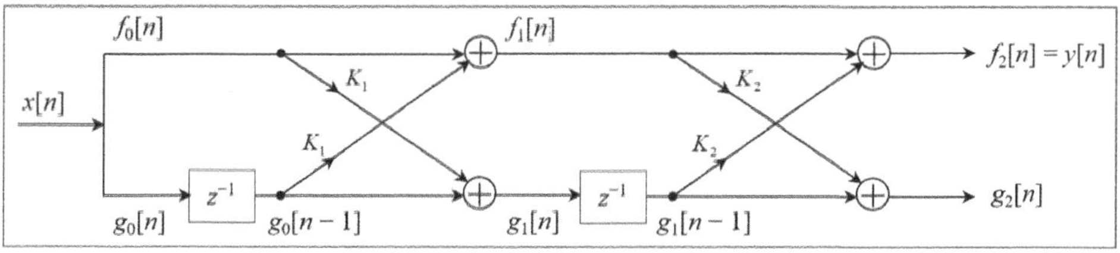

Figure 11.10. Two-stage lattice filter.

Now, consider a FIR filter for which $m = 2$. The output for this case from the direct form structure (Eq. (11.22)) is

$$y[n] = x[n] + \alpha_2[1]x[n-1] + \alpha_2[2]x[n-2]. \tag{11.26}$$

By cascading two single-stage lattice structures as shown in Fig. 11.10, it is possible to obtain the same output as in Eq. (11.26). Indeed, the output from the first stage is,

$$f_1[n] = x[n] + K_1 x[n-1], \tag{11.27}$$

$$g_1[n] = K_1 x[n] + x[n-1]. \tag{11.28}$$

The output from the second stage is,

$$f_2[n] = f_1[n] + K_2 g_1[n-1], \tag{11.29}$$

$$g_2[n] = K_2 f_1[n] + g_1[n-1]. \tag{11.30}$$

By substituting for $f_1[n]$ and $g_1[n]$, we can show that

$$f_2[n] = x[n] + K_1(1 + K_2)x[n-1] + K_2 x[n-2]. \tag{11.31}$$

This equation is identical to the output of the direct-form FIR filter as given by Eq. (11.26), if we equate the coefficients as

$$\alpha_2[2] = K_2, \quad \alpha_2[1] = K_1(1 + K_2) \tag{11.32}$$

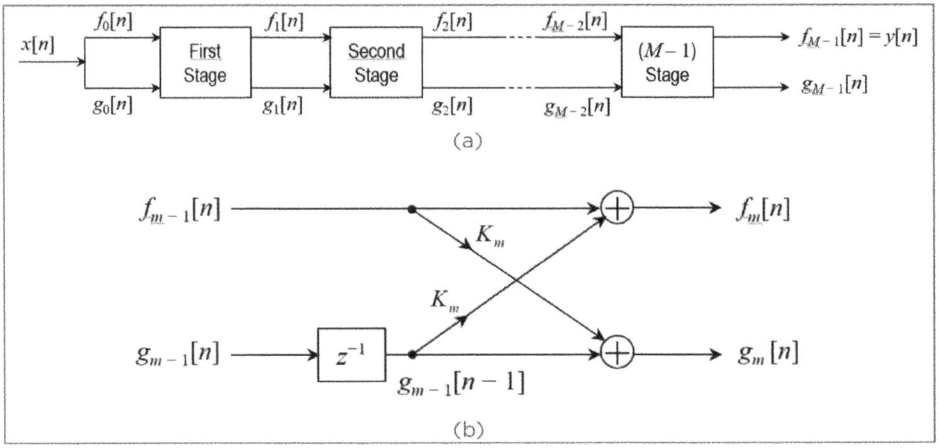

Figure 11.11. $(M-1)$-stage lattice filter.

or, equivalently,

$$K_2 = \alpha_2[2], \quad K_1 = \frac{\alpha_2[1]}{(1+\alpha_2[2])} \tag{11.33}$$

Thus, the reflection coefficients K_1 and K_2 of the lattice filter can be obtained from the coefficients $\{\alpha_m[k]\}$ of the direct-form realization.

Continuing like this, by induction, we can show the equivalence between an mth order direct-form FIR filter and an m-order of m-stage lattice filter. The lattice filter is generally described by the following set of order-recursive equations (for $m = 1, 2, \ldots, M-1$),

$$f_0[n] = g_0[n] = x[n], \tag{11.34}$$

$$f_m[n] = f_{m-1}[n] + K_m g_{m-1}[n-1], \tag{11.35}$$

$$g_m[n] = K_m f_{m-1}[n] + g_{m-1}[n-1]. \tag{11.36}$$

Then the output of the $(M-1)$-stage lattice filter corresponds to the output of a $(M-1)$-order FIR filter, that is,

$$y[n] = f_{M-1}[n]. \tag{11.37}$$

Figure 11.11(a) shows a $(M-1)$-stage lattice filter in block diagram and Fig. 11.11(b) shows one of the stages in the block diagram, where the computations are specified by Eq. (11.35) and Eq. (11.36).

As a consequence of the equivalence between an FIR filter and a lattice filter, the output $f_m[n]$ of an $m-$stage filter can be expressed as

$$f_m[n] = \sum_{k=0}^{m} \alpha_m[k]x[n-k], \quad \alpha_m[0] = 1 \tag{11.38}$$

Since Eq. (11.38) is a convolution sum, the z-transform relationship is

$$F_m(z) = A_m(z)X(z) \tag{11.39}$$

or

$$A_m(z) = \frac{F_m(z)}{X(z)} = \frac{F_m(z)}{F_0(z)} \tag{11.40}$$

The other output component from the lattice, namely, $g_m[n]$, can also be expressed in the form of a convolution sum as in Eq. (11.38), by choosing another set of coefficients $\{\beta_m[k]\}$.

From Eq. (11.28), we note that the filter coefficients for the lattice filter that produces $f_1[n]$ are $\{1, K_1\} = \{1, \alpha_1[1]\}$ while the coefficients for the filter with output $g_1[n]$ are $\{K_1, 1\} = \{\alpha_1[1], 1\}$. That means they are in *reverse order.* If we consider the two-stage filter, with the output given by Eq. (11.30), we find that $g_2[n]$ can be expressed in the form,

$$g_2[n] = \alpha_2[2]x[n] + \alpha_2[1]x[n-1] + x[n-2], \tag{11.41}$$

where, the filter coefficients are $\{\alpha_2[2], \alpha_2[1], 1\}$ and the coefficients of the filter that produces the output $f_2[n]$ are $\{1, \alpha_2[1], \alpha_2[2]\}$. Here again, the filter coefficients are in reverse order. From this development, for an m-stage lattice filter, we have

$$g_m[n] = \sum_{k=0}^{m} \beta_m[k]x[n-k]. \tag{11.42}$$

where the filter coefficients

$$\beta_m[k] = \alpha_m[m-k], \quad k = 0, 1, \ldots, m \tag{11.43}$$

with $\beta_m[m] = 1$.

In the context of linear prediction, let us assume that the data $x[n], x[n-1], \ldots, x[n-m+1]$ is used to linearly predict the signal value $x[n-m]$ by the use of a linear filter with coefficients $\{-\beta_m[k]\}$. Then the predicted value is

$$\hat{x}[n-m] = -\sum_{k=0}^{m-1} \beta_m[k]x[n-k]. \tag{11.44}$$

Since the data run in reverse order through the predictor, the prediction performed in Eq. (11.44) is called the *backward prediction.* In contrast, the FIR filter with system function $A_m(z)$ is called a *forward predictor.*

In the z-transform domain, Eq. (11.42) becomes

$$G_m(z) = B_m(z)X(z) \tag{11.45}$$

or,

$$B_m(z) = \frac{G_m(z)}{X(z)} \tag{11.46}$$

where $B_m(z)$ represents the system function of the FIR filter with coefficients $\{\beta_m[k]\}$. That is,

$$B_m(z) = \sum_{k=0}^{m} \beta_m[k]z^{-k} \tag{11.47}$$

Since $\beta_m[n] = \alpha_m[m-k]$, Eq. (11.47) may be expressed as

$$\begin{aligned} B_m(z) &= \sum_{k=0}^{m} \alpha_m[m-k]z^{-k} \\ &= \sum_{l=0}^{m} \alpha_m[l]z^{l-m} \\ &= z^{-m}\sum_{l=0}^{m} \alpha_m[l]z^{l} \\ &= z^{-m}A_m(z^{-1}) \end{aligned} \tag{11.48}$$

This relationship implies that the zeros of the FIR filter with system function $B_m(z)$ are simply the reciprocals of the zeros of $A_m(z)$. Hence $B_m(z)$ is called the reciprocal or *reverse* polynomial of $A_m(z)$.

Now that we have established this interesting relationships between the direct-form FIR filter and the lattice structure, let us return to the recursive lattice equations (11.34) through (11.36) and transfer them to the z-domain. Thus we have (for $m = 1, 2, \ldots, M-1$),

$$F_0(z) = G_0(z) = X(z) \tag{11.49}$$

$$F_m(z) = F_{m-1}(z) + K_m z^{-1}G_{m-1}(z), \tag{11.50}$$

$$G_m(z) = K_m F_{m-1}(z) + z^{-1}G_{m-1}(z). \tag{11.51}$$

If we divide each equation by $X(z)$, we get the desired results in the form (for $m = 1, 2, \ldots, M-1$),

$$A_0(z) = B_0(z) = 1, \tag{11.52}$$

$$A_m(z) = A_{m-1}(z) + K_m z^{-1} B_{m-1}(z), \tag{11.53}$$

$$B_m(z) = K_m A_{m-1}(z) + z^{-1} B_{m-1}(z). \tag{11.54}$$

In the matrix form,

$$\begin{bmatrix} A_m(z) \\ B_m(z) \end{bmatrix} = \begin{bmatrix} 1 & K_m \\ K_m & 1 \end{bmatrix} \begin{bmatrix} A_{m-1}(z) \\ z^{-1} B_{m-1}(z) \end{bmatrix}. \tag{11.55}$$

Conversion of lattice coefficients to direct-form filter coefficients: The direct form FIR filter coefficients $\{\alpha_m[k]\}$ can be obtained from the lattice coefficients $\{K_i\}$ by using the following relations (for $m = 1, 2, \ldots, M-1$):

$$A_0(z) = B_0(z) = 1, \tag{11.56}$$

$$A_m(z) = A_{m-1}(z) + K_m z^{-1} B_{m-1}(z), \tag{11.57}$$

$$B_m(z) = z^{-m} A_m(z^{-1}). \tag{11.58}$$

The solution is obtained recursively, beginning with $m = 1$. Thus we obtain a sequence of $(M-1)$ FIR filters, one for each value of m.

Conversion of direct-form filter coefficients to lattice coefficients: Suppose that we are given the FIR filter coefficients for the direct-form realization or, equivalently, the polynomial $A_m(z)$, and we wish to determine the corresponding lattice filter parameters $\{K_i\}$. For the m-stage lattice, we immediately obtain the parameter $K_m = \alpha_m[m]$. To obtain K_{m-1}, we need the polynomials $A_{m-1}(z)$ since, in general, K_m is obtained form the polynomial $A_m(z)$ for $m = M-1, M-2, \ldots, 1$. Consequently, we need to compute the polynomials $A_m(z)$ starting from $m = M-1$ and "stepping down" successively to $m = 1$. The desired recursive relation for the polynomials is easily determined from Eq. (11.53) and Eq. (11.54) as,

$$\begin{aligned} A_m(z) &= A_{m-1}(z) + K_m z^{-1} B_{m-1}(z) \\ &= A_{m-1}(z) + K_m \left[B_m(z) - K_m A_{m-1}(z)\right]. \end{aligned} \tag{11.59}$$

If we solve for $A_{m-1}(z)$, we get,

$$A_{m-1}(z) = \frac{A_m(z) - K_m B_m(z)}{1 - K_m^2}, \quad m = M-1, M-2, \ldots, 1. \tag{11.60}$$

Thus, we compute all lower-degree polynomials $A_m(z)$ beginning with $A_{M-1}(z)$ and obtain the desired lattice coefficients from the relation $K_m = \alpha_m[m]$. We observe that the procedure works only if $|K_m| \neq 1$ for $m = 1, 2, \ldots, M-1$.

Estimation of the Reflection coefficients: From the step-down recursive equation (Eq. (11.60)), it is easy to obtain a formula for recursively estimating the reflection coefficients, K_m, beginning with $m = M-1$ and stepping down to $m = 1$. For $m = M-1, M-2, \ldots, 1$, we have

$$K_m = \alpha_m[m], \quad \alpha_{m-1}[0] = 1 \tag{11.61}$$

$$\alpha_{m-1}[k] = \frac{\alpha_m[k] - K_m \beta_m[k]}{1 - K_m^2} \tag{11.62}$$

$$= \frac{\alpha_m[k] - \alpha_m[m]\alpha_m[m-k]}{1 - \alpha_m^2[m]}, \quad 1 \leq k \leq m-1. \tag{11.63}$$

As mentioned above, the recursive equation in (11.62) breaks down if any lattice parameters $|K_m| = 1$. If this occurs, it is indicative of the fact that the polynomial $A_{m-1}(z)$ has a root on the unit circle. Such a root can be factored out from $A_{m-1}(z)$ and the iterative process in Eq. (11.62) is carried out for the reduced order system.

11.3 Structure for IIR Systems

We consider four different IIR system structures described by the difference equation (Eq. (11.1)) or equivalently, by the system function (Eq. (11.2)). We are going to discuss the four main realization structures of the IIR systems, that are:

- Direct-form structures,
- Cascade-form structures,
- Parallel-form structures, and
- Lattice structures

11.3.1 Direct-Form Structures

The rational system function in Eq. (11.2) that characterizes an IIR system can be viewed as two systems in cascade, that is,

$$H(z) = H_1(z)H_2(z), \tag{11.64}$$

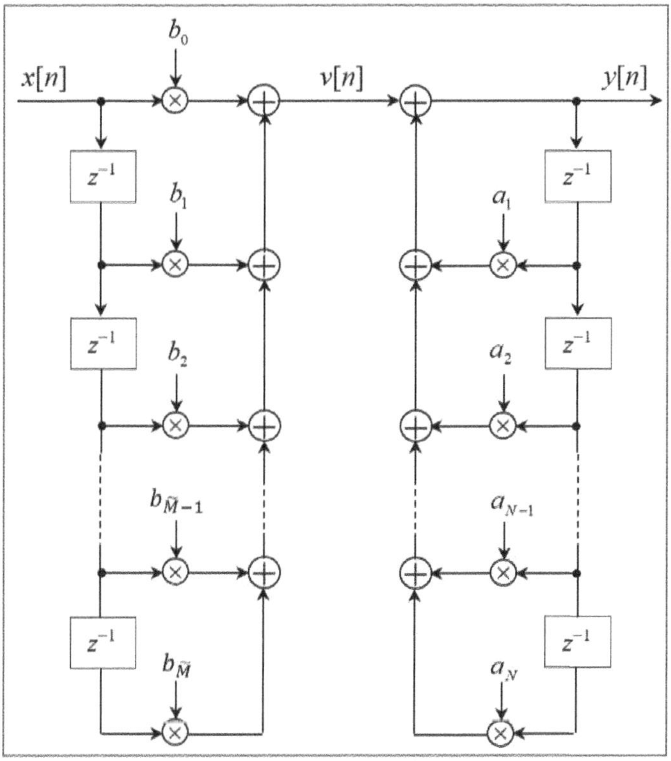

Figure 11.12. Direct Form I realization of IIR system.

where $H_1(z)$ is an all-zero system which consists of the zeros of $H(z)$, and $H_2(z)$ is an all-pole system consisting of the poles of $H(z)$. That is,

$$H_1(z) = \sum_{k=0}^{M-1} b_k z^{-k} = \sum_{k=0}^{\tilde{M}} b_k z^{-k}, \tag{11.65}$$

and

$$H_2(z) = \frac{1}{1 + \sum_{k=1}^{N} a_k z^{-k}}. \tag{11.66}$$

In Eq. (11.65), $M - 1$ in the summation is replaced with $\tilde{M}$ for notational simplicity in the subsequent sections. Since $H_1(z)$ is an FIR system, its direct-form realization was discussed already in the Section 11.2.1 (see Fig. 11.1). By attaching the second system in cascade with $H_1(z)$, we obtain the *Direct Form I* realization of the IIR system and it is depicted in Fig. 11.12. This realization requires $\tilde{M} + N + 1$ multiplications, $\tilde{M} + N$ additions, and $\tilde{M} + N$ memory locations.

If the all-pole filter ($H_2(z)$) is placed before the all-zero filter ($H_1(z)$), a more compact structure is obtained as explained below.

If we consider a 1st order system, Eq. (11.1), becomes

$$y[n] = -a_1 y[n-1] + b_0 x[n] + b_1 x[n-1], \tag{11.67}$$

which is realized as in Fig. 11.13(a) (Direct Form I structure). This realization uses separate delays (memory) for both the input and output signal samples. As we explained here, this system can be considered as the cascade of two LTI systems. The first system is a non-recursive system (all-zero) described by

$$v[n] = b_0 x[n] + b_1 x[n-1], \tag{11.68}$$

whereas the second system is a recursive (all-pole) system described by

$$y[n] = -a_1 y[n-1] + v[n]. \tag{11.69}$$

It is known that interchanging the order of the cascaded LTI systems does not change the overall response of the system [8]. So, by interchanging/swapping the all-zero and all-pole systems, we obtain an alternative realization of the system in Eq. (11.67). The resulting structure of the system is shown in Fig. 11.13(b). From this figure, we obtain the following two difference equations

$$w[n] = -a_1 w[n-1] + x[n], \tag{11.70}$$

$$y[n] = b_0 w[n] + b_1 w[n-1], \tag{11.71}$$

which provide an alternative algorithm for computing the output of the system described by Eq. (11.67). A close observation of Fig. 11.13(b) reveals that the two memory elements contain the same input $w[n]$ and hence the same output $w[n-1]$. Hence, these two memory elements are replaced by a singe unit as shown in Fig. 11.13(c). Compared to the *Direct Form I* structure, the new structure is more efficient in terms of the memory requirements. It is called the *Direct Form II* structure. This idea can be extended to the general LTI system described in Eq. (11.1) and is shown in Fig. 11.14. This structure requires $\tilde{M} + N + 1$ multiplications, $\tilde{M} + N$ additions, and maximum of $\{\tilde{M}, N\}$ memory locations. Since this structure minimizes the number of memory units, it is also known as *canonical structure*.

11.3.2 Cascade-Form Structure

By assuming $N \geq \tilde{M}$, the IIR system in Eq. (11.2) can be factored into a cascade of second-order subsystems, such that $H(z)$ can be expressed as,

$$H(z) = \prod_{k=1}^{K} H_k(z) \tag{11.72}$$

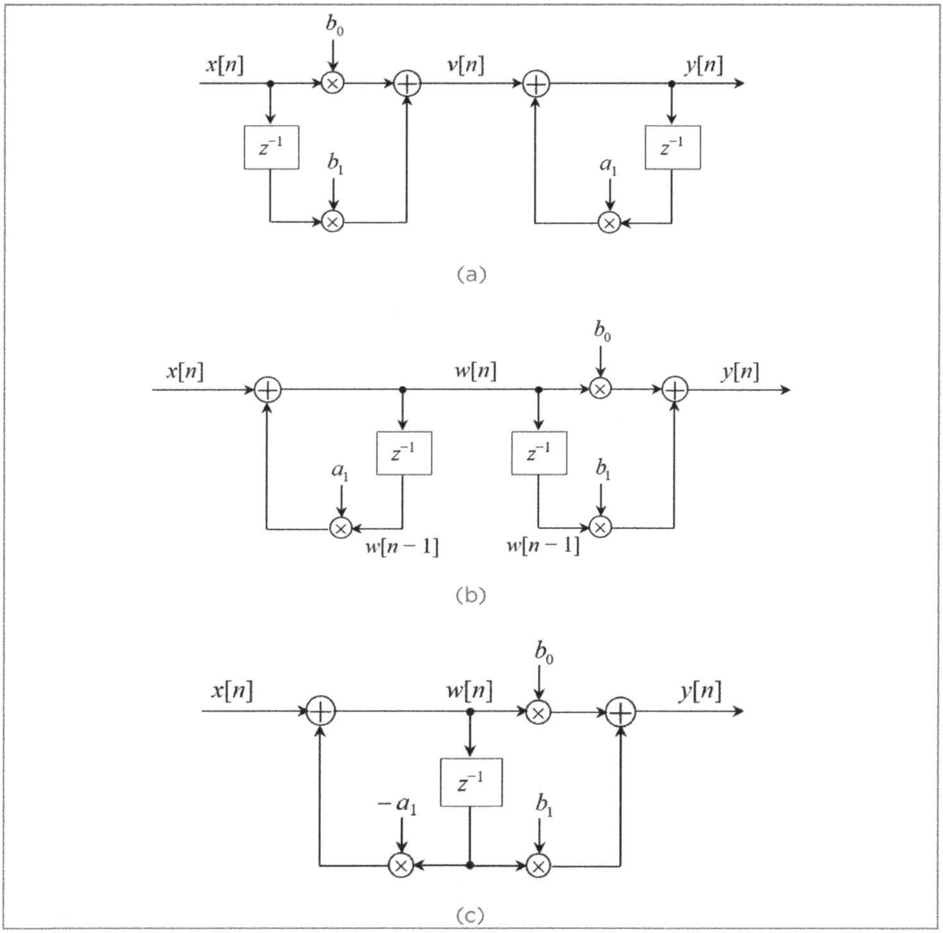

Figure 11.13. Steps in converting from the Direct Form I realization in (a) to the Direct Form II realization in (c).

where K is the integer part of $(N+1)/2$. $H_k(z)$ has the general form

$$H_k(z) = \frac{b_{k0} + b_{k1}z^{-1} + b_{k2}z^{-2}}{1 + a_{k1}z^{-1} + a_{k2}z^{-2}}. \tag{11.73}$$

As in the case of FIR systems based on a cascade form realizations, the parameter b_0 can be distributed equally among the K filter sections.

The coefficients $\{a_{ki}\}$ and $\{b_{ki}\}$ in the second-order subsystems are real. This implies that in forming the subsystems or quadratic factors in Eq. (11.73),

- We should group together a pair of complex-conjugate poles and a pair of complex-conjugate zeros.

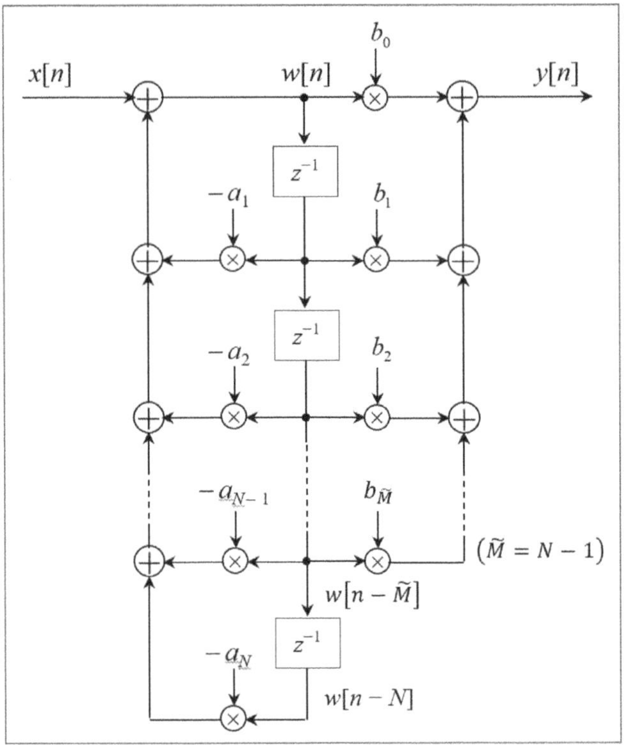

Figure 11.14. Direct Form II realization of IIR system ($M = N - 1$).

- The pairing of two complex-conjugate poles with a pair of complex-conjugate zeros or real-valued zeros to form the sub-systems can be done arbitrarily.
- Any two real-valued zeros can be paired together to form a quadratic factor and, likewise, any two real-valued poles can be paired together to form the quadratic factor.

This means that the roots of the numerator as well as the denominator of Eq. (11.73) may consist of either a pair of real roots or a pair of complex conjugate roots.

If $N > \tilde{M}$, some of the subsystems have either $b_{k2} = 0$, or $b_{k1} = 0$, or $b_{k2} = b_{k1} = 0$ for some k. Furthermore, if N is odd, one of the subsystems, say $H_k(z)$, must have $a_{k2} = 0$, so that the sub-system is of first order. To preserve the modularity in the implementation of $H(z)$, it is often preferable to use the basic second-order subsystems in the cascade structure and have some zero-valued coefficients in some of the sub-systems.

The general form of the cascade structure is illustrated in Fig. 11.15. If we use the Direct Form II structure for each of the sub-systems, the computational algorithm for realizing the IIR system is described by the following set of equations

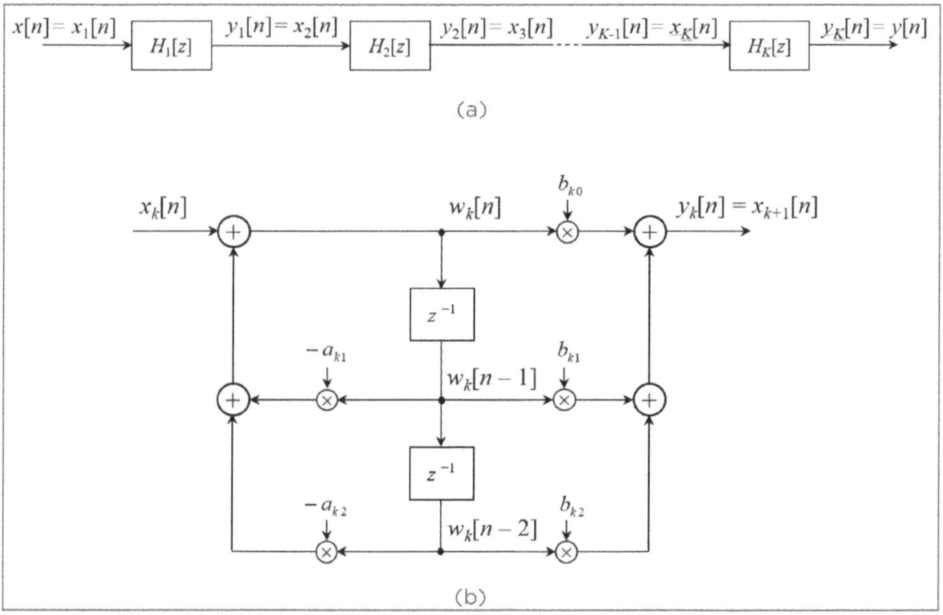

Figure 11.15. Cascade structure of second-order system (a) and realization of each second-order section (b).

(for $k = 1, 2, \ldots, K$):

$$y_0[n] = x[n], \tag{11.74}$$

$$w_k[n] = -a_{k1}w_k[n-1] - a_{k2}w_k[n-2] + y_{k-1}[n], \tag{11.75}$$

$$y_k[n] = b_{k0}w_k[n] + b_{k1}w_k[n-1] + b_{k2}w_k[n-2], \tag{11.76}$$

$$y[n] = y_K[n]. \tag{11.77}$$

11.3.3 Parallel-Form Structure

This structure of the IIR systems is developed based on the partial fraction expansion of $H(z)$. By assuming $N = \tilde{M}$, the partial fraction expansion can be written as

$$H(z) = C + \sum_{k=1}^{N} \frac{A_k}{1 - p_k z^{-1}}, \tag{11.78}$$

where $\{p_k\}$ are the poles, $\{A_k\}$ are the residues, and $C = \frac{b_N}{a_N}$. The structure implied by Eq. (11.78) is shown in Fig. 11.16. It consists of a parallel bank of single-pole

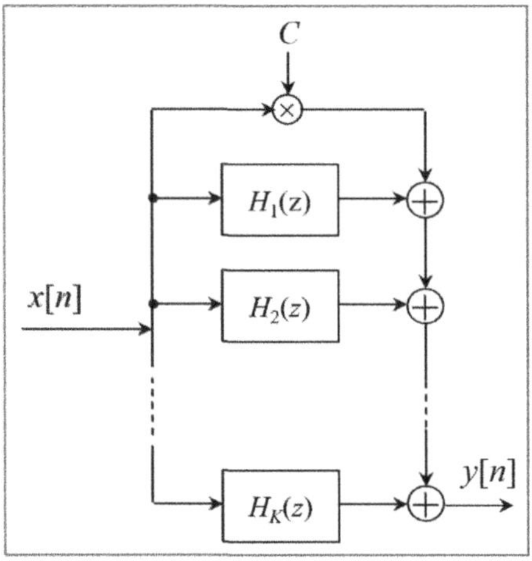

Figure 11.16. Parallel structure of IIR system.

filters. In general, some of the poles of $H(z)$ may be complex valued and the corresponding A_k's are also complex valued. To avoid multiplications by complex numbers, we can combine complex conjugate poles to form two-pole sub-systems. In addition, we can combine, in arbitrary manner, pairs of real-valued poles to form two-pole sub-systems. Each of these sub-systems has the form

$$H_k(z) = \frac{b_{k0} + b_{k1}z^{-1}}{1 + a_{k1}z^{-1} + a_{k2}z^{-2}}. \tag{11.79}$$

The overall function can now be expressed as

$$H(z) = C + \sum_{k=1}^{K} H_k(z). \tag{11.80}$$

The Direct Form II structure of the individual second-order sections which are the basic building blocks for $H(z)$ is shown in Fig. 11.17. The parallel form realization of the IIR system is described by (for $k = 1, 2, \ldots, K$):

$$w_k[n] = -a_{k1}w_k[n-1] - a_{k2}w_k[n-2] + x[n], \tag{11.81}$$

$$y_k[n] = b_{k0}w_k[n] + b_{k1}w_k[n-1], \tag{11.82}$$

$$y[n] = Cx[n] + \sum_{k=1}^{K} y_k[n]. \tag{11.83}$$

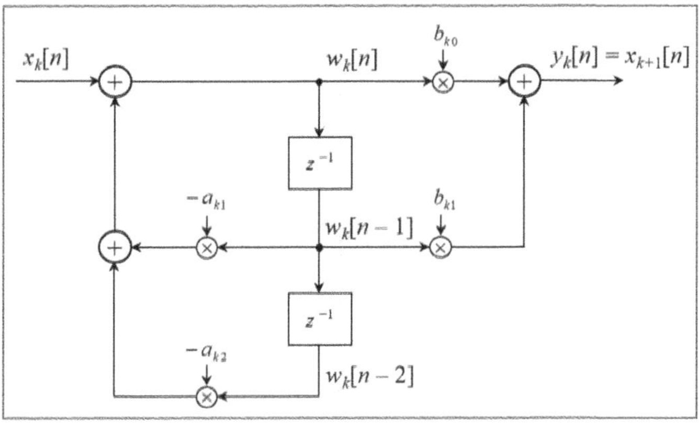

Figure 11.17. Structure of second-order section in a parallel IIR system realization.

11.3.4 Lattice Structure for IIR Systems

Let us begin with an all-pole system with the system function,

$$H(z) = \frac{1}{1 + \sum_{k=1}^{N} a_N[k] z^{-1}} = \frac{1}{A_N(z)}. \tag{11.84}$$

The direct form realization of this system is illustrated in Fig. 11.18. The corresponding difference equation for the IIR system is

$$y[n] = -\sum_{k=1}^{N} a_N[k] y[n-k] + x[n]. \tag{11.85}$$

It is interesting to note that if we interchange the roles of input and output (i.e., interchange $x[n]$ and $y[n]$), we obtain

$$x[n] = -\sum_{k=1}^{N} a_N[k] x[n-k] + y[n], \tag{11.86}$$

or, equivalently,

$$y[n] = x[n] + \sum_{k=1}^{N} a_N[k] x[n-k]. \tag{11.87}$$

We note that this equation describes an FIR system having the system function $H(z) = A_N(z)$, while the system described by the difference equation in Eq. (11.85) represents an IIR system with the system function $H(z) = \frac{1}{A_N(z)}$. That means, one system can be obtained from the other simply by interchanging the roles of input and output.

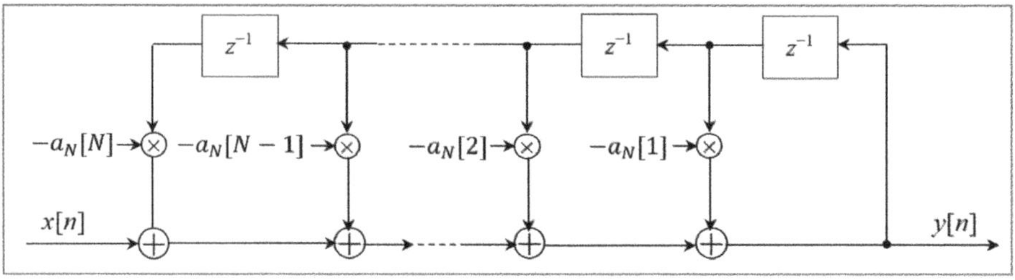

Figure 11.18. Direct form realization of an all-pole system.

Based on this, we shall use the all-zero (FIR) lattice described in Section 11.2.4 to obtain a lattice structure for the all-pole IIR system by interchanging the roles of input and output. First, we take the all-zero lattice filter illustrated in Fig. 11.11 and then redefine the input as

$$x[n] = f_N[n], \tag{11.88}$$

and the output as

$$y[n] = f_0[n]. \tag{11.89}$$

These are exactly the opposite of the definitions of the all-zero lattice filter. These definitions, dictate that the quantities $\{f_m[n]\}$ be computed in descending order [i.e., $f_N[n]$, $f_{N-1}[n]$, ...]. This computations can be accomplished by rearranging the recursive equation (Eq. (11.35)) and thus solving for $f_{m-1}[n]$ in terms of $f_m[n]$, that is,

$$f_{m-1}[n] = f_m[n] - K_m g_{m-1}[n-1], \quad m = N, N-1, \ldots, 1 \tag{11.90}$$

Equation (11.36) for $g_m[n]$ remains unchanged. The resulting set of equations are (for $m = N, N-1, \ldots, 1$):

$$f_N[n] = x[n], \tag{11.91}$$

$$f_{m-1}[n] = f_m[n] - K_m g_{m-1}[n-1], \tag{11.92}$$

$$g_m[n] = K_m f_{m-1}[n] + g_{m-1}[n-1], \tag{11.93}$$

$$y[n] = f_0[n] = g_0[n]. \tag{11.94}$$

The corresponding structure is shown in Fig. 11.19.

11.4 Signal Flow Graphs and Transposed Structures

A signal flow graph provides an alternative, but equivalent graphical representation to a block diagram structure that we have been using to illustrate various

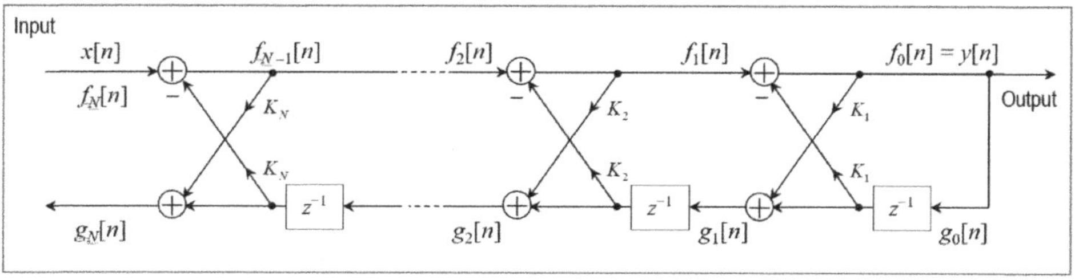

Figure 11.19. Lattice structure for an all-pole IIR system.

system realizations. The basic elements of a flow graph are *branches* and *nodes*. A signal flow graph is basically a set of directed branches that connect at nodes. By definition, the signal out of a branch is equal to the branch gain (system function) times the signal into the branch. Furthermore, the signal at a node of a flow graph is equal to the sum of the signals from all branches connecting to the node.

11.4.1 Transposed Structure of IIR Systems

To illustrate these basic notions, let us consider a two-pole and two-zero IIR system illustrated in block diagram form in Fig. 11.20(a). This block diagram can be converted to the signal flow graph as shown in Fig. 11.20(b). We note that the flow graph contains 5 nodes labeled 1 to 5. Two of the nodes (1,3) are summing nodes (i.e., they contain adding units) and the other three nodes representing branching points. Branch transmittances are indicated for the branches in the flow graph. A delay is indicated by the branch transmittance z^{-1}. When the branch transmittance is unity, it is left unlabelled. The input to the system originates at a *source node* and the output signal is extracted at a *sink node*.

One technique that is useful in deriving new structures for FIR and IIR systems stems from the *transposition* or *flow-graph reversal theorem*. This theorem states that if we reverse the directions of all branch transmittances and interchange the input and output in the flow graph, the system function remains unchanged. The resulting structure is called a *transposed structure* or *transposed form*.

For example, the transposition of the signal flow graph in Fig. 11.20(b) is illustrated in Fig. 11.21(a). The corresponding block diagram realization of the transposed form is shown in Fig. 11.21(b). It is interesting to note that the transposition of the original flow graph resulted in branching nodes becoming adding nodes, and vice versa.

Let us apply this theorem to a Direct Form II structure. First, we reverse all the signal flow directions in Fig. 11.14 with $(M = N)$. Second, we change nodes into adders, adders into nodes, and finally, we interchange the input and the output.

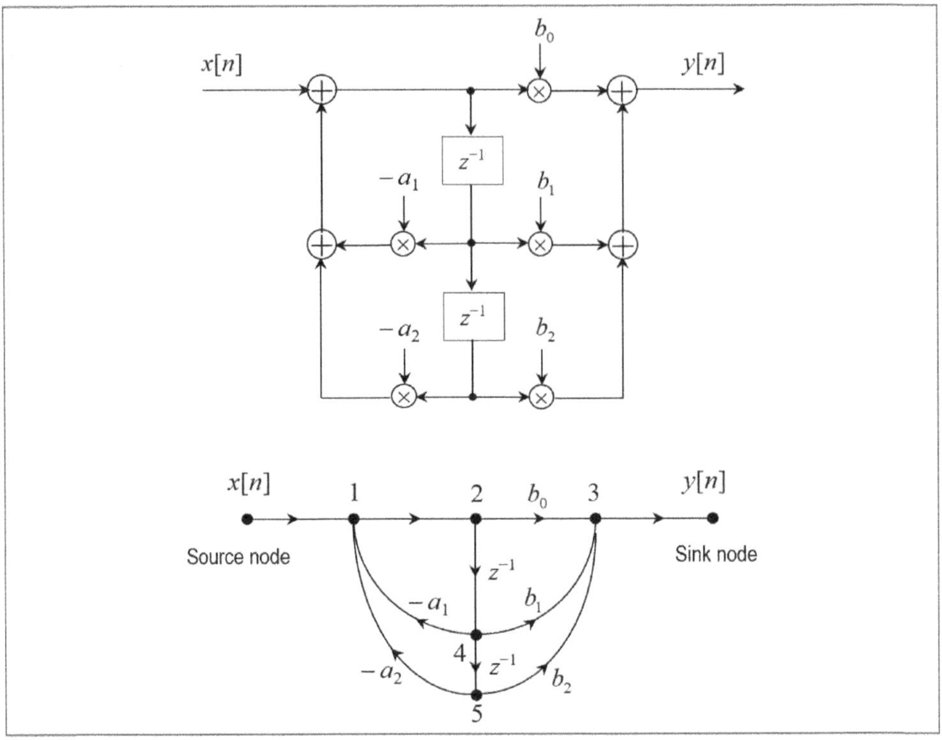

Figure 11.20. Second-order filter structure (a) and its signal flow graph (b).

The resulting transposed Direct Form II structure is shown in Fig. 11.22. This can be redrawn as in Fig. 11.23. which shows the input on the left and the output on the right.

The transposed Direct Form II realization can be described by the set of difference equations (for $k = 1, 2, \ldots, N-1$ and assuming $\tilde{M} = N$):

$$y[n] = w_1[n-1] + b_0 x[n], \tag{11.95}$$

$$w_k[n] = w_{k+1}[n-1] - a_k y[n] + b_k x[n], \tag{11.96}$$

$$w_N[n] = b_N x[n] - a_N y[n]. \tag{11.97}$$

It is also clear from observing Fig. 11.23 that the above set of difference equations is equivalent to a single difference equation as follows

$$y[n] = -\sum_{k=1}^{N} a_k y[n-k] + \sum_{k=0}^{\tilde{M}} b_k x[n-k] \tag{11.98}$$

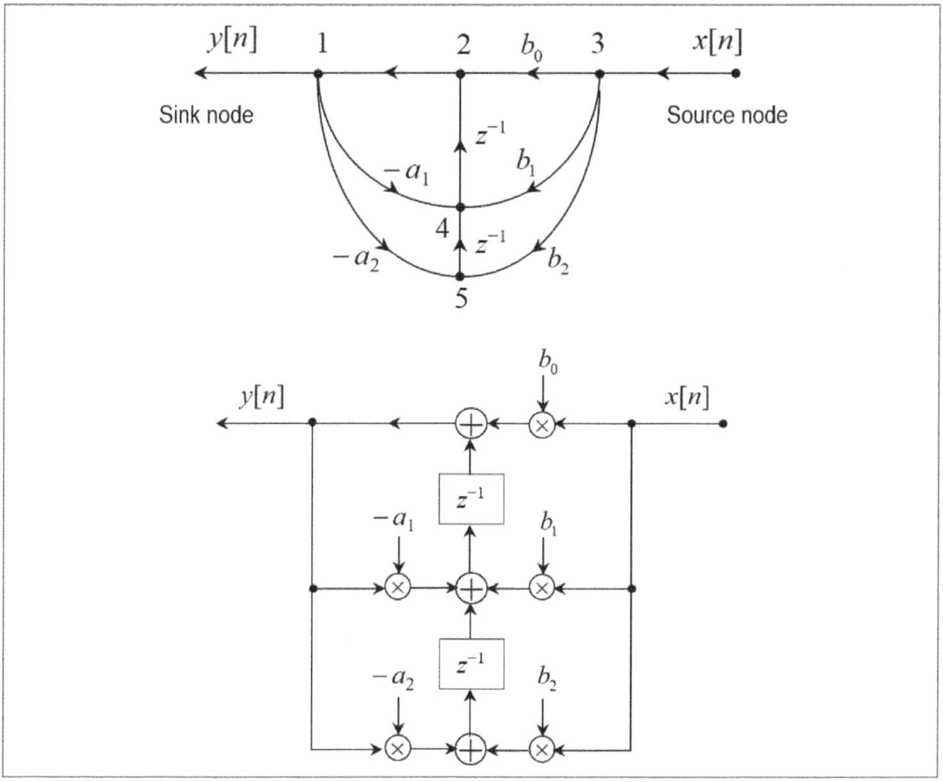

Figure 11.21. Signal flow graph (a) Its filter structure (b).

Finally, we observe that the transposed Direct Form II structure require the same amount of computational complexity (multiplications, additions, and memory units) as the original Direct Form II structure.

11.4.2 Transposed Structure of FIR Systems

Although our discussion has been concerned with the general IIR systems, it is interesting to note that an FIR system obtained by setting $a_k = 0, \;\; k = 1, 2, \ldots, N$ in Eq. (11.98), also has a transposed direct form structure as illustrated in Fig. 11.24. This realization may be described by the set of difference equations (for $k = M-1, M-2, \ldots, 1$):

$$w_{\tilde{M}}[n] = b_{\tilde{M}}x[n] \tag{11.99}$$

$$w_k[n] = w_{k+1}[n-1] + b_k x[n] \tag{11.100}$$

$$y[n] = w_1[n-1] + b_0 x[n] \tag{11.101}$$

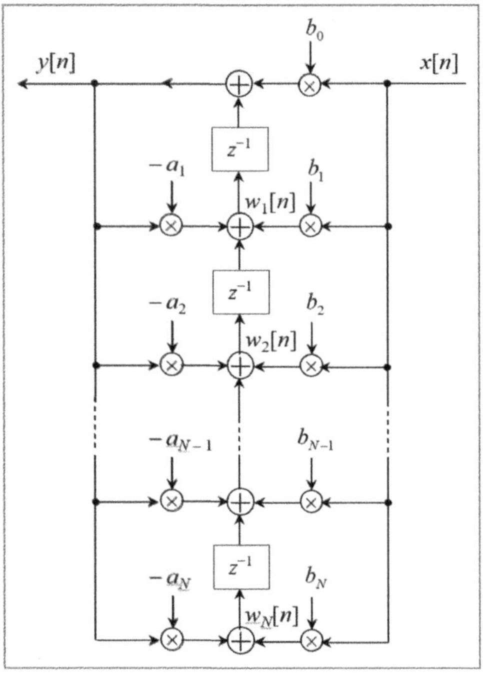

Figure 11.22. Transposed direct-form II structure.

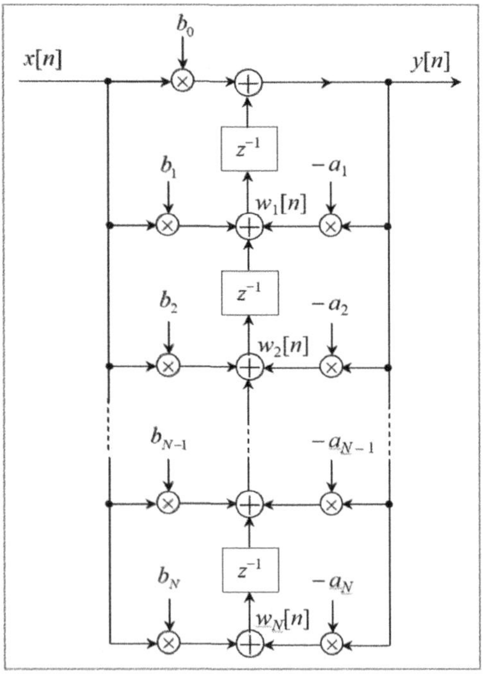

Figure 11.23. Transposed direct-form II structure.

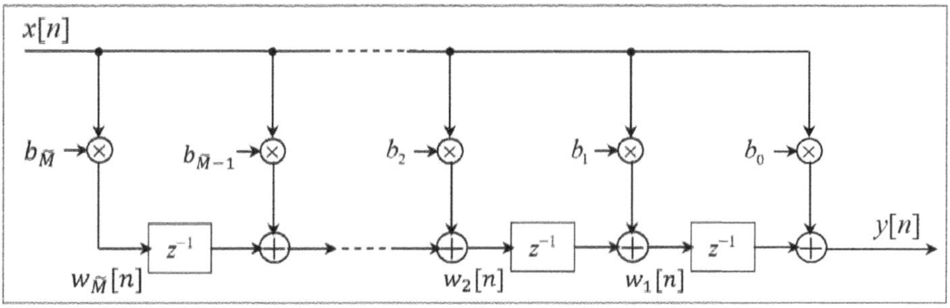

Figure 11.24. Transposed FIR structure.

11.5 Concluding Remarks

In this chapter, various schemes for the implementation of FIR and IIR systems were discussed in detail. Alternate structures were also discussed. They are:

- The structures for FIR systems include:
  - Direct form structure
  - Cascade-form structure
  - Frequency sampling structure, and
  - Lattice structure
- For IIR systems, the structures were:
  - Direct-form I structure
  - Direct-form II structure or Canonical structure
  - Parallel-form structure, and
  - Lattice structure.
- Transposed structures for FIR as well as IIR systems.

Exercises

11.1. A second order digital filter is implemented using Direct Form II structure as shown in Fig. P11.1. Here a_0, a_1, a_2, b_1, and b_2 are the filter coefficients, $x[n]$ is the input and $y[n]$ is the output.

(a) Determine the difference equation relating the input and the output and the frequency response of the filter.

(b) Write down the pseudo-code for implementing the filter in Question (a). Comment on the implementation complexity.

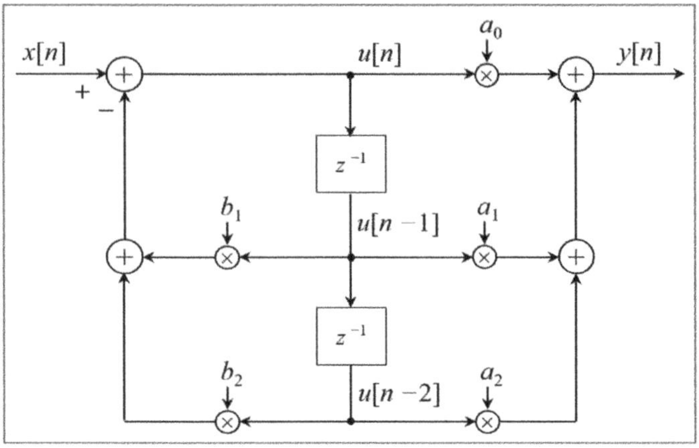

Figure P11.1. Direct form II structure of 2nd order digital filter.

11.2. The transfer function of a second order filter is given by:

$$H(f) = \frac{a_0 + a_1 e^{-j2\pi f T_s} + a_2 e^{-j4\pi f T_s}}{1 + b_1 e^{-j2\pi f T_s} + b_2 e^{-j4\pi f T_s}},$$

where the filter coefficients a_0, a_1, a_2, b_1, and b_2 are real numbers, and T_s is the sampling interval. The input and output signal of the filter are $x[n]$ and $y[n]$, respectively.

(a) Write down the system function $H(z)$ of the filter. Write down the difference equation that describes the operation of the filter. Draw the block diagram of the filter using Direct Form II.

(b) In the following, we will consider a simplified version of the filter for which $a_1 = 0$ and $a_2 = 0$. Furthermore, the sampling interval is $T_s = 250\mu s$, $b_1 = -2r\cos(\beta)$ and $b_2 = r^2$, where $0 < r \leq 1$, and $0 \leq \beta \leq \pi$. The impulse response of the simplified filter can be written as:

$$h[n] = a_0 \frac{\sin([n+1]\beta)}{\sin(\beta)} r^n,$$

and is valid when $n \geq 0$. Find the pair of values a_0 and β, for which the amplification of the filter at $f = 1$kHz is 1, and the amplification for frequencies 0 and 2kHz is the same (i.e., $|H(0)| = |H(2\text{kHz})|$).

(c) Find the impulse response ($h[n]$) of the filter for $\beta = 0$. Examine whether the filter is stable for all values of r in the given interval, when $\beta = 0$.

(d) The signal $x[n]$ is applied at the input of the filter. $x[n]$ is the sampled version of the periodic analog signal $x(t)$ given by:

$$x(t) = d_1 \cos(2\pi f_0 t) + d_2 \cos(4\pi f_0 t),$$

where f_0, d_1, and d_2 are constants. Find which frequency components are present in the output signal $y[n]$ within the frequency range $0 - 2\text{kHz}$, when $f_0 = 9.5\text{kHz}$.

11.3. A digital filter (recursive) is implemented using Direct Form II as shown in Fig. P11.3. The filter coefficients a and b are constants, and the input and the output signals are denoted as $x[n]$ and $y[n]$, respectively. The sampling interval is denoted as T_s. The symbol z^{-1} in the figure represents a unit sample delay.

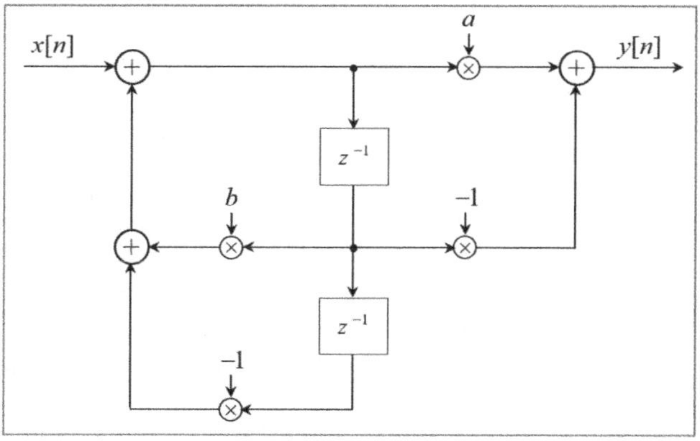

Figure P11.3.

(a) Draw the block diagram of the filter implemented in Direct Form I. Write down the difference equation that describes the operation of the filter. Find the system function $H(z)$ and the transfer function $H(f)$ expressed as functions of a, b, and T_s.

(b) For the rest of the problem, we assume that $x[n]$ is a real digital signal, and that a and b are given by

$$a = e^{j\frac{2\pi m}{N}},$$
$$b = 2\cos\left(\frac{2\pi m}{N}\right) = e^{j\frac{2\pi m}{N}} + e^{-j\frac{2\pi m}{N}},$$

where m and N are integer constants. The output signal $y[n]$ is complex, i.e., $y[n] = y_R[n] + jy_I[n]$.

Find the complex impulse response $h[n]$ of the filter (use the complex expression for b).

(c) The Discrete Fourier transform (DFT) can be realized as a filter bank. The mth coefficient of the DFT (calculated at digital frequency $f_m = \frac{m}{N}$, where $0 \leq m \leq N-1$) is the output of a filter with an impulse responde $h_m[n]$. Show the first N values of $h[n]$ have a simple relation to $h_m[n]$.

11.4. This problem considers a filter, which approximates an ideal low-pass filter with an impulse response (unit sample response) $h[n]$ and a transfer function $H(f) = |H(f)|e^{j\theta(f)}$, where $\theta(f) = \mathsf{arg}(H(f))$. The non-zero values of the impulse response are given in the table below. The impulse response is equal to zero ($h[n] = 0$) for all other time instances, n.

n	1	3	5	7	8	9	11	13	15
$h[n]$	$-\frac{1}{128}$	$\frac{1}{32}$	$-\frac{11}{128}$	$\frac{5}{16}$	$\frac{1}{2}$	$\frac{5}{16}$	$-\frac{11}{128}$	$\frac{1}{32}$	$-\frac{1}{128}$

The width of the filter's pass-band is approximately $\frac{f_g}{2}$, where $f_g = \frac{f_s}{2} = \frac{1}{2T_s}$. The two constants f_s and T_s are the sampling frequency and sampling interval, respectively. Within its pass-band, the filter has no zeros.

(a) Find $H(f)$ at the three frequencies 0, $\frac{f_g}{2}$, and f_g. Explain why the phase, $\mathsf{arg}(H(f))$ of the transfer function of the filter is linear below $f = \frac{f_g}{2}$ and sketch $\mathsf{arg}(H(f))$ in this region.

11.5. The operation of a digital filter can be described by the following difference equation:

$$y[n] = a_0 x[n] + a_2 x[n-2] + b_1 y[n-1],$$

where $x[n]$ is the input signal to the filter and $y[n]$ is its output signal. a_0, a_2, and b_1 are real constants. The sampling interval for the digital signal is T_s.

(a) Draw the block diagram of the filter implemented in Direct Form I and in Direct Form II. Then show that the impulse response of the filter is:

$$h[n] = \begin{cases} 0 & \text{for } n < 0 \\ a_0 & \text{for } n = 0 \\ a_0 b_1 & \text{for } n = 1 \\ a_0 b_1^n + a_2 b_1^{n-2} & \text{for } n \geq 2. \end{cases}$$

(b) Write down the condition under which the filter is stable. Write down the transfer function $H(f)$ of the filter.

In the following question, the sampling interval is chosen to be $T_s = 500\mu s$. The following digital signal is applied at the input of the filter:

$$x[n] = \sin(2\pi f_1 n T_s),$$

where $f_1 = 3.5$kHz. The filter coefficients a_0, a_2, and b_1 are set to be $a_0 = a_2 = b_1 = \frac{1}{2}$ for the next question.

(c) Give the placement and the amplitudes of the components of $y[n]$ in the amplitude spectrum for frequencies in the range 0 to 2kHz.

11.6. In this problem, we will consider a series of identical filters whose inputs and outputs are cascaded as shown in Fig. P11.6(a), and the digital input and output signals are $x[n]$ and $y[n]$, respectively.

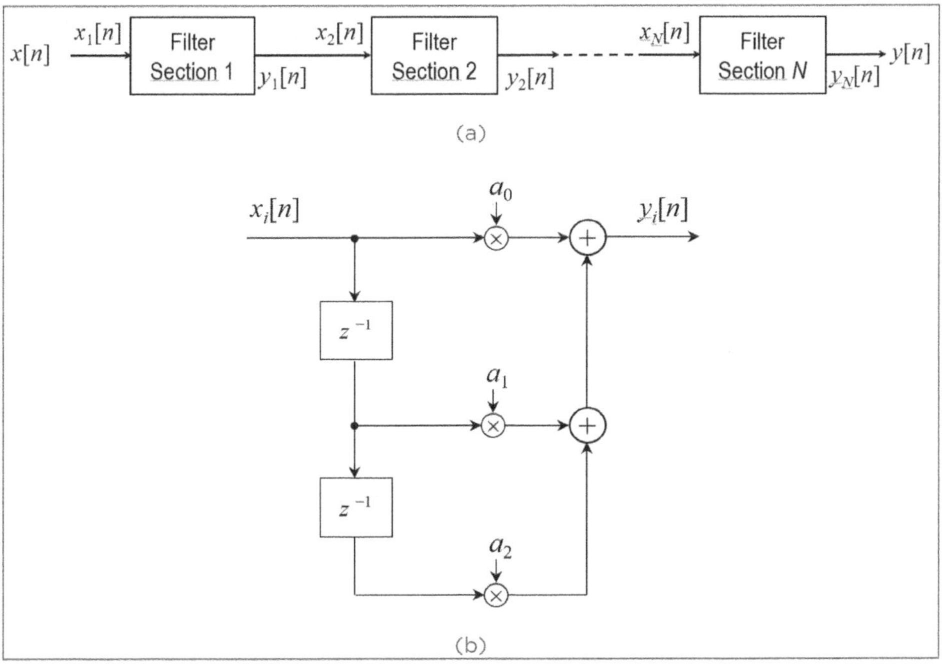

Figure P11.6.

The input signal to the ith filter is denoted as $x_i[n]$ and its output signal, $y_i[n]$. The input signal to the ith stage $x_i[n]$ is the output from the $(i-1)^{th}$ stage $y_{i-1}[n]$, i.e., $x_i[n] \equiv y_{i-1}[n]$. Figure P11.6(b) shows the block diagram of a single section (i^{th} section) from the filter, where a_0, a_1, and a_2 are real constants, and T_s is the sampling interval.

(a) For the first question, the cascade filter consists of N identical sections, where N is a positive integer number. Give an expression for the impulse response $h_i[n]$ and the transfer function $H_i(f)$ of a single section expressed through the filter coefficients a_0, a_1, a_2, and T_s. Express the combined impulse response $h[n]$ through $h_i[n]$ and the combined transfer function $H(f)$ through $H_i(f)$.

(b) Here, the number of sections N is chosen to be 2. Write down the impulse response $h[n]$ of the total filter expressed as a function of a_0, a_1, and a_2. Draw a block diagram of the (total) filter realized in Direct Form I.

(c) The signal $x[n]$ at the input of the filter comes from an analog to digital converter (ADC). At the input of the ADC, the following periodic signal is applied:

$$x(t) = d_0 \cos(2\pi f_0 t) + d_1 \cos(2\pi f_1 t),$$

where d_0, d_1, f_0, and f_1 are real constants. The conversion to the digital signal can be assumed to be exact (without quantization errors). The sampling frequency is 1.5kHz, and $f_0 = 9.5$kHz, and $f_1 = 2.3$kHz. Give the frequencies of those frequency components that are present in the frequency range from -1.5kHz to 3kHz.

11.7. In this problem, we will consider a digital low-pass filter described by the following difference equation:

$$y[n] = a_N x[n-N] + \sum_{q=0}^{N-1} \left[a_q x[n-q] + a_q x[n-2N+q]\right],$$

where N is a positive integer number, and a_q and a_N are real numbers. The amplitude of the transfer function of the filter is shown in Fig. P11.7. Here, $f_g = \frac{1}{2T_s}$, where T_s is the sampling interval of the signal $x[n]$.

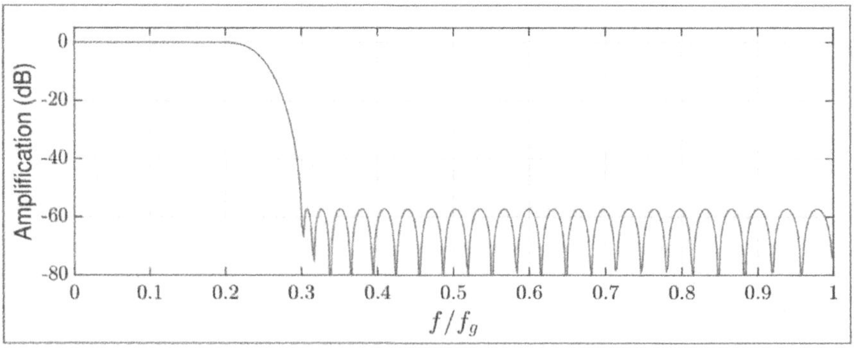

Figure P11.7.

(a) Explain why the impulse response of the filter is symmetric around $n = N$ and draw a block diagram of the filter realized in Direct Form I. Sketch the phase of the transfer function of the filter for $|f| < 0.2 f_g$, when $N = 32$.

(b) Using the given low-pass filter as a basis, design a new complex filter (with complex coefficients), which lets the frequency components from $0.3 f_g$ to $0.7 f_g$ pass with an unaltered amplitude and suppresses the components at negative frequencies. Write down the impulse response of the complex filter expressed as a function of f_g, N, a_q, and a_N. Sketch the amplitude of the transfer function of the filter in the frequency range from $-2 f_g$ to $2 f_g$ in the same way as shown in Fig. P11.7.

11.8. In a "talking dictionary" system, the "lattice" filter shown in Fig. P11.8 is used to generate a synthetic speech signal $y[n]$ from an input excitation $x[n]$ created from the acoustic template of the word to be spoken.

(a) Derive the transfer function of the filter and write a pseudo-code programme for implementing the filter directly (at the expense of 2 multiply operations per sample).

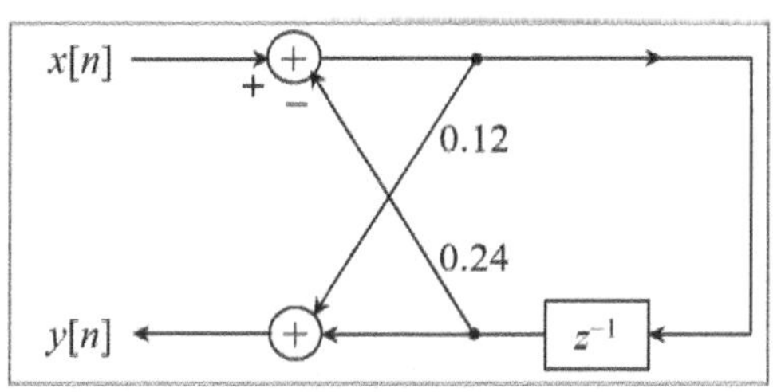

Figure P11.8.

(b) The direct implementation of the lattice filter in Fig. P11.8 requires 2 multiplication operations per sample. Since multiplication is difficult to implement in hardware and is a time-consuming operation, a creative engineer attempts to implement the filter using just 1 multiply operation per sample. Can this be done? How would you propose to do this? Rewrite the programme if necessary.

11.9. Consider the block diagram representation of a causal linear time invariant (LTI) system shown in Fig. P11.9. Here $x[n]$ and $y[n]$ are the input and output, respectively of the system. In the figure, the symbol z^{-1} denotes a unit sample delay and $v[n]$ is an intermediate variable.

(a) Derive the transfer function $H(z)$ for the given system and draw the direct form-I implementation structure of the system.

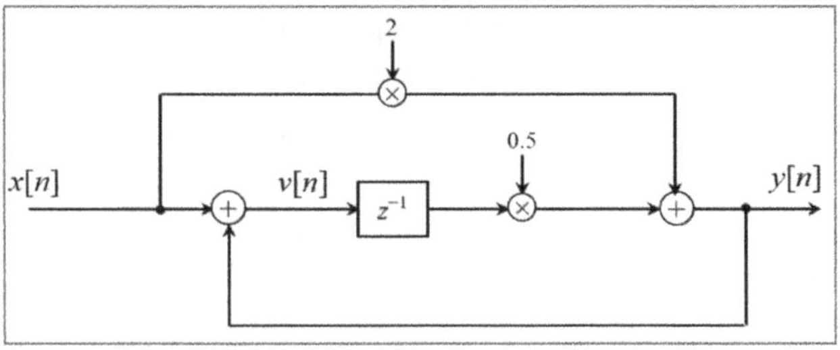

Figure P11.9.

(b) Reconfigure the system implementation structure in the form of a standard canonic (Direct Form-II) structure and compare its implementation complexity to that of the Direct Form I implementation scheme.

(c) Sketch the pole-zero distribution of the system in the Z-domain and comment on its stability.

(d) Derive an expression for the impulse response $h[n]$ of the system and plot the first five samples (i.e., the numerical values of $h[n]$ for $n = 0, 1, 2, 3,$ and 4) as a function of n.

11.10. Consider the block schematic of a linear time-invariant (LTI) digital system as shown in Fig. P11.10. Here $x[n]$ and $y[n]$ are the input and output, respectively of the system. $h[n]$ is the impulse response of the system.

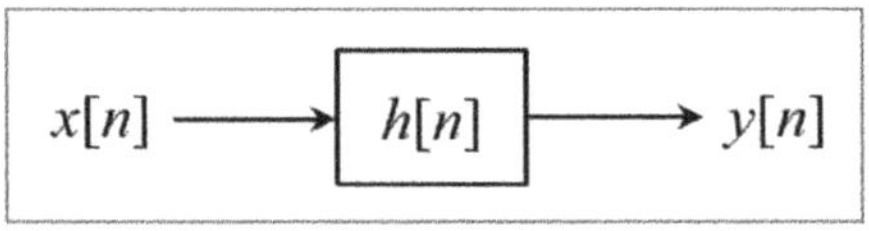

Figure P11.10.

The system function $H(z) = \frac{Y(z)}{X(z)}$ is given by,

$$H(z) = \frac{1 - 0.36z^{-2}}{1 + 0.1z^{-1} - 0.72z^{-2}},$$

where z^{-1} represents a unit sample delay and $X(z)$ and $Y(z)$ are the z-transforms of the input and output, respectively.

(a) Draw the "transposed Direct Form II" structure for the given system.

(b) Determine whether the system is stable or not.

(c) Assume that the overall system ($H(z)$) is realized by cascading two first-order sub-systems ($H_1(z)$ and $H_2(z)$). Draw the respective Direct Form I structures for $H_1(z)$ and $H_2(z)$.
(d) Derive the impulse response $h[n]$ of the overall system.

11.11. Consider the block diagram representation of the cascade of three causal first-order linear time-invariant (LTI) systems shown in Fig. P11.11.

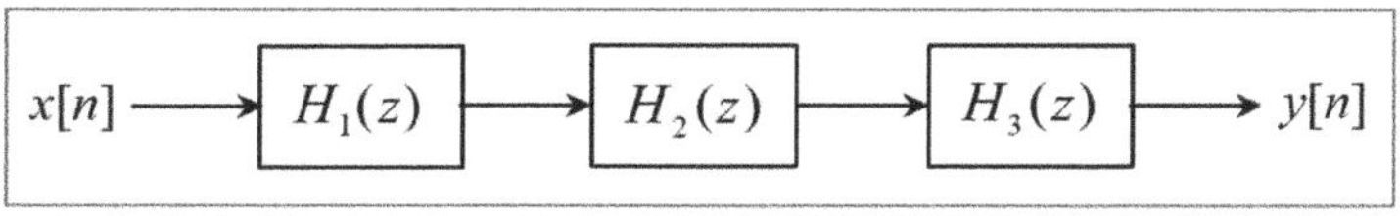

Figure P11.11.

The system functions for the sub-systems are:

$$H_1(z) = \frac{2+0.1z^{-1}}{1+0.4z^{-1}}, \quad H_2(z) = \frac{3+0.2z^{-1}}{1-0.3z^{-1}}, \quad H_3(z) = \frac{1}{1-0.2z^{-1}}.$$

(a) Determine the transfer function, $H(z)$, of the overall system as a ratio of two polynomials in z^{-1}.
(b) Determine the difference equation characterizing the overall system.
(c) Sketch the pole-zero distribution of $H(z)$ in the z-domain and comment on its stability.
(d) Develop a parallel form realization of the overall system with each parallel section containing a single pole.
(e) Derive the impulse response $h[n]$ of the overall system in a closed form.

11.12. Consider the block diagram representation of a second-order filter shown in Fig. P11.12. Here $x[n]$ and $y[n]$ are the input and output, respectively of the filter. The parameters α, β, and γ are scalars. The symbol z^{-1} denotes unit sample delay.

(a) Derive the transfer function $H(z)$ for the given filter and draw the direct form-I implementation structure of the filter.
(b) Reconfigure the filter in the form of a standard second-order canonic (Direct Form-II) structure and compare its implementation complexity to that of the Direct Form I implementation scheme.
(c) Sketch the pole-zero distribution of the filter in the Z-domain and comment on its stability (Assume $\alpha = 0.5$, $\beta = 2$, and $\gamma = 1$).
(d) Derive the impulse response $h[n]$ of the filter.

11.13. Consider the block diagram representation of a second-order filter shown in Fig. P11.13. Here $x[n]$ and $y[n]$ are the input and output,

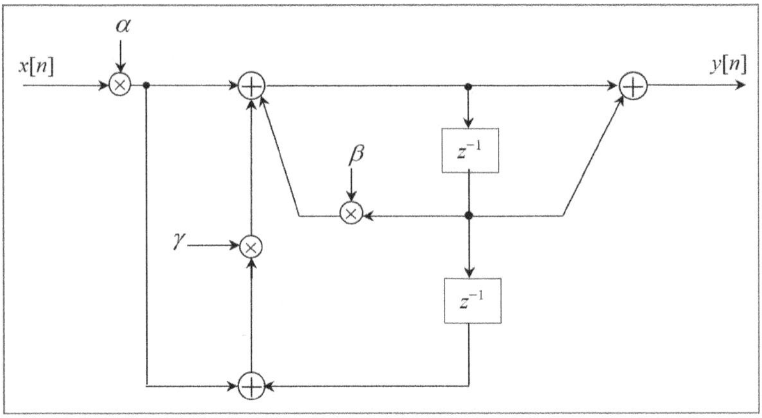

Figure P11.12.

respectively of the filter. The system function $H(z)$ ($z-$transform of $h[n]$) of the filter is given by,

$$H(z) = \frac{1 + 2z^{-1} + z^{-2}}{1 - 0.75z^{-1} + 0.125z^{-2}}.$$

The symbol z^{-1} in $H(z)$ denotes a unit sample delay.

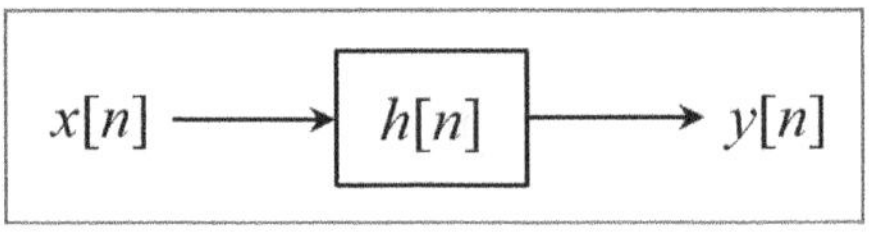

Figure P11.13.

(a) Determine the difference equation that relates the output $y[n]$ to the input $x[n]$. Draw the Direct Form I implementation structure of the filter.

(b) Reconfigure the filter in the form of a standard second-order canonic (Direct Form II) structure and compare its implementation complexity to the Direct Form I implementation scheme.

(c) The system function $H(z)$ for the given second order filter can be considered as the cascade of two first order recursive systems, $H_1(z)$ and $H_2(z)$, respectively. From $H(z)$, identify the first order recursive systems $H_1(z)$ and $H_2(z)$ and draw the cascade form structure for $H(z)$. Make sure to use Direct Form II structures for $H_1(z)$ and $H_2(z)$.

(d) As an alternative to factoring the numerator and denominator polynomials in $H(z)$, this rational function can be written as a partial fraction expansion resulting in the parallel form realization of the second order system. Derive and draw the parallel form structure of $H(z)$ with first order sub-sections.

(e) By making appropriate assumptions (initial rest), derive the expression for the impulse response ($h[n]$) of the filter.

11.14. Consider the second-order filter depicted by the block diagram in Fig. P11.14. Here $x[n]$ and $y[n]$ are the input and output, respectively of the filter and α_1 and α_2 are the filter coefficients. The symbol z^{-1} in the figure denotes a unit sample delay. The transfer function and impulse response of the filter are denoted by $H(z)$ and $h[n]$, respectively. The sampling interval T_s of the digital signal is assumed to be normalized to unity.

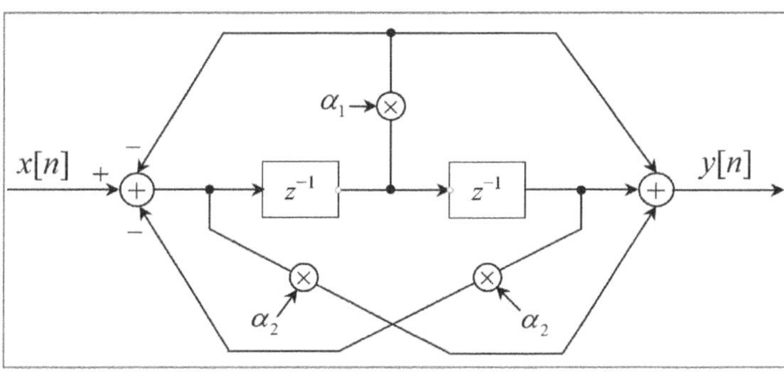

Figure P11.14.

(a) Derive the expression for the transfer function $H(z)$ of the filter. Comment on the type of the filter (i.e., low-pass, high-pass, all-pass, band-pass, or band-stop etc.) based on its frequency response.

(b) Reconfigure the filter in the form of a standard second-order canonic (Direct Form II) structure and compare its implementation complexity to the realization given in Fig. P11.14.

In the following, we will assume that $\alpha_1 = -\frac{3}{4}$ and $\alpha_2 = \frac{1}{8}$.

(c) Sketch the pole-zero distribution of the filter in the Z−domain and comment on the stability of the filter.

(d) By making appropriate assumptions (initial rest), write out the expressions (computation not required) for the first six values of the impulse response of the filter (i.e., $h[n]$ for $n = 0, 1, \ldots, 5$).

11.15. A digital filter ($H(z)$) is constructed by cascading two simple filters with transfer functions $H_1(z) = \frac{1}{2}(1 - z^{-2})$ and $H_2(z) = \frac{1}{1-\frac{1}{4}z^{-1}-\frac{1}{8}z^{-2}}$.

(a) Derive the transfer function $H(z)$ of the resulting filter and draw its pole-zero diagram.
(b) Comment on the stability of the filter $H(z)$
(c) Draw the Direct Form II realization of the filter $H(z)$
(d) Compute the first six values of the impulse response ($h[n]$) of the filter.

11.16. A digital filter is described by the following difference equation

$$y[n] = \frac{1}{2}y[n-1] + x[n]$$

where $y[n]$ is the output and $x[n]$ is the input to the system.

(a) Derive the transfer function $H(z)$ of the resulting filter and draw its pole-zero diagram.
(b) Draw the Direct Form II realization of the system and comment on its stability.

DOI: 10.1561/9781680839791.ch12

Chapter 12

Discrete Fourier Transform

In many practical signal processing applications, Discrete Fourier Transform (DFT) is the most important transform method used to perform Fourier analysis. Mathematically, DFT converts a finite input sequence of equally spaced samples of a signal into a sequence (same length) of equally spaced samples of the DTFT, which is a complex valued function of frequency. The interval at which the DTFT is sampled is the reciprocal of the length of the input sequence. An inverse DFT (IDFT) is a Fourier series, using the DTFT samples as coefficients of complex sinusoids at the corresponding DTFT frequencies. It has the same sample values as the original input sequence. The DFT, therefore, provides the frequency domain representation of the original input sequence. If the original sequence spans all the non-zero values of a signal, its DTFT is continuous (and periodic), and the DFT provides discrete samples of one cycle of the signal. If the original sequence is one cycle of a periodic signal, the DFT provides all the non-zero values of one DTFT cycle.

So far in this book, we have discussed three different frequency-domain representations for time-domain signals:

- Continuous-Time Fourier Series (CTFS; Chapter 3): To determine the spectra of continuous-time periodic signals. These signals result in discrete-frequency spectra.
- Continuous-Time Fourier Transform (CTFT; Chapter 4): To determine the spectra of continuous-time non-periodic signals. These signals result in

continuous-frequency spectra. The CTFS can be considered as a special case of CTFT for periodic signals.

- Discrete-Time Fourier Transform (DTFT; Chapter 9): To determine the spectra of discrete-time non-periodic (and periodic) signals. These signals result in continuous-frequency spectra which are periodic.

In this chapter, the objective is *the computation of the Fourier spectra of finite-duration sampled signals*. This arises from the fact that the signal measured in any practical application will be finite in duration. We are expected to use this finite-duration record of the signal to make inferences on the spectrum of the original signal (e.g., speech or bio-medical signals), which may be infinite in duration (in reality). Therefore, it is of interest to know how the finite data length of the measured record affects the computed spectrum.

The following points will be discussed in detail in this chapter.

- How does computing the spectrum at discrete values of frequency (i.e., sampling in frequency domain) affect the signal in time domain?
- Frequency-domain representation of periodic discrete-time signals.
- Frequency-domain representation of finite-duration discrete-time signals.
- The relevance of periodicity in time domain to computation of the spectra of discrete-time finite-duration signals.
- What precautions should be taken to guarantee that the computed spectrum is close to the actual spectrum?

12.1 Practical Aspects of Spectrum Computation

There are mainly two limitations associated with practical computation of spectra.

- The signal is recorded for a finite duration in time. This effect can be examined using the *method of windows*.
- The spectrum is computed for discrete values of the frequency. This is equivalent to *sampling the spectrum in frequency domain*.

12.1.1 Spectra of Windowed Signals

Let $x_a(t)$ be a continuous-time signal of very large duration. Further, let $x[n]$ denote the samples of this signal, taken at the rate of $f_s = \frac{1}{T_s}$ samples/second (i.e., T_s is the sampling period). Thus, we have

$$x[n] = x_a(nT_s), \quad \text{for} \quad -\infty < n < \infty. \tag{12.1}$$

The measured record of samples (data) in a particular experiment is given by

$$y[n] = x[n] \qquad \text{for} \quad n = 0, 1, \ldots, N-1. \tag{12.2}$$

In other words, the signal is observed for a finite duration of time from $t = 0$ to $t = [N-1]T_s$. Therefore, we can express the measured signal as

$$y[n] = x[n]w[n] = x_a(nT_s)w[n] \tag{12.3}$$

where $w[n]$ is a window function defined as

$$w[n] = \begin{cases} 1 & \text{for } n = 0, 1, \ldots, N-1 \\ 0 & \text{otherwise.} \end{cases} \tag{12.4}$$

Thus, we find that

$$y[n] = \begin{cases} x[n] & \text{for } n = 0, 1, \ldots, N-1 \\ 0 & \text{otherwise.} \end{cases} \tag{12.5}$$

Using Eq. (12.3), we can express the spectrum (DTFT) of $y[n]$ as

$$Y(\tilde{f}) = X(\tilde{f}) \odot W(\tilde{f}) = \int_{-0.5}^{0.5} X(u)W(\tilde{f} - u)du \tag{12.6}$$

where $\odot$ denotes 'circular convolution' (also known as 'periodic convolution'), $W(\tilde{f})$ is the DTFT of $w[n]$ and is given by

$$\begin{aligned} W(\tilde{f}) &= \sum_{n=-\infty}^{\infty} w[n]e^{-j2\pi \tilde{f} n} = \sum_{n=0}^{N-1} w[n]e^{-j2\pi \tilde{f} n} \\ &= \frac{1 - e^{-j2\pi \tilde{f} N}}{1 - e^{-j2\pi \tilde{f}}} = e^{-j\pi(N-1)\tilde{f}} \left(\frac{\sin(\pi \tilde{f} N)}{\sin(\pi \tilde{f})} \right). \end{aligned} \tag{12.7}$$

Clearly, we will get $Y(\tilde{f}) = X(\tilde{f})$ only if $W(\tilde{f}) = \delta(\tilde{f})$, which will happen only when $N \to \infty$. Thus, we find that *the finite data length causes the computed spectrum, $Y(\tilde{f})$, to be different from the actual spectrum, $X(\tilde{f})$.*

12.1.2 Sampling in Frequency-Domain: Derivation of DFT

Here, we will derive the expression for the DFT of a sampled signal.

The spectrum, $Y(\tilde{f})$, of the measured data is given by[1]

$$Y(\tilde{f}) = \sum_{n=0}^{N-1} y[n]e^{-j2\pi \tilde{f} n}, \qquad 0 \le \tilde{f} < 1.0 \tag{12.8}$$

Even though the frequency, $\tilde{f}$, varies continuously in the range [0, 1.0], in practice, we evaluate the spectrum, $Y(\tilde{f})$, only at discrete values of the frequency, given by

$$\left\{0,\ \Delta\tilde{f},\ 2\Delta\tilde{f},\ \cdots,\ (N-2)\Delta\tilde{f},\ (N-1)\Delta\tilde{f}\right\} \tag{12.9}$$

where $\Delta\tilde{f} = \frac{1}{N}$.

Define $Y_k = Y(k\Delta\tilde{f}) = Y\left(\frac{k}{N}\right)$. Then, we can obtain from Eq. (12.8)

$$Y_k = \sum_{n=0}^{N-1} y[n]e^{-j2\pi\left(\frac{k}{N}\right)n}, \tag{12.10}$$

for $k = 0, 1, \ldots, N-1$. Since $e^{-j2\pi \frac{k}{N} n}$ is periodic in k with period N (i.e., $e^{-j2\pi \frac{k+N}{N} n} = e^{-j2\pi \frac{k}{N} n}$), we see that Y_k is periodic with period N (i.e., $Y_{k+N} = Y_k$).

Let $l \in \{0, 1, 2, \ldots, N-1\}$, i.e., l is an integer in the range $0, 1, \cdots, N-1$. Then, consider the following[2]:

$$\begin{aligned}
\sum_{k=0}^{N-1} Y_k e^{j2\pi \frac{k}{N} l} &= \sum_{k=0}^{N-1}\left[\sum_{n=0}^{N-1} y[n]e^{-j2\pi\left(\frac{k}{N}\right)n}\right] e^{j2\pi \frac{k}{N} l} \\
&= \sum_{n=0}^{N-1} y[n] \sum_{k=0}^{N-1} e^{j2\pi \frac{k}{N}[l-n]}
\end{aligned}$$

1. In Eq. (12.8), for the sake of convenience, we use the period defined by $0 \le \tilde{f} < 1.0$ instead of $-0.5 \le \tilde{f} < 0.5$ for our analysis.

2. Note the following.

$$\begin{aligned}
\sum_{k=0}^{N-1} e^{j2\pi \frac{k}{N}[l-n]} &= N \qquad \text{if } l = n \\
&= \frac{1 - e^{j2\pi \frac{N}{N}[l-n]}}{1 - e^{j2\pi \frac{1}{N}[l-n]}} = 0 \qquad \text{if } l \ne n. \\
\Rightarrow \quad \sum_{k=0}^{N-1} e^{j2\pi \frac{k}{N}[l-n]} &= N\delta[l-n].
\end{aligned}$$

$$= \sum_{n=0}^{N-1} y[n]\,[N\delta[l-n]] = Ny[l]$$

$$\Rightarrow \quad y[n] = \frac{1}{N}\sum_{k=0}^{N-1} Y_k e^{j2\pi \frac{k}{N}n} \tag{12.11}$$

for $n = 0, 1, \ldots, N-1$. Since $e^{-j2\pi \frac{k}{N}n}$ is also periodic in n with period N (i.e., $e^{-j2\pi \frac{k}{N}[n+N]} = e^{-j2\pi \frac{k}{N}n}$), we see that $y[n]$ is periodic with period N (i.e., $y[n+N] = y[n]$). We have the following remarks.

- *Sampling of the spectrum causes the time-domain signal to be periodic* (i.e., $y[n+N] = y[n]$). *The period is given by the inverse of the sampling period used for sampling the spectrum* (i.e., $\frac{1}{\Delta \tilde{f}}$). In other words, *the DFT treats the finite-duration signal* $\{y[0], y[1], y[2], \cdots, y[N-1]\}$ *as one period of a periodic signal with period* N.
- But *the original measured signal record (i.e.,* $y[n]$ *for* $n = 0, 1, \ldots, N-1$*) is not periodic!* That is, *discretizing the frequency axis (i.e.,* $\tilde{f} = 0, 1/N, 2/N, \ldots, [N-1]/N$*) has the effect of making the input signal* $y[n]$ *appear as though it is a periodic discrete-time signal with period* N.
- The above observed property is the dual of the sampling theorem[3] we studied in Chapter 7. We saw in Chapter 7 that *sampling in time domain results in periodicity in frequency domain.* The derivation above shows that *sampling in frequency domain results in periodicity in time domain.*
- Equations (12.10) and (12.11) define the DFT pair of equations. *Equation (12.10) defines the forward DFT,* and *Eq. (12.11) defines the inverse DFT (IDFT).*
- *For a finite-duration signal,* $y[n]$, *of length* N *samples, its DFT is obtained by evaluating its DTFT (defined in Chapter 9) at frequency resolution* $1/N$. That is,

  $$Y_k = Y(\tilde{f})\Big|_{\tilde{f}=k/N} = Y\left(\frac{k}{N}\right), \tag{12.12}$$

  where $Y(\tilde{f})$ denotes the DTFT of $y[n]$. The resulting DFT spectrum, Y_k, is periodic with period N.

---

3. If you wish to explore the meaning of *aliasing in time domain caused by sampling in frequency domain,* use $\Delta\tilde{f} = 1/M$ in Eq. (12.9). The frequency grid is then given by $\{0,\ \Delta\tilde{f},\ 2\Delta\tilde{f},\ \cdots,\ (M-2)\Delta\tilde{f},\ (M-1)\Delta\tilde{f}\}$ and M will be the period in time and frequency domains. Then, aliasing will occur in the time domain if $M < N$. For example, let $N = 5$ and $M = 3$. Then, one period of the aliased signal (i.e., left-side of Eq. (12.11)) will be $\{y[0] + y[3], y[1] + y[4], y[2]\}$.

Since there is a close connection between DFT and periodic discrete-time signals, we shall next explore the frequency-domain representation of discrete-time periodic signals.

12.2 Discrete Fourier Series (DFS)

In Chapter 3, we studied Fourier series as a frequency-domain representation of continuous-time periodic signals. In this section, we explore a similar representation for discrete-time periodic signals.

Let $x[n]$ for $n = 0, 1, \ldots, N$, be one period of a discrete-time periodic signal with period N. Then, using Eqs. (12.10) and (12.11), we can obtain the DFT equations

$$X_k = \sum_{n=0}^{N-1} x[n] e^{-j2\pi \frac{k}{N} n}, \tag{12.13}$$

$$x[n] = \frac{1}{N} \sum_{k=0}^{N-1} X_k e^{j2\pi \frac{k}{N} n}, \tag{12.14}$$

for $k = 0, 1, 2, \ldots, N-1$ and $n = 0, 1, 2, \ldots, N-1$. We have the following remarks.

- We have

$$x[n+N] = x[n] \qquad \text{and} \qquad X_{k+N} = X_k. \tag{12.15}$$

- The periodicity in time domain imposed by the IDFT computation in Eq. (12.14) is naturally satisfied by $x[n]$, since we chose $x[n]$ to be a periodic signal with period N.
- Equation (12.14) is the Fourier series representation of the periodic signal $x[n]$ using N discrete-time complex sinusoids at frequencies 0, $1/N$, $2/N$, $\cdots$, $(N-1)/N$. Therefore, *Eq. (12.13) is known as the forward DFS* and *Eq. (12.14) as the inverse DFS.*
- Because $e^{j2\pi \frac{k+N}{N} n} = e^{j2\pi \frac{k}{N} n}$, *we need only N complex sinusoids with frequencies k/N, $k = 0, 1, 2, \ldots, N-1$, in the Fourier series representation (Eq. (12.14)).*
- *We call Eqs. (12.13) and (12.14) as DFS equations when the original $x[n]$ is periodic. We call these as DFT equations when the original $x[n]$ is not periodic (aperiodic).*

12.2.1 Derivation of DFS

Let $x_a(t)$ be a continuous-time periodic signal with period $T_p = 1/f_p$, where f_p is its fundamental frequency. Therefore, using Fourier series, we can express $x_a(t)$ as

$$x_a(t) = \sum_{k=-\infty}^{\infty} \tilde{X}_k e^{j2\pi k f_p t} \tag{12.16}$$

where $\tilde{X}_k$ are the Fourier series coefficients. It is easy to see that we can construct a discrete-time periodic signal, $x[n]$, with period N from $x_a(t)$ by sampling $x_a(t)$ with sampling rate $f_s = Nf_p$. That is,

$$x[n] = x_a(nT_s) \qquad \text{with} \quad T_s = \frac{1}{f_s} = \frac{1}{Nf_p} \tag{12.17}$$

where T_s is the sampling period. Using Eq. (12.16), we can express $x[n]$ as

$$\begin{aligned} x[n] = x_a(nT_s) &= \sum_{k=-\infty}^{\infty} \tilde{X}_k e^{j2\pi k f_p n T_s} \\ &= \sum_{k=-\infty}^{\infty} \tilde{X}_k e^{j2\pi \frac{k}{N} n}, \end{aligned} \tag{12.18}$$

since $f_p T_s = 1/N$. Express k as $k = lN + m$, where l ranges from $-\infty$ to $+\infty$ and $m = 0, 1, 2, \ldots, N-1$. Then,

$$e^{j2\pi \frac{k}{N} n} = e^{j2\pi \frac{lN+m}{N} n} = e^{j2\pi l n} e^{j2\pi \frac{m}{N} n} = e^{j2\pi \frac{m}{N} n}.$$

Using these, we can express the right side of Eq. (12.18) as

$$\begin{aligned} \sum_{k=-\infty}^{\infty} \tilde{X}_k e^{j2\pi \frac{k}{N} n} &= \sum_{l=-\infty}^{\infty} \sum_{m=0}^{N-1} \tilde{X}_{lN+m} e^{j2\pi \frac{lN+m}{N} n}, \\ &= \sum_{m=0}^{N-1} \left[\sum_{l=-\infty}^{\infty} \tilde{X}_{lN+m} \right] e^{j2\pi \frac{m}{N} n}, \\ &= \frac{1}{N} \sum_{m=0}^{N-1} X_m e^{j2\pi \frac{m}{N} n}, \end{aligned} \tag{12.19}$$

where

$$X_m = N \sum_{l=-\infty}^{\infty} \tilde{X}_{lN+m}. \tag{12.20}$$

Substituting Eq. (12.19) in Eq. (12.18), we get

$$x[n] = \frac{1}{N}\sum_{k=0}^{N-1} X_k e^{j2\pi \frac{k}{N} n}, \tag{12.21}$$

for $n = 0, 1, 2, \ldots, N-1$. Further, we have

$$\begin{aligned}
\sum_{n=0}^{N-1} x[n]e^{-j2\pi \frac{l}{N} n} &= \sum_{n=0}^{N-1}\left[\frac{1}{N}\sum_{k=0}^{N-1} X_k e^{j2\pi\left(\frac{k}{N}\right)n}\right] e^{-j2\pi \frac{l}{N} n}, \\
&= \frac{1}{N}\sum_{k=0}^{N-1} X_k \sum_{n=0}^{N-1} e^{j2\pi \frac{n}{N}[k-l]}, \\
&= \frac{1}{N}\sum_{k=0}^{N-1} X_k \left[N\delta[k-l]\right] = X_l,
\end{aligned}$$

$$\Rightarrow \quad X_k = \sum_{n=0}^{N-1} x[n]e^{-j2\pi \frac{k}{N} n}, \tag{12.22}$$

for $k = 0, 1, \ldots, N-1$. *Equation (12.22) is the forward DFS* and *Eq. (12.21) is the inverse DFS.* Comparing Eqs. (12.22) and (12.21) with Eqs. (12.13) and (12.14), respectively, we again see that *the Fourier series representation of a discrete-time periodic signal is same as the DFT of one period of this signal.*

12.2.2 Periodicity and Circular Shift of DFT

Recall that DFT makes the finite-length signal $x[n]$ of N samples to appear periodic with period N. That is,

$$x[n + mN] = x[n] \qquad \text{for any integer} \quad m \tag{12.23}$$

with $n = 0, 1, 2, \ldots, N-1$. This is illustrated in Figure 12.1(a) which shows the periodically repeated version of the signal $x[n]$ for $N = 6$. Figure 12.1(b) illustrates another way of showing the periodic signal $x[n]$. Each period is arranged with N equally spaced points on a circle, in the counter clockwise direction. Therefore, the different periods of the signal, wrap around this circle, as illustrated for three periods in Fig. 12.1.

Taking a closer look at Fig. 12.1 will reveal that *circularly shifting a N-point signal $\{x[0], x[1], \cdots, x[N-1]\}$ is equivalent to linearly shifting its periodic version.* For example, let $\tilde{x}[n] = x[n-2]$. Then, using Fig. 12.1(a), one period of $\tilde{x}[n]$ for $n = 0, 1, \ldots, N-1$ is given by $\{x[-2], x[-1], x[0], \cdots, x[3]\}$ which is same

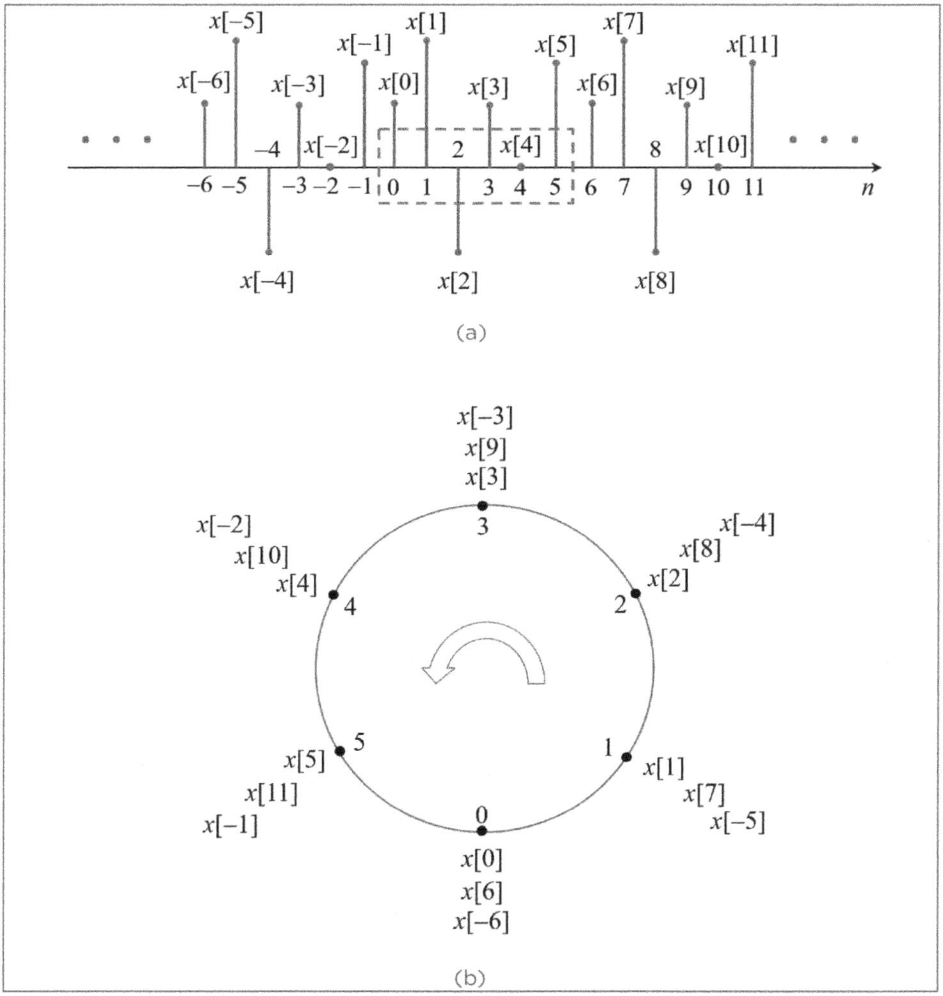

Figure 12.1. Illustration of the time-domain periodicity caused by DFT for a N-point signal record $x[n]$ with $N = 6$. (a) linear representation and (b) circular representation.

as $\{x[4], x[5], x[0], \cdots, x[3]\}$. This can also be obtained from Figure 12.1(b) by starting from the index '4' $[n - 2 + N]$ with $n = 0$, $N = 6$) and completing one circle in the counterclockwise direction.

12.3 Properties of DFT and DFS

We shall now discuss some properties of DFT and DFS. Since DFT and DFS are essentially the same, we shall only refer to DFT in this section.

Let $x[n]$ and $y[n]$ be two discrete-time signals of length N with DFTs X_k and Y_k, respectively. That is,

$$X_k = \sum_{n=0}^{N-1} x[n]e^{-j2\pi \frac{k}{N}n}, \qquad Y_k = \sum_{n=0}^{N-1} y[n]e^{-j2\pi \frac{k}{N}n}$$

$$x[n] = \frac{1}{N}\sum_{k=0}^{N-1} X_k e^{j2\pi \frac{k}{N}n}, \qquad y[n] = \frac{1}{N}\sum_{k=0}^{N-1} Y_k e^{j2\pi \frac{k}{N}n}$$

for $n = 0, 1, 2, \ldots, N-1$ and $k = 0, 1, 2, \ldots, N-1$.

12.3.1 Linearity

Let $p[n] = \alpha x[n] + \beta y[n]$ where α and β are two scalars. Then, we can easily show that

$$P_k = \alpha X_k + \beta Y_k. \tag{12.24}$$

Thus, *DFT is a linear transform.*

12.3.2 Time Shifting

Let $p[n] = x[n - n_0]$ where $n_0 \in \{0, 1, \ldots, N-1\}$. Then, we get (with $m = n - n_0$)

$$\begin{aligned} P_k &= \sum_{n=0}^{N-1} x[n-n_0]e^{-j2\pi \frac{k}{N}n} = \sum_{m=-n_0}^{N-1-n_0} x[m]e^{-j2\pi \frac{k}{N}[m+n_0]} \\ &= e^{-j2\pi \frac{k}{N}n_0} \sum_{m=-n_0}^{N-1-n_0} x[m]e^{-j2\pi \frac{k}{N}m} \\ &= e^{-j2\pi \frac{k}{N}n_0} X_k \end{aligned} \tag{12.25}$$

since $x[m]e^{-j2\pi \frac{k}{N}m}$ is periodic with period N. Thus, as in earlier chapters, the only effect of shifting a signal in time domain is a linear phase change in the frequency domain (no change in magnitude spectrum).

12.3.3 Time Reversal

Let $p[n] = x[-n]$. Then, we get (with $m = N - n$)

$$P_k = \sum_{n=0}^{N-1} x[-n]e^{-j2\pi \frac{k}{N}n} = \sum_{n=0}^{N-1} x[N-n]e^{-j2\pi \frac{k}{N}n}$$

$$= \sum_{m=1}^{N} x[m]e^{-j2\pi \frac{k}{N}[N-m]} = \sum_{m=1}^{N} x[m]e^{j2\pi \frac{k}{N}m}$$
$$= X_{-k}. \tag{12.26}$$

12.3.4 Frequency Shifting

Let $p[n] = e^{j2\pi \frac{l}{N}n}x[n]$ with $l \in \{0, 1, \ldots, N-1\}$. Then, we get

$$P_k = \sum_{n=0}^{N-1} e^{j2\pi \frac{l}{N}n}x[n]e^{-j2\pi \frac{k}{N}n} = \sum_{n=0}^{N-1} x[n]e^{-j2\pi \frac{k-l}{N}n}$$
$$= X_{k-l}. \tag{12.27}$$

12.3.5 Circular Convolution

Let $p[n]$ denote the 'circular convolution' of $x[n]$ and $y[n]$ given by

$$p[n] = x[n] \odot y[n] = \sum_{l=0}^{N-1} x[l]y[n-l]. \tag{12.28}$$

Then, we get (with $m = n - l$)

$$P_k = \sum_{n=0}^{N-1} \left[\sum_{l=0}^{N-1} x[l]y[n-l] \right] e^{-j2\pi \frac{k}{N}n}$$
$$= \sum_{l=0}^{N-1} x[l] \sum_{m=-l}^{N-1-l} y[m]e^{-j2\pi \frac{k}{N}[m+l]}$$
$$= \sum_{l=0}^{N-1} x[l]e^{-j2\pi \frac{k}{N}l} \sum_{m=-l}^{N-1-l} y[m]e^{-j2\pi \frac{k}{N}m}$$
$$= X_k Y_k. \tag{12.29}$$

Thus, in the case of DFT, circular convolution in time domain results in multiplication in frequency domain. This is unlike in CTFT and DTFT where linear convolution[4] in time domain results in multiplication in frequency domain.

---

4. Difference between linear and circular convolutions: Let $a[n]$ and $b[n]$ be N-point data sequences which are zero for n outside $\{0, 1, \ldots, N-1\}$. The linear convolution between $a[n]$ and $b[n]$ is given by

$$p_l[n] = a[n] \otimes b[n] = \sum_{l=0}^{N-1} a[l]b[n-l]$$

Similarly, we can show that multiplication in time domain results in circular convolution of DFTs. That is,

$$q[n] = x[n]y[n] \tag{12.30}$$

$$\Rightarrow \quad Q_k = \frac{X_k \odot Y_k}{N} = \frac{1}{N}\sum_{l=0}^{N-1} X_l Y_{k-l}. \tag{12.31}$$

12.3.6 Real Symmetric Signals

If $x[n]$ is real, i.e., $x[n] = x^*[n]$, then

$$X_k^* = X_{-k} = X_{N-k}. \tag{12.32}$$

If $x[n]$ is real and even-symmetric, i.e., $x[n] = x[-n] = x^*[n]$, then the DFT X_k will be purely real:

$$X_k = X_{-k} = X_k^* = X_{N-k}. \tag{12.33}$$

If $x[n]$ is real and odd-symmetric, i.e., $x[n] = -x[-n] = x^*[n]$, then the DFT X_k will be purely imaginary:

$$X_k = -X_{-k} = -X_k^* = -X_{N-k}. \tag{12.34}$$

12.4 DFT of Sinusoidal Signals

Consider the continuous-time signal $x_a(t)$ obtained by adding two complex sinusoids with amplitudes $\{A_1, A_2\}$ and frequencies $\{f_1, f_2\}$ as

$$x_a(t) = A_1 e^{j2\pi f_1 t} + A_2 e^{j2\pi f_2 t}. \tag{12.35}$$

Let us sample this signal at rate f_s samples/second such that there is no aliasing in the frequency domain. This can be ensured by choosing[5] $f_s > \max\{2|f_1|, 2|f_2|\}$.

---

where $a[n] = 0$ and $b[n] = 0$ if $n \notin \{0, 1, \ldots, N-1\}$. The circular convolution between $a[n]$ and $b[n]$ is given by

$$p_2[n] = a[n] \odot b[n] = \sum_{l=0}^{N-1} \tilde{a}[l]\tilde{b}[n-l]$$

where $\tilde{a}[n]$ and $\tilde{b}[n]$ are obtained by periodically extending $a[n]$ and $b[n]$, respectively. In writing Eq. (12.28), it is assumed that $x[n]$ and $y[n]$ are periodic because of DFT.

5. For complex signals, the sampling frequency for no aliasing can be smaller than twice the maximum frequency in the signal. For example, if $f_1 = -50$ Hz and $f_2 = 120$ Hz, then it is easy to verify that any sampling rate $f_s > 170$ samples/second is sufficient to guarantee no aliasing. This is because the spectra of complex signals may not be symmetric.

Then, we get the sampled signal as (with sampling period $T_s = \frac{1}{f_s}$)

$$x[n] = x_a(nT_s) = A_1 e^{j2\pi \tilde{f}_1 n} + A_2 e^{j2\pi \tilde{f}_2 n} \tag{12.36}$$

where $\tilde{f}_1 = \frac{f_1}{f_s}$ and $\tilde{f}_2 = \frac{f_2}{f_s}$. Let us examine the DFT of this signal $x[n]$ for a N-sample record.

$$\begin{aligned} X_k &= \sum_{n=0}^{N-1} \left[A_1 e^{j2\pi \tilde{f}_1 n} + A_2 e^{j2\pi \tilde{f}_2 n} \right] e^{-j2\pi \frac{k}{N} n} \\ &= A_1 \sum_{n=0}^{N-1} e^{j2\pi \left(\tilde{f}_1 - \frac{k}{N}\right) n} + A_2 \sum_{n=0}^{N-1} e^{j2\pi \left(\tilde{f}_2 - \frac{k}{N}\right) n}. \end{aligned} \tag{12.37}$$

Consider the following cases.

Case 1: The frequencies f_1 and f_2 are such that

$$\frac{k_1}{N} = \frac{f_1}{f_s} \quad \text{and} \quad \frac{k_2}{N} = \frac{f_2}{f_s} \tag{12.38}$$

where k_1 and k_2 belong to $\{0, 1, \ldots, N-1\}$. In this case, we get the DFT from Eq. (12.37) as

$$X_k = NA_1 \delta[k - k_1] + NA_2 \delta[k - k_2] \tag{12.39}$$

This implies that

$$X_k = \begin{cases} NA_1 & \text{if } k = k_1 \\ NA_2 & \text{if } k = k_2 \\ 0 & \text{otherwise.} \end{cases} \tag{12.40}$$

Case 2: The frequencies f_1 and f_2 do not satisfy Eq. (12.38) for any k_1, k_2 in $\{0, 1, \ldots, N-1\}$. In this case, we get the DFT from Eq. (12.37) as

$$X_k = A_1 \frac{1 - e^{j2\pi \left(N\tilde{f}_1 - k\right)}}{1 - e^{j2\pi \left(\tilde{f}_1 - \frac{k}{N}\right)}} + A_2 \frac{1 - e^{j2\pi \left(N\tilde{f}_2 - k\right)}}{1 - e^{j2\pi \left(\tilde{f}_2 - \frac{k}{N}\right)}}$$

$$= A_1 e^{j\pi[N-1]\left(\tilde{f}_1 - \frac{k}{N}\right)} \frac{\sin\left(\pi(k - N\tilde{f}_1)\right)}{\sin\left(\pi(k/N - \tilde{f}_1)\right)}$$

$$+A_2 e^{j\pi[N-1]\left(\tilde{f}_2 - \frac{k}{N}\right)} \frac{\sin\left(\pi(k - N\tilde{f}_2)\right)}{\sin\left(\pi(k/N - \tilde{f}_2)\right)}. \tag{12.41}$$

Case 3: The frequency f_1 satisfies Eq. (12.38) and f_2 does not satisfy Eq. (12.38). In this case, we get the DFT from Eq. (12.37) as

$$X_k = NA_1\delta[k - k_1]$$

$$+A_2 e^{j\pi[N-1]\left(\tilde{f}_2 - \frac{k}{N}\right)} \frac{\sin\left(\pi(k - N\tilde{f}_2)\right)}{\sin\left(\pi(k/N - \tilde{f}_2)\right)}. \tag{12.42}$$

Based on the above, we can infer the following.

- The DFT will result in the expected spectrum of two discrete lines at $\tilde{f}_1$ and $\tilde{f}_2$ only if the frequencies of the sinusoids, (f_1, f_2), and the choice of the sampling frequency, (f_s), and data length, N, satisfy the condition given by Eq. (12.38).
- This condition ensures that the normalized frequencies of the sinusoids correspond to two of the frequencies (k/N) at which the DFT is evaluated.
- Since Eq. (12.38) implies that $N = \frac{k_1 f_s}{f_1} = \frac{k_2 f_s}{f_2}$, this is also equivalent to saying that the data length N contains an integer number of periods of the signal $x_a(t)$. This is because Eq. (12.38) is equivalent to the following:

$$NT_s = k_1 T_1 = k_2 T_2 \tag{12.43}$$

where $T_1 = 1/f_1$ and $T_2 = 1/f_2$ are the periods of the two sinusoids, and NT_s is the time duration over which the signal is observed and sampled.

12.5 Concluding Remarks

We conclude this chapter with the following remarks.

- DFT is a tool for computing the spectrum of signals in practical applications.

- DFT computes the spectrum at discrete frequencies, which are equally spaced, based on finite data records. The spacing between the frequencies are given by $1/N$ where N is the length of the data record.
- The sampling of the spectrum by DFT causes the signal to appear periodic in time domain with period N, even though the original signal record is not periodic.
- DFT for a signal record $\{x[n]\}$ of N samples is the same as the DFS of a periodic signal which is a periodic extension of the signal record $\{x[n]\}$. In other words, *the DFT treats the finite-duration signal* $\{x[0], x[1], x[2], \cdots, x[N-1]\}$ *as one period of a periodic signal with period* N.
- Because of the periodicity in time domain caused by DFT, multiplication of the DFTs of two finite-length sequences becomes equivalent to their circular convolution in time domain. Similarly, multiplication of the data sequences in time domain becomes equivalent to circular convolution of their DFTs.
- The DFT of a sinusoidal signal will indicate the actual frequencies of the sinusoids only if the normalized frequencies of the sinusoids fall exactly on the frequencies at which the DFT is evaluated.

Exercises

12.1. Let us consider the use of DFT to compute the autocorrelation of the complex-valued sequence $x[n]$, that is,

$$r_{xx}[m] = \frac{1}{N} \sum_{n=0}^{N-m-1} x^*[n]x[n+m], \quad m \geq 0$$

Suppose the size M of the DFT is much smaller than that of the data length N. Specifically, assume that $N = KM$.

(a) Determine the steps needed to section $x[n]$ and compute $r_{xx}[m]$ for $-(M/2)+1 \leq m \leq (M/2)-1$, by using $4K$ $M-$point DFTs and one $M-$point IDFT.

(b) Now, consider the following three sequences $x_1[n], x_2[n]$, and $x_3[n]$, each of duration M. Let the sequence $x_1[n]$ and $x_2[n]$ have arbitrary values in the range $0 \leq n \leq (M/2) - 1$, but be zero for $(M/2) \leq n \leq M-1$. the sequence $x_3[n]$ is defined as

$$x_3[n] = \begin{cases} x_1[n], & 0 \leq (M/2) - 1 \\ x_2\left[n - \frac{M}{2}\right] & (M/2) \leq n \leq M-1 \end{cases}$$

Determine a simple relationship among the $M-$point DFTs $X_1[k]$, $X_2[k]$ and $X_3[k]$.

(c) By using the result in part (b), show how the computation of the DFTs in part (a) can be reduced in number from $4K$ to $2K$.

12.2. (a) Find the 89 point DFT of $x[n] = e^{j2\pi \frac{15}{89} n}$.

(b) Find the 89 point DFT of $x[n] = e^{j2\pi \frac{15.7}{89} n}$.

(c) The 128 point DFT of $e^{j2\pi f_1 n}$ has a magnitude of

$$|X[k]| = \left| \frac{\sin\left[\pi\left(\frac{k}{128} - \tilde{f}_1\right) 128\right]}{\sin\left[\pi\left(\frac{k}{128} - \tilde{f}_1\right)\right]} \right|$$

Find $|X[k]|$ for $\tilde{f}_1 = \frac{15}{128}$ and $\frac{15.5}{128}$.

(d) An analog signal $x_a(t) = e^{j\pi 100t}$ is digitized at 500 Hz sampling rate to generate the discrete-time signal $x[n]$. Use $N-$ point DFT to sketch the magnitude spectrum for $N = 4$, 6, and 8.

12.3. A sampling rate of f_s Hz is used to generate the discrete-time signal $x[n]$ from the analog signal $x_a(t) = e^{-2t} u_a(t)$, where $u_a(t)$ denoting the unit step function. Find the spectra of $x[n]$ using the $N-$ point DFT. What happens when N is large?

12.4. (a) The 6-point DFT $X[0], X[1], \ldots, X[5]$ of $x[n]$ is given by

$$X[k] = \sum_{n=0}^{5} x[n] e^{-j2\pi \frac{nk}{6}}, \quad k = 0, 1, \ldots, 5$$

Ignoring the range for k, show that $X[k]$ is periodic and has a period of 6.

(b) The 7-point DFT of $x[n]$ is 1, 2, 3, 4, 5, 6, 7. Find the 7-point DFT of $y[n] = x[n] e^{j2\pi \frac{2}{7} n}$.

12.5. (a) Consider the 6-point DFT $X[0], X[1], \ldots, X[5]$ of a real signal $x[n]$. Relate $X[5]$ to $X[1]$ and $X[4]$ to $X[2]$. Show that $X[0]$ and $X[3]$ are real.

(b) The 6-point DFT $X[0], X[1], \ldots, X[5]$ of a real signal $x[n]$ satisfy $X[5] = X^*[1]$, $X[4] = X^*[2]$, $X[0] =$ real and $X[3] =$ real. Sketch the DFT spectrum.

12.6. (a) Sketch the 6-point DFT spectrum of $x[n] = 2\cos\left(2\pi \frac{n}{6}\right)$.

(b) Find the 9-point DFT of $x[n] = 2\cos(2\pi 0.2n) + 3\sin(2\pi 0.3n)$.

12.7. Find the DFT of $x[n]$, which is given by 4, 2, -2, 0, 2, -2.

12.8. In this problem, let $X[k]$ be the N-point DFT (Discrete Fourier Transform) of the real-valued N-point sequence $x[n]$, $0 \leq n \leq N-1$.

We define a $2N$-point sequence $y[n]$ as

$$y[n] = \begin{cases} x\left[\frac{n}{2}\right] & \text{for } n \text{ even} \\ 0 & \text{for } n \text{ odd.} \end{cases}$$

(a) Express the $2N$-point DFT $Y[k]$ of $y[n]$ in terms of $X[k]$ (*Hint*: Use the appropriate "twiddle factor" to derive the DFT. For example, the twiddle factor W_N is defined as $W_N = e^{-j\frac{2\pi}{N}}$).)

In the following, it is assumed that the N-point DFT $X[k]$ of $x[n]$ is bandlimited. That is, $X[k] = 0$ for $k_0 \leq k \leq N - k_0$, where $k_0 < N$ is a positive integer constant. We insert $(L-1)N$ more zeros in the middle of $X[k]$ to obtain the following LN-point DFT:

$$X'[k] = \begin{cases} X[k] & \text{for } \; 0 \leq k \leq k_0 - 1 \\ 0 & \text{for } \; k_0 \leq k \leq LN - k_0 \\ X[k + N - LN] & \text{for } \; LN - k_0 + 1 \leq k \leq LN - 1, \end{cases}$$

where L is a positive integer constant.

(b) Show that $Lx'[Ln] = x[n], \quad 0 \leq n \leq N-1$, where $x'[n]$ is the IDFT (inverse DFT) of $X'[k]$ (*Hint*: Use the appropriate "twiddle factor" to derive the DFT and IDFT).

In the following, it is assumed that $N = 7$ and $x[n] = 1$ for $n = 0, 1, \ldots, N-1$. The $z-$transform $X(z)$ of $x[n]$ is sampled at 5 points on the unit circle to obtain the corresponding 5 DFT coefficients $X'[k], \; k = 0, 1, \ldots, 4$. This is obtained by substituting $z = e^{j\frac{2\pi}{5}}$ in the expression for $X(z)$.

(c) Determine the IDFT $x'[n]$ of $X'[k]$, compare it with the original sequence $x[n]$ and comment on the results.

12.9. Consider the continuous-time signal, $x_a(t) = 2\cos(80\pi t) + 2\sin(90\pi t)$. We want to use discrete Fourier transform (DFT) for examining the spectrum of this signal.

(a) For this signal, what is the minimum possible sampling frequency that results in alias-free spectrum?

(b) Derive the relation between sampling frequency and DFT-size N so that the DFT spectrum is free from spectral leakage (i.e., the DFT spectrum contains exactly four non-zero values corresponding to the frequencies present in $x_a(t)$). (*Note*: You may use the results available for complex sinusoids to derive this.)

(c) If the sampling frequency is chosen as 91 samples/second, what is the minimum possible value for N so that the DFT spectrum is free from spectral leakage?

(d) Assume that the signal $x_a(t)$ is sampled at the rate of 100 samples/second. Consider the following two cases. In Case 1, you are given 50 samples. In Case 2, you are given 555 samples. Will there be spectral leakage if we compute DFT in these two cases? In which of these cases you prefer to use a non-rectangular window while computing the DFT spectrum? Justify your answer.

DOI: 10.1561/9781680839791.ch13

Chapter 13

Fast Fourier Transform

Fast Fourier transforms (FFT) are widely used algorithms in digital signal processing for applications in several fields including almost all engineering disciplines, science, and mathematics. The basic ideas for the FFT algorithms were proposed by Cooley and Tukey in 1965 [13]. In 1994, Gilbert Strang described the FFT as *the most important numerical algorithm of our lifetime.*[1] In an article that was published in 2000 in the IEEE magazine on Computing in Science and Engineering, FFT algorithm was listed as one of the top 10 Algorithms of the 20th century.[2]

In the last chapter, the DFT and IDFT algorithms were derived. It can be seen that for finding the DFT of a sequence of length N, it requires N^2 multiplications, which is quite high as N increases. Fast Fourier transform (FFT) is an algorithm that computes the DFT of a sequence, or its IDFT with as low as $\frac{N}{2}\log_2 N$ multiplications. This means that the FFT algorithm(s) are much faster than the DFT algorithm, especially when the length of the input sequence N increases.

In this chapter, the fundamental steps in the development of a Radix-2 FFT algorithm is described in detail.

1. G.Strang, "Wavelets", American Scientist, vol. 82, no. 3, 1994, pp. 250–255. JSTOR, www.jstor.org/stable/29775194

2. J. Dongarra and F. Sullivan, "Guest Editors Introduction to the top 10 algorithms", IEEE Magazine on Computing in Science & Engineering. vol. 2, No. 1, pp. 22–23, 2000.

13.1 Fast Fourier Transform (FFT)

The FFT is a computationally efficient approach to compute the DFT spectrum. For example, the computation of a N-point DFT (i.e., $X_k, k = 0, 1, 2, \ldots, N-1$) will require N^2 complex multiplications (and $N(N-1)$ complex additions). This can be quite costly for large values of N.

The principle behind FFT can be explained using a 15-point DFT example shown later in this section. Since the computational complexity of DFT is proportional to the square of the number of samples, the basic principle behind FFT is to transform the problem of computing a N-point DFT into that of computing DFTs of smaller sizes ('Divide -and-conquer' approach), and using these smaller-sized DFTs to obtain the required N-point DFT.

13.1.1 Direct Computation of DFT

Given a sequence of data $\{x[n]\}$ of length N, its N-point DFT sequence $\{X[k]\}$ (complex valued) is calculated according to,

$$X[k] = \sum_{n=0}^{N-1} x[n] W_N^{kn}, \quad 0 \leq k \leq N-1 \tag{13.1}$$

where

$$W_N = e^{-j\frac{2\pi}{N}} \tag{13.2}$$

is the *twiddle factor.*

In general, the sequence $\{x[n]\}$ is also assumed to be complex valued. The IDFT becomes,

$$x[n] = \frac{1}{N} \sum_{k=0}^{N-1} X[k] W_N^{-nk}, \quad 0 \leq n \leq N-1. \tag{13.3}$$

Since the DFT and IDFT involves basically the same type of computations, the method to efficiently compute DFT also applies to IDFT as well.

It can be seen that for the N point DFT, it requires N^2 complex multiplications ($4N^2$ real multiplications) and $N^2 - N$ complex additions ($2N(2N-1) = 4N^2 - 2N$ real additions). This makes the direct computation of DFT highly inefficient especially when N is large. By exploiting the symmetry ($W_N^{k+N/2} = -W_N^k$) and periodicity ($W_N^{k+N} = W_N^k$) properties of the phase factor W_N, an efficient algorithm (known collectively as FFT) can be derived.

13.1.2 Divide-and-Conquer Approach

This approach is based on the decomposition of an N-point DFT into successively smaller DFTs. Assuming that N is not a prime number, it is possible to factor N as the product of two integers as,

$$N = LM \tag{13.4}$$

The assumption that N is not a prime number is not restrictive, since we can pad any sequence with zeros to ensure a factorization as above.

The sequence $\{x[n]\}$, $0 \leq n \leq N-1$, can be stored either in a one-dimensional array indexed by n or as a two-dimensional array indexed by the row index, l ($0 \leq l \leq L-1$) and column index, m ($0 \leq m \leq M-1$) as shown in the following Tables 13.1 and 13.2.

Thus, the sequence $\{x[n]\}$ can be stored in a rectangular array in the following two different ways, each of which depends on the mapping of index n to the indices $[l, m]$.

If we select the mapping,

$$n = Ml + m, \tag{13.5}$$

it leads to the arrangement in which the first row consists of the first M elements of $x[n]$, the second row consists of the next M elements of $x[n]$, and so on, as shown in the Table 13.3 (row-wise).

Table 13.1. Storing of the data in one-dimensional array.

n	0	1	2	$\cdots$	$N-1$
x	$x[0]$	$x[1]$	$x[2]$	$\cdots$	$x[N-1]$

Table 13.2. Two-dimensional data array for storing the data sequence.

$l \setminus m$	0	1	2	$\cdots$	$M-1$
0	$x[0,0]$	$x[0,1]$	$x[0,2]$	$\cdots$	$x[0,M-1]$
1	$x[1,0]$	$x[1,1]$	$x[1,2]$	$\cdots$	$x[1,M-1]$
2	$x[2,0]$	$x[2,1]$	$x[2,2]$	$\cdots$	$x[2,M-1]$
$\vdots$	$\vdots$	$\vdots$	$\vdots$	$\cdots$	$\vdots$
$L-1$	$x[L-1,0]$	$x[L-1,1]$	$x[L-1,2]$	$\cdots$	$x[L-1,M-1]$

Table 13.3. Row-wise arrangements for the data arrays.

$l \setminus m$	0	1	$\cdots$	$M-1$
0	$x[0]$	$x[1]$	$\cdots$	$x[M-1]$
1	$x[M]$	$x[M+1]$	$\cdots$	$x[2M-1]$
2	$x(2M)$	$x[2M+1]$	$\cdots$	$x[3M-1]$
$\vdots$	$\vdots$	$\vdots$	$\cdots$	$\vdots$
$L-1$	$x[[L-1]M]$	$x[[L-1]M+1]$	$\cdots$	$x[LM-1]$

Table 13.4. Column-wise arrangements for the data arrays.

$l \setminus m$	0	1	$\cdots$	$M-1$
0	$x[0]$	$x[L]$	$\cdots$	$x[[M-1]L]$
1	$x[1]$	$x[L+1]$	$\cdots$	$x[[M-1]L+1]$
2	$x[2]$	$x[L+2]$	$\cdots$	$x[[M-1]L+2]$
$\vdots$	$\vdots$	$\vdots$	$\cdots$	$\vdots$
$L-1$	$x[L-1]$	$x[2L-1]$	$\cdots$	$x[LM-1]$

On the other hand, the mapping

$$n = l + mL \tag{13.6}$$

stores the first L elements of $x[n]$ in the first column, the next L elements in the second column, and so on, as shown in the Table 13.4 (column-wise):

A similar arrangement can be used to store the computed DFT values as well. In particular, the mapping is from the index k to a pair of indices $[p, q]$, where $0 \le p \le L-1$ and $0 \le q \le M-1$. If we select the mapping $k = Mp + q$, the DFT ($X[k]$) is stored *row-wise* and if the mapping is $k = qL + p$, then the DFT is stored *column-wise.*

Now, suppose that $x[n]$ is mapped into the rectangular array $x[l, m]$ and $X[k]$ is mapped into the corresponding rectangular array $X[p, q]$. Then the DFT can be expressed as a double sum over the elements of the rectangular array multiplied by the corresponding phase vectors. To be specific, let us adopt a column-wise mapping for $x[n]$ and row-wise mapping for the DFT. Then,

$$X[p, q] = \sum_{m=0}^{M-1} \sum_{l=0}^{L-1} x[l, m] W_N^{(Mp+q)(mL+l)}. \tag{13.7}$$

However, $W_N^{Nmp} = 1$, $W_N^{mqL} = W_{N/L}^{mq} = W_M^{mq}$, and $W_N^{Mpl} = W_{N/M}^{pl} = W_L^{pl}$. With these simplifications, the above equation can be expressed as

$$X[p,q] = \sum_{l=0}^{L-1} \left\{ W_N^{lq} \left[\sum_{m=0}^{M-1} x[l,m] W_M^{mq} \right] \right\} W_L^{lp}. \tag{13.8}$$

This expression involves the computation of DFTs of length M and L. To see this, let us subdivide the computations into three steps:

- First, we compute the M-point DFTs

$$F[l,q] = \sum_{m=0}^{M-1} x[l,m] W_M^{mq}, \quad 0 \le q \le M-1, \tag{13.9}$$

  for each of the rows $l = 0, 1, \ldots, L-1$.
- Second, compute a new rectangular array $G[l,q]$ defined as

$$G[l,q] = W_N^{lq} F[l,q] \tag{13.10}$$

  for $0 \le l \le L-1$ and $0 \le q \le M-1$.
- Finally, compute the L-point DFTs

$$X[p,q] = \sum_{l=0}^{L-1} G[l,q] W_L^{lp}, \tag{13.11}$$

  for each column $q = 0, 1, \ldots, M-1$, of the array $G[l,q]$.

By looking at the procedure, one may feel that the outlined procedure is more complex than the direct computation of the DFT. However, by evaluating the computational complexity of Eq. (13.8), this doubt can be cleared. The first step involves the computation of L DFTs, each of M points (LM^2 complex multiplications and $LM(M-1)$ complex additions). The second step requires LM complex multiplications. The final step in the computation requires ML^2 complex multiplications and $ML(L-1)$ complex additions. So, there are a total of $N(M+L+1)$ complex multiplications and $N(M+L-2)$ complex additions, where $N = ML$. The number of complex multiplications has been reduced from N^2 to $N(M+L+1)$ and the number of complex additions has been reduced from $N(N-1)$ to $N(M+L-2)$. For example, suppose $N = 1000$ and we select $L = 2$ and $M = 500$. Then instead of having to perform 10^6 complex multiplications, this approach reduces to 503, 000 complex multiplications (a reduction of $\approx$50%). The number of additions is also reduced by a factor of 2.

When N is a highly composite number, that is, N can be factored into a product of prime numbers of the form

$$N = r_1 r_2 \dots r_\nu \tag{13.12}$$

then the decomposition above can be repeated $(\nu - 1)$ more times. This procedure results in smaller DFTs, which, in turn, leads to a more efficient computational algorithm.

Example 1: Given the input $x[n]$ of length 15, find the 15 point DFT of the sequence.

The given input sequence is $x[0], x[1], \dots, x[14]$. $N = 15$. Since $N = 5\times 3 = 15$, we select $L = 5$ and $M = 3$ so that the 15 point sequence is stored column-wise as follows:

Row 1: $x[0,0] = x[0]$ $x[0,1] = x[5]$ $x[0,2] = x[10]$

Row 2: $x[1,0] = x[1]$ $x[1,1] = x[6]$ $x[1,2] = x[11]$

Row 3: $x[2,0] = x[2]$ $x[2,1] = x[7]$ $x[2,2] = x[12]$

Row 4: $x[3,0] = x[3]$ $x[3,1] = x[8]$ $x[3,2] = x[13]$

Row 5: $x[4,0] = x[4]$ $x[4,1] = x[9]$ $x[4,2] = x[14]$

Now, we compute the 3-point DFTs for each of the five rows giving us the following 5×3 array:

$$\begin{matrix} F[0,0] & F[0,1] & F[0,2] \\ F[1,0] & F[1,1] & F[1,2] \\ F[2,0] & F[2,1] & F[2,2] \\ F[3,0] & F[3,1] & F[3,2] \\ F[4,0] & F[4,1] & F[4,2] \end{matrix}$$

The next step is to multiply each of the terms $F[l,q]$ by the phase factors $W_N^{lq} = W_{15}^{lq}$, $0 \le l \le 4$ and $0 \le q \le 2$. This results in the following 5×3 array:

Column 1	Column 2	Column 3
$G[0,0]$	$G[0,1]$	$G[0,2]$
$G[1,0]$	$G[1,1]$	$G[1,2]$
$G[2,0]$	$G[2,1]$	$G[2,2]$
$G[3,0]$	$G[3,1]$	$G[3,2]$
$G[4,0]$	$G[4,1]$	$G[4,2]$

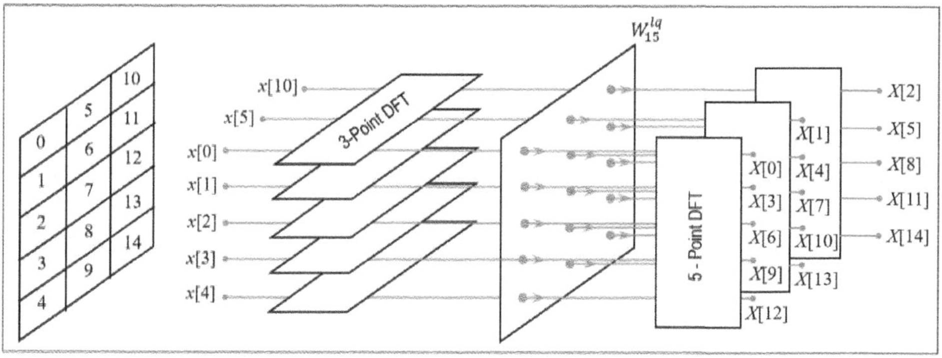

Figure 13.1. Computation of $N = 15$-point DFT by means of 3-point and 5-point DFTs.

The final step is to compute the 5-point DFTs for each of the three columns. This gives the desired values of the DFT in the form

$$
\begin{array}{lll}
X[0,0] = X[0] & X[0,1] = X[1] & X[0,2] = X[2] \\
X[1,0] = X[3] & X[1,1] = X[4] & X[1,2] = X[5] \\
X[2,0] = X[6] & X[2,1] = X[7] & X[2,2] = X[8] \\
X[3,0] = X[9] & X[3,1] = X[10] & X[3,2] = X[11] \\
X[4,0] = X[12] & X[4,1] = X[13] & X[4,2] = X[14]
\end{array}
$$

Figure 13.1 illustrates the computational steps. It may be noted that the input sequence is in shuffled order, whereas the output sequence is in normal order. Input sequence order: $x[0], x[5], x[10], x[1], x[6], x[11], x[2], x[7], x[12], x[3], x[8], x[13], x[4], x[9], x[14]$.

To summarize, the steps involved in the algorithm are:

Algorithm 1:

- Store the signal column-wise.
- Compute the M-point DFT of each row.
- Multiply the resulting array by the phase factors W_N^{lq}.
- Compute the L-point DFT of each column.
- Read the resulting array row-wise.

Algorithm 2:

- Store the signal row-wise.
- Compute the L-point DFT of each column.
- Multiply the resulting array by the phase factors W_N^{pm}.
- Compute the M-point DFT of each row.
- Read the resulting array column-wise.

13.2 Radix-2 FFT Algorithm

In the previous sub-section, we have shown that the N-point DFT can be efficiently computed by factoring N as the product of prime numbers $r_1, r_2, \ldots, r_\nu$. One of the special case is when $r_1 = r_2 = \ldots = r_\nu \equiv r$, so that $N = r^\nu$. In such a case, the DFTs are of the same size (r) and the computation of the N-point DFT has a regular pattern. Here, the number r is called the *radix* of the FFT algorithm. When $r = 2$, the resulting FFT algorithm is termed the radix-2 FFT algorithm.

Let us consider the computation of the $N = 2^\nu$ point DFT by divide-and-conquer approach. We select $M = N/2$ and $L = 2$. This selection results in splitting the N point data sequence into two $N/2$ point data sequences $f_1[n]$ and $f_2[n]$, corresponding to the even-numbered and odd-numbered samples of $x[n]$, respectively as

$$f_1[n] = x[2n], \quad n = 0, 1, \ldots, \frac{N}{2} - 1, \tag{13.13}$$

$$f_2[n] = x[2n+1]. \tag{13.14}$$

Here $f_1[n]$ and $f_2[n]$ are obtained by decimating $x[n]$ by a factor of 2 and hence the resulting FFT algorithm is called the *decimation-in-time* algorithm.

Now, the N-point DFT can be expressed in terms of the DFTs of the decimated sequences as follows:

$$\begin{aligned} X[k] &= \sum_{n=0}^{N-1} x[n] W_N^{kn}, \quad k = 0, 1, \ldots, N-1, \\ &= \sum_{n\ even} x[n] W_N^{kn} + \sum_{n\ odd} x[n] W_N^{kn}, \\ &= \sum_{m=0}^{N/2-1} x[2m] W_N^{2mk} + \sum_{m=0}^{N/2-1} x[2m+1] W_N^{k[2m+1]}. \end{aligned} \tag{13.15}$$

But $W_N^2 = W_{N/2}$. With this substitution, the above Eq. becomes

$$\begin{aligned} X[k] &= \sum_{m=0}^{N/2-1} f_1[m] W_{N/2}^{mk} + W_N^k \sum_{m=0}^{N/2-1} f_2[m] W_{N/2}^{mk}, \\ &= F_1[k] + W_N^k F_2[k], \quad k = 0, 1, \ldots, N-1. \end{aligned} \tag{13.16}$$

where $F_1[k]$ and $F_2[k]$ are the $N/2$-point DFTs of $f_1[n]$ and $f_2[n]$, respectively.

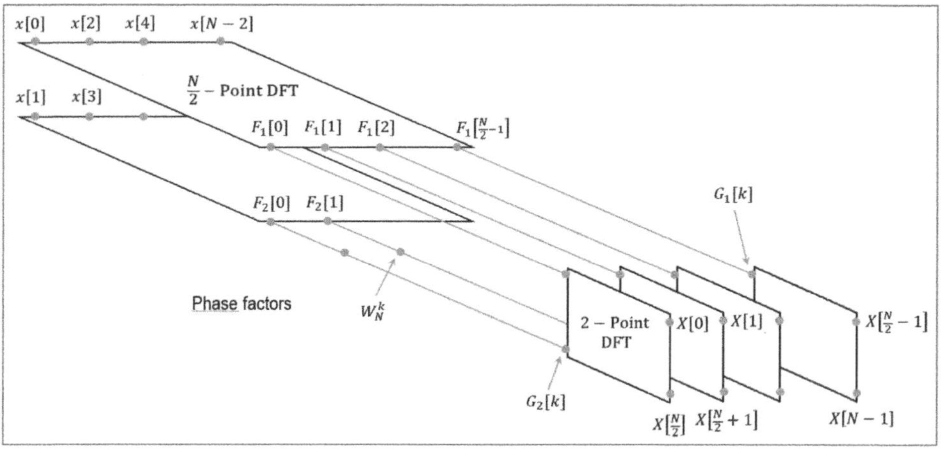

Figure 13.2. First step in the decimation-in-time algorithm.

Since $F_1[k]$ and $F_2[k]$ are periodic with period $N/2$, we have $F_1[k + N/2] = F_1[k]$ and $F_2[k + N/2] = F_2[k]$. In addition, the factor $W_N^{k+N/2} = -W_N^k$. Hence, Eq. (13.16) can be expressed as (for $k = 0, 1, \ldots, \frac{N}{2} - 1$),

$$X[k] = F_1[k] + W_N^k F_2[k], \tag{13.17}$$

$$X\left[k + \frac{N}{2}\right] = F_1[k] - W_N^k F_2[k]. \tag{13.18}$$

It can be seen that to compute the N-point DFT of $x[n]$, by this process, the number of complex multiplications has been reduced to $2(N/2)^2 + N/2 = N^2/2 + N/2$ compared to N^2 (for direct computation); a reduction of a factor of ≈ 2 for large N.

Figure 13.2 illustrates the first step in the *decimation-in-time* algorithm.

Having performed the decimation-in-time once, we can repeat the process for each of the sequences $f_1[n]$ and $f_2[n]$. Thus $f_1[n]$ would result in two $N/4$ point sequences. The number of complex multiplications to compute $F_1[k]$ and $F_2[k]$ is now reduced to $N^2/4 + N/2$. To compute $X[k]$ using $F_1[k]$ and $F_2[k]$, we require another $N/2$ complex multiplications. Consequently, the number of multiplications is reduced again by a factor of 2 to $N^2/4 + N$.

This process can be repeated until the resulting sequences are reduced to one-point sequences. For $N = 2^\nu$, this decimation can be performed $\nu = \log_2 N$ times. Thus the total number of complex multiplications reduces to $(N/2)\log_2 N$. The number of complex additions is $N \log_2 N$. A comparison of the computational complexity for the direct computation of the DFT versus the FFT algorithm for different N is provided in Table 13.5.

Table 13.5. Comparison of the computational complexity for the direct computation of the DFT versus the FFT algorithm.

N	N^2 (DFT)	$\frac{N}{2}\log_2 N$ (FFT)	Improvement factor
4	16	4	4.0
8	64	12	5.3
64	4096	192	21.3
256	65536	1024	64.0
1024	1048576	5120	204.8

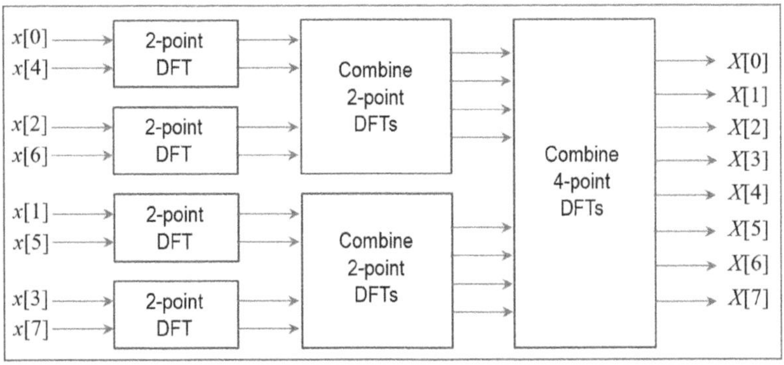

Figure 13.3. Three stages in the computation of an $N = 8$-point DFT.

For illustration, Fig. 13.3 depicts the computation of an $N = 8$-point DFT. This is achieved in three stages, beginning with 4 two-point DFTs, then 2 four-point DFTs and finally, one 8-point DFT. The combination of the smaller DFTs to form the larger DFT is illustrated in Fig. 13.4.

Observe that the basic computation performed at every stage, is to take two complex numbers, say the pair $[a, b]$, multiply b with W_N^r, and then add and subtract the product from a to form the two new complex numbers (A, B). This basic computation is called the *butterfly* (Fig. 13.5) structure because the flow-graph resembles a butterfly.

Each butterfly involves one complex multiplication and two complex additions. For $N = 2^\nu$, there are $N/2$ butterflies per stage of the computation process and $\log_2 N$ stages, which results in $(N/2)\log_2 N$ complex multiplications and $N \log_2 N$ complex additions.

An important observation is concerned with the order of the input data sequence after it is decimated $(\nu - 1)$ times. Consider the case where $N = 8$. After the first decimation, we get the sequence order as $x[0], x[2], x[4], x[6], x[1], x[3], x[5], x[7]$. After the second decimation, the order becomes $x[0], x[4], x[2], x[6], x[1], x[5], x[3], x[7]$. This shuffling of the input data sequence

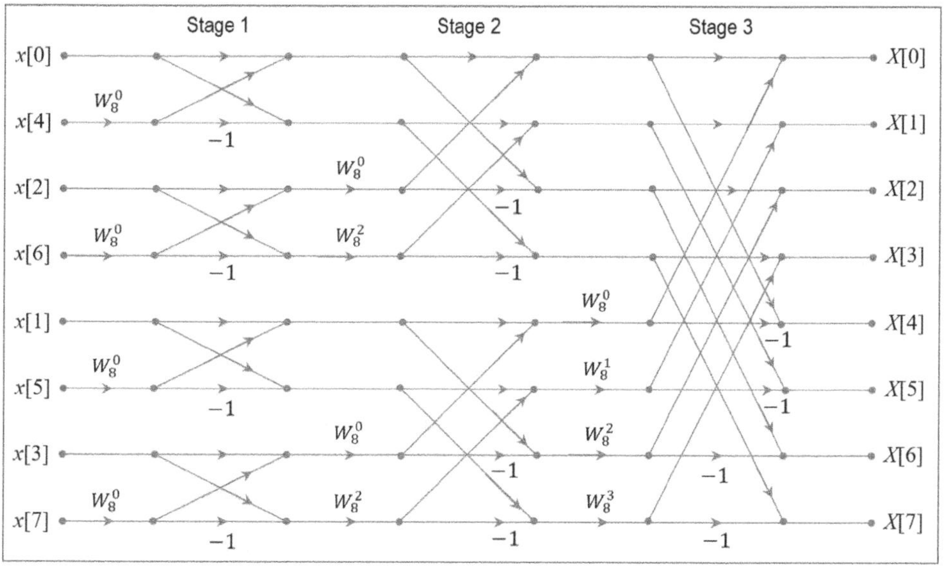

Figure 13.4. Eight-point decimation-in-time FFT algorithm.

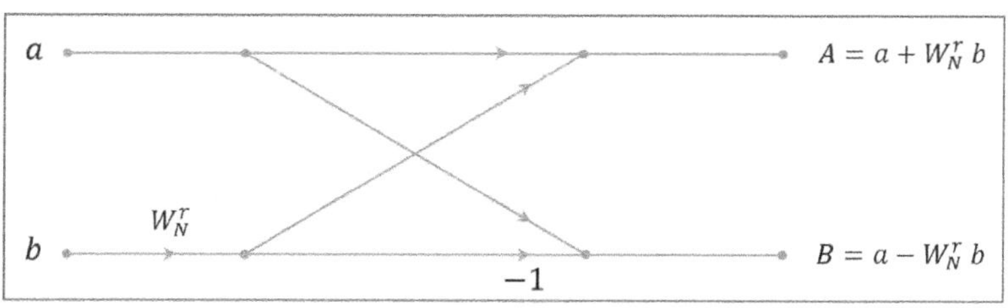

Figure 13.5. Basic butterfly computation in the decimation-in-time FFT algorithm.

has a well-defined order as can be seen from Fig. 13.6 and Table 13.6. By expressing the data index n in binary form, we note that the order of the decimated data sequence is easily obtained by reading the binary representation in the reverse order.

Another important radix-2 FFT algorithm, called the *decimation-in-frequency algorithm*, is obtained by using the divide-and-conquer approach with the choice of $M = 2$ and $L = N/2$. This choice implies a column-wise storage of the input data sequence [5].

§ Radix-4 FFT Algorithms: *When the number of data points (N) in the DFT is a power of 4 (i.e., $N = 4^\nu$), we can of course always use a radix-2 algorithm for the computation. However, for this case it is more efficient to employ a radix-4 FFT algorithm.*

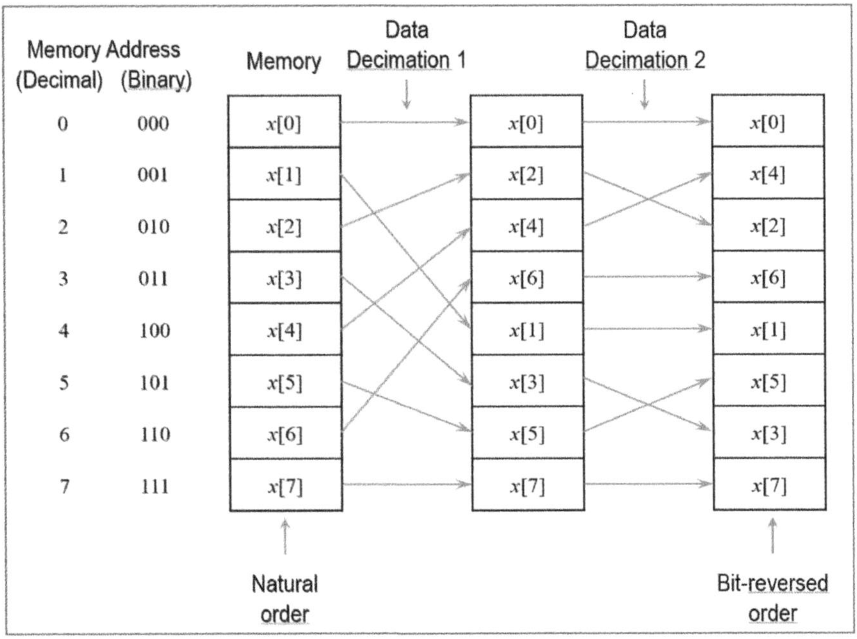

Figure 13.6. Shuffling of the data and bit reversal.

Table 13.6. Shuffling of the data and bit reversal.

Original order [n]	$n_2n_1n_0$	$n_1n_0n_2$	$n_0n_1n_2$	Shuffled order
0	000	000	000	0
1	001	010	100	4
2	010	100	010	2
3	011	110	110	6
4	100	001	001	1
5	101	011	101	5
6	110	101	011	3
7	111	111	111	7

13.2.1 Applications of FFT Algorithms

Basically, the FFT algorithm is used as an efficient means to compute the DFT and IDFT of sequences. Other applications include linear filtering, correlation, and spectrum analysis [5].

13.3 Concluding Remarks

We conclude this chapter with the following remarks.

- The fast Fourier transform is a computationally efficient means of computing DFT, requiring only $\frac{N}{2}\log_2 N$ multiplications, compared with N^2 required for direct computation.

Exercises

13.1. To efficiently compute the 256 point FFT of two real signas $x[n]$ and $y[n]$, we form $w[n] = x[n] + jy[n]$ and compute $W[k]$. Can $X[3]$, $X[253]$, $Y[3]$ and $Y[253]$ be found from $W[3]$ and $W[253]$?

13.2. To compute the 64-point FFT of $x[0], \ldots, x[63]$, we divide $x[n]$ into four sequences:

$$x_0[n] = x[0], x[4], \ldots$$
$$x_1[n] = x[1], x[5], \ldots$$
$$x_2[n] = x[2], x[6], \ldots$$
$$x_3[n] = x[3], x[7], \ldots$$

Can the 64-point FFT of $X[k]$ be found from combining the 16-point FFTs of these sequences?

13.3. Dual Tone Frequency Modulation (DTMF) is used in touch-tone keypad. Pressing '1' gives two real sinusoids at 697 and 1209 Hz. Pressing other keys give two sinusoids at other frequencies. Consider a simpler two key single tone system. '1' gives a 506 Hz and '2' gives a 1215 Hz tone. We need to detect the keys or tones every 0.01 seconds or so. We need to sample at 8 kHz as the telephone speech has a bandwidth of up to about 4 kHz.

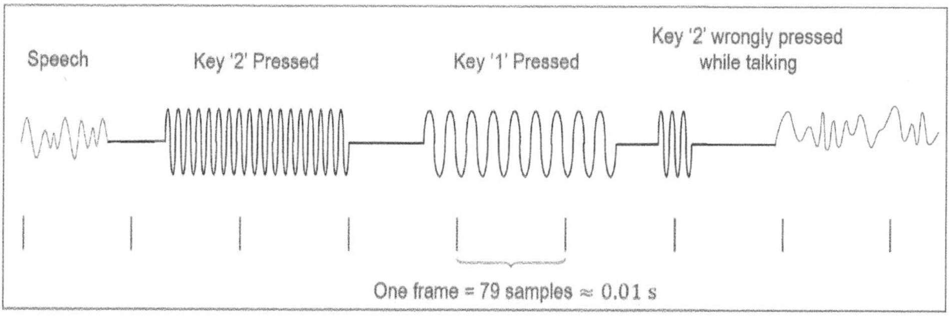

Figure P13.3.

(a) If DFT is to be used in this case, what should the DFT size be?
(b) Suppose a 79 point DFT is used in Question (a) above. Sketch the DFT spectrum for the signal shown in Fig. P13.3 below and indicate whether the keys will be detected.

DOI: 10.1561/9781680839791.ch14

Chapter 14

Design of Digital Filters

A digital filter is a LTI discrete-time system that is realized using finite precision arithmetic. The design of digital filters involves three basic and independent steps:

- The specification of the desired properties of the system.
- The approximation of these specifications using a causal discrete-time system.
- The realization of these specifications using finite precision arithmetic.

In the filter design process, we determine the coefficients of a causal FIR or IIR filter that closely approximates the desired frequency response specifications. The issue of which type of filter to design, FIR or IIR, depends on the application as well as the specifications of the desired frequency response. In practice, FIR filters are employed in problems where there is a requirement of linear phase characteristics within the pass-band of the filter. If such requirements are not specified, either an IIR or FIR filter may be employed. However, as a general rule, an IIR filter has lower side-lobes in the stop-band than the FIR filter having the same number of parameters. For this reason, if some phase distortion is either tolerable or unimportant, an IIR filter is preferable, primarily because of its fewer parameters for implementation. Further it require less memory and has lower computational complexity.

14.1 Characteristics of Practical Frequency-Selective Filters

Figure 14.1 illustrates the desired frequency response characteristics of different types of filters. Ideal filters are non-causal and hence physically unrealizable for real-time signal processing applications. Causality implies that the frequency response characteristics $H(\omega)$ of the filter cannot be zero, except at a finite set of points in the frequency range. In addition, $H(\omega)$ cannot have an infinitely sharp cut-off from pass-band to stop-band, that is $H(\omega)$ cannot drop from unity to zero abruptly. In this chapter, we limit our discussions to the design of the class of LTI systems specified by the difference equation

$$y[n] = -\sum_{k=1}^{N} a_k y[n-k] + \sum_{k=0}^{M-1} b_k x[n-k], \tag{14.1}$$

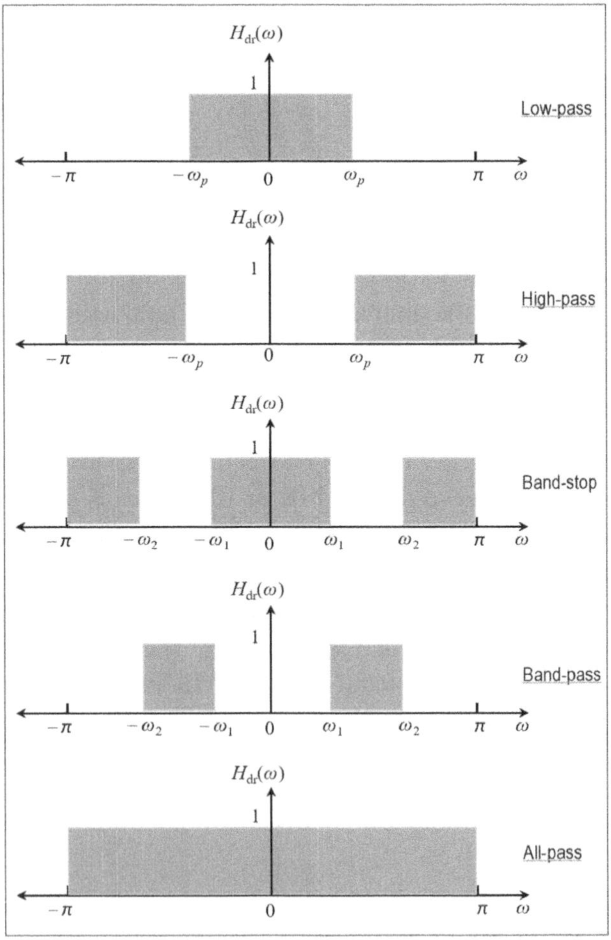

Figure 14.1. Ideal frequency response characteristics of different types of filters.

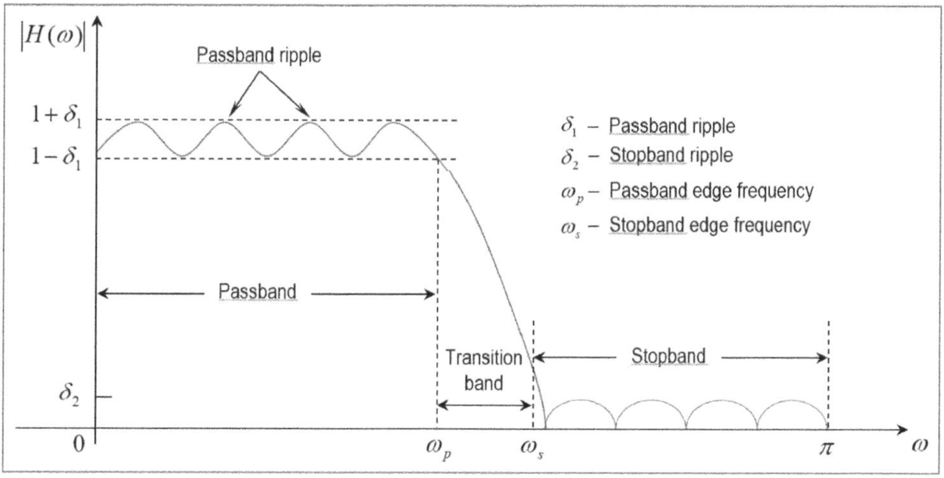

Figure 14.2. Magnitude characteristics of physically realizable filters.

which are causal and physically realizable. Such systems have a frequency response

$$H(\omega) = \frac{\sum_{k=0}^{M-1} b_k e^{-j\omega k}}{1 + \sum_{k=1}^{N} a_k e^{-j\omega k}}. \tag{14.2}$$

Although the ideal filter characteristics are desirable, they are not absolutely necessary in most practical applications. If we relax these conditions, it is possible to realize causal filters that approximate the ideal filters. The magnitude characteristics of such a filter (low-pass) is shown in Fig. 14.2. In the pass-band and stop-band of the filter, you may note that the magnitude response is deviated from the ideal characteristics. These deviations are specified by the pass-band (denoted as δ_1) and stop-band (denoted as δ_2) ripples. It may also be noted that there exist a transition band ($\omega_s - \omega_p$) in a practical filter where as it is absent in an ideal filter.

In any filter design problem, we can specify, (i) the maximum tolerable pass-band ripple, (ii) the maximum tolerable stop-band ripple, (iii) pass-band edge frequency (ω_p), and (iv) the stop-band edge frequency (ω_s). Based on these specifications, we can select the parameters $\{a_k\}$ and $\{b_k\}$ in the frequency response characteristics given by Eq. (14.2), which best approximates the desired specifications. The degree to which $H(\omega)$ approximates the specifications depends on the criterion used in the selection of the filter coefficients $\{a_k\}$ and $\{b_k\}$ as well as on N and M.

14.2 Design of FIR Filters

In this section, we discuss three different methods for the design of linear phase FIR digital filters. They are:

- Windowing
- Frequency Sampling
- Optimization Method

14.2.1 Symmetric and Anti-symmetric FIR Filters

An FIR filter of length M with input $x[n]$ and output $y[n]$ can be described by the difference equation

$$y[n] = b_0x[n] + b_1x[n-1] + \ldots + b_{M-1}x[n-M+1] = \sum_{k=0}^{M-1} b_k x[n-k], \tag{14.3}$$

where $\{b_k\}$ is the set of filter coefficients. Alternatively, the output of the filter can be described as the convolution of the input and the filter's impulse response $h[n]$. Thus we have,

$$y[n] = \sum_{k=0}^{M-1} h[k]x[n-k]. \tag{14.4}$$

Here, the lower and upper limits of the convolution sum reflect the causality and finite-duration characteristics of the filter. Clearly Eq. (14.3) and Eq. (14.4) are identical in their form and hence $b_k = h[k],\ k = 0, 1, \ldots, M-1$. The filter can also be characterized by its system function

$$H(z) = \sum_{k=0}^{M-1} h[k]z^{-k}. \tag{14.5}$$

The roots of the above polynomial constitutes the *zeros* of the filter.

An FIR filter has linear phase if it satisfies the condition

$$h[n] = \pm h[M-1-n], \quad n = 0, 1, \ldots, M-1. \tag{14.6}$$

When the symmetry (+) and anti-symmetry (−) conditions are incorporated in Eq. (14.5), we have,

$$H(z) = h[0] + h[1]z^{-1} + \ldots + h[M-1]z^{-[M-1]} \tag{14.7}$$

$$\Longrightarrow H(z) = \begin{cases} z^{-[M-1]/2}\,\{h_{odd}\} & \text{for } M \text{ odd} \\ z^{-[M-1]/2}\,\{h_{even}\} & \text{for } M \text{ even} \end{cases} \tag{14.8}$$

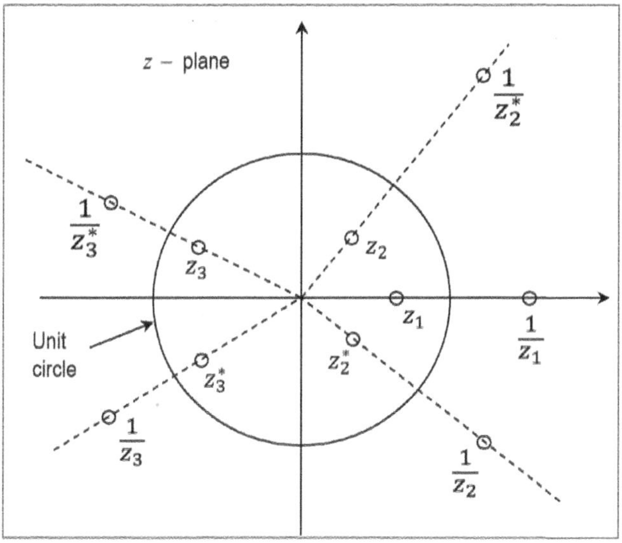

Figure 14.3. Symmetry of zero locations for a linear-phase FIR filter.

where

$$h_{odd} = h\left[\frac{M-1}{2}\right] + \sum_{n=0}^{\frac{M-3}{2}} h[n]\left[z^{\left(\frac{M-1-2n}{2}\right)} \pm z^{-\left(\frac{M-1-2n}{2}\right)}\right], \quad (14.9)$$

$$h_{even} = \sum_{n=0}^{\frac{M}{2}-1} h[n]\left[z^{\left(\frac{M-1-2n}{2}\right)} \pm z^{-\left(\frac{M-1-2n}{2}\right)}\right]. \quad (14.10)$$

Now, if we substitute z^{-1} for z in Eq. (14.5) and multiply both sides of the equation by $z^{-[M-1]}$, we get

$$z^{-[M-1]}H(z^{-1}) = \pm H(z). \quad (14.11)$$

This implies that the roots of the polynomial $H(z)$ are identical to the roots of the polynomial $H(z^{-1})$. That is, the roots of $H(z)$ must occur in reciprocal pairs. That is if z_1 is a root or zero of $H(z)$, then $1/z_1$ is also a root. Furthermore, if the impulse response $h[n]$ of the filter is real, then the complex valued roots must occur in conjugate pairs. Hence if z_1 is a complex valued root, then z_1^* is also a root. $H(z)$ also has a root at $1/z_1^*$. Figure 14.3 shows all these for the case of a linear phase FIR filter.

The frequency response characteristics $H(\omega)$ of the linear-phase FIR filters are obtained by evaluating Eq. (14.8) on the unit circle.

When $h[n] = h[M-1-n]$ (i.e., symmetric linear phase FIR filter): $H(\omega)$ can be expressed as

$$H(\omega) = H_r(\omega)e^{-j\omega\left[\frac{M-1}{2}\right]}, \tag{14.12}$$

where $H_r(\omega)$ is a real function of ω and can be expressed as

$$H_r(\omega) = h\left[\frac{M-1}{2}\right] + 2\sum_{n=0}^{\frac{M-3}{2}} h[n]\cos\omega\left[\frac{M-1}{2} - n\right], \tag{14.13}$$

when M is odd.

$$H_r(\omega) = 2\sum_{n=0}^{\frac{M}{2}-1} h[n]\cos\omega\left[\frac{M-1}{2} - n\right], \tag{14.14}$$

when M is even.

The phase characteristics of the filter for both odd and even M are

$$\Theta(\omega) = \begin{cases} -\omega\left[\frac{M-1}{2}\right], & \text{if } H_r(\omega) > 0 \\ -\omega\left[\frac{M-1}{2}\right] + \pi, & \text{if } H_r(\omega) < 0. \end{cases} \tag{14.15}$$

When $h[n] = -h[M-1-n]$ (i.e., anti-symmetric linear phase FIR filter): The unit sample response is anti-symmetric. For M odd, the centre point of the anti-symmetric $h[n]$ is $n = \frac{M-1}{2}$. Hence,

$$h\left[\frac{M-1}{2}\right] = 0. \tag{14.16}$$

However, if M is even, each term in $h[n]$ has a matching term of opposite sign. The frequency response of this anti-symmetric FIR filter is

$$H(\omega) = H_r(\omega)e^{j\left[-\omega\left[\frac{M-1}{2}\right] + \frac{\pi}{2}\right]}, \tag{14.17}$$

where

$$H_r(\omega) = 2\sum_{n=0}^{\frac{M-3}{2}} h[n]\sin\omega\left[\frac{M-1}{2} - n\right], \tag{14.18}$$

when M is odd and

$$H_r(\omega) = 2 \sum_{n=0}^{\frac{M}{2}-1} h[n] \sin \omega \left[\frac{M-1}{2} - n \right], \tag{14.19}$$

when M is even. The corresponding phase characteristics of the the filter are

$$\Theta(\omega) = \begin{cases} \frac{\pi}{2} - \omega \left[\frac{M-1}{2} \right], & \text{if} \;\; H_r(\omega) > 0 \\ \frac{3\pi}{2} - \omega \left[\frac{M-1}{2} \right], & \text{if} \;\; H_r(\omega) < 0 \end{cases} \tag{14.20}$$

These general frequency response formulas can be used to design linear-phase FIR filters with symmetric and anti-symmetric impulse responses. The number of filter coefficients that specify the frequency response for different types of filters is shown in Table 14.1 below:

Table 14.1. Number of filter coefficients that specify the frequency response of different types of FIR filters.

Filter type	M odd	M even
Symmetric	$\frac{M+1}{2}$	$\frac{M}{2}$
Anti-symmetric	$\frac{M-1}{2}$, $h\left[\frac{M-1}{2}\right] = 0$	$\frac{M}{2}$

14.2.2 Design of Linear-Phase FIR Digital Filters Using Windowing

We begin with the desired frequency response specification $H_d(\omega)$ and determine the corresponding unit sample response $h_d[n]$. We have ($H_d(\omega) \leftrightarrow h_d[n]$),

$$H_d(\omega) = \sum_{n=0}^{\infty} h_d[n] e^{-j\omega n}, \tag{14.21}$$

where

$$h_d[n] = \frac{1}{2\pi} \int_{-\pi}^{\pi} H_d(\omega) e^{j\omega n}. \tag{14.22}$$

So, given $H_d(\omega)$, the unit sample response $h_d[n]$ can be evaluated using Eq. (14.22). In general, this $h_d[n]$ is infinite in duration and must be truncated at some point, say at $n = M - 1$, to provide a finite length (order M) FIR filter. This truncation is equivalent to multiplying $h_d[n]$ by a finite "rectangular window", defined as

$$w[n] = \begin{cases} 1 & \text{for } n = 0, 1, \ldots, M-1 \\ 0 & \text{otherwise} \end{cases} \tag{14.23}$$

Thus the unit sample response of the FIR filter becomes

$$h[n] = h_d[n]w[n] \tag{14.24}$$

which means that,

$$h[n] = \begin{cases} h_d[n] & \text{for } n = 0, 1, \ldots, M-1 \\ 0 & \text{otherwise} \end{cases} \tag{14.25}$$

We have to see the effect of this multiplication on the desired frequency response $H_d(\omega)$. To see this, recall the multiplication in time domain is equivalent to the convolution of $H_d(\omega)$ and $W(\omega) = \sum_{n=0}^{M-1} w[n]e^{-j\omega n}$ (Fourier transform of $w[n]$) in the frequency domain. Thus,

$$H(\omega) = \frac{1}{2\pi}\int_{-\pi}^{\pi} H_d(\nu)W(\omega - \nu)d\nu \tag{14.26}$$

We have

$$\begin{aligned} W(\omega) &= \sum_{n=0}^{M-1} e^{-j\omega n} \\ &= \frac{1 - e^{-j\omega M}}{1 - e^{-j\omega}} \\ &= e^{-j\omega[M-1]/2}\left(\frac{\sin(\omega M/2)}{\sin(\omega/2)}\right). \end{aligned} \tag{14.27}$$

This function has a magnitude response

$$|W(\omega)| = \frac{|\sin(\omega M/2)|}{|\sin(\omega/2)|}, \quad -\pi \le \omega \le \pi \tag{14.28}$$

and a piece-wise linear phase

$$\Theta(\omega) = \begin{cases} -\omega\left[\frac{M-1}{2}\right] & \text{when } \sin(\omega M/2) \ge 0 \\ -\omega\left[\frac{M-1}{2}\right] + \pi, & \text{when } \sin(\omega M/2) < 0. \end{cases} \tag{14.29}$$

The magnitude response of the window function is shown in Fig. 14.4 for $M = 31$ and $M = 61$. The width of the main lobe is $\frac{4\pi}{M}$. Hence as M increases, the

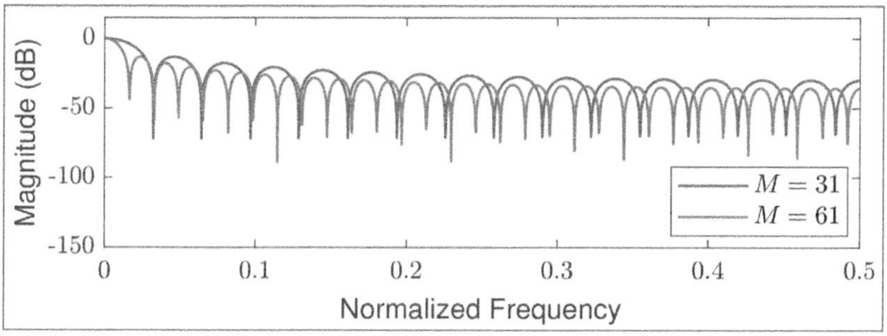

Figure 14.4. Frequency response for rectangular window of length (a) $M = 31$, (b) $M = 61$.

main lobe becomes narrower. However, the side-lobes of $|W(\omega)|$ are relatively high and remain unaffected by an increase in M. In fact, even though the width of each side-lobe decreases with an increase in M the height of each side-lobe increases with an increase in M in such a manner that the area under each side-lobe remains invariant to changes in M. This characteristic behaviour is not evident from Fig. 14.4 because $W(\omega)$ has been normalized by M such that the normalized peak values of the side lobes remain invariant to an increase in M. The characteristics of the rectangular window plays a significant role in determining the frequency response of the truncated unit impulse response. Specifically, the convolution of $H_d(\omega)$ with $W(\omega)$ has the effect of smoothing $H_d(\omega)$. As M increases, $W(\omega)$ becomes narrower, and the smoothing is reduced. On the other hand, the large side lobes of $W(\omega)$ result in some undesirable ringing[1] effects in the FIR filter frequency response $H(\omega)$ and also in relatively large side lobes of $H(\omega)$. These undesirable effects can be minimized by using other types of windows that do not contain abrupt discontinuities in their time-domain characteristics, and have correspondingly low side lobes in their frequency domain characteristics.

Several popular non-rectangular window types are listed in Table 14.2. Figure 14.5 illustrates the time-domain characteristics of those windows for length $M = 61$. The frequency response characteristics of the most commonly used non-rectangular windows, such as, Hanning, Hamming, and Blackman windows for different window lengths are illustrated in Figs.14.6 through Fig. 14.8.

All of these window functions have significantly lower side-lobes compared to the rectangular window. However, for the same value of M, the width of the main lobe is also wider compared to the rectangular window. As a result, these window functions provide more smoothing and as a result, the transition region in the FIR filter is wider. To reduce the width of this transition region, we can simply increase

1. Gibb's phenomenon discussed earlier.

Table 14.2. Time domain description of popular window functions of length M.

Name of Window	Time-domain Sequence ($w[n]$)
Kaiser	$\dfrac{I_0\left[\alpha\sqrt{\left(\frac{M-1}{2}\right)^2-\left(n-\frac{M-1}{2}\right)^2}\right]}{I_0\left[\alpha\left(\frac{M-1}{2}\right)\right]}$
Bartlett	$1-\frac{2\lvert n-\frac{M-1}{2}\rvert}{M-1}$
Blackman	$0.42-0.5\cos\frac{2\pi n}{M-1}+0.08\cos\frac{4\pi n}{M-1}$
Tukey	$\frac{1}{2}\left[1+\cos\left(\frac{n-(1+a)(M-1)/2}{(1-a)(M-1)/2}\pi\right)\right]\ \alpha(M-1)/2\le\left\lvert n-\frac{M-1}{2}\right\rvert\le\frac{M-1}{2}$
Hamming	$0.54-0.46\cos\frac{2\pi n}{M-1}$
Hanning	$\frac{1}{2}\left(1-\cos\frac{2\pi n}{M-1}\right)$
Triangular	$1-\frac{2\lvert n-\frac{M-1}{2}\rvert}{M-1}$
Gaussian	$\exp\left(-\frac{1}{2}\left(\frac{n-M/2}{\sigma M/2}\right)^2\right);\ 0\le n\le M$ and $\sigma\le 0.5$
Parzen	$w_0\left(n-\frac{M}{2}\right),\ 0\le n\le M$
	$w_0(n)=\begin{cases}1-6\left(\frac{n}{(M+1)/2}\right)^2\left(1-\frac{\lvert n\rvert}{(M+1)/2}\right), & 0\le n\le\frac{M+1}{4}\\ 2\left(1-\frac{\lvert n\rvert}{(M+1)/2}\right)^3, & \frac{M+1}{4}\le\lvert n\rvert\le\frac{M+1}{2}\end{cases}$
Bohman[1]	$\sin\left(\frac{\pi n}{M}\right)=\cos\left(\frac{\pi n}{M}-\frac{\pi}{2}\right),\ 0\le n\le M.$
Chebyshev	$\frac{1}{M+1}\sum_{k=0}^{M}W_0(k)e^{j\frac{2\pi k(n-M/2)}{M+1}},\quad 0\le n\le M,$
	$W_0(k)=\dfrac{T_M\left(\beta\cos\left(\frac{\pi k}{M+1}\right)\right)}{T_M(\beta)},\quad 0\le k\le M,$
	$T_n(x)=\begin{cases}\cos\left(n\cos^{-1}(x)\right) & \text{if } -1\le x\le 1,\\ \cosh\left(n\cosh^{-1}(x)\right) & \text{if } x\ge 1,\\ (-1)^n\cosh\left(n\cosh^{-1}(-x)\right) & \text{if } x\le -1.\end{cases}$

[1]This window function is also known as the half-sine window or half-cosine window.

the length of the window, which results in higher order filter. Table 14.3 summarizes the important frequency domain features of the various window functions.

Example: Suppose that we want to design a symmetric low-pass linear phase FIR filter having the following (desired) frequency response:

$$H_d(\omega)=\begin{cases}1e^{-j\omega[M-1]/2} & \text{for } 0\le\lvert\omega\rvert\le\omega_c\\ 0 & \text{otherwise}\end{cases}$$

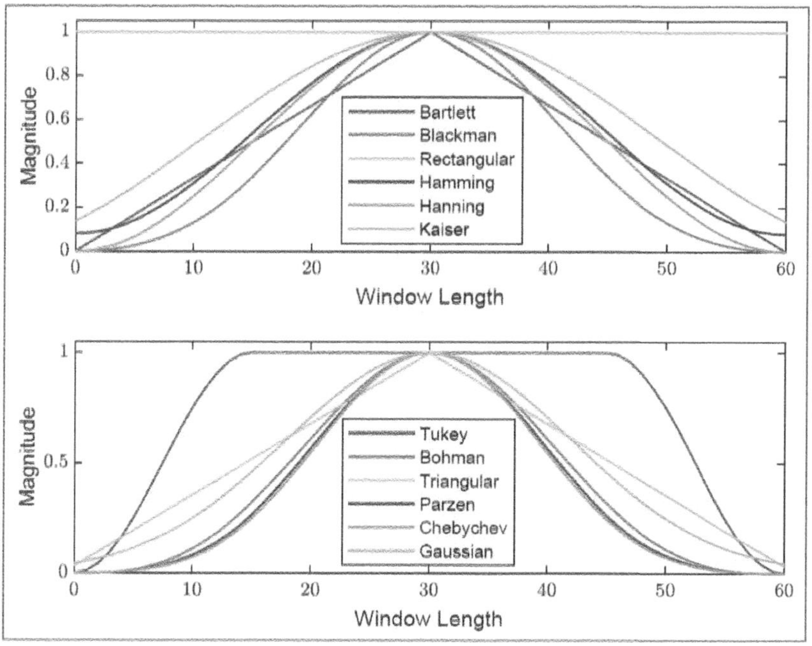

Figure 14.5. Shapes of different types of windows.

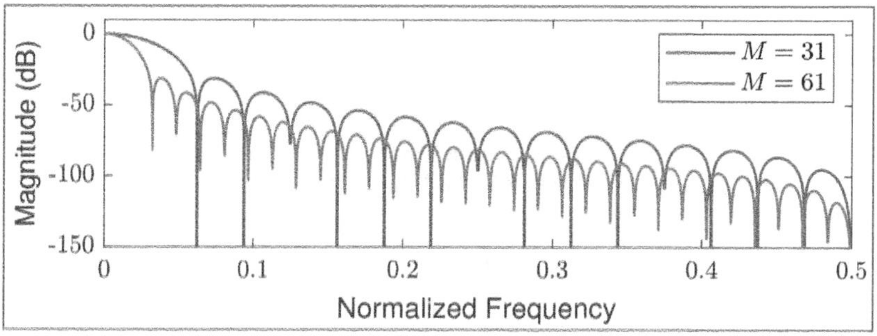

Figure 14.6. Frequency response for Hanning window of length (a) $M = 31$, (b) $M = 61$.

A delay of $[M-1]/2$ units is incorporated into $H_d(\omega)$ in anticipation of forcing the filter to be of length M. The corresponding sample response is

$$
\begin{aligned}
h_d[n] &= \frac{1}{2\pi}\int_{-\omega_c}^{\omega_c} e^{j\omega\left[n-\frac{M-1}{2}\right]}d\omega \\
&= \frac{\sin\omega_c\left[n-\frac{M-1}{2}\right]}{\pi\left[n-\frac{M-1}{2}\right]}, \quad n \neq \frac{M-1}{2}.
\end{aligned}
\tag{14.30}
$$

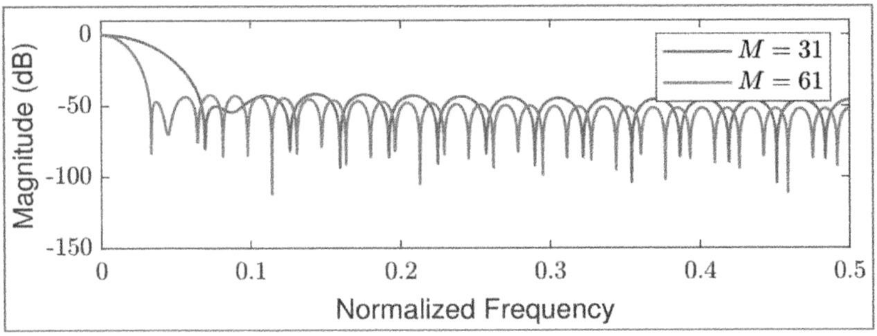

Figure 14.7. Frequency response for Hamming window of length (a) $M = 31$, (b) $M = 61$.

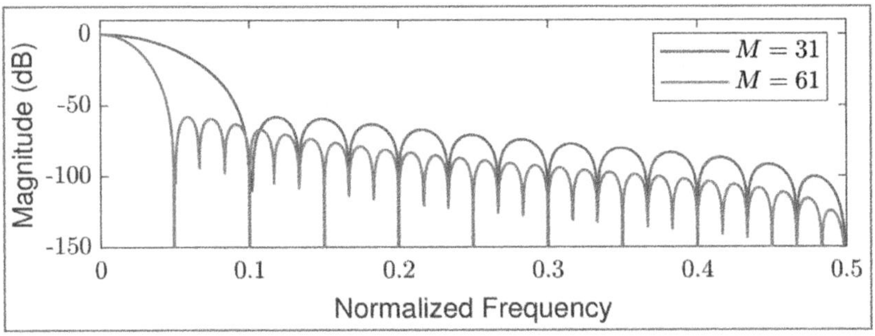

Figure 14.8. Frequency response for Blackman window of length (a) $M = 31$, (b) $M = 61$.

Table 14.3. Important frequency characteristics of most commonly referred types of windows.

Window	Approx. transition width of main lobe	Peak side lobe (dB)
Rectangular	$\frac{4\pi}{M}$	-13
Bartlett	$\frac{8\pi}{M}$	-25
Hanning	$\frac{8\pi}{M}$	-31
Hamming	$\frac{8\pi}{M}$	-41
Blackman	$\frac{12\pi}{M}$	-57

Clearly, $h_d[n]$ is non-causal and infinite duration. If we multiply $h_d[n]$ by the rectangular window (of length M), we obtain a FIR filter of length M having the unit sample response

$$h[n] = \frac{\sin \omega_c \left[n - \frac{M-1}{2}\right]}{\pi \left[n - \frac{M-1}{2}\right]}, \quad 0 \leq n \leq M-1, \ n \neq \frac{M-1}{2}. \tag{14.31}$$

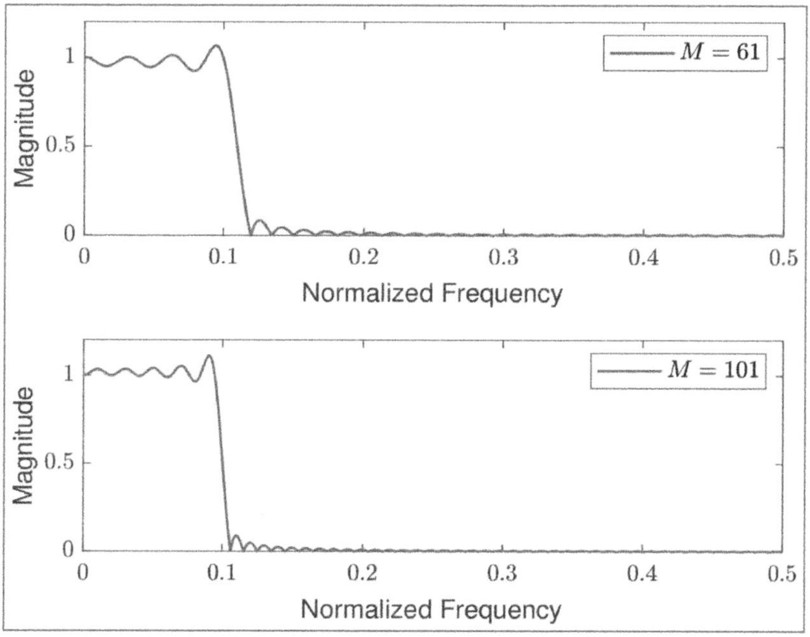

Figure 14.9. Low-pass FIR filter designed with rectangular window: (a) $M = 61$ and (b) $M = 101$.

If M is selected to be odd, the value of $h[n]$ at $n = [M - 1]/2$ is

$$h\left[\frac{M-1}{2}\right] = \frac{\omega_C}{\pi}. \tag{14.32}$$

The magnitude of the frequency response $H(\omega)$ of this low pass filter (with cut off $\omega_c = 0.1$) for $M = 61$ and $M = 101$ is illustrated in Fig. 14.9. The relatively large oscillations (ripples) near the band edge is due to the Gibbs phenomenon (more when use the rectangular window). As the filter length increases, the transition band decreases and the transition from pass band to stop band becomes sharper.

For a comparison, the magnitude responses of the same filter designed using different types of windows (Rectangular, Hamming, Blackman, and Kaiser) with a filter length, $M = 61$, are shown in Figs. 14.10(a)–(d). It may be observed that the design with rectangular window provides the sharp cut off characteristics however, its side lobe levels are the highest among the 4 characteristics. The width of the main lobe of the filter characteristics for the designs with non-rectangular windows (Figs. 14.10(b)–(d)) are higher than that with the rectangular window. On the other hand, the side lobes, as you can see, are much lower for these designs compared to the rectangular window. Gibb's phenomenon is clearly visible in the case of rectangular window (Fig. 14.10(a)).

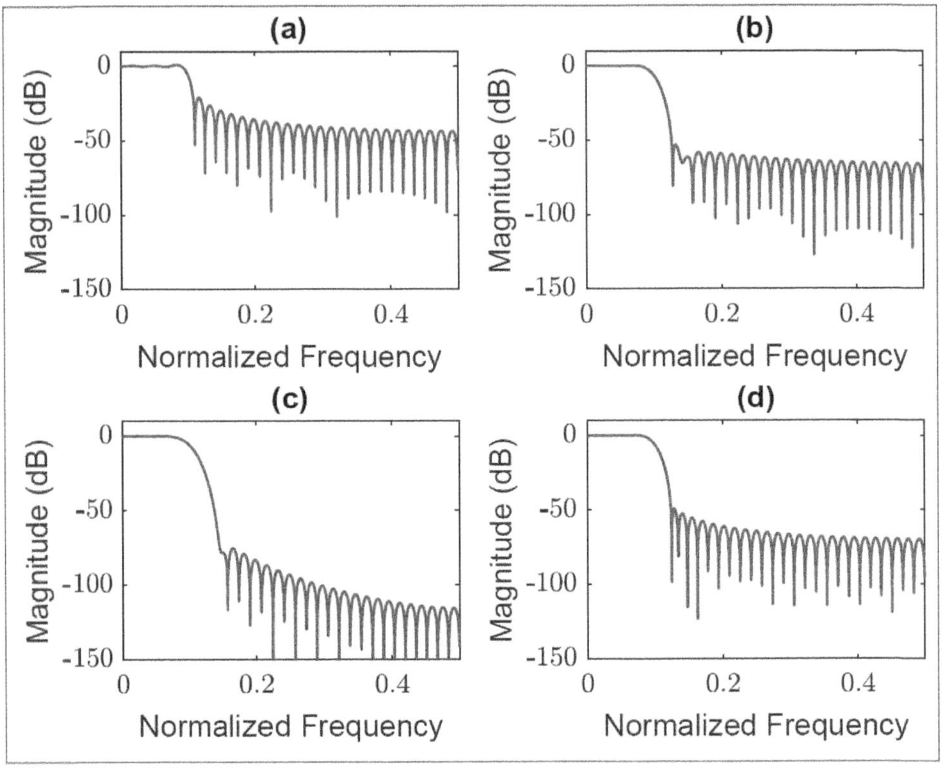

Figure 14.10. Low-pass FIR filter designed with different types of windows of length $M = 61$: (a) Rectangular window, (b) Hamming window, (c) Blackman window, and (d) Kaiser window.

14.2.3 Design of Linear-Phase FIR Filters by the Frequency Sampling Method

Remember the frequency sampling structure of the FIR systems which we have discussed earlier in Section 11.2.3. In this method of FIR filter design, we specify the desired frequency response $H_d(\omega)$ at a set of M equally spaced frequencies, namely,

$$\omega_k = \frac{2\pi k}{M}, \tag{14.33}$$

where $k = 0, 1, \ldots, \frac{M-1}{2}$ for odd values of M, $k = 0, 1, \ldots, \frac{M}{2} - 1$ for even values of M and solve for the impulse response $h[n]$ of the FIR filter. To reduce side lobes, the frequency specifications in the transition band may be optimized (numerically) by using linear programming techniques.

To simplify the computations, we can exploit the symmetry property of the sampled frequency response. Let us begin with the desired frequency response of the

FIR filter,

$$H_d(\omega) = \sum_{n=0}^{M-1} h[n]e^{-j\omega n}. \tag{14.34}$$

Suppose that we specify the frequency response of the filter at the frequencies given by Eq. (14.33). Then from Eq. (14.34), we get (for $k = 0, 1, \ldots, M-1$),

$$\begin{aligned} H_d[k] &\equiv H_d\left(\frac{2\pi k}{M}\right), \\ &\equiv \sum_{n=0}^{M-1} h[n]e^{-j\frac{2\pi k}{M}n}. \end{aligned} \tag{14.35}$$

Taking the inverse of Eq. (14.35) will give $h[n]$. Multiplying both sides of Eq. (14.35) by $e^{j\frac{2\pi k}{M}m}$, $m = 0, 1, \ldots, M-1$ and taking the sum over $k = 0, 1, \ldots, M-1$, the right-hand side of this equation reduces to $Mh[m]$. Thus we obtain (for $n = 0, 1, \ldots, M-1$),

$$h[n] = \frac{1}{M}\sum_{k=0}^{M-1} H_d[k]e^{j\frac{2\pi k}{M}n}. \tag{14.36}$$

Note that Eq. (14.35) is the DFT of the sequence $\{h[n]\}$ and Eq. (14.36) is its IDFT.

Since $h[n]$ is real, we can easily show that the frequency samples $H_d[k]$ satisfy the symmetry condition

$$H_d[k] = H_d^*[M-k]. \tag{14.37}$$

This symmetry condition along with the symmetry of $h[n]$ can be used to reduce the frequency specifications from M to $(M+1)/2$ points for odd values of M and $M/2$ points for even values of M. This reduces the computational requirement considerably. In particular, if Eq. (14.17) is simplified at the frequencies, $\omega_k = \frac{2\pi k}{M}$, $k = 0, 1, \ldots, M-1$, we obtain,

$$H_d[k] = H_r\left(\frac{2\pi k}{M}\right)e^{j\left[\beta\frac{\pi}{2} - \frac{2\pi k}{2M}[M-1]\right]}, \tag{14.38}$$

where $\beta = 0$ when $\{h[n]\}$ is symmetric and $\beta = 1$ when $\{h[n]\}$ is anti-symmetric. A simplification occurs by defining a set of real frequency samples $\{G[k]\}$

$$G[k] = (-1)^k H_r\left(\frac{2\pi k}{M}\right), \quad k = 0, 1, \ldots, M-1. \tag{14.39}$$

We use Eq. (14.39) in Eq. (14.38) to eliminate $H_r(\omega_k)$. Thus we obtain,

$$H_d[k] = G[k]e^{j\pi(-1)k}e^{j\left[\beta\frac{\pi}{2}-\frac{2\pi k}{2M}[M-1]\right]}. \tag{14.40}$$

Now the symmetric condition for $H_d[k]$ in Eq. (14.37) translates into a corresponding symmetry condition for $G[k]$, which can be exploited by substituting into Eq. (14.36), to simplify the expressions for the FIR filter impulse response $\{h[n]\}$ for the two cases $\beta = 0$ and $\beta = 1$. The results are summarized below:

<u>When $h[n] = h[M-1-n]$ (symmetric: $\beta = 0$):</u>

$$H_d[k] = G[k]e^{j\frac{\pi k}{M}}, \quad k = 0, 1, \ldots, M-1$$

$$G[k] = (-1)^k H_r\left(\frac{2\pi k}{M}\right); \quad G[k] = -G[M-k]$$

M odd:

$$h[n] = \frac{1}{M}\left\{G[0] + 2\sum_{k=1}^{\frac{M-1}{2}} G[k]\cos\frac{2\pi k}{M}\left(n+\frac{1}{2}\right)\right\}. \tag{14.41}$$

M even:

$$h[n] = \frac{1}{M}\left\{G[0] + 2\sum_{k=1}^{\frac{M}{2}-1} G[k]\cos\frac{2\pi k}{M}\left(n+\frac{1}{2}\right)\right\}. \tag{14.42}$$

<u>When $h[n] = -h[M-1-n]$ (anti-symmetric: $\beta = 1$)):</u>

$$H_d[k] = G[k]e^{j\frac{\pi}{2}}e^{j\frac{\pi k}{M}}, \quad k = 0, 1, \ldots, M-1$$

$$G[k] = (-1)^k H_r\left(\frac{2\pi k}{M}\right); \quad G[k] = G[M-k]$$

M odd:

$$h[n] = -\frac{2}{M}\sum_{k=1}^{\frac{M-1}{2}} G[k]\sin\frac{2\pi k}{M}\left(n+\frac{1}{2}\right). \tag{14.43}$$

M even:

$$h[n] = \frac{1}{M}\left\{(-1)^{n+1}G\left(\frac{M}{2}\right) - 2\sum_{k=1}^{\frac{M}{2}-1} G[k]\sin\frac{2\pi k}{M}\left(n+\frac{1}{2}\right)\right\}. \tag{14.44}$$

Example: Determine the coefficients of a linear-phase FIR filter of length $M = 15$ which has a symmetric unit sample response and frequency response that satisfies the conditions

$$H_r\left(\frac{2\pi k}{15}\right) = \begin{cases} 1 & \text{for } k = 0, 1, 2, 3 \\ 0.4 & \text{for } k = 4 \\ 0 & \text{for } k = 5, 6, 7 \end{cases} \tag{14.45}$$

Since $h[n]$ is symmetric, we use the corresponding formula from the four cases described earlier. In this case,

$$G[k] = (-1)^k H_r\left(\frac{2\pi k}{15}\right), \quad k = 0, 1, \ldots, 7 \tag{14.46}$$

The result of this computation is

$$\begin{aligned} h[0] &= -0.014112893 = h[14] \\ h[1] &= -0.001945309 = h[13] \\ h[2] &= 0.04000004 = h[12] \\ h[3] &= 0.01223454 = h[11] \\ h[4] &= -0.09138802 = h[10] \\ h[5] &= -0.01808986 = h[9] \\ h[6] &= 0.3133176 = h[8] \\ h[7] &= 0.52 = h[7] \end{aligned}$$

The frequency response characteristic of this filter is shown in Fig. 14.11. We should emphasize that $H_r(\omega)$ is exactly equal to the values given by the specifications above at $\omega_k = \frac{2\pi k}{15}$.

14.2.4 Design of Optimum Equiripple Linear-Phase FIR Filters

The window method and the frequency sampling method are relatively simple techniques for the design of linear-phase FIR filters. However, they have some disadvantages which may render them undesirable for some applications. They are:

- Lack of precise control of the critical frequencies such as ω_p and ω_s. We have to accept whatever values we obtain after the design.
- We cannot specify both δ_1 and δ_2 ripple factors simultaneously. Either we have to specify $\delta_1 = \delta_2$ in the window method or we can optimize only δ_2 in the frequency sampling method.
- The approximation errors between the ideal and the actual responses is not uniformly distributed in the frequency bands. It is higher near the band

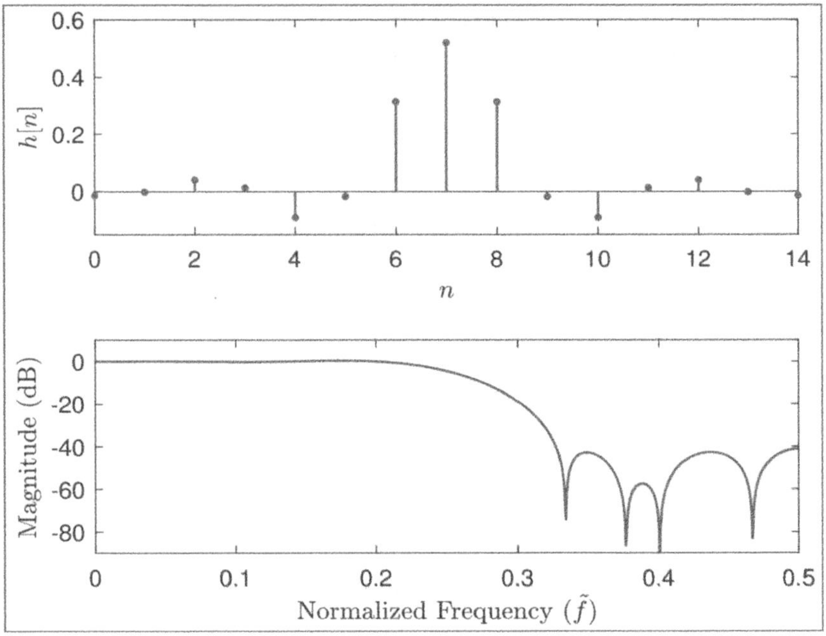

Figure 14.11. Impulse response and frequency response of the linear-phase FIR filter (Example 2).

edges and lower in frequency regions away from band edges. By distributing the error uniformly, we can obtain a lower-order filter satisfying the same specifications.

To overcome these problems, an optimization method, named as the *Parks–McClellan algorithm* for the design of FIR filters have been proposed.[2] The goal of the algorithm is to minimize the error in the pass and stop bands by utilizing the Chebyshev approximation. It has become a standard method for FIR filter design.

The optimization method described in this section is formulated as a Chebyshev approximation problem. In this method, the weighted approximation error between the desired frequency response and the actual frequency response is spread evenly across the pass-band and the stop-band of the filter by minimizing the maximum error.

To describe the design procedure, let us consider the design of a low pass linear-phase FIR filter with a pass-band edge frequency ω_p and stop-band edge

2. J.H. McClellan and T.W. Parks, "A personal history of the Parks-McClellan algorithm", *IEEE Signal Processing Magazine*, vol. 22, No. 2, pp. 82–86, 2005.

frequency ω_s. From the general specifications given in Fig. 14.2, we can write

$$1 - \delta_1 \leq H_r(\omega) \leq 1 + \delta_1, \quad |\omega| \leq \omega_p, \tag{14.47}$$

and

$$-\delta_2 \leq H_r(\omega) \leq \delta_2, \quad |\omega| > \omega_s. \tag{14.48}$$

Here, δ_1 and δ_2 represents the pass-band and stop-band ripples, respectively. The remaining filter parameter is its length M.

Let us focus on the four different cases that result in a linear-phase FIR filter. To derive a standard (compact) objective function for all these four cases, the following manipulations are suggested:

- <u>**Case 1**: $h[n] = h[M-1-n]$ (symmetric) and M odd:</u> In this case, the real valued frequency response characteristics is

$$H_r(\omega) = h\left[\frac{M-1}{2}\right] + 2\sum_{n=0}^{\frac{M-3}{2}} h[n]\cos\omega\left[\frac{M-1}{2} - n\right]. \tag{14.49}$$

If we let $k = \left[\frac{M-1}{2}\right] - n$ and define a new set of filter parameters $\{a[k]\}$ as

$$a[k] = \begin{cases} h\left[\frac{M-1}{2}\right] & \text{for } k = 0, \\ 2h\left[\frac{M-1}{2} - k\right] & \text{for } k = 1, 2, \ldots, \frac{M-1}{2}, \end{cases} \tag{14.50}$$

then Eq. (14.49) reduces to the compact form

$$H_r(\omega) = \sum_{k=0}^{\frac{M-1}{2}} a[k]\cos\omega k. \tag{14.51}$$

- <u>**Case 2**: $h[n] = h[M-1-n]$ (symmetric) and M even:</u>

$$H_r(\omega) = 2\sum_{n=0}^{\frac{M}{2}-1} h[n]\cos\omega\left[\frac{M-1}{2} - n\right] \tag{14.52}$$

By letting $k = \left[\frac{M}{2}\right] - n$ and defining a new set of filter parameters $\{b[k]\}$ as

$$b[k] = 2h\left[\frac{M}{2} - k\right], \quad k = 1, 2, \ldots, M/2 \tag{14.53}$$

Then, Eq. (14.52) becomes

$$H_r(\omega) = \sum_{k=1}^{\frac{M}{2}} b[k] \cos\omega \left[k - \frac{1}{2}\right] \tag{14.54}$$

This can be further modified to

$$H_r(\omega) = \cos\frac{\omega}{2} \sum_{k=0}^{\frac{M}{2}-1} \tilde{b}[k] \cos\omega k, \tag{14.55}$$

where

$$\tilde{b}[0] = \frac{1}{2}b[1] \tag{14.56}$$

$$\tilde{b}[k] = 2b[k] - \tilde{b}[k-1], \quad k = 1, 2, \ldots, \frac{M}{2} - 2 \tag{14.57}$$

$$\tilde{b}\left[\frac{M}{2} - 1\right] = 2b\left[\frac{M}{2}\right]. \tag{14.58}$$

- <u>**Case 3**: $h[n] = -h[M-1-n]$ (anti-symmetric) and M odd:</u>

$$H_r(\omega) = 2\sum_{n=0}^{\frac{M-3}{2}} h[n] \sin\omega \left[\frac{M-1}{2} - n\right] \tag{14.59}$$

By setting $k = \left[\frac{M-1}{2}\right] - n$ and define a new set of filter parameters $\{c[k]\}$ as

$$c[k] = 2h\left[\frac{M-1}{2} - k\right], \quad k = 1, 2, \ldots, [M-1]/2 \tag{14.60}$$

Then Eq. (14.59) becomes

$$H_r(\omega) = \sum_{k=1}^{\frac{M-1}{2}} c[k] \sin\omega k \tag{14.61}$$

This is further modified to

$$H_r(\omega) = \sin\omega \sum_{k=0}^{\frac{M-3}{2}} \tilde{c}[k] \cos\omega k \tag{14.62}$$

where

$$\tilde{c}\left[\frac{M-3}{2}\right] = c\left[\frac{M-1}{2}\right], \tag{14.63}$$

$$\tilde{c}\left[\frac{M-5}{2}\right] = 2c\left[\frac{M-3}{2}\right], \tag{14.64}$$

$$\vdots = \vdots$$

$$\tilde{c}[k-1] - \tilde{c}[k+1] = 2c[k], \quad 2 \le k \le \frac{M-5}{2}, \tag{14.65}$$

$$\tilde{c}[0] + \frac{1}{2}\tilde{c}[2] = c[1]. \tag{14.66}$$

- **Case 4**: $h[n] = -h[M-1-n]$ (anti-symmetric) and M even:

$$H_r(\omega) = 2\sum_{n=0}^{\frac{M}{2}-1} h[n]\sin\omega\left[\frac{M-1}{2} - n\right] \tag{14.67}$$

A change of summation index by setting $k = \frac{M}{2} - n$ combined with a definition of a new set of filter parameters $\{d[k]\}$ according to

$$d[k] = 2h\left[\frac{M}{2} - k\right], \quad k = 1, 2, \ldots, \frac{M}{2} \tag{14.68}$$

This results in

$$H_r(\omega) = \sum_{k=1}^{\frac{M}{2}} d[k]\sin\omega\left[k - \frac{1}{2}\right] \tag{14.69}$$

This can be further modified to

$$H_r(\omega) = \sin\frac{\omega}{2}\sum_{k=0}^{\frac{M}{2}-1} \tilde{d}[k]\cos\omega k \tag{14.70}$$

where,

$$\tilde{d}\left[\frac{M}{2} - 1\right] = 2d\left[\frac{M}{2}\right], \tag{14.71}$$

$$\tilde{d}[k-1] - \tilde{d}[k] = 2d[k], \quad 2 \le k \le \frac{M}{2} - 1, \tag{14.72}$$

$$\tilde{d}[0] - \frac{1}{2}\tilde{d}[1] = d[1]. \tag{14.73}$$

We note that the rearrangements in Case 2, Case 3, and Case 4 have allowed us to express $H_r(\omega)$ in a more compact form as

$$H_r(\omega) = Q(\omega)P(\omega), \tag{14.74}$$

where $Q(\omega)$ and $P(\omega)$ for each case is shown in Table 14.4 below.

Table 14.4. Filter types and the corresponding $Q(\omega)$ and $P(\omega)$ functions.

Filter type	$Q(\omega)$	$P(\omega)$
Case 1	1	$\sum_{k=0}^{[M-1]/2} a[k] \cos \omega k$
Case 2	$\cos \frac{\omega}{2}$	$\sum_{k=0}^{(M/2)-1} \tilde{b}[k] \cos \omega k$
Case 3	$\sin \omega$	$\sum_{k=0}^{(M-3)/2} \tilde{c}[k] \cos \omega k$
Case 4	$\sin \frac{\omega}{2}$	$\sum_{k=0}^{(M/2)-1} \tilde{d}[k] \cos \omega k$

It may be noted that here $P(\omega)$ has a common form

$$P(\omega) = \sum_{k=0}^{L} \alpha[k] \cos \omega k, \tag{14.75}$$

with $\{\alpha[k]\}$ representing the parameters of the filter which are linearly related to the unit sample response $h[n]$ of the FIR filter and the upper limit in the summation L is different for different cases as can be seen from the table.

In addition to the common framework given above, we also define the real-valued desired frequency response $H_{dr}(\omega)$ (unity in the pass-band and zero in the stop-band. See Fig. 14.1 for the illustration of $H_{dr}(\omega)$ for different types filters.) and the weighting function $W(\omega)$ on the approximation error. This $W(\omega)$ allows us to choose the relative size of the errors in the different frequency bands. In particular, it is convenient to normalize $W(\omega)$ to unity in the stop-band and set $W(\omega) = \delta_2/\delta_1$ in the pass-band, that is,

$$W(\omega) = \begin{cases} \frac{\delta_2}{\delta_1} & \omega \text{ in the pass-band} \\ 1 & \omega \text{ in the stop-band} \end{cases} \tag{14.76}$$

The weighted approximation error is defined as

$$\begin{aligned} E(\omega) &= W(\omega)\left[H_{dr}(\omega) - H_r(\omega)\right] \\ &= W(\omega)\left[H_{dr}(\omega) - Q(\omega)P(\omega)\right] \\ &= W(\omega)Q(\omega)\left[\frac{H_{dr}(\omega)}{Q(\omega)} - P(\omega)\right] \\ &= \hat{W}(\omega)\left[\hat{H}_{dr}(\omega) - P(\omega)\right] \end{aligned} \tag{14.77}$$

for all four different types of linear-phase FIR filters.

Given the error function $E(\omega)$, the Chebyshev approximation problem is to determine the filter parameters $\{\alpha[k]\}$ that minimizes the maximum absolute value of $E(\omega)$ over the frequency bands in which the approximation is to be performed. In mathematical terms, this can be written as

$$\min_{\text{over}[\alpha[k]]}\left[\max_{\omega\in S}|E(\omega)|\right] = \min_{\text{over}[\alpha[k]]}[xx], \tag{14.78}$$

where

$$xx = \max_{\omega\in S}\left|\hat{W}(\omega)\left[\hat{H}_{dr}(\omega) - \sum_{k=0}^{L}\alpha[k]\cos\omega k\right]\right| \tag{14.79}$$

In Eq. (14.78), S is the set (disjoint union) of frequency bands over which the optimization is to be performed. This set consists of the pass-bands and stop-bands of the desired filter.

The solution to the optimization problem in Eq. (14.78) is given by Parks and McClellan (1972) by applying the *alternation*[3] theorem to the theory of Chebyshev approximation.

To elaborate on the alternation theorem, let us consider the design of a low-pass filter with pass-band $0 \le \omega \le \omega_p$ and stop-band $\omega_s \le \omega \le \pi$. Since the desired frequency response $H_{dr}(\omega)$ and the weighting function $W(\omega)$ are piece-wise constant, we have,

$$\begin{aligned} \frac{dE(\omega)}{d\omega} &= \frac{d}{d\omega}\{W(\omega)\left[H_{dr}(\omega) - H_r(\omega)\right]\} \\ &= -\frac{dH_r(\omega)}{d\omega} = 0 \end{aligned} \tag{14.80}$$

3. Alternation Theorem: Let S be a compact subset of the interval $[0, \pi)$. A necessary and sufficient condition for $P(\omega)$ in Eq. (14.75) to be unique, best weighted Chebyshev approximation to $\tilde{H}_{dr}(\omega)$ in S, is that the error function $E(\omega)$ exhibit at least $L+2$ extremal frequencies in S. That is, there must exist at least $L+2$ frequencies $\{\omega_i\}$ in S such that $\omega_1 < \omega_2 < \ldots < \omega_{L+2}$, $E(\omega_i) = -E(\omega_{i+1})$, and $|E(\omega_i)| = \max_{\omega\in S}|E(\omega)|$, $i = 1, 2, \ldots, L+2$.

Consequently, the frequencies $\{\omega_i\}$ corresponding to the peaks of $E(\omega)$ also corresponds to peaks at which $H_r(\omega)$ meets the error tolerance. Since $H_r(\omega)$ is a trigonometric polynomial of degree L, for Case 1, for example,

$$
\begin{aligned}
H_r(\omega) &= \sum_{k=0}^{L} \alpha[k] \cos \omega k \\
&= \sum_{k=0}^{L} \alpha[k] \left[\sum_{n=0}^{k} \beta_{nk} (\cos \omega)^n \right] \\
&= \sum_{k=0}^{L} \alpha'[k] (\cos \omega)^k. \qquad (14.81)
\end{aligned}
$$

It follows that $H_r(\omega)$ can have at most $L-1$ local maxima and minima in the open interval $0 < \omega < \pi$. In addition, $\omega = 0$ and $\omega = \pi$ are usually extrema of $H_r(\omega)$ and, also, of $E(\omega)$. Therefore, $H_r(\omega)$ has at most $L+1$ extremal frequencies. Furthermore, the band edge frequencies ω_p and ω_s are also extrema of $E(\omega)$ since $|E(\omega)|$ is maximum at $\omega = \omega_p$ and $\omega = \omega_s$. Hence there are at most $L+3$ extremal frequencies in $E(\omega)$ for the unique, best approximation of the ideal low-pass filter. On the other hand, the alternation theorem states that there are at least $L+2$ extremal frequencies in $E(\omega)$. Thus the error function for the low-pass filter design has either $L+3$ or $L+2$ extrema.

The alternation theorem guarantees a unique solution for the optimization problem in Eq. (14.78). At the desired extremal frequencies $\{\omega_n\}$, we have the set of equations,

$$
\hat{W}(\omega_n)[\hat{H}_{dr}(\omega_n) - P(\omega_n)] = [-1]^n \delta, \quad n = 0, 1, \ldots, L+1 \qquad (14.82)
$$

where δ is the maximum value of the error function $E(\omega)$. If we select $W(\omega)$ by Eq. (14.76), it follows that $\delta = \delta_2$. The set of linear equations in Eq. (14.82) can be rearranged as

$$
P(\omega_n) + \frac{[-1]^n \delta}{\hat{W}(\omega_n)} = \hat{H}_{dr}(\omega_n), \quad n = 0, 1, \ldots, L+1 \qquad (14.83)
$$

where

$$
P(\omega_n) = \sum_{k=0}^{L} \alpha[k] \cos \omega_n k \qquad (14.84)
$$

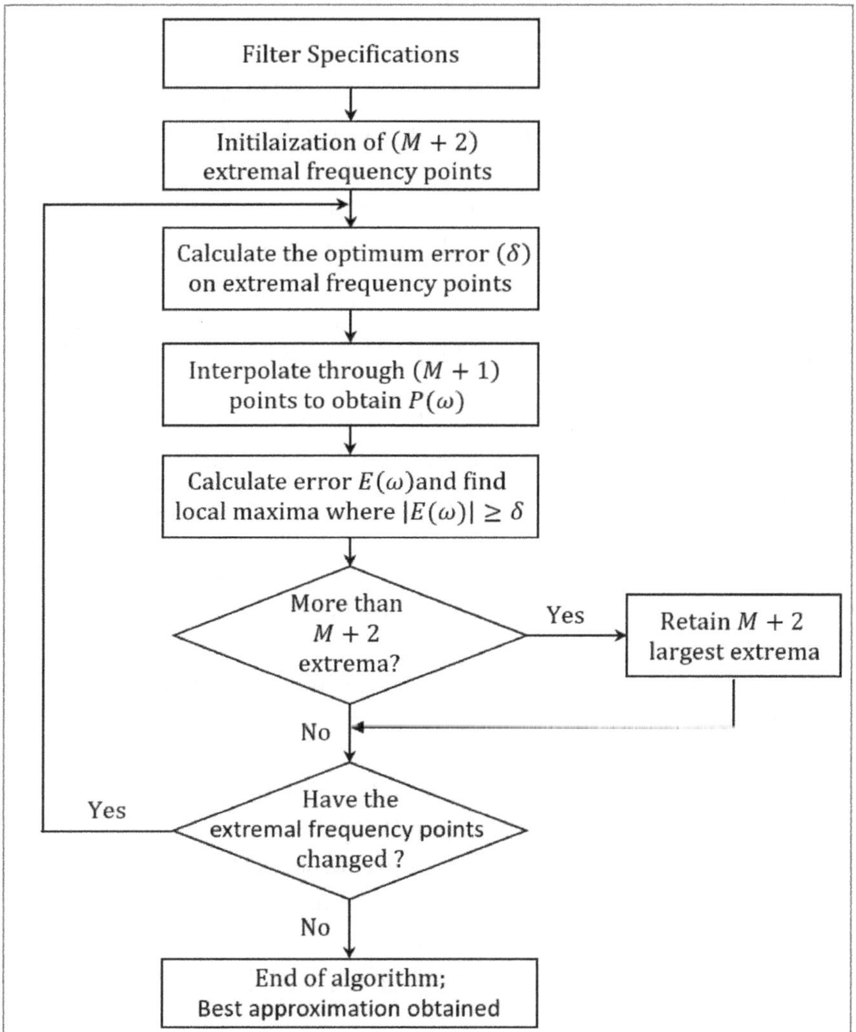

Figure 14.12. Flowchart of Remez algorithm.

Here, $\{\alpha[k]\}$ and δ are the parameters to be determined by solving the $L+1$ equations (Eq. (14.83)). Initially we do not know either the set of extremal frequencies $\{\omega_n\}$ or the parameters $\{\alpha[k]\}$ and δ. An iterative approach,[4] known as re *Remez Exchange Algorithm* can be used to solve for the parameters [5]. In this algorithm, by guessing the set of extremal frequencies, determine $P(\omega)$ and δ and then compute the error function $E(\omega)$. From this $E(\omega)$, we determine another set of $L+2$ extremal frequencies and repeat the process iteratively until it converges to

4. Rabiner et al. (1975) have proposed a more efficient method.

the optimal set of extremal frequencies. Figure 14.12 shows the flowchart of Remez algorithm.

A computer programme written by Parks and McClellan is available for designing linear-phase FIR filters of different types using the optimization method just described in this sub-section.

The Remez programme requires a number of input parameters which determine the filter characteristics. The following parameters must be specified:

- NFILT: Filter length (M)
- JTYPE: Type of filter (1 (multiple passband/stopband), 2 (differentiator), 3 (Hilbert transformer))
- NBANDS: Number of frequency bands (2–10)
- LGRID: Grid density for interpolating the error function $E(\omega)$ (default value is 16)
- EDGE: Frequency bands specified by lower and upper cutoff frequencies (up to maximum of 10 bands)
- FX: Array of maximum size 10 that specifies the desired frequency response in each band
- WTX: Array of maximum size 10 that specifies the weight function in each band

Example: Design a low-pass filter of length $M = 61$ with a pass-band frequency $f_p = 0.1$ and a stop-band frequency $f_s = 0.15$.

The low pass filter is a two band filter with pass band edge frequencies (0, 0.1) and stop-band edge frequencies (0.15, 0.5). The desired response is (1,0) and the weight function is arbitrarily selected as (1,1).

NFILT $= 61$, JTYPE $= 1$, NBANDS $= 2(0.0, 0.1, 0.15, 0.5)$, FX $= 1.0, 0.0$, WTX $= 1.0, 1.0$

The impulse response and the frequency response of the designed filter are shown in Fig. 14.13. The resulting filter has a stop band attenuation of −56dB and a pass band ripple of 0.0135 dB.

If we increase the length of the filter to 101, keeping all the other parameters same, the resulting filter has an impulse response and frequency response as shown in Fig. 14.14. Here, the stop-band attenuation is −85dB and the pass-band ripple has reduced to 0.00046 dB.

14.3 Design of IIR Filters

IIR filters have infinite-duration impulse responses. It is easy to match those to analog filters which generally have infinitely-long impulse response. There are several

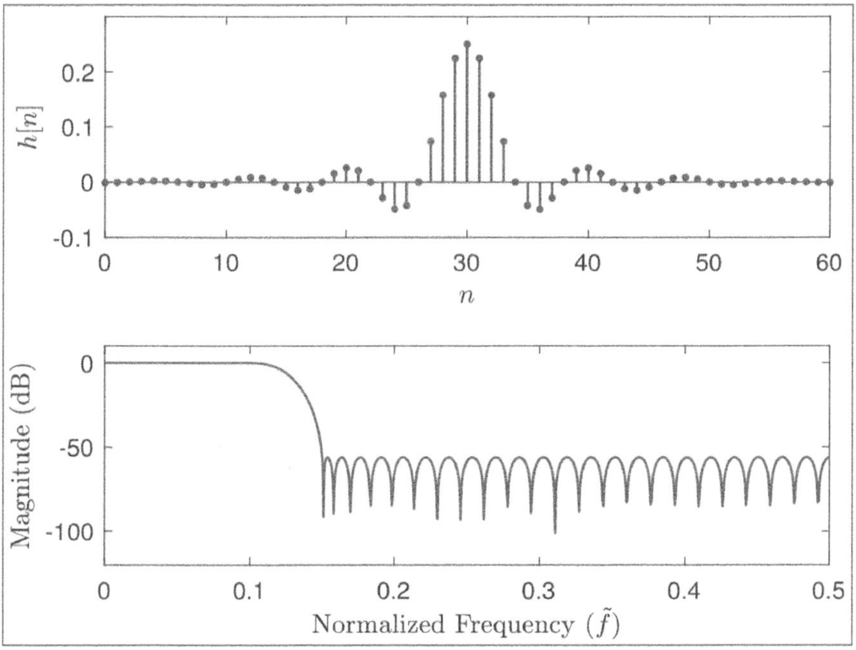

Figure 14.13. Impulse response and frequency response of the designed filter in Example.

methods that can be used for the design of IIR filters. The techniques described in this section are all based on converting an analog filter into a digital filter. Here, three such schemes are discussed. They are:

- Design by approximation of derivatives
- Design by impulse invariance (preserves the shape of the impulse response)
- Design by bilinear transformation (preserves the system function representation)

14.3.1 Design of IIR Filters From Analog Filters

Analog filter design is a mature and well-developed field. We begin the design of a digital filter in the analog domain and then convert the design into the digital domain by complex-valued mapping.

An analog filter can be described by its system function,

$$H_a(s) = \frac{B(s)}{A(s)} = \frac{\sum_{k=0}^{M} \beta_k s^k}{\sum_{k=0}^{N} \alpha_k s^k} \tag{14.85}$$

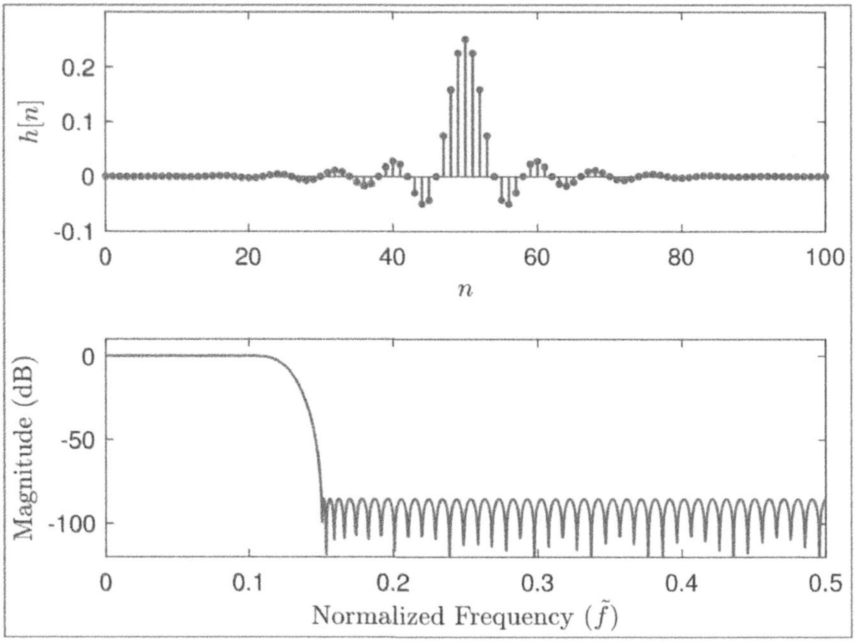

Figure 14.14. Impulse response and frequency response of the designed filter in Example.

where, α_k and β_k are the filter coefficients, or by its impulse response, which is related to $H_a(s)$ by the Laplace transform

$$H_a(s) = \int_{-\infty}^{\infty} h(t)e^{-st}dt. \tag{14.86}$$

An analog filter having the rational system function $H_a(s)$ can be described by the linear constant coefficient differential equation as

$$\sum_{k=0}^{N} \alpha_k \frac{d^k y(t)}{dt^k} = \sum_{k=0}^{M} \beta_k \frac{d^k x(t)}{dt^k} \tag{14.87}$$

where $x(t)$ and $y(t)$ are the input and output, respectively of the system.

Each of these three equivalent characterization of an analog filter leads to alternative methods for converting them into the digital domain.

We recall that an analog LTI system with system function $H_a(s)$ is stable if all its poles lie in the left half of the s-plane. Accordingly, if the conversion technique is to be effective, it should posses the following desirable properties:

- The $j\Omega$ axis in the s-plane should map into the unit circle in the z-plane. This means that the frequencies in s and z domains are directly related.

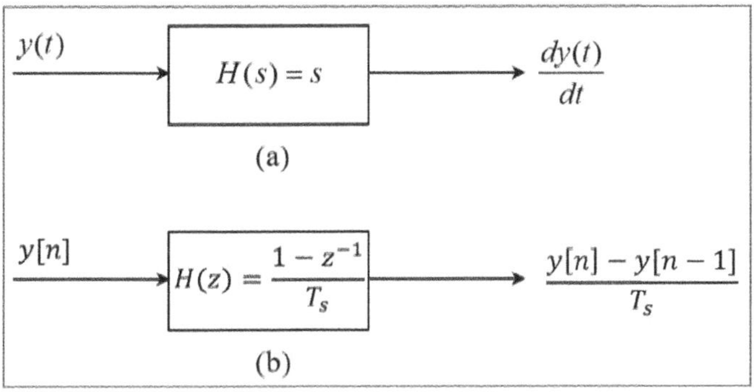

Figure 14.15. Frequency domain equivalent.

- The LHP of the s-plane should map into the interior of the unit circle in the z-plane. Thus a stable analog filter will be converted into a stable digital filter.

In the design of IIR filters, we shall specify the desired filter characteristics for the magnitude response only and accepts the phase response that is obtained from the design methodology.

14.3.2 IIR Filter Design by Approximation of Derivatives

It is known that a differential equation can be approximated by an equivalent difference equation. This approach is used to find the equivalent difference equation representation of Eq. (14.87). The derivative $\frac{dy(t)}{dt}$ at time instant $t = nT_s$ is approximated by the difference equation as follows:

$$\begin{aligned}\left.\frac{dy(t)}{dt}\right|_{t=nT_s} &= \frac{y(nT_s) - y(nT_s - T_s)}{T_s}\\ &= \frac{y[n] - y[n-1]}{T_s} \qquad (14.88)\end{aligned}$$

where T_s is the sampling interval and $y[n] \equiv y(nT_s)$. The analog differentiator with its output $\frac{dy(t)}{dt}$ has a system function $H(s) = s$, while its digital counterpart has the system function $H(z) = \frac{1-z^{-1}}{T_s}$ (see Fig. 14.15). The frequency domain equivalent for the relationship in Eq. (14.88) is

$$s = \frac{1 - z^{-1}}{T_s}. \qquad (14.89)$$

The second derivative is replaced by the second difference equation as below:

$$\left.\frac{d^2y(t)}{dt^2}\right|_{t=nT_s} = \frac{y[n] - 2y[n-1] + y[n-2]}{T_s^2} \tag{14.90}$$

In the frequency domain, it is equivalent to

$$s^2 = \frac{1 - 2z^{-1} + z^{-2}}{T_s^2} = \left(\frac{1 - z^{-1}}{T_s}\right)^2. \tag{14.91}$$

Similarly, the kth derivative of $y(t)$ results in the equivalent frequency domain relationship

$$s^k = \left(\frac{1 - z^{-1}}{T_s}\right)^k. \tag{14.92}$$

Hence the system function for the digital IIR filter obtained as a result of the approximation of the derivatives by finite difference is

$$H(z) = H_a(s)|_{s=\frac{1-z^{-1}}{T_s}} \tag{14.93}$$

Implications: The mapping from s-plane to z-plane as given in Eq. (14.89) or equivalently,

$$z = \frac{1}{1 - sT_s}. \tag{14.94}$$

If we substitute $s = j\Omega$ in the above equation, we get

$$z = \frac{1}{1 + \Omega^2 T_s^2} + j\frac{\Omega T_s}{1 + \Omega^2 T_s^2}. \tag{14.95}$$

As Ω varies from $-\infty$ to ∞, the corresponding locus of points in the z-plane is a circle of radius 0.5 and with centre at $(0.5, 0)$, as illustrated in Fig. 14.16.

It can be shown that the mapping in Eq. (14.94) takes points on the LHP of the s-plane into corresponding points inside this circle in the z-plane, and points in the RHP of the s-plane are mapped into points outside this circle. So, though the mapping preserves the stability, the range of frequencies are limited and as a result, the mapping is restricted to the design of low-pass and band-pass filters having relatively small resonant frequencies. It is not possible, for example, to transform a high-pass analog filter into a corresponding high-pass digital filter.

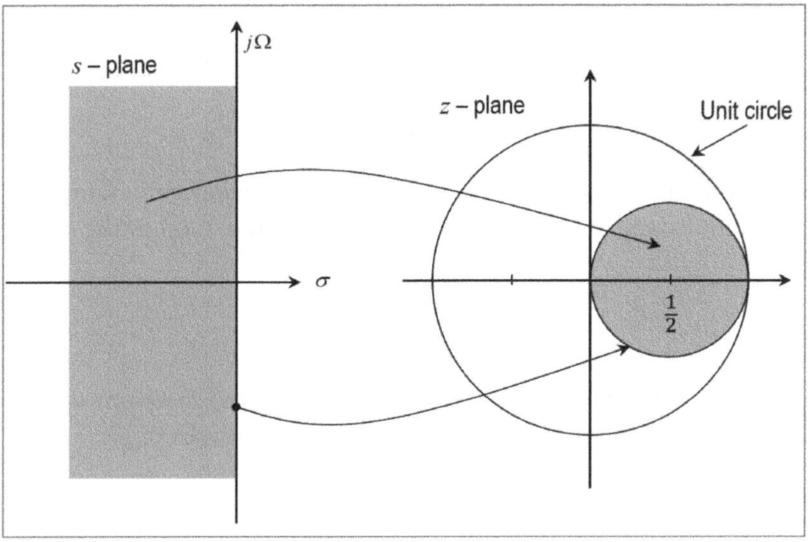

Figure 14.16. The mapping from s-plane to z-plane.

To overcome this problem, a Lth order difference equation has been proposed as follows:

$$\left.\frac{dy(t)}{dt}\right|_{t=nT_s} = \sum_{k=1}^{L} \alpha_k \frac{y(nT_s + kT_s) - y(nT_s - kT_s)}{T_s}, \tag{14.96}$$

where α_k are a set of parameters that can be chosen to optimize the approximation. The resulting mapping is

$$s = \frac{1}{T_s} \sum_{k=1}^{L} \alpha_k (z^k - z^{-k}). \tag{14.97}$$

When $z = e^{j\omega}$, we have

$$s = j\frac{2}{T_s} \sum_{k=1}^{L} \alpha_k \sin \omega k, \tag{14.98}$$

which is purely imaginary. Thus,

$$\Omega = \frac{2}{T_s} \sum_{k=1}^{L} \alpha_k \sin \omega k, \tag{14.99}$$

is the resulting mapping between the two frequency variables. By proper choice of the α_k's, it is possible to map the $j\Omega$-axis into the unit circle. Furthermore, the points in the LHP of the s-plane is mapped into the unit circle. The problem here is the selection of the α_k's which is in general a difficult one.

14.3.3 IIR Filter Design by Impulse Invariance

In this method, the objective is to design an IIR filter having a unit sample response $h[n]$ that is the sampled version of the impulse response $h_a(t)$ of the analog filter. That is,

$$h[n] \equiv h_a(nT_s), \quad n = 0, 1, \ldots \tag{14.100}$$

where T_s is the sampling interval. Using the rule of the spectrum of a sampled signal, the spectral relationship of Eq. (14.100) can be shown to be

$$H(f) = f_s \sum_{k=-\infty}^{\infty} H_a(f - kf_s) \tag{14.101}$$

where f is the analog frequency, $f_s = 1/T_s$ is the sampling frequency. As you can see, a scaled version (scaled by f_s) of the original spectrum repeats on either side of the frequency axis at a rate of f_s Hz.

Equation (14.101) can be written as a function of the normalized frequency, $\tilde{f} = \frac{f}{f_s}$ as follows:

$$H(\tilde{f}) = f_s \sum_{k=-\infty}^{\infty} H_a\left(\left(\tilde{f} - k\right) f_s\right) \tag{14.102}$$

Equation (14.102) is also equivalently

$$H(\omega) = f_s \sum_{k=-\infty}^{\infty} H_a((\omega - 2\pi k) f_s), \tag{14.103}$$

where $\omega = 2\pi \tilde{f}$. This equation can be re-written as,

$$H(\Omega T_s) = \frac{1}{T_s} \sum_{k=-\infty}^{\infty} H_a\left(\Omega - \frac{2\pi k}{T_s}\right), \tag{14.104}$$

as the analog and digital frequencies are related by $\omega = \Omega T_s$ or $e^{j\omega} = e^{j\Omega T_s}$. Since $z = e^{j\omega}$ on the unit circle, $s = j\Omega$ on the imaginary axis, we have the mapping $z = e^{sT_s}$.

Figure 14.17 depicts the frequency responses of a low-pass analog filter and the corresponding digital filter. It is noted that the sampling interval T_s is to be chosen

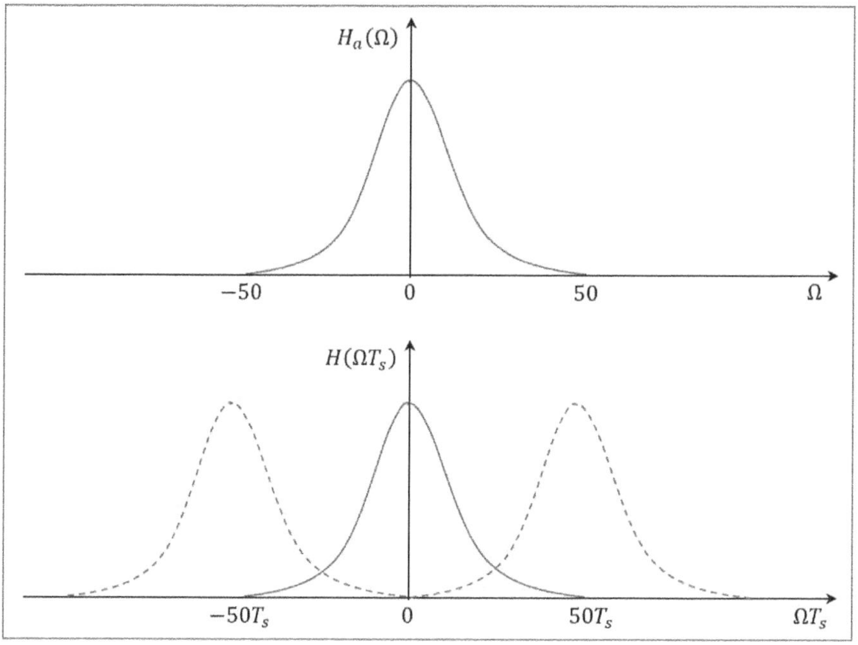

Figure 14.17. Frequency response $H_a(\Omega)$ of the analog filter and the corresponding digital filter with aliasing.

small enough to avoid the effects of aliasing. It is also clear that the impulse invariance method is inappropriate for designing high-pass filters due to the spectrum aliasing that results form the sampling process.

To investigate the mapping of points between the two planes, we rely on a generalization of Eq. (14.104) which relates the z-transform of $h[n]$ to the Laplace transform of $h_a(t)$. The relationship is

$$H(z)|_{z=e^{sT_s}} = \frac{1}{T_s} \sum_{k=-\infty}^{\infty} H_a\left(s - j\frac{2\pi k}{T_s}\right) \tag{14.105}$$

where

$$H(z) = \sum_{n=0}^{\infty} h[n]z^{-n} \tag{14.106}$$

$$H(z)|_{z=e^{sT_s}} = \sum_{n=0}^{\infty} h[n]e^{-sT_s n} \tag{14.107}$$

Note that when $s = j\Omega$, Eq. (14.105) reduces to Eq. (14.104). Let us consider the mapping of points from s-plane to the z-plane implied by the relation $z = e^{sT_s}$. If we substitute $s = \sigma + j\Omega$ and express the complex variable z in polar form, we get

$$re^{j\omega} = e^{\sigma T_s} e^{j\Omega T_s} \tag{14.108}$$

Clearly, we must have

$$r = e^{\sigma T_s} \tag{14.109}$$

$$\omega = \Omega T_s \tag{14.110}$$

Correspondingly, $\sigma < 0$ implies that $0 < r < 1$ and $\sigma > 0$ implies $r > 1$. When $\sigma = 0$, we have $r = 1$. Therefore, the LHP of the s-plane is mapped inside the unit circle in z-plane and the RHP in s-plane is mapped to the outside of the unit circle in the z-plane.

Also, the $j\Omega$-axis is mapped onto the unit circle in z domain. However, the mapping of the $j\Omega$-axis onto the unit circle is not one-to-one. Since ω is unique over the range $(-\pi, \pi)$, the mapping $\omega = \Omega T_s$ implies that the interval $-\frac{\pi}{T_s} \leq \Omega \leq \frac{\pi}{T_s}$ maps into the corresponding values of $-\pi \leq \omega \leq \pi$. Furthermore, the frequency interval $\frac{\pi}{T_s} \leq \Omega \leq 3\frac{\pi}{T_s}$ also maps into the interval $-\pi \leq \omega \leq \pi$ and, in general, so does the interval $(2k-1)\frac{\pi}{T_s} \leq \Omega \leq (2k+1)\frac{\pi}{T_s}$, when k is an integer. This means that the mapping from the analog frequency Ω to the frequency variable ω in the digital domain is many-to-one, which reflects the effect of aliasing due to sampling. Figure 14.18 illustrates this.

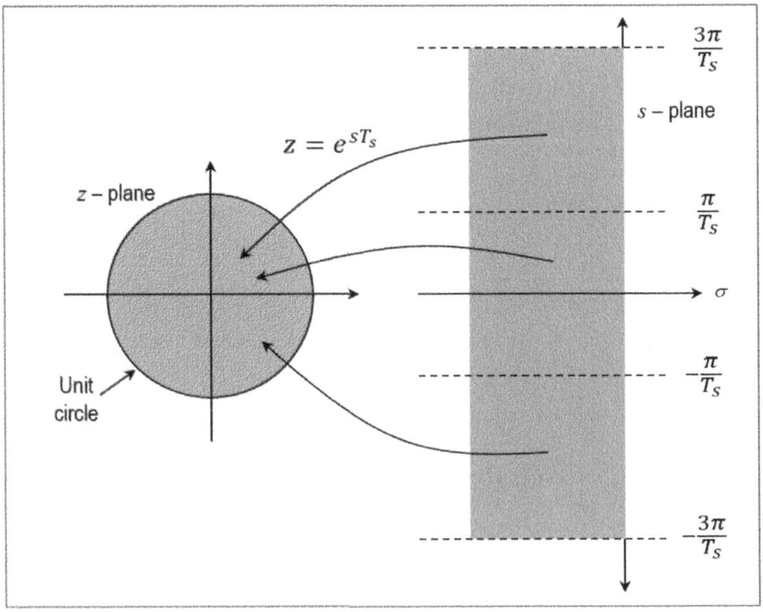

Figure 14.18. The mapping in impulse invariance method.

To explore further the effect of the impulse invariance method on the characteristics of the resulting filter, let us express the system function of the analog filter in partial fraction form. On the assumption that the poles of the analog filter are distinct, we can write

$$H_a(s) = \sum_{k=1}^{N} \frac{c_k}{s - p_k}, \tag{14.111}$$

where, p_k's are the poles of the analog filter and c_k's are the residues. Hence,

$$h_a(t) = \sum_{k=1}^{N} c_k e^{p_k t}, \quad t \geq 0. \tag{14.112}$$

If we sample $h_a(t)$ periodically at $t = nT_s$, we have

$$\begin{aligned} h[n] &= h_a(nT_s) \\ &= \sum_{k=1}^{N} c_k e^{p_k n T_s}. \end{aligned} \tag{14.113}$$

Now with the substitution of Eq. (14.113), the system function of the resulting digital IIR filter becomes,

$$\begin{aligned} H(z) &= \sum_{n=0}^{\infty} h[n] z^{-n} \\ &= \sum_{k=1}^{N} c_k \sum_{n=0}^{\infty} \left(e^{p_k T_s} z^{-1}\right)^n \end{aligned} \tag{14.114}$$

The inner sum in Eq. (14.114) converges because $p_k < 0$ and yields

$$\sum_{n=0}^{\infty} \left(e^{p_k T_s} z^{-1}\right)^n = \frac{1}{1 - e^{p_k T_s} z^{-1}} \tag{14.115}$$

Therefore the system function of the digital filter is

$$H(z) = \sum_{k=1}^{N} \frac{c_k}{1 - e^{p_k T_s} z^{-1}}. \tag{14.116}$$

The poles of the digital filter is

$$z_k = e^{p_k T_s}, \quad k = 1, 2, \ldots, N. \tag{14.117}$$

Although the poles are mapped according to Eq. (14.117), the zeros in the two domains do not satisfy the same relationship. Therefore the impulse invariance method does not correspond to the simple mapping of the points given by $z = e^{sT_s}$.

14.3.3.1 Design Procedure

Given the filter specifications (ω_p, ω_s, etc.), we have to follow the following steps:

- Choose T_s and determine the analog frequencies

$$\Omega_p = \frac{\omega_p}{T_s} \tag{14.118}$$

and

$$\Omega_s = \frac{\omega_s}{T_s} \tag{14.119}$$

- Design an analog filter $H_a(s)$ using the specifications (Butterworth, Chebyschev, or elliptic).
- Using partial fraction approach, find

$$H_a(s) = \sum_{k=1}^{N} \frac{c_k}{s - p_k} \tag{14.120}$$

- Transform the analog poles p_k into digital poles $e^{p_k T_s}$ to obtain the digital filter

$$H(z) = \sum_{k=1}^{N} \frac{c_k}{1 - e^{p_k T_s} z^{-1}}. \tag{14.121}$$

Example: Transform $H_a(s) = \frac{s+1}{s^2+5s+6}$ into a digital filter using the impulse invariant technique in which $T_s = 0.1$.

We have

$$H_a(s) = \frac{s+1}{s^2+5s+6} = \frac{2}{s+3} - \frac{1}{s+2}, \tag{14.122}$$

by partial fraction approach.

Poles are $p_1 = -3$ and $p_2 = -2$. We get $H(z)$ as

$$\begin{aligned} H(z) &= \frac{2}{1 - e^{-3T_s} z^{-1}} - \frac{1}{1 - e^{-2T_s} z^{-1}} \\ &= \frac{1 - 0.8966z^{-1}}{1 - 1.5595z^{-1} + 0.6065z^{-2}} \end{aligned} \tag{14.123}$$

Note: The advantages of the impulse invariant mapping are that it is a stable design and that frequencies Ω and ω are linearly related ($\omega = \Omega T_s$). But the disadvantage is that we should expect some aliasing of the analog frequency response, and in some cases this aliasing is unacceptable. Hence, this design method is useful only when the analog filter is essentially band-limited to a low-pass or band-pass filter.

14.3.4 IIR Filter Design by the Bilinear Transformation

The two methods ((i) method of approximation of derivatives and (ii) the impulse invariance method) of the IIR filter designs discussed have severe limitation in that they are appropriate only for low-pass filters and a limited class of band-pass filters. The bilinear method overcomes this limitation.

- The bilinear transformation is a conformal mapping that transforms the $j\Omega$-axis onto the unit circle in the z-domain only once, thus avoiding the aliasing of frequency components.
- All points in the LHP of the s-plane are mapped inside the unit circle and all points in the RHP of the s-plane are mapped into corresponding points outside the unit circle in the z-plane.
- This method preserves the system function representation as well.

Let us consider an analog linear filter with system function

$$H(s) = \frac{b}{s+a} \tag{14.124}$$

This system is also characterized by the differential equation

$$\frac{dy(t)}{dt} + ay(t) = bx(t). \tag{14.125}$$

Instead of using the difference equation approach, suppose that we integrate the derivative and approximate the integral by the trapezoidal formula [9]. Thus

$$y(t) = \int_{t_0}^{t} y'(t)dt + y(t_0), \tag{14.126}$$

where $y'(t)$ denotes the 1st derivative of $y(t)$. By applying the trapezoidal formula at $t = nT_s$ and $t_0 = nT_s - T_s$, the approximation of Eq. (14.126) becomes

$$y(nT_s) = \frac{T_s}{2}[y'(nT_s) + y'(nT_s - T_s)] + y(nT_s - T_s). \tag{14.127}$$

Now the differential equation in (14.125) evaluated at $t = nT_s$ yields

$$y'(nT_s) = -ay(nT_s) + bx(nT_s). \tag{14.128}$$

We use Eq. (14.128) to substitute for the derivative in Eq. (14.127) and thus obtain a difference equation for the equivalent discrete-time system. With $y[n] \equiv y(nT_s)$ and $x[n] \equiv x(nT_s)$, we get

$$\left(1 + \frac{aT_s}{2}\right) y[n] - \left(1 - \frac{aT_s}{2}\right) y[n-1] = \frac{bT_s}{2}(x[n] + x[n-1]) \tag{14.129}$$

The z-transform of this difference equation is

$$\left(1 + \frac{aT_s}{2}\right) Y(z) - \left(1 - \frac{aT_s}{2}\right) z^{-1} Y(z) = \frac{bT_s}{2}(1 + z^{-1})X(z) \tag{14.130}$$

Hence, the system function of the equivalent digital filter is

$$H(z) = \frac{Y(z)}{X(z)} = \frac{\left(\frac{bT_s}{2}\right)(1 + z^{-1})}{1 + \left(\frac{aT_s}{2}\right) - \left(1 - \frac{aT_s}{2}\right) z^{-1}} \tag{14.131}$$

or, equivalently

$$H(z) = \frac{b}{\frac{2}{T_s}\left(\frac{1-z^{-1}}{1+z^{-1}}\right) + a} \tag{14.132}$$

Comparing Eqs. (14.124) and (14.132), we get the mapping from s-plane to the z-plane as

$$s = \frac{2}{T_s}\left(\frac{1 - z^{-1}}{1 + z^{-1}}\right). \tag{14.133}$$

This is called the *bilinear transformation.*

14.3.4.1 Characteristics of the bilinear transformation

To investigate the characteristics of the transformation, let $z = re^{j\omega}$ and $s = \sigma + j\Omega$. Then Eq. (14.133) can be expressed as

$$s = \frac{2}{T_s}\left(\frac{r^2 - 1}{1 + r^2 + 2r\cos\omega} + j\frac{2r\sin\omega}{1 + r^2 + 2r\cos\omega}\right) \tag{14.134}$$

From this, we get,

$$\sigma = \frac{2}{T_s}\left(\frac{r^2 - 1}{1 + r^2 + 2r\cos\omega}\right), \tag{14.135}$$

$$\Omega = \frac{2}{T_s}\left(\frac{2r\sin\omega}{1 + r^2 + 2r\cos\omega}\right). \tag{14.136}$$

From these equations, we note that if $r < 1$, then $\sigma < 0$, and if $r > 1$, then $\sigma > 0$. Hence, the LHP in s-plane maps into the interior of the unit circle in the z-plane and the RHP in s-plane maps to the outside of the unit circle. When $r = 1$, then $\sigma = 0$ and

$$\Omega = \frac{2}{T_s}\left(\frac{\sin\omega}{1 + \cos\omega}\right) \tag{14.137}$$

$$= \frac{2}{T_s}\tan\left(\frac{\omega}{2}\right) \tag{14.138}$$

or, equivalently

$$\omega = 2\tan^{-1}\left(\frac{\Omega T_s}{2}\right). \tag{14.139}$$

This relationship between the frequency variables in the two domains is called "frequency warping" and is illustrated in Fig. 14.19. We observe that the entire range in Ω is mapped only once into the range $-\pi \leq \omega \leq \pi$. However, the mapping is highly non-linear. We observe a frequency compression or *frequency warping* due to the non-linearity of the *arctangent* function. It is also interesting to note that the bilinear transformation maps the point $s = \infty$ into the point $z = -1$. Consequently, the single-pole low-pass filter in Eq. (14.124), which has a zero at $s = \infty$, results in a digital filter that has a zero at $z = -1$.

14.3.4.2 Consequences of Frequency Warping

It is easy to verify that the bilinear transform gives a one-to-one, order-preserving, conformal mapping between the analog frequency axis $s = j\Omega$ and the digital

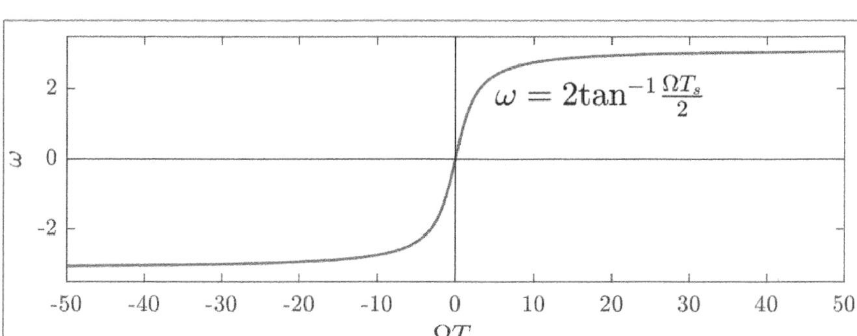

Figure 14.19. Relation between ω and Ω in bilinear transformation.

frequency axis $z = e^{j\Omega T_s}$, where T_s is the sampling interval. Therefore, the amplitude response takes on exactly the same values over both frequency axes, with the only problem being a frequency warping such that equal increments along the unit circle in the z-plane correspond to larger and larger bandwidths along the $j\Omega$ axis in the s-plane. The frequency warping is certainly unavoidable in any one-to-one map because the analog frequency axis ($j\Omega$) is infinite while the digital frequency axis (unit circle) is finite. The relation between the analog and digital frequency axes can be seen in Figs. 14.19 and 14.20 as well as in the following equations.

$$\Omega_p = \frac{2}{T_s} \tan\left(\frac{\omega_p}{2}\right), \tag{14.140}$$

and

$$\Omega_s = \frac{2}{T_s} \tan\left(\frac{\omega_s}{2}\right). \tag{14.141}$$

This is illustrated in Fig 14.20 below.

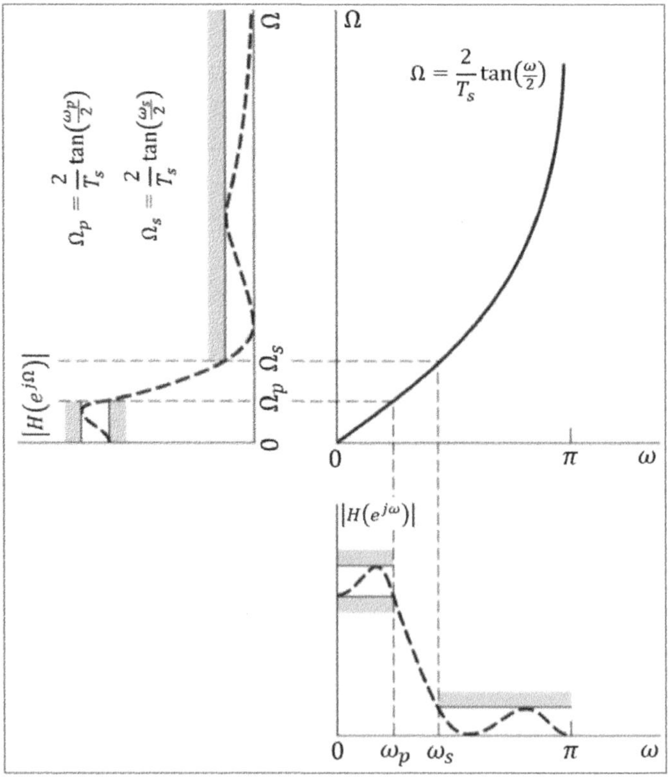

Figure 14.20. Consequence of frequency warping.

Here, the good news is that we do not have to worry about frequency aliasing. On the other hand, the bad news is that we have to account for frequency warping when we start from a discrete-time (digital) filter specification.

14.3.4.3 Design Procedure

Given the filter specifications (ω_p, ω_s, etc.), to design the digital filter, we have the following steps:

- Choose T_s (usually set as 1) and determine the analog frequencies

$$\Omega_p = \frac{2}{T_s}\tan\left(\frac{\omega_p}{2}\right) \tag{14.142}$$

and

$$\Omega_s = \frac{2}{T_s}\tan\left(\frac{\omega_s}{2}\right) \tag{14.143}$$

- Design an analog filter $H_a(s)$ using the specifications (Butterworth, Chebyschev, or Elliptic) to meet the specifications.
- Finally, set

$$H(z) = H_a\left(\frac{2}{T_s}\frac{1-z^{-1}}{1+z^{-1}}\right) \tag{14.144}$$

Example: Transform $H_a(s) = \frac{s+1}{s^2+5s+6}$ into a digital filter using the bilinear transformation technique. Choose $T_s = 1$.

Using the transformation, we have

$$H(z) = H_a\left(\frac{2}{T_s}\left.\frac{1-z^{-1}}{1+z^{-1}}\right|_{T_s=1}\right) \tag{14.145}$$

$$= H_a\left(2\frac{1-z^{-1}}{1+z^{-1}}\right) \tag{14.146}$$

$$= \frac{2\frac{1-z^{-1}}{1+z^{-1}}+1}{\left(2\frac{1-z^{-1}}{1+z^{-1}}\right)^2+5\left(2\frac{1-z^{-1}}{1+z^{-1}}\right)+6} \tag{14.147}$$

Simplifying,

$$H(z) = \frac{3+2z^{-1}-z^{-2}}{20+4z^{-1}} = \frac{0.15+0.1z^{-1}-0.05z^{-2}}{1+0.2z^{-1}}. \tag{14.148}$$

Once $H(z)$ is obtained, all the analysis, as well as its implementation schemes of the filter can be performed.

14.4 Concluding Remarks

In this chapter, the most commonly used methods for the design of both FIR and IIR filters were discussed in detail. They are:

- For the design of FIR filters
  - Windowing method,
  - Frequency sampling method, and
  - Optimization method.
- For IIR Filters
  - Approximation of derivatives method,
  - Impulse invariance method, and
  - Bilinear transformation method.

Exercises

14.1. This problem deals with the design of IIR (infinite impulse response) digital filter from respective analog filter whose transfer function is given as

$$H(s) = \frac{s}{s^2 + 400s + 2 \times 10^5}$$

where $s = \sigma + j\Omega$ is the Laplace variable.

(a) Using the method of approximation of derivatives, design a digital IIR filter (that is, determine the system function $H(z)$ and the corresponding difference equation relating the input, $x[n]$ and the output, $y[n]$ of the filter) to simulate the analog filter given by $H(s)$ above. Use a sampling rate of 1 kHz (that is, $f_s = 1000$).

(b) Using the method of bilinear transformation, design a digital IIR filter (that is, determine the system function $H(z)$ and the corresponding difference equation relating the input, $x[n]$ and the output, $y[n]$ of the filter) to simulate the analog filter given by $H(s)$ above. Use the same sampling rate of $f_s = 1000$ here as well.

(c) Compare and comment on the IIR system functions derived using these two methods.

14.2. Use the frequency sampling method to design an FIR filter whose frequency responses goes through the points as given in table below:

k	0	1	2	3	4	5	6	7
$H\left(e^{j\frac{2\pi k}{8}}\right)$	0	1	j	0	0	0	$-j$	1

14.3. Using bilinear transformation, design a digital low-pass Butterworth filter with a 3dB cut-off frequency of 2kHz and minimum attenuation of 30dB at 4.25kHz for a sampling rate of 10kHz.

DOI: 10.1561/9781680839791.ch15

Chapter 15

Random Signals

The type of signals that we have been dealing with in the previous chapters are *deterministic* in nature, i.e., signals which may be described by mathematical expressions and whose behaviour does not contain any uncertainty or randomness. Typical examples of deterministic signals are $2\cos(100\pi t + \pi/4)$, $u(t-2)$, $\delta(t) - \delta(t-5)$, $e^{-2t}u(t)$, $e^{-4|t|}$, *etc.*

In this chapter, we study the class of signals known as *random signals*. Unlike deterministic signals, the *random signals* have some randomness or uncertainty associated with them. The study of random signals is very important since many of the signals in the real world are random in nature.

In this chapter, the following points are discussed in detail.

- What causes randomness in signals?
- How to characterize or describe random signals?
- Meaning of statistics (or, averages) such as mean, variance, power, correlation, power spectrum, etc.
- Linear filtering of random signals.
- Concept of white noise.

15.1 Introduction

In engineering, we often encounter signals that do not have a precise mathematical description, since they develop as random functions of time. Sometimes this random behaviour may be caused by a single variable, but often it is the consequence of interactions of many random variables. An example is the thermal noise which is generated by the random movement of electrons in an electric conductor. Another example would be the response of brain cells (neurons) to external stimulus, like auditory/visual/tactile, etc. In other cases, the causes of randomness are not clear and a description is not possible, but the signal is characterized by means of measurements only.

A random time function may be a desired signal (information), such as an *audio* or *video* signal, or it may be an unwanted signal that is unintentionally added to a desired signal and disturbs the desired signal. The desired signal can be considered as a random signal and the unwanted signal the noise. However, the latter often does not behave like noise in the classical sense, but it is more like interference. This means that it is an information bearing signal but undesired. In all cases, a description of the signals is required in order to be able to analyse its impact on the performance of the system under study. The random character as a function of time makes the signals difficult to describe and the same hold for signal processing or filtering. Nevertheless, there is a need for characterizing these signals by a few deterministic parameters that enable the system to assess the performance of the system.

Consider the following examples of random signals:

- The signals received by the antenna in a radio/TV receiver is the transmitted signal mixed with noise, where the noise is random in nature.
- When we speak into a microphone, the signal that enters the microphone is not only our speech but also the random noise that is present in the acoustic environment.
- If you say your name several times, the speech signal that you produce each time is not exactly the same. There is some randomness associated with the speech production mechanism in your vocal tract, even though we hear you saying the same name again and again.
- The output of a sinusoidal signal generator is given by

$$x(t) = A\cos(2\pi f_0 t + \theta). \tag{15.1}$$

The amplitude A and frequency f_0 can be set by us and are hence deterministic. On the other hand, the phase θ is random and it can be anywhere in the interval $[0, 2\pi]$. Each time we switch on the signal generator, the starting phase of the signal is different and unpredictable. This randomness in the phase makes $x(t)$ a random signal.

Thus, there is a lot of randomness in the real world around us.

15.2 Stochastic Process

In probability theory, a random variable x is a rule that assigns a number to every outcome of an experiment, such as, for example, rolling a die. This random variable is associated with a sample space S, such that according to a well-defined procedure to each event s in the sample space, a number is assigned to x and is denoted by $x(s)$. For stochastic process, on the other hand, a time function $x(t, s)$ is assigned to every outcome in the sample space. Within the framework of the experiment, the family (or ensemble) of all possible functions that can be realized is called a *stochastic process*. A specific *realization* out of this family is denoted as $x_n(t)$ and is called a *sample function* of the stochastic process. In general, the subscript n is omitted in representing a sample function. Figure 15.1 shows a few sample functions that are supposed to form an ensemble.

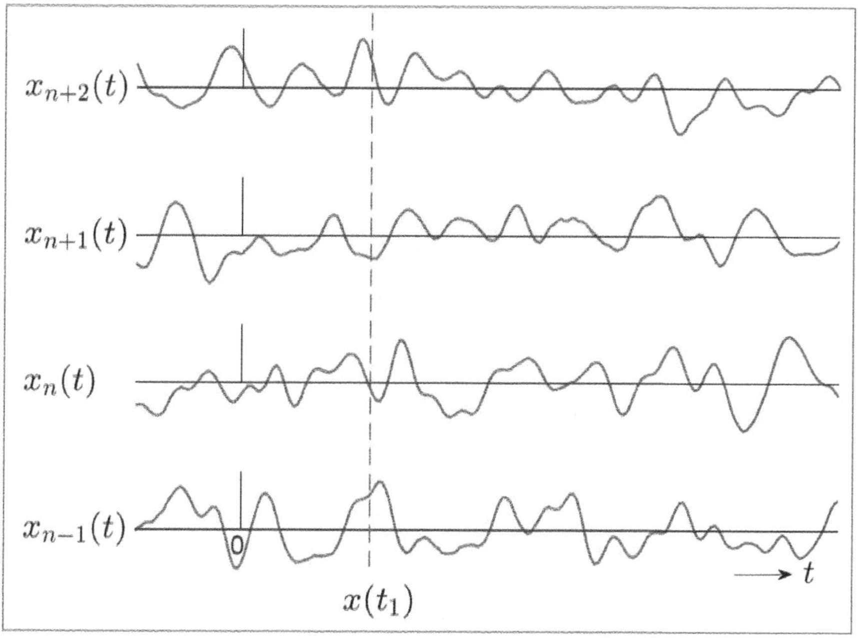

Figure 15.1. A few sample functions of a stochastic process.

This figure gives an example of a finite number of possible realizations, but an ensemble may consist of infinite number of realizations. A stochastic process produces a random variable that arises from giving t, a fixed value with s being variable. This means that the random variable $x(t_1, s) = x(t_1)$ is found by considering the family of realizations at the fixed point in time t_1 (see Fig. 15.1). Instead of $x(t_1)$ we may use the notation x_1. The random variable x_1 describes the statistical properties of the process at the instant of time t_1. The *expectation* of x_1 (denoted as $E[x_1]$) is called the *ensemble mean* or the *expected value* or the *mean* of the stochastic process (at the instant of time t_1). Since t_1 is an arbitrarily chosen time, the mean of the process will in general may not be same, i.e., it may have different values at different time instances. Finally, a stochastic process may represent a single number by giving both t and s fixed values.

The phrase *stochastic process* may therefore have four different interpretations as:

- A family (or ensemble) of time functions. Both t and s are variables.
- A single time function called a sample function or a realization of the stochastic process. Then t is a variable and s is fixed.
- A random variable; t is fixed and s is variable.
- A single number; both t and s are fixed.

Different classes of stochastic processes may be distinguished. They are classified on the basis of the characteristics of the realization values of the process x and the time parameter t. Both can be either *continuous* or *discrete*, in any combination. Then we have the following four classes:

- Continuous stochastic process: Both the values of $x(t)$ and t are continuous.
- Discrete-time processes: The values of $x(t)$ are continuous, whereas time t is discrete. These processes are also called continuous random sequences.
- Discrete stochastic processes: Here the values of $x(t)$ are discrete but the time axis is continuous.
- Discrete random sequence: Both the values of $x(t)$ and t are discrete.

15.2.1 Continuous Stochastic Processes

For this class of processes, it is assumed that in principle the following holds:

$$-\infty < x(t) < \infty \quad \text{and} \quad -\infty < t < \infty \tag{15.2}$$

An example of this class is already given by Figure 15.1. This could be an ensemble of realizations of a thermal noise process produced by a resistor.

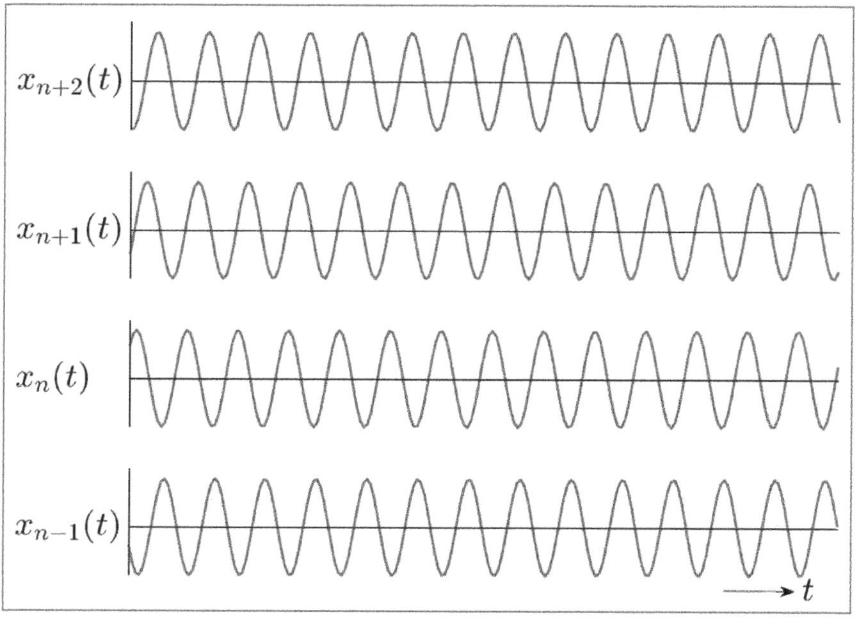

Figure 15.2. Ensemble of sample functions of the stochastic process $A\cos(2\pi f_0 t + \theta)$, with θ uniformly distributed in the interval $(0, 2\pi]$.

Another example is the process described by the following equation:

$$x(t) = A\cos(2\pi f_0 t + \theta). \tag{15.3}$$

with A, and f_0 constant and θ a random variable with a uniform probability density function on the interval $(0, 2\pi]$. In this example, the number of realizations is in fact uncountable, as θ assumes continuous values. An ensemble of sample functions is shown in Fig. 15.2.

15.2.2 Discrete-Time Processes (Continuous Random Sequences)

As an example of this type of process, we can imagine sampling the process that is shown in Fig. 15.1. Let us suppose that to this process ideal sampling is applied at equidistant points in time with sampling period T_s (the ideal sampling means that the values at T_s are replaced by delta functions of amplitude $x(nT_s)$). The process that is produced in this way is shown in Fig. 15.3, where the sample values are presented by means of the length of the samples.

Another important example of the discrete-time process is the so-called *Poisson process*, where there are no equidistant samples in time but the processes produce samples at random points in time.

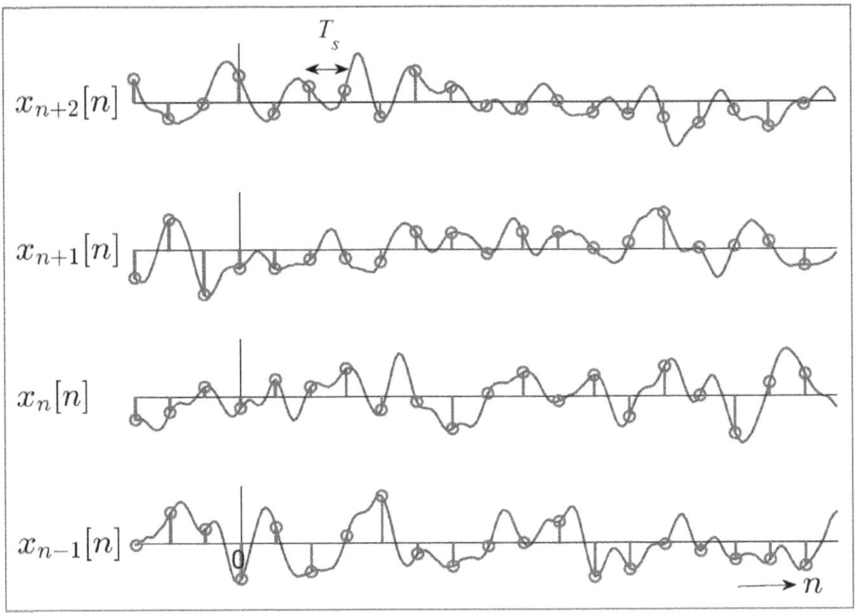

Figure 15.3. Example of a discrete-time stochastic process.

15.2.3 Discrete Stochastic Processes

Two examples are presented here:

Example 1: The ensemble or realizations consists of a set of constant time functions. According to the outcome of an experiment, one of these constants may be chosen. This experiment can be, for example, the rolling of a die. In this case, the number of realizations can be six ($n = 6$), equal to the number of faces of a die. Each of the outcomes, $s \in \{1, 2, 3, 4, 5, 6\}$ has a one-to-one correspondence to one of these numbered constant functions of time. The ensemble is depicted in Fig. 15.4.

Example 2: Another important stochastic process is the random data (bits) signal. It is produced by many data sources and is described by

$$x(t) = \sum_n A_n p(t - nT - \phi). \tag{15.4}$$

where $\{A_n\}$ are the data bits that are randomly chosen from the set $A_n \in \{+1, -1\}$. The rectangular pulse $p(t)$ of width T serves as the carrier of the information. Now ϕ is supposed to be uniformly distributed in the bit interval $(0, T]$, so that all data sources of the family have the same bit period, but these periods are not synchronized. The ensemble is shown in Fig. 15.5.

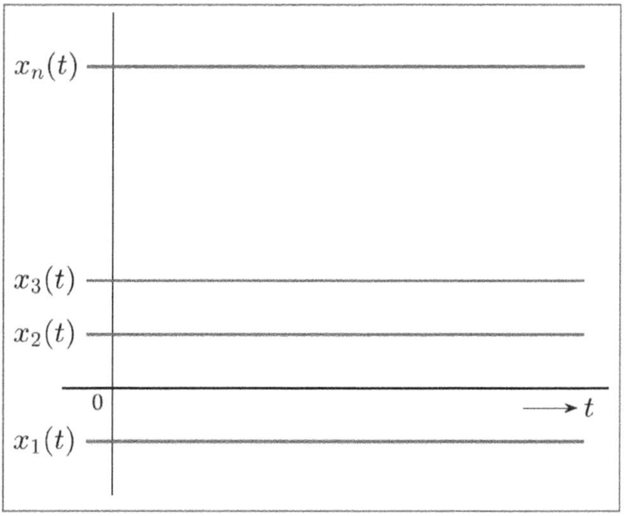

Figure 15.4. Ensemble of sample functions of the stochastic process constituted by a number of constant time functions.

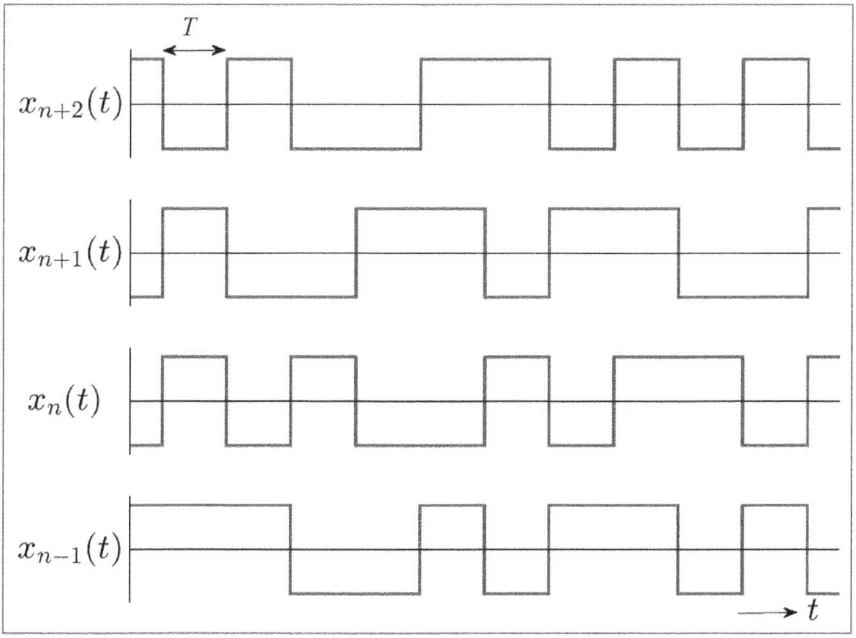

Figure 15.5. Ensemble of sample functions of the stochastic process $x(t) = \sum_n A_n p(t - nT - \phi)$, with ϕ uniformly distributed in the interval $(0, T]$.

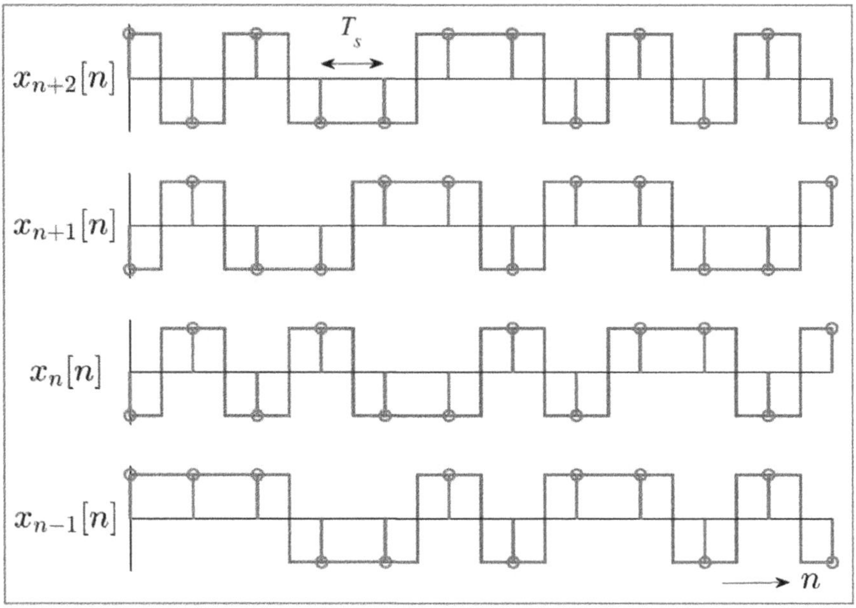

Figure 15.6. Example of a discrete random sequence.

15.2.4 Discrete Random Sequences

The discrete random sequence can be imagined to result from sampling a discrete stochastic process. Figure 15.6 shows the result of sampling the random data signal from Example 2.

15.3 Deterministic Function vs Stochastic Process

The concept of the stochastic process does not conflict with the theory of deterministic functions. It should be recognized that a deterministic function can be considered as a special case of a stochastic process. This is elucidated by considering the example given in Eq. (15.3). If the random variable θ is given the probability density function $f_\theta(\theta) = \delta(\theta)$, then the stochastic process reduces to the function $\cos(2\pi f_0 t)$. The given probability density function is actually a discrete one with a single outcome.

In fact, the ensemble of the process reduces in this case to a family comprising merely one member. This is a general rule; when the probability density function of a stochastic process that is governed by a single random variable consists of a single delta function, then a deterministic function results. This way of generalization avoids the often confusing discussion on the difference between a deterministic function on the one hand and a stochastic process on the other hand. In view of the consideration presented here, they can actually be considered as members of the same class, namely the class of stochastic process.

The random processes are characterized by statistical measures, which are described in the following sections.

15.4 Probability, Cumulative Distribution Function and Probability Density Function

The set of possible outcomes of a random experiment is called the *sample space*. An element in the sample space is called a *sample point*. Each outcome of a random experiment corresponds to a sample point. Subset of the sample space are called events, and events consisting of a single element (sample point) are called elementary events.

Probability is a number, associated with the events according to some appropriate probability law. The probability assigned to the event s from the sample space S ($s \in S$) is denoted as $P(s)$ and has a value between 0 and 1 ($0 \leq P(s) \leq 1$). In order to be a valid probability assignment, the following three axioms must be satisfied:

- $P(s) \geq 0$ for every event $s \in S$.
- $P(s) = 0$ for the certain event S.
- For any two mutually exclusive events s_1 and s_2, $P(s_1 \cup s_2) = P(s_1) + P(s_2)$.

In signal processing applications, it is the probabilistic description of the random variable, rather than the statistical characterization of events in the sample space, that is generally of interest. It is therefore more convenient to have a probability law assigned to the random variable itself. For a fixed value of the time parameter (say t_1), the *cumulative probability distribution function (CDF)* or, for short, the *distribution function* of the random variable x_1 is defined by

$$F_x(x_1; t_1) \triangleq P\{x(t_1) \leq x_1\}. \tag{15.5}$$

From this notation, it follows that F_x may be a function of the value of t_1 that has been chosen.

For two random variables $x_1 = x(t_1)$ and $x_2 = x(t_2)$, the 2nd order joint probability distribution function can be derived as the two-dimensional extension of the the above equation as follows:

$$F_x(x_1, x_2; t_1, t_2) \triangleq P\{x(t_1) \leq x_1, x(t_2) \leq x_2\}. \tag{15.6}$$

Similarly, this can be extended to the Nth-order joint cumulative probability distribution function as

$$F_x(x_1, \ldots, x_N; t_1, \ldots, t_N) \triangleq P\{x(t_1) \leq x_1, \ldots, x(t_N) \leq x_N\}. \tag{15.7}$$

The corresponding (joint) *probability density functions (PDF)* are found by *taking the derivatives respectively of the above equations* as shown below:

$$f_x(x_1; t_1) \triangleq \frac{\partial F_x(x_1; t_1)}{\partial x_1}, \tag{15.8}$$

$$f_x(x_1, x_2; t_1, t_2) \triangleq \frac{\partial^2 F_x(x_1, x_2; t_1, t_2)}{\partial x_1 \partial x_2}, \tag{15.9}$$

$$f_x(x_1, \ldots, x_N; t_1, \ldots, t_N) \triangleq \frac{\partial^N F_x(x_1, \ldots, x_N; t_1, \ldots, t_N)}{\partial x_1, \ldots, \partial x_N}. \tag{15.10}$$

Statistical Independence: Two processes $x(t)$ and $y(t)$ are called statistically independent if the set of random variables $\{x(t_1), x(t_2), \ldots, x(t_N)\}$ is independent of the set of random variables $\{y(t_1'), y(t_2'), \ldots, y(t_M')\}$, for each arbitrary choice of the time parameters $\{t_1, t_2, \ldots, t_N; t_1', t_2', \ldots, t_M'\}$. Independence implies that the joint probability density function can be factored in the following way

$$\begin{aligned} &f_{x,y}(x_1, \ldots, x_N; y_1, \ldots, y_M; t_1, \ldots, t_N; t_1', \ldots, t_M') \\ &= f_x(x_1, \ldots, x_N; t_1, \ldots, t_N).f_y(y_1, \ldots, y_M; t_1', \ldots, t_M'). \end{aligned} \tag{15.11}$$

Thus, the joint probability density function of two independent processes is written as the product of the two marginal probability density functions.

15.5 Description of Random Signals

One of the main issues in the study of random signals is: *how to describe or characterize random signals?* This problem arises because of the randomness associated with such signals. For example, if the speech signal you produce is different each time you speak your name, how can we present a single waveform as the signal corresponding to your name? Each record of the signal is similar to one another, and yet different, with some unpredictable variability associated with each record. So, we may consider taking an *average of several records* and present that as the speech signal corresponding to your name.

Consider another example of waiting to board a bus that is scheduled to arrive at 8.30a.m. daily. If we keep track of the exact arrival time of the bus each day, we will find that there is some amount of variability associated with the arrival time. If we observe the arrival pattern of the bus for several days, we can estimate the *average arrival time* and the *associated variability* (e.g., ±3 minute, ±30 seconds, *etc.* with respect to the average). The variability in the arrival time depends on many factors such as driver's mood, condition of the bus, weather, traffic flow *etc.* These factors are clearly random in nature and these give rise to the randomness in the arrival time of the bus. In summary, we can say the following.

§ *Associated with every random signal, there are some physical phenomena which cause randomness. While these physical phenomena may not be observable, what we can observe and measure are the effects causes by them, which are the random signals.*

The above examples show that we have to use *averages* to characterize the features of random signals. Averages are also called *statistics.* Just as we need to observe the arrival time on several days for estimating the average arrival time, *we need to observe several copies of a random signal before we can estimate the averages associated with it.*

Two of the most commonly used statistical averages are *mean* and *correlation,* which are *the first and the second order statistics,* respectively, of a random process. In practical applications, very often, these two averages are sufficient to make conclusions or assessments about the observed random signals. So, we will look into the definition and meaning of these averages.

- *Mean of a Random Process*: This is the most simple of all the averages. Let $x(t)$ be an analog random signal. Its mean, denoted by m_x, is the average value of $x(t)$ and is given by

$$m_x = E[x(t)] = \int_c x f_x(x; t) dx, \tag{15.12}$$

  where $f_x(x; t)$ is the PDF of the random variable $x(t)$, $E[\cdot]$ denotes the statistical expectation operator, and c defines the complete range of the random variable $x(t)$. Observe that the mean may be a function of the time index t.

  Linearity of Mean: It is easy to verify from Eq. (15.12) that mean is a linear operator. If $x(t)$ and $y(t)$ are two random signals, then for any two scalars α and β,

$$E[\alpha x(t) + \beta y(t)] = \alpha E[x(t)] + \beta E[y(t)]. \tag{15.13}$$

  Thus, *the mean of a linear combination of signals is equal to the same linear combination of the means of the respective signals.*
- *Auto correlation of a Random Process*: The autocorrelation function (ACF) of a stochastic process is defined as the correlation $E[x_1 x_2]$ of the two random

variables $x_1 = x(t_1)$ and $x_2 = x(t_2)$. It is defined as

$$\gamma_{xx}(t_1, t_2) = E[x(t_1)x^*(t_2)] = \int_C \int_C x_1 x_2^* f_x(x_1, x_2; t_1, t_2) dx_1 dx_2 \tag{15.14}$$

where $f_x(x_1, x_2; t_1, t_2)$ is the joint PDF of the random variables x_1 and x_2. Observe that the autocorrelation may be a function of the time indices t_1 and t_2.

Substituting $t_1 = t$ and $t_2 = t_1 + \tau$, Eq. (15.14) becomes

$$\gamma_{xx}(t, t + \tau) = E[x(t)x^*(t + \tau)]. \tag{15.15}$$

- *Cross-correlation function (CCF) between two random processes*: It is defined as

$$\gamma_{xy}(t, t + \tau) = E[x(t)y^*(t + \tau)]. \tag{15.16}$$

§ *Prove the following:*

$$\gamma_{xx}(t_2, t_1) = \gamma_{xx}^*(t_1, t_2) \tag{15.17}$$

and

$$\gamma_{yx}(t_2, t_1) = \gamma_{xy}^*(t_1, t_2). \tag{15.18}$$

15.6 Stationary Processes

We can say that a stochastic process is stationary if its statistical properties do not depend on the time parameter. There are several types of stationarity like *strict-sense stationarity, wide-sense stationarity*, etc.

15.6.1 First-Order Stationary Processes

A stochastic process is called first-order stationary process if the first-order PDF is independent of time. That is,

$$f_x(x_1; t_1) = f_x(x_1; t_1 + \tau) \tag{15.19}$$

holds for all values of τ. As a result of this property, the mean m_x of such process is constant (i.e., it is independent of time).

15.6.2 Second-Order Stationary Processes

A stochastic process is called second-order stationary process if the two-dimensional PDF

$$f_x(x_1, x_2; t_1, t_2) = f_x(x_1, x_2; t_1 + \tau, t_2 + \tau), \tag{15.20}$$

for all values of τ. It is easy to verify that the above equation is only a function of the time difference $t_2 - t_1$ and does not depend on the absolute time. In order to gain that insight, put $\tau = -t_1$. A process that is second-order stationary is first-order stationary as well, since the second-order joint PDF uniquely determines the lower order (first-order in this case) PDF.

15.6.3 N^{th}-Order Stationary Processes

By extending the reasoning from the last subsection to N random variables, we arrive at the Nth-order stationary process. For such processes, the Nth-order joint PDF is independent of time shift. i.e.,

$$f_x(x_1, \dots, x_N; t_1, \dots, t_N) = f_x(x_1, \dots, x_N; t_1 + \tau, \dots, t_N + \tau) \quad (15.21)$$

for all values of τ. Such processes are stationary for all orders $k \leq N$. An Nth-order stationary process where N can have an arbitrary large value is called a *strict-sense stationary process.* A random process $\{x(t)\}$ is said to be *stationary in the strict-sense* if *all* of its statistical averages or characteristics do not change with a shift in the time origin. That is, the joint PDF of $\{x(t_1), x(t_2), \dots, x(t_N)\}$ depends only on the differences between the time indices $\{t_1, t_2, \dots, t_N\}$ and *not* on the actual value of each index.

15.6.4 Wide-Sense Stationary (WSS) Processes

Since the two-dimensional joint PDF for a second-order stationary process depends only on the time difference ($\tau = t_2 - t_1$), the autocorrelation function will also be a function of the difference τ. Then Eq. (15.15) can be written as

$$\gamma_{xx}(t, t + \tau) = \gamma_{xx}(\tau). \quad (15.22)$$

The mean and the autocorrelation function of a stochastic process are often the most important characterizing features. It becomes easier if these two quantities do not change in absolute time. A second order stationary process guarantees this independence but at the same time places severe demands on the process. Therefore, we define a broader class of stochastic process, the so called wide sense stationary process. So, a random process $\{x(t)\}$ is said to be *stationary in the wide-sense* (or, *weakly stationary*) if its mean and autocorrelation do not change with a shift in the time origin. That is, it satisfies the following conditions:

$$E[x(t)] = m_x = \text{constant} \quad (15.23)$$

$$E[x(t)x(t + \tau)] = \gamma_{xx}(\tau). \quad (15.24)$$

It is clear that a second-order stationary process is also wide-sense stationary. The converse, however, is not necessarily true.

15.6.5 Jointly WSS Random Processes

The random processes $\{x(t)\}$ and $\{y(t)\}$ are said to be *jointly stationary in the wide sense* if they are individually WSS and their cross-correlation $\gamma_{xy}(t, t+\tau)$ depends only on τ not on the absolute time parameter. Therefore, we have

$$\gamma_{xy}(t, t+\tau) = E[x(t)y(t+\tau)] = \gamma_{xy}(\tau). \tag{15.25}$$

15.6.6 Properties of $\gamma_{xx}(\tau)$

If a process is at least WSS, its autocorrelation function exhibits the following properties:

- $|\gamma_{xx}(\tau)| \leq \gamma_{xx}(0)$. This means that $|\gamma_{xx}(\tau)|$ attains its maximum value when $\tau = 0$.

  To prove this property, let us consider the expression

$$\begin{aligned} E[\{x(t) \pm x(t+\tau)\}^2] &= E[x^2(t) + x^2(t+\tau) \pm 2x(t)x(t+\tau)] \\ &= 2\{\gamma_{xx}(0) \pm \gamma_{xx}(\tau)\} \end{aligned} \tag{15.26}$$

  Since the expectation $E[\{x(t) \pm x(t+\tau\}^2]$ is taken over the squared value of a certain random variable, this expectation should be ≥ 0. Thus the proof.
- Symmetry property ($\gamma_{xx}(-\tau) = \gamma_{xx}(\tau)$). i.e., $\gamma_{xx}(\tau)$ is an even function of τ.

  To prove this property, substitute $t' = t + \tau$ in the definition of the autocorrelation function. Then,

$$\begin{aligned} \gamma_{xx}(\tau) &= E[x(t)x(t+\tau)] = E[x(t'-\tau)x(t')] \\ &= E[x(t')x(t'-\tau)] = \gamma_{xx}(-\tau) \end{aligned} \tag{15.27}$$

- $\gamma_{xx}(0) = E[x^2(t)] = \sigma_x^2 + m_x^2$. This property follows immediately from the definition of $\gamma_{xx}(\tau)$ by inserting $\tau = 0$ (Eq. (15.24)). It may also be noted that

$$\gamma_{xx}(0) = E[x^2(t)] = p_x = \sigma_x^2 + m_x^2, \tag{15.28}$$

  where p_x is the power of the random process $x(t)$. σ_x^2 is the variance.
- If $x(t)$ has no periodic component, then $\gamma_{xx}(\tau)$ comprises a constant term which is equal to $E[x(t)]^2$. i.e., $\lim_{|\tau|\to\infty} \gamma_{xx}(\tau) = E[x(t)]^2 = m_x^2$.

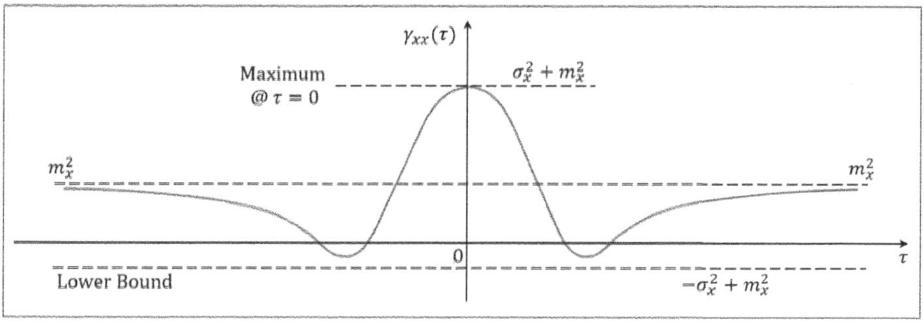

Figure 15.7. Example of an ACF.

Proof: From a physical point of view, most processes have the property that the random variables $x(t)$ and $x(t+\tau)$ are independent when $\tau \to \infty$. From the definition of the autocorrelation, it follows that

$$\begin{aligned}\lim_{\tau\to\infty} \gamma_{xx}(\tau) &= \lim_{\tau\to\infty} E[x(t)x(t+\tau)] \\ &= E[x(t)]E[x(t+\tau)] \\ &= m_x m_x = m_x^2 \end{aligned} \quad (15.29)$$

Figure 15.7 shows a typical ACF with the properties just discussed. The negative values of the ACF in this figure show that for these values, the shift τ of the expected value $E\{x(t+\tau)x(t)\}$ is negative. It means that with a separation τ, $x(t+\tau)$ and $x(t)$ are expected to have opposite signs. Like large positive values of the ACF, large negative values also indicate a strong relationship between the values of $x(t)$. The lower bound, $-\sigma_x^2 + m_x^2$ is not reached in this example. The properties such as the maximum value of the ACF, lower bound and the symmetry hold for all stationary random signals. The uncorrelation at very high values of τ holds when distantly separated values do not correlate.

- If $x(t)$ has a periodic component, then $\gamma_{xx}(\tau)$ will comprise a periodic component as well, and which has the same periodicity.

Proof: Periodic processes may be decomposed into cosine and sine components according to the Fourier analysis. It therefore suffices to consider the autocorrelation function of one such component:

$$E[\cos(\omega t - \theta)\cos(\omega t + \omega\tau - \theta)] = \frac{1}{2}E[\cos(\omega\tau) + \cos(2\omega t + \omega\tau - 2\theta)] \quad (15.30)$$

Since our considerations are limited to WSS, the autocorrelation function should be independent of the absolute time t, and thus the expectation of the

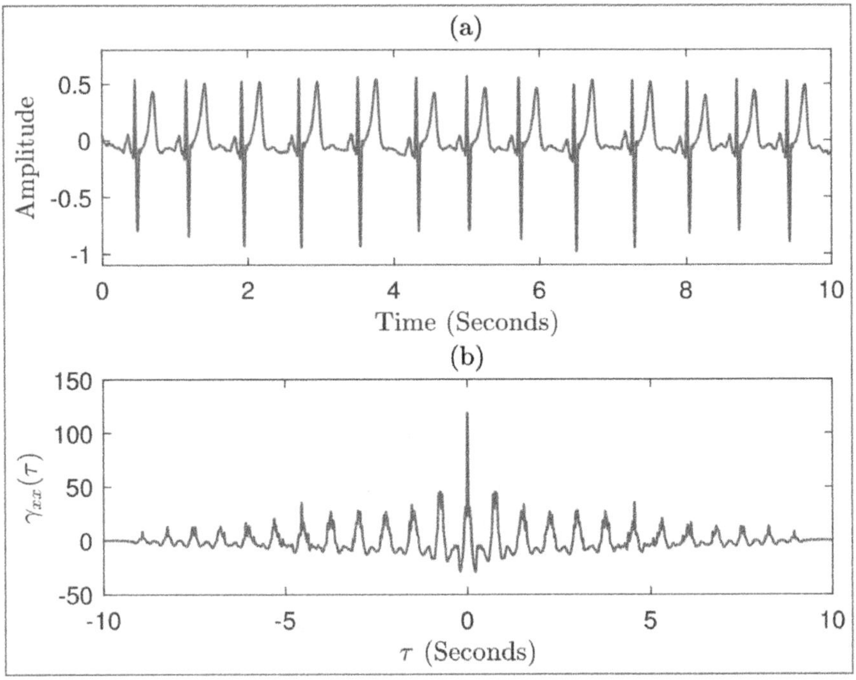

Figure 15.8. Typical ECG waveform (a) and its ACF (biased estimate) (b).

last term of the latter expression should be zero. Thus only the term comprising $\cos(\omega\tau)$ remains after taking the expectation, which proves this property.

Figure 15.8 shows an example ECG waveform (Fig. 15.8(a)) for 10 seconds and its corresponding ACF (estimated) (Fig. 15.8(b)). It can be seen that the maximum of the ACF happens at $\tau = 0$. Because of the repetitive (not exactly periodic) nature of the ECG waveforms, the ACF also shows some kind of repetitiveness. The Matlab function `xcorr` is used to find a biased estimate of the ACF of the ECG signal.

15.6.7 Properties of $\gamma_{xy}(\tau)$

If two processes $x(t)$ and $y(t)$ are jointly WSS, then the cross-correlation function exhibits the following properties:

- $\gamma_{xy}(-\tau) = \gamma_{yx}(\tau)$.

  Proof: In the definition of the cross-correlation function, replace $t - \tau$ by t' and after some manipulations, we get

$$\begin{aligned}\gamma_{xy}(-\tau) &= E[x(t)y(t-\tau)] = E[x(t'+\tau)y(t')] \\ &= E[y(t')x(t'+\tau)] = \gamma_{yx}(\tau)\end{aligned} \tag{15.31}$$

- $|\gamma_{xy}(\tau)| \leq \sqrt{\gamma_{xx}(0)\gamma_{yy}(0)}$.

  To prove this property, we consider the expectation of the process $\{x(t) + cy(t+\tau)\}^2$, where c is a constant; i.e., we investigate

$$E[\{x(t) + cy(t+\tau)\}^2] = E[X] = \gamma_{xx}(0) + c^2\gamma_{yy}(0) + 2c\gamma_{xy}(\tau). \tag{15.32}$$

  where

$$\begin{aligned} E[X] &= E[x^2(t) + c^2y^2(t+\tau) + 2cx(t)y(t+\tau)], \\ &= E[x^2(t)] + c^2E[y^2(t+\tau)] + 2cE[x(t)y(t+\tau)]. \end{aligned} \tag{15.33}$$

  This latter expression is a quadratic form in c, and since it is the expectation of a quantity squared, this quantity can never be less than 0. As a result, the discriminant cannot be positive; i.e.,

$$\gamma_{xy}^2(\tau) - \gamma_{xx}(0)\gamma_{yy}(0) \leq 0. \tag{15.34}$$

  This implies that,

$$|\gamma_{xy}(\tau)| \leq \sqrt{\gamma_{xx}(0)\gamma_{yy}(0)}. \tag{15.35}$$

- $|\gamma_{xy}(\tau)| \leq \frac{1}{2}[\gamma_{xx}(0) + \gamma_{yy}(0)]$.

  Proof: This property is a consequence of the well-known fact that the arithmetic mean of two positive numbers is always greater than or equal to their geometric mean.

Example: Based on the correlation function, it is possible to measure the distance of an object from a transmitter. This example explains the scheme in detail. Consider the system shown in Fig. 15.9 where a signal source produces a random signal, being a realization of a stochastic process. Let us assume that the process is WSS. The signal is applied to a transmitter that produces a wave in a transmission medium; let it be an acoustic wave or an electromagnetic wave denoted by $x(t)$. Let us further assume that the transmitted wave strikes a distant object which reflects (partly) the wave. Then the reflected wave will travel backwards to the position of the measuring equipment. The receiver at the measuring equipment receives this signal, denoted as $y(t)$. These signals $x(t)$ and $y(t)$ can be applied to a correlator that estimates the cross-correlation function $\gamma_{xy}(\tau)$ given by,

$$\gamma_{xy}(\tau) = E[x(t)y(t+\tau)] = E[x(t)\alpha x(t-T+\tau)] = \alpha\gamma_{xx}(\tau - T) \tag{15.36}$$

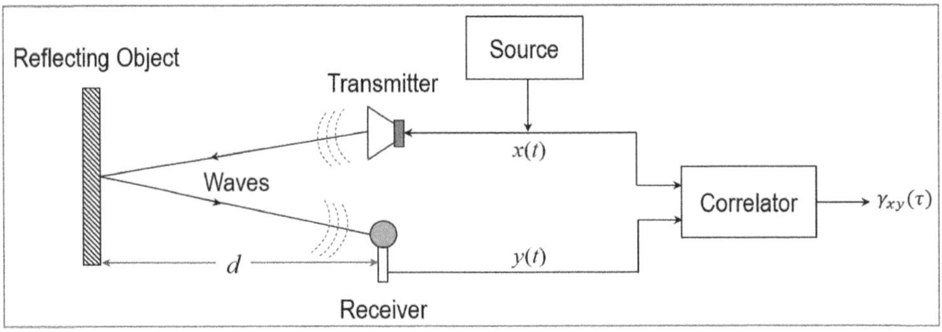

Figure 15.9. Set up for measuring a distance based on the correlation function.

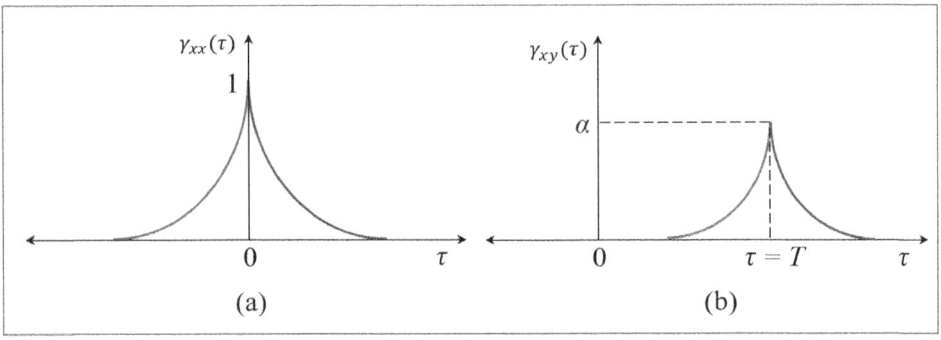

Figure 15.10. (a) The autocorrelation function of the transmitted signal $x(t)$ and (b) the measured cross-correlation function (between the transmitted signal, $x(t)$ and the received signal, $y(t)$) of the distance measuring set up.

where $y(t) = \alpha x(t-T)$ is the delayed and attenuated version of $x(t)$, T is the total travel time. Figure 15.10 shows the autocorrelation and cross-correlation functions. We will be able to estimate the distance (d) of the object from the transmitter if we know the propagation velocity (v) of the wave in the medium and the time, T using the relation, $2d = v \times T$, where $2d$ is the distance travelled during the time period of T units.

15.6.8 Independent, Uncorrelated, and Orthogonal Processes

The random processes $\{x(t)\}$ and $\{y(t)\}$ are said to be

- *Independent* if the random variables $\{x(t_1), x(t_2), \ldots, x(t_m)\}$ are mutually independent[1] of the random variables $\{y(t_1'), y(t_2'), \ldots, y(t_m')\}$ for all m.
- *Uncorrelated* if $E[x(t)y(t+\tau)] = m_x.m_y$ for all t and τ.

1. The set of random variables $x_1, x_2, \ldots, x_m$ are said to be independent if their joint PDF is separable. That is, $f_{x_1x_2\cdots x_m}(x_1, x_2, \ldots,, x_m) = f_{x_1}(x_1)f_{x_2}(x_2)\cdots f_{x_m}(x_m)$.

- *Orthogonal* if $\gamma_{xy}(t, t+\tau) = 0$ for all t and τ.

Thus, zero-mean orthogonal processes are also uncorrelated.

§ *Examine why independent processes are also uncorrelated, while uncorrelated processes need not be independent.*

15.6.9 Covariance Functions

Correlations of random processes after removing the mean are called covariances. For stationary processes, the auto- and cross-covariances are given by

$$C_{xx}(\tau) = E\left[(x(t) - m_x)(x(t+\tau) - m_x)\right] = \gamma_{xx}(\tau) - m_x^2. \tag{15.37}$$

$$C_{xy}(\tau) = E\left[(x(t) - m_x)(y(t+\tau) - m_y)\right] = \gamma_{xy}(\tau) - m_x m_y. \tag{15.38}$$

For random signals, the signal samples become less correlated as they become more and more separated in time. Therefore,

$$\lim_{\tau\to\infty} \gamma_{xx}(\tau) = m_x^2 \quad \Rightarrow \quad \lim_{\tau\to\infty} C_{xx}(\tau) = 0, \tag{15.39}$$

provided the process $x(t)$ does not have a periodic component.

The symmetry properties of correlation and covariance functions are useful in signal analysis:

$$\gamma_{xx}(\tau) = \gamma_{xx}(-\tau), \quad C_{xx}(\tau) = C_{xx}(-\tau) \tag{15.40}$$

$$\gamma_{xy}(\tau) = \gamma_{yx}(-\tau), \quad C_{xy}(\tau) = C_{yx}(-\tau). \tag{15.41}$$

Further (when $\tau = 0$),

$$\gamma_{xx}(0) = E\left[|x(t)|^2\right] = p_x \tag{15.42}$$

$$C_{xx}(0) = E\left[|x(t) - m_x|^2\right] = \sigma_x^2 \tag{15.43}$$

$$|\gamma_{xx}(\tau)| \leq \gamma_{xx}(0) \qquad \text{for all } \tau. \tag{15.44}$$

where, p_x is the average power of $x(t)$. Note that σ_x^2 is the variance of $x(t)$. When the mean of the random process is zero, the *correlation functions are same as the covariance functions.*

15.7 Physical Interpretation of Process Parameters

In the previous sections, stochastic processes have been described from a mathematical point of view. In practice, we want to relate these descriptions to physical concepts such as signal, represented, for example, by a voltage or a current. In these cases, the following physical interpretations are connected to the parameters of the stochastic processes:

- The mean $m_x = E[x(t)]$ is proportional to the d.c. component of the signal
- The squared mean value $m_x^2 = \{E[x(t)]\}^2$ is proportional to the power in the d.c. component ($p_{x(DC)}$) of the signal.
- The mean squared value $E[x^2(t)]$ is proportional to the total average power of the signal.
- The variance $\sigma_x^2 = E[x^2(t)] - m_x^2$ is proportional to the power in the time varying components of the signal, i.e., the a.c. power.
- The standard deviation σ_x is the square root of the mean squared value of the time-varying components, i.e., the root-mean-square (r.m.s) value.

15.8 Gaussian Random Processes

We will now briefly describe some properties of Gaussian[2] random processes which are very useful in the study of signal processing and communication systems.

Definition: The (real-valued) process $\{x(t)\}$ is said to be a Gaussian process if it has the following PDF:

$$f_x(x) = \frac{1}{\sigma_x\sqrt{2\pi}} \exp\left[-\frac{(x-m_x)^2}{2\sigma_x^2}\right]. \tag{15.45}$$

When considering two jointly Gaussian random variables x and y, the joint PDF becomes,

$$f_{xy}(x, y) = \frac{1}{2\pi\sigma_x\sigma_y\sqrt{1-\rho^2}} \exp\left\{\frac{-1}{2(1-\rho^2)}[xx]\right\} \tag{15.46}$$

where,

$$[xx] = \left[\frac{(x-m_x)^2}{\sigma_x^2} - \frac{2\rho(x-m_x)(y-m_y)}{\sigma_x\sigma_y} + \frac{(y-m_y)^2}{\sigma_y^2}\right] \tag{15.47}$$

and ρ is the correlation coefficient defined by,

$$\rho = \frac{E\left[(x-m_x)(y-m_y)\right]}{\sigma_x\sigma_y}. \tag{15.48}$$

2. More details on Gaussian PDF can be found in Appendix B.

For N jointly Gaussian random variables, the random vector $\mathbf{x} = \{x_1, x_2, \cdots, x_N\}^T$ has a *jointly Gaussian* PDF as shown below:

$$f_{x_1,x_2,\ldots,x_N}(\mathbf{x}) = \frac{1}{\sqrt{(2\pi)^N |\mathbf{C}_x|}} \exp\left(-\frac{(\mathbf{x}-\mathbf{m}_x)^T \mathbf{C}_{xx}^{-1}(\mathbf{x}-\mathbf{m}_x)}{2}\right) \tag{15.49}$$

where $\mathbf{C}_{xx} = E\left[(\mathbf{x}-\mathbf{m}_x)(\mathbf{x}-\mathbf{m}_x)^T\right]$ and $\mathbf{m}_x = E[\mathbf{x}]$ are the auto-covariance matrix and mean, respectively, of $\mathbf{x}$, and $|\mathbf{C}_{xx}|$ denotes the determinant of $\mathbf{C}_{xx}$. Further,

$$\mathbf{C}_{xx} = E[\mathbf{x}\mathbf{x}^T] - \mathbf{m}_x\mathbf{m}_x^T = \Gamma_{xx} - \mathbf{m}_x\mathbf{m}_x^T \tag{15.50}$$

where Γ_{xx} is the autocorrelation matrix of $\mathbf{x}$.

15.8.1 Properties of Gaussian Random Processes

- For Gaussian random processes, the *knowledge of the mean and autocorrelation function gives a complete description of the process.* This is because the joint PDF of a Gaussian process has only the mean vector and autocorrelation matrix as the two parameters.
- If a Gaussian process $\{x(t)\}$ is passed through a LTI system, the output process $\{y(t)\}$ will also be a Gaussian process. This is because *linear combinations of Gaussian random variables is also Gaussian.*
- For Gaussian random processes, *wide-sense stationarity and strict-sense stationarity are equivalent.* This follows from the fact that Gaussian PDF uses only up to second order statistics (mean and autocorrelation).
- For Gaussian random processes, *uncorrelatedness and independence are equivalent.*

  This is because the exponent $(\mathbf{x}-\mathbf{m}_x)^T\mathbf{C}_{xx}^{-1}(\mathbf{x}-\mathbf{m}_x)$ in the joint PDF becomes $\sum_{i=1}^{N} \mathbf{C}_{xx}(i,i)(x_i - m_{x_i})^2$ when the process is uncorrelated. This results in the joint PDF becoming a product of the N marginal PDFs.

15.9 Complex Stochastic Processes

A complex stochastic process is defined by,

$$z(t) = x(t) + jy(t) \tag{15.51}$$

with $x(t)$ and $y(t)$ are real stochastic processes. Such a process is said to be stationary if $x(t)$ and $y(t)$ are jointly stationary. Expectation and autocorrelation function of a complex stochastic process is defined as

$$E[z(t)] = E[x(t) + jy(t)] = E[x(t)] + jE[y(t)] \tag{15.52}$$

and

$$\gamma_{zz}(t, t+\tau) = E[z(t)z^*(t+\tau)] \tag{15.53}$$

where $*$ indicates the complex conjugate.

For the autocovariance function, the above Eq. (15.53) can be used, where $z(t)$ is replaced with $z(t) - E[z(t)]$. This yields

$$C_{zz}(t, t+\tau) = \gamma_{zz}(t, t+\tau) - E[z(t)]E^*[z(t+\tau)] \tag{15.54}$$

The cross-correlation function of two complex processes $p(t)$ and $q(t)$ reads

$$\gamma_{pq}(t, t+\tau) = E[p(t)q^*(t+\tau)] \tag{15.55}$$

and the cross-covariance function is found from Eq. (15.56) below:

$$C_{pq}(t, t+\tau) = \gamma_{pq}(t, t+\tau) - E[p(t)]E^*[q(t+\tau)]. \tag{15.56}$$

One may wonder why the correlation functions of complex processes are defined in the way it has been done in Eqs. (15.53) and (15.55). The explanation for this arises from an engineering point of view; the given expressions of the correlation functions evaluated for $\tau = 0$ have to result in the expectation of the squared process for real processes.

15.10 Ergodic Processes

Estimation of ensemble averages of a random process requires a large number of realizations of the process. This is not feasible since we usually have only one realization of the process in any practical situation. Using one realization, all that we can obtain are time averages. Therefore, if the process is ergodic, we can replace the ensemble averages by time averages.

A stationary random process is said to be *ergodic* if its *ensemble averages are equal to time averages*. It has to satisfy the following conditions:

$$\tilde{m}_x = \overline{x(t)} = E[x(t)] = m_x \tag{15.57}$$

$$\tilde{\gamma}_{xx}(\tau) = \overline{x(t)x(t+\tau)} = E[x(t)x(t+\tau)] = \gamma_{xx}(\tau) \tag{15.58}$$

where, $\tilde{m}_x = \overline{x(t)}$ and $\tilde{\gamma}_{xx}(\tau) = \overline{x(t)x(t+\tau)}$ are the time averages of the mean and autocorrelation, respectively of the ergodic random process $x(t)$. By assuming ergodicity, a number of important statistical properties (such as the mean and autocorrelation) of a process may be estimated from the observation of a single available realization.

15.10.1 Mean

For an ergodic process $x(t)$, the time averaged mean is given by

$$\tilde{m}_x = \overline{x(t)} = \lim_{T\to\infty} \frac{1}{T}\int_{-T/2}^{T/2} x(t)dt, \tag{15.59}$$

where T is the observation period.

15.10.2 Correlation

Correlation and power spectrum are other two closely related statistics of random signals. We will introduce correlation for the ergodic process in this subsection and power spectrum in the next subsection.

Cross-Correlation: The time averaged *cross-correlation* between two random signals $x(t)$ and $y(t)$ is defined as

$$\begin{aligned}\tilde{\gamma}_{xy}(\tau) &= \overline{x^*(t)y(t+\tau)} \\ &= \lim_{T\to\infty} \frac{1}{T}\int_{-T/2}^{T/2} x^*(t)y(t+\tau)dt\end{aligned} \tag{15.60}$$

where τ is a time shift. That is, $\tilde{\gamma}_{xy}(\tau)$ is a measure of the similarity between $x(t)$ and $y(t+\tau)$. In several applications, we would like to know the value of τ at which the cross-correlation $\tilde{\gamma}_{xy}(\tau)$ becomes maximum.[3] In other words, $\tilde{\gamma}_{xy}(\tau)$ is a measure of how much information does $x(t)$ have about $y(t+\tau)$. If $x(t)$ and $y(t)$ are independent signals, then there is no correlation or dependence between them. Hence, for independent or uncorrelated signals, we can write,

$$\begin{aligned}\tilde{\gamma}_{xy}(\tau) &= \overline{x^*(t)y(t+\tau)} = \overline{x(t)}^* \cdot \overline{y(t+\tau)} \\ &= \overline{x(t)}^* \cdot \overline{y(t)} = \tilde{m}_x^* \cdot \tilde{m}_y\end{aligned} \tag{15.61}$$

where $\tilde{m}_y$ is the mean of $y(t)$. We have the following remarks.

- If $\tilde{\gamma}_{xy}(\tau) = \tilde{m}_x^* \cdot \tilde{m}_y$ for all values of τ, then we say that the signals $x(t)$ and $y(t)$ are *uncorrelated signals*.

3. For example, when we transmit a signal $s(t)$, let the received signal be $y(t) = s(t-\tau_0) + n(t)$ where τ_0 is a time-delay caused by the channel and $n(t)$ is the channel noise. At the receiver, we may want to know the value of τ_0. To determine this, we can compute the cross-correlation $\tilde{\gamma}_{sy}(\tau) = \overline{s^*(t)y(t+\tau)}$. Then, we will find that $|\tilde{\gamma}_{sy}(\tau)|$ will have a peak when $\tau = \tau_0$.

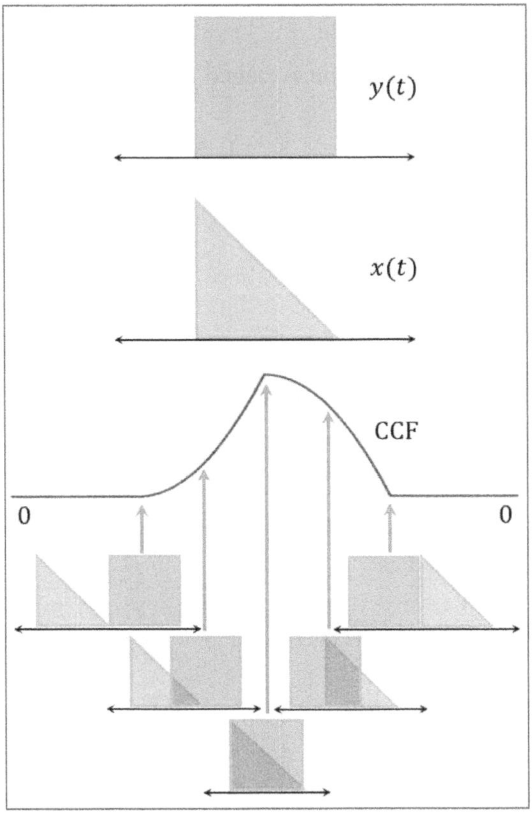

Figure 15.11. Illustration of the steps involved in estimating the CCF between two finite duration functions.

- If $x(t)$ and $y(t)$ are uncorrelated signals with either $\tilde{m}_x = 0$ or $\tilde{m}_y = 0$, then we have

$$\tilde{\gamma}_{xy}(\tau) = 0 \qquad \text{for all } \tau \tag{15.62}$$

$$\text{Power}\,[\alpha x(t) + \beta y(t)] = |\alpha|^2 p_x + |\beta|^2 p_y \tag{15.63}$$

where α and β are two scalars, and p_x and p_y are the powers of $x(t)$ and $y(t)$, respectively. With $\alpha = 1 = \beta$, Eq. (15.63) becomes,

$$\text{Power}\,[x(t) + y(t)] = p_x + p_y. \tag{15.64}$$

That is, *power of a sum of uncorrelated zero-mean signals is equal to the sum of the individual powers.*

Figure 15.11 illustrates the process of estimation of cross-correlation function (CCF) between two functions $x(t)$ and $y(t)$.

Autocorrelation: When $y(t) = x(t)$, we get from Eq. (15.60)

$$\tilde{\gamma}_{xx}(\tau) = \overline{x^*(t)x(t+\tau)} = \lim_{T\to\infty} \frac{1}{T}\int_{-T/2}^{T/2} x^*(t)x(t+\tau)dt \qquad (15.65)$$

and this is called the *autocorrelation of* $x(t)$. Thus, $\tilde{\gamma}_{xx}(\tau)$ *is a measure of the similarity of* $x(t)$ *with its shifted version.* That is, it is a measure of the redundancy (or, self-similarity) within $x(t)$. Based on this interpretation, we can state the following.

- Autocorrelation $\tilde{\gamma}_{xx}(\tau)$ becomes maximum when $\tau = 0$, since the maximum self-similarity occurs when we do not shift the signal. Further, note from Eq. (15.65) that

$$\tilde{\gamma}_{xx}(0) = \overline{x^*(t)x(t)} = \overline{|x(t)|^2} = p_x = \text{Power.} \qquad (15.66)$$

  Thus, power of the signal is equal to autocorrelation with no time shift (i.e., $\tau = 0$). Thus, we have

$$|\tilde{\gamma}_{xx}(\tau)| \le \tilde{\gamma}_{xx}(0) \qquad \text{for all } \tau. \qquad (15.67)$$

- In general, autocorrelation or self-similarity decreases as τ becomes large. So, from Eq. (15.61), we get

$$\tilde{\gamma}_{xx}(\tau) = |\tilde{m}_x|^2 = |\text{Mean}|^2 \qquad \text{for large } \tau. \qquad (15.68)$$

The steps involved in the estimation of the ACF of a signal, $x(t)$ is illustrated in Fig. 15.12.

15.10.3 Variance and Power

Power and variance are closely related to each other and hence we shall consider them together in this subsection. *The power of a random signal* $x(t)$ *is the average of the magnitude-square of the signal,* i.e., it is the mean-square value $\overline{|x(t)|^2}$ of the

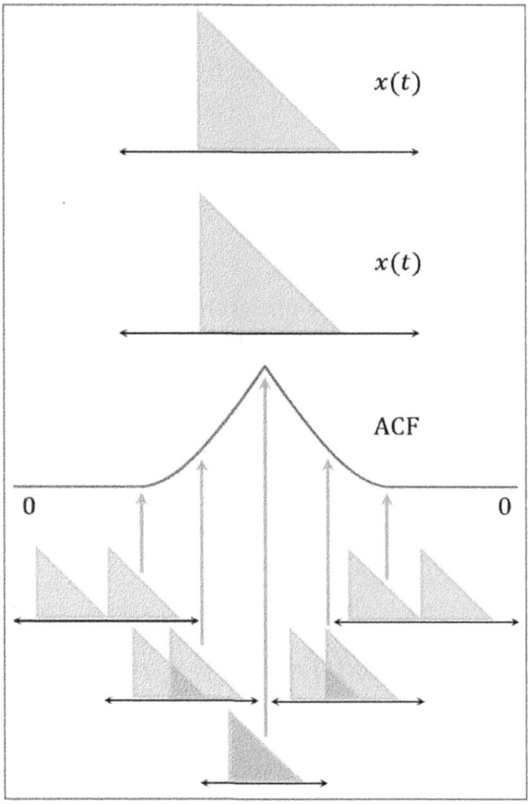

Figure 15.12. Illustration of the steps involved in estimating the ACF of a finite duration function.

signal. Denoting the power by $\tilde{p}_x$, we get

$$\text{Power,} \quad \tilde{p}_x = \overline{|x(t)|^2} = \lim_{T \to \infty} \frac{1}{T} \int_{-T/2}^{T/2} |x(t)|^2 dt. \tag{15.69}$$

The *variance (denoted by $\tilde{\sigma}_x^2$) of a random signal is the power of the signal after removing the mean*, i.e., it is the power of $x(t) - \tilde{m}_x$. So, we get

$$\text{Variance,} \ \tilde{\sigma}_x^2 = \overline{|x(t) - \tilde{m}_x|^2} = \lim_{T \to \infty} \frac{1}{T} \int_{-T/2}^{T/2} |x(t) - \tilde{m}_x|^2 \, dt. \tag{15.70}$$

Variance is a measure of the spread in the signal amplitude with respect to its mean.

Relation between Power and Variance: Note that

$$|x(t) - \tilde{m}_x|^2 = |x(t)|^2 - \tilde{m}_x \cdot x^*(t) - \tilde{m}_x^* \cdot x(t) + |\tilde{m}_x|^2 \tag{15.71}$$

where $x^*(t)$ and $\tilde{m}_x^*$ are complex conjugates of $x(t)$ and $\tilde{m}_x$, respectively. Since $\tilde{m}_x$ is independent of t, and the mean of $x^*(t)$ is $\tilde{m}_x^*$, we get

$$\overline{\tilde{m}_x \cdot x^*(t)} = \tilde{m}_x \cdot \overline{x^*(t)} = \tilde{m}_x \cdot \overline{x(t)}^* = |\tilde{m}_x|^2 \tag{15.72}$$

$$\overline{\tilde{m}_x^* \cdot x(t)} = |\tilde{m}_x|^2, \qquad \overline{|\tilde{m}_x|^2} = |\tilde{m}_x|^2. \tag{15.73}$$

Using Eqn. (15.71) in Eq. (15.70), we can express the variance as

$$\text{Variance, } \sigma_x^2 = \overline{|x(t) - \tilde{m}_x|^2} = \overline{|x(t)|^2} - |\tilde{m}_x|^2 \tag{15.74}$$

$$\Rightarrow \quad \text{Variance} = \text{Power} - |\text{Mean}|^2. \tag{15.75}$$

Eq. (15.10.3) shows that *power and variance of a signal are equal if its mean is zero.* Thus, for signals with zero mean, we have

$$\text{Variance} = \text{Power} = \text{Mean-square value} = \overline{|x(t)|^2}. \tag{15.76}$$

It is easy to verify that the power and variance of the sinusoidal random signal in Eq. (15.1) are equal and is given by $\frac{A^2}{2}$.

Because of the squaring operation in Eqs. (15.69) and (15.70), the power and variance are not linear operators. That is,

$$\text{Power}\,[\alpha x_1(t) + \beta x_2(t)] \neq \alpha \text{Power}\,[x_1(t)] + \beta \text{Power}\,[x_2(t)]. \tag{15.77}$$

Similar equation holds for variance also.

§ *Examine if ergodocity implies stationarity, or vice versa?*

Two stochastic processes $x(t)$ and $y(t)$ are jointly ergodic if the individual processes are ergodic and if the time-averaged cross-correlation function equals the statistical cross-correlation. That is,

$$\tilde{\gamma}_{xy}(\tau) = \overline{x(t)y(t+\tau)} = E[x(t)y(t+\tau)] = \gamma_{xy}(\tau) \tag{15.78}$$

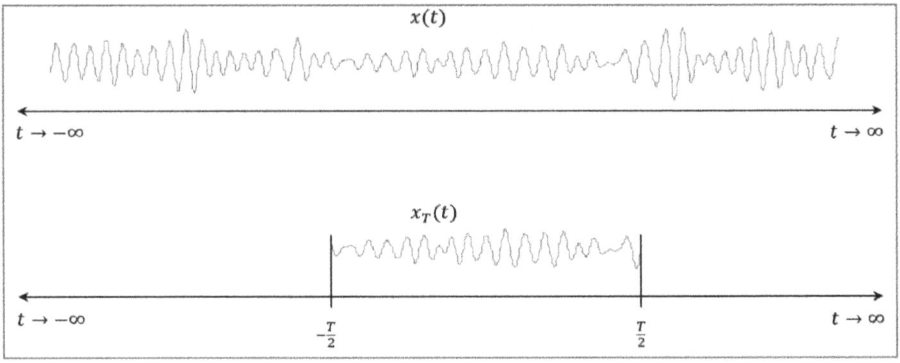

Figure 15.13. Signal $x(t)$ and its truncated version $x_T(t)$.

If we have only one realization of a random process $x(t)$, we can assume it to be ergodic (unless otherwise specified) for further analysis where its time averages becomes ensemble averages. This means that $\tilde{\gamma}_{xx}(\tau) = \gamma_{xx}(\tau)$, $\tilde{m}_x = m_x$, etc. Hence, in the rest of the book, the time averages are replaced with ensemble averages.

15.11 Power Spectrum

The statistics that we have considered so far (i.e., mean, power, variance, and correlation) are related to the time-domain description of the random signal. While mean, power and variance are constants, correlation is a function of the time-shift τ. On the other hand, power spectrum (or power spectral density (PSD)) is a frequency-domain statistic. The power spectrum of the random signal $x(t)$ is defined as,

$$\Gamma_{xx}(f) = \lim_{T\to\infty} \frac{1}{T} |X_T(f)|^2 \tag{15.79}$$

$$\text{where} \quad X_T(f) = \int_{-\infty}^{\infty} x_T(t) e^{-j2\pi f t} dt = \int_{-T/2}^{T/2} x(t) e^{-j2\pi f t} dt. \tag{15.80}$$

Here, $x_T(t)$ is nothing but $x(t)$ truncated to the interval $[-\frac{T}{2}, \frac{T}{2}]$ as shown in Fig. 15.13.

The *power spectrum* $\Gamma_{xx}(f)$ *gives the distribution of the power in* $x(t)$ *with frequency*. For example, the power of $x(t)$ in the frequency band $[f_1, f_2]$ is given by the integral $\int_{f_1}^{f_2} \Gamma_{xx}(f)df$. So, *the total power in* $x(t)$ can be expressed as

$$\text{Power}, \ p_x = \int_{-\infty}^{\infty} \Gamma_{xx}(f)df = \gamma_{xx}(0) = \overline{|x(t)|^2}. \tag{15.81}$$

Equation (15.81) shows the various ways (time-domain and frequency-domain) of expressing the total power in $x(t)$.

Wiener-Khinchin Theorem: This theorem relates the autocorrelation and power spectrum. For stationary random processes, the power spectral density (PSD) $\Gamma_{xx}(f)$ and autocorrelation function are Fourier transforms of each other. That is, we have

$$\Gamma_{xx}(f) = \int_{-\infty}^{\infty} \gamma_{xx}(\tau)e^{-j2\pi f\tau}d\tau \tag{15.82}$$

$$\text{or} \quad \gamma_{xx}(\tau) = \int_{-\infty}^{\infty} \Gamma_{xx}(f)e^{j2\pi f\tau}df. \tag{15.83}$$

Thus, to determine the power spectrum, it is enough that we determine the autocorrelation and then find its Fourier transform. This approach can be used to estimate power spectrum of measured random signals.

Some features of the PSD are listed below.

- The PSD $\Gamma_{xx}(f)$ is always real and non-negative.
- The PSD of a real-valued random process is even symmetric. That is, $\Gamma_{xx}(f) = \Gamma_{xx}(-f)$. This may not be true for complex-valued processes.
- $\text{Im}\{\Gamma_{xx}(f)\} \equiv 0$. Here, $\text{Im}\{.\}$ is defined as the imaginary part of the quantity within the braces.
- The quantity $\frac{1}{2\pi}\Gamma_{xx}(f)df$ is the average contribution of the frequency components in the range $[f, f + df]$ to the total power. We also note that

$$\gamma_{xx}(0) = E\left[|x(t)|^2\right] = \frac{1}{2\pi}\int_{-\pi}^{\pi} \Gamma_{xx}(f)df. \tag{15.84}$$

The Cross-Power Spectrum: The cross-power spectrum can be defined as the Fourier transform of the cross-correlation function.

$$\Gamma_{xy}(f) = \int_{-\infty}^{\infty} \gamma_{xy}(\tau)e^{-j2\pi f\tau}d\tau \tag{15.85}$$

$$\text{or} \quad \gamma_{xy}(\tau) = \int_{-\infty}^{\infty} \Gamma_{xy}(f)e^{j2\pi f\tau}df. \tag{15.86}$$

15.12 Linear Filtering of Random Signals

In practical applications, we often need to filter random signals. For example, in communication systems, we need to filter the signal at the transmitter to ensure that the transmitted signal does not exceed the allocated bandwidth. At the receiver, we need to filter the received signal to reject noise and interferences that are outside the signal bandwidth of interest, as well as to shape the signal according to some requirements. Similarly, the measured signals in all practical applications are first passed through some filters to appropriately condition the signal.

> § *In the design and analysis of systems (e.g., transmitter, receiver, front-end filters, etc.), it is required to know how the various statistics (e.g., mean, power, variance, correlation, power spectrum) change due to these filtering operations.*

Let $y(t)$ be the output of the filter with impulse response $h(t)$, when the input is $x(t)$. Further, let $y_T(t)$ be the output when the input is $x_T(t)$, where $x_T(t)$ is as defined in Section 15.11. Then, we get

$$y(t) = x(t) \otimes h(t) = \int_{-\infty}^{\infty} h(t')x(t-t')dt' \tag{15.87}$$

$$y_T(t) = x_T(t) \otimes h(t), \qquad Y_T(f) = H(f)X_T(f). \tag{15.88}$$

Using the definitions of mean, power and correlation, we shall now examine the various statistics of the output signal[4] $y(t)$.

<u>**Mean**</u>: We get the mean of $y(t)$ as

$$\begin{aligned} m_y = \overline{y(t)} &= \overline{\int_{-\infty}^{\infty} h(t')x(t-t')dt'} = \int_{-\infty}^{\infty} h(t')\overline{x(t-t')}dt' \\ &= m_x \int_{-\infty}^{\infty} h(t')dt' \end{aligned} \tag{15.89}$$

since mean $m_x = \overline{x(t)} = \overline{x(t-t')}$ is independent of time.

4. What is important in Section 15.12 is the meaning of the results. The derivation of the formulas may be omitted.

Power Spectrum: We get the power spectrum of $y(t)$ as

$$\begin{aligned}\Gamma_{yy}(f) &= \lim_{T\to\infty}\frac{1}{T}|Y_T(f)|^2 = \lim_{T\to\infty}\frac{1}{T}|H(f)X_T(f)|^2 \\ &= |H(f)|^2 \lim_{T\to\infty}\frac{1}{T}|X_T(f)|^2 = |H(f)|^2\Gamma_{xx}(f)\end{aligned} \tag{15.90}$$

since $H(f)$ is independent of T and $\Gamma_{xx}(f)$ is the power spectrum of the input signal $x(t)$.

Autocorrelation: Using Eq. (15.83), we can get the autocorrelation of $y(t)$ by taking the inverse Fourier transform of power spectrum $\Gamma_{yy}(f)$. Since multiplication in frequency domain is equivalent to convolution in time domain, we get

$$\begin{aligned}\gamma_{yy}(\tau) &= \mathcal{F}^{-1}\left[\Gamma_{yy}(f)\right] = \mathcal{F}^{-1}\left[|H(f)|^2\Gamma_{xx}(f)\right] \\ &= \mathcal{F}^{-1}\left[|H(f)|^2\right] \otimes \mathcal{F}^{-1}\left[\Gamma_{xx}(f)\right] = r_{hh}(\tau)\otimes\gamma_{xx}(\tau)\end{aligned} \tag{15.91}$$

where $r_{hh}(\tau)$ is the inverse Fourier transform of[5] $|H(f)|^2$. Please note that $h(t)$, the impulse response of the filter is a deterministic process and its ACF is denoted as $r_{hh}(\tau)$.

Power: We get the power of $y(t)$ as

$$p_y = \gamma_{yy}(0) = \int_{-\infty}^{\infty}\Gamma_{yy}(f)df = \int_{-\infty}^{\infty}|H(f)|^2\Gamma_{xx}(f)df. \tag{15.92}$$

Cross-Correlation: Using Eq. (15.87), we get the cross-correlation $\gamma_{xy}(\tau)$ between input $x(t)$ and output $y(t)$ as

$$\begin{aligned}\gamma_{xy}(\tau) &= \overline{x^*(t)y(t+\tau)} = \overline{x^*(t)\int_{-\infty}^{\infty}x(t+\tau-t')h(t')dt'} \\ &= \int_{-\infty}^{\infty}\overline{x^*(t)x(t+\tau-t')}h(t')dt' \\ &= \int_{-\infty}^{\infty}\gamma_{xx}(\tau-t')h(t')dt' = \gamma_{xx}(\tau)\otimes h(\tau).\end{aligned} \tag{15.93}$$

5. Note that $|H(f)|^2 = H(f)H^*(f)$, and the Fourier transform of $h^*(-t)$ is $H^*(f)$. Therefore, we get $r_{hh}(\tau) = \mathcal{F}^{-1}\left[|H(f)|^2\right] = \mathcal{F}^{-1}[H(f)]\otimes\mathcal{F}^{-1}\left[H^*(f)\right] = h(t)\otimes h^*(-t)$.

Thus, Eqs. (15.89) to (15.93) show that *the output statistics can be obtained from the input statistics and knowledge of the filter.*

15.13 White Noise

White noise (denoted as $n(t)$) is a random signal whose power spectrum is flat over all the frequencies.[6] That is, we have

$$\Gamma_{nn}(f) = K \qquad \text{for all } f, \tag{15.94}$$

where, K is just a scalar variable which specifies the value of the power spectrum at each frequency. Thus, *all the frequencies have equal weight in the power spectrum of white noise.* Taking inverse Fourier transform of $\Gamma_{nn}(f)$, we get the autocorrelation of white noise as

$$\gamma_{nn}(\tau) = K\delta(\tau) \tag{15.95}$$

$$\Rightarrow \quad \gamma_{nn}(\tau) = 0 \qquad \text{for all } \tau \neq 0. \tag{15.96}$$

Thus, the autocorrelation of white noise is zero for all τ except $\tau = 0$, i.e., *white noise is uncorrelated even for an arbitrarily small time shift.* For this reason, *white noise is said to be a completely uncorrelated signal (or, highly random signal).* In view of Eq. (15.68), Eq. (15.96) implies that *the mean of white noise must be zero.*

Eq. (15.94) implies that *the power of white noise is infinity.* Clearly, such a signal cannot exist in practice. Nevertheless, white noise is used extensively in the development and analysis of several practical systems (e.g., communication systems which assume the channel noise to be white) since *the bandwidth of the noise is much much larger than the bandwidth of the underlying desired signal. So, as far as the signal is concerned, the noise appears like a white noise, even though it is not truly white.*

White Noise Processes: The random process $\{n[n]\}$ is said to be a white noise process if its PSD is flat over all the frequencies, i.e., if $\Gamma_{nn}e^{jw}$ is a constant for $f \in [-\pi, \pi]$. Therefore, its autocorrelation is given by,

$$\gamma_{nn}[k] = \frac{1}{2\pi}\int_{-\pi}^{\pi}(constant)e^{-j\omega k}d\omega \tag{15.97}$$

6. The thermal noise caused by a resistor has a power spectrum that is flat over a really wide bandwidth.

This means that,

$$\gamma_{nn}[k] = \begin{cases} 0 & \text{if } k \neq 0 \\ \sigma_n^2 & \text{if } k = 0. \end{cases} \tag{15.98}$$

Therefore, we may also say that $\{n[n]\}$ is a white noise process if its mean is zero, and $n[n]$ and $n[m]$ are uncorrelated for $n \neq m$. It is called *strictly white* if its mean is zero, and $n[n]$ and $n[m]$ are independent for $n \neq m$.

§ *It is often assumed that the noise observed in most of the communications channels is white and Gaussian. Is this justifiable?*
§ *In discrete-time processing, we use discrete-time white noise in place of continuous-time white noise. Examine the relation between these two noises.*

15.14 Properties of ACF

Some of the important properties of ACFs for different types (random, periodic and aperiodic) of signals ($x(t)$) is provided in Table 15.1 below.

Table 15.1. Important properties of ACF.

Random Signals	Periodic Signals	Aperiodic Signals
$\gamma_{xx}(\tau) = \gamma_{xx}(-\tau)$	$r_{xx}(\tau) = r_{xx}(-\tau)$	$r_{xx}(\tau) = r_{xx}(-\tau)$
$\lvert\gamma_{xx}(\tau)\rvert \leq \gamma_{xx}(0)$	$\lvert r_{xx}(\tau)\rvert \leq r_{xx}(0)$	$\lvert r_{xx}(\tau)\rvert \leq r_{xx}(0)$
$\lim_{\tau\to\infty} \gamma_{xx}(\tau) = m_x^2 = p_{x(DC)}$	$r_{xx}(\tau) = r_{xx}(\tau + T_p)$	$\lim_{\tau\to\infty} r_{xx}(\tau) = x_\infty^2$
$p_x = \gamma_{xx}(0) = m_x^2 + \sigma_x^2$	$p_x = r_{xx}(0)$	$p_x = r_{xx}(0)$
$\gamma_{yy}(\tau) = r_{hh}(\tau) \otimes \gamma_{xx}(\tau)$	$r_{yy}(\tau) = r_{hh}(\tau) \otimes r_{xx}(\tau)$	$r_{yy}(\tau) = r_{hh}(\tau) \otimes r_{xx}(\tau)$

In the Table 15.1, T_p is the period of the periodic signal. $p_{x(DC)}$ is the power of the d.c component of the signal. Similar properties exists for discrete-time signals as well.

15.15 Properties of CCF

Some of the important properties of CCFs for different types (random, periodic, and aperiodic) of signals ($x(t)$ and $y(t)$) is provided in Table 15.2 below.

Table 15.2. Important properties of CCFs.

Random Signals	Periodic Signals	Aperiodic Signals
$\gamma_{xy}(\tau) = \gamma_{xy}(-\tau)$	$r_{xy}(\tau) = r_{xy}(-\tau)$	$r_{xy}(\tau) = r_{xy}(-\tau)$
$\gamma_{xy}(\tau) \leq \sqrt{\gamma_{xx}(0)\gamma_{yy}(0)}$	$r_{xy}(\tau) \leq \sqrt{r_{xx}(0)r_{yy}(0)}$	$r_{xy}(\tau) \leq \sqrt{r_{xx}(0)r_{yy}(0)}$
$2\lvert\gamma_{xy}(\tau)\rvert \leq \gamma_{xx}(0) + \gamma_{yy}(0)$	$2\lvert r_{xy}(\tau)\rvert \leq r_{xx}(0) + r_{yy}(0)$	$2\lvert r_{xy}(\tau)\rvert \leq r_{xx}(0) + r_{yy}(0)$
$\lim_{\tau\to\infty} \gamma_{xy}(\tau) = m_x m_y$	$r_{xy}(\tau) = r_{xy}(\tau + T_p)$	$\lim_{\tau\to\infty} r_{xy}(\tau) = x_\infty y_\infty$
$\gamma_{xy}(\tau) = m_x m_y$	$r_{xy}(\tau) = 0, \quad \forall\tau$	$r_{xy}(\tau) = x_\infty y_\infty$
uncorrelated	if no common frequencies	uncorrelated
$\gamma_{yx}(\tau) = h(\tau) \otimes \gamma_{xx}(\tau)$	$r_{yx}(\tau) = h(\tau) \otimes r_{xx}(\tau)$	$r_{yx}(\tau) = h(\tau) \otimes r_{xx}(\tau)$

15.16 Properties of Power Spectra

Some of the important properties of power spectra for different types (random, periodic, and aperiodic) of signals ($x(t)$) is provided in Table 15.3 below.

Table 15.3. Important properties of Power Spectra.

Random Signals	Periodic Signals	Aperiodic Signals
$\Gamma_{xx}(f) = \Gamma_{xx}(-f)$	$S_{xx}(k) = S_{xx}(-k)$	$S_{xx}(f) = S_{xx}(-f)$
$p_x = \int_{-\infty}^{\infty} \Gamma_{xx}(f)df$	$p_x = \sum_{k=-\infty}^{\infty} S_{xx}(k)$	$p_x = \int_{-\infty}^{\infty} S_{xx}(f)df$
$p_x = \int_{-1/2}^{1/2} \Gamma_{xx}(\tilde{f})d\tilde{f}$	$p_{x(DC)} = S_{xx}(0)$	
$\Gamma_{xx}(f) = \lim_{T_p\to\infty} \frac{1}{T_p}$	$S_{xx}(k) = \lvert X(k)\rvert^2 \geq 0$	$S_{xx}(f) = \lvert X(f)\rvert^2 \geq 0$
$E\left[\lvert X(f)\rvert^2\right] \geq 0$		
$X(f) = \int_0^{T_p} x(t)e^{-j2\pi ft}dt$		
Const. real spec., $m_x^2\delta(f)$	Real discrete spec at $f = \frac{k}{T_p}$	Const. real spec., $x_\infty^2\delta(f)$
$\Gamma_{yy}(f) = \lvert H(f)\rvert^2\Gamma_{xx}(f)$	$S_{yy}(k) = \lvert H(k/T_p)\rvert^2 S_{xx}(k)$	$S_{yy}(f) = \lvert H(f)\rvert^2 S_{xx}(f)$

15.17 Properties of Cross-Power Spectra

Some of the important properties of cross-power spectra for different types (random, periodic, and aperiodic) of signals ($x(t)$ and $y(t)$) is provided in Table 15.4 below.

Table 15.4. Important properties of Cross-Power Spectra.

Random Signals	Periodic Signals	Aperiodic Signals
$\Gamma_{xy}(f) = \Gamma^*_{yx}(f)$	$S_{xy}(k) = S^*_{yx}(k)$	$S_{xy}(f) = S^*_{yx}(f)$
$\Gamma_{xy}(f) = \Gamma^*_{xy}(-f)$	$S_{xy}(k) = S^*_{xy}(-k)$	$S_{xy}(f) = S^*_{xy}(-f)$
$2\lvert\Gamma_{xy}(f)\rvert \leq \Gamma_{xx}(f) + \Gamma_{yy}(f)$	$2\lvert S_{xy}(k)\rvert \leq S_{xx}(k) + S_{yy}(k)$	$2\lvert S_{xy}(f)\rvert \leq S_{xx}(f) + S_{yy}(f)$
$\lvert\Gamma_{xy}(f)\rvert \leq \sqrt{\Gamma_{xx}(f)\Gamma_{yy}(f)}$	$S_{xy}(k) = 0, \ \forall k$ if no common frequencies	$\lvert S_{xy}(f)\rvert \leq \sqrt{S_{xx}(f)S_{yy}(f)}$
$\Gamma_{xy}(f) = \lim_{T_p \to \infty} \frac{1}{T_p} E\left[X(f)Y^*(f)\right]$	$S_{xy}(k) = X(k)Y^*(k)$	$S_{xy}(f) = X(f)Y^*(f)$
$X(f) = \int_0^{T_p} x(t)e^{-j2\pi f t}dt$		
Const. complex spec., $m_x m_y \delta(f)$	Discrete complex spec at $f = \frac{k}{T_p}$	Const. complex spec., $x_\infty y_\infty \delta(f)$
$\Gamma_{yx}(f) = H(f)\Gamma_{xx}(f)$	$S_{yx}(k) = H(k/T_p)S_{xx}(k)$	$S_{yx}(f) = H(f)S_{xx}(f)$

15.18 Concluding Remarks

We can make the following concluding remarks based on the material presented in this chapter.

- The signals that we encounter in many of the real-world practical applications are random in nature.
- Characterization of random signals are done by means of specifying averages (or statistics) such as mean, power, correlation, etc.
- When a random signal is applied to the input of a linear filter, the statistics of the output random signal can be computed from the knowledge of the input statistics and the filter impulse/frequency response.
- The power spectrum of white noise is flat over all the frequencies. Therefore, white noise is a completely uncorrelated signal.

Exercises

15.1. Starting with the continuous time-average definition of the autocorrelation function (ACF), prove that the Fourier Transform of the ACF is the power spectral density (PSD) of the signal.

15.2. A signal $x(t)$ is transmitted through a channel. The received signal $y(t)$ is scaled, shifted, and noisy version of $x(t)$ given as $y(t) = \alpha x(t - t_0) + \eta(t)$, where α is a scale factor, t_0 is the time delay and $\eta(t)$ is the noise. Assume that the noise process has zero mean and is statistically independent of the signal process, and that all process are stationary.

   (a) Derive expressions for the mean and the ACF of $y(t)$ in terms of the statistics of $x(t)$ and $\eta(t)$.

15.3. Derive an expression for the ACF of the signal $x(t) = \sin(\omega_0 t)$. Use the time-average definition of the ACF. From the ACF, derive an expression for the PSD of the signal. Show all steps.

15.4. A rhythmic episode of a theta (θ) wave in an EEG (electroencephalogram - brain signals) signal is approximated by a researcher to be a sine wave of frequency 5Hz. The signal is sampled at 100 Hz.

   (a) Draw a schematic representation of the ACF of the episode for delays up to 0.5 seconds. Label the time axis in samples and in seconds.

   (b) Draw a schematic representation of the PSD of the episode. Label the frequency axis in Hz.

15.5. A digital random signal $x_1[n]$ gets the values of 0 and 1 with equal probability equal to 0.5. The signal $x_1[n]$ is assumed to be ergodic. The samples of the signal $x_1[n]$ and $x_1[m]$ are uncorrelated when $n \neq m$.

(a) Find and sketch the one dimensional probability density function $f_x(x_1)$ of the signal and calculate the mean and the power of the signal.

(b) Show that the autocorrelation function (ACF) of $x_1[n]$ is

$$\gamma_{x_1x_1}[k] = \begin{cases} c & \text{for } \ k = 0 \\ d & \text{otherwise,} \end{cases}$$

and find the constants c and d. Sketch one period of the power spectrum of the signal.

(c) The digital random signal $x_0[n]$ is derived from $x_1[n]$ so that $x_0[n]$ gets the value 1 when $x_1[n] = 0$, and the value 0 when $x_1[n] = 1$.

The cross-correlation $\gamma_{x_1x_0}[k]$ between $x_1[n]$ and $x_0[n]$ is given by

$$\gamma_{x_1x_0}[k] = \begin{cases} 0 & \text{for } \ k = 0 \\ d & \text{otherwise.} \end{cases}$$

Sketch one period of the cross power spectrum $\Gamma_{x_1x_0}(\tilde{f})$, and find the cross-correlation function $\gamma_{x_0x_1}[k]$ between $x_0[n]$ and $x_1[n]$. Finally, find the ACF and the power spectrum of the digital random signal $y[n] = x_1[n] - x_0[n]$.

15.6. Consider the analog random signal $x(t)$ with a rectangular probability density function. The ACF of $x(t)$ is given by,

$$\gamma_{xx}(\tau) = A\delta(\tau),$$

where A is a positive real constant.

The signal $x(t)$ is filtered by a filter whose impulse response is given by

$$h(t) = \begin{cases} \frac{a}{T} & \text{for } \ 0 \leq t \leq T \\ d & \text{otherwise,} \end{cases}$$

where a and T are positive real constants. The result of the filtration is the random signal $y(t)$.

(a) Find the mean value $E\{x(t)\}$ of $x(t)$ and the transfer function $H(f)$ of $h(t)$ expressed in terms of a and T. Find an expression for the ACF of $h(t)$ expressed in a and T.

(b) Find the ACF $\gamma_{yy}(\tau)$ of $y(t)$ and its power spectrum $\Gamma_{yy}(f)$ expressed by a, A and T.
(c) What is the probability density function that $y(t)$ can be assumed to have, and find the mean and the standard deviation of $y(t)$ expressed by a, A and T.
(d) The analog signal $y(t)$ is converted to a digital signal after the filtration with an ideal low-pass filter, whose transfer function is given by:

$$H_{lp}(f) = \begin{cases} 1 & \text{for } |f| < 4/T \\ d & \text{otherwise.} \end{cases}$$

The noise generated in the sampling process is assumed to have negligible power, and unimportant in our further considerations. Suggest the minimum possible sampling frequency expressed in terms of T, so that the analog and the digital signals are equivalent. Sketch the power spectrum ($\Gamma_{yy}(\tilde{f})$) of the sampled signal. Give the amplitude of the spectrum at $f = 0$ expressed by A, a and T. Note the zeros in the power spectrum for $|f| < \frac{12}{T}$ for the chosen sampling frequency.

15.7. The autoregressive (AR) process generator consists of a digital generator of white noise followed by a time-invariant recursive digital filter as shown in Fig. P15.7.

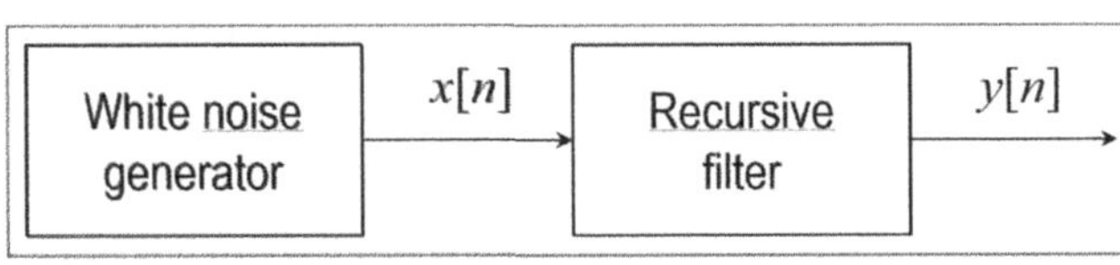

Figure P15.7.

The output from the white noise generator is denoted by $x[n]$ and has a power p_x and a mean value $E\{x[n]\} = 0$. The ACF and power spectrum of $x[n]$ are denoted by $\gamma_{xx}[k]$ and $\Gamma_{xx}(\tilde{f})$, respectively. The output $y[n]$ of the digital filter is found by:

$$y[n] = x[n] + \sum_{q=1}^{Q} b_q y[n-q],$$

where Q is a positive integer number, and b_q are real constants. The impulse response of the filter is denoted by $h[n]$, its transfer function by $H(\tilde{f})$ and the sampling interval by T_s. The ACF and power spectrum of $y[n]$ are denoted by $\gamma_{yy}[k]$ and $\Gamma_{yy}(\tilde{f})$, respectively.

(a) Draw a block diagram of the filter implemented in direct form II, and write down $H(\tilde{f})$ expressed by b_q and T_s. Show that the filter is causal, and that $h[0] = 1$.

(b) Write down $\Gamma_{xx}(\tilde{f})$ expressed by p_x, and sketch $\gamma_{xx}[k]$. Find $\gamma_{yy}[k]$ and p_y expressed by p_x and $h[n]$. Determine $E\{y[n]\}$.

(c) Because the filter is time invariant, we can shift in time both the input and the output so that we get:

$$y[n+k] = x[n+k] + \sum_{q=1}^{Q} b_q y[n+k-q],$$

where k is an arbitrary integer number. Show that

$$\gamma_{xy}[k] = \gamma_{xx}[k] + \sum_{q=1}^{Q} b_q \gamma_{xy}[k-q],$$

where $\gamma_{xy}[k]$ is the CCF between $x[n]$ and $y[n]$. Explain why $\gamma_{xy}[k] = 0$ for $k < 0$, and find $\gamma_{xy}[k]$ when $Q = 1$.

(d) Show that

$$\gamma_{yy}[k] = \gamma_{xy}[-k] + \sum_{q=1}^{Q} b_q \gamma_{yy}[k-q],$$

and hereby derive $\gamma_{yy}[k]$ expressed by $\gamma_{xy}[k]$ and $h[n]$. Write a system (set) of equations, from which b_q can be found, when $\gamma_{yy}[k]$ is known.

15.8. In this problem, we will consider an analog random ergodic signal $x(t)$. $x(t)$ has a normally distributed probability density function $f_x(x)$ given by

$$f_x(x) = \frac{k_2}{\sqrt{\pi}} e^{-k_2^2(x-k_1)^2}.$$

The ACF $\gamma_{xx}(\tau)$ of $x(t)$ is:

$$\gamma_{xx}(\tau) = p_x e^{-2\pi f_x|\tau|},$$

where k_1, k_2, and f_x are real constants, and p_x is the power of the signal.

(a) Find k_1 and then express k_2 through p_x. Write down the power density spectrum $\Gamma_{xx}(f)$ of the signal expressed through p_x and $f_x(x)$ (Do not solve any Fourier integrals.)

(b) $x(t)$ is applied at the input of a frequency independent amplifier (same amplification for all frequencies). The output of the amplifier is denoted as $y(t)$ and is given by:

$$y(t) = \alpha x(t) + z(t),$$

where α is a positive real constant and $z(t)$ is a new random signal, which reflects the (for example) the overloading (no need to worry about this term as far as this problem is concerned) of the amplifier. Find the ACF $\gamma_{yy}(\tau)$ of $y(t)$ and the CCF $\gamma_{xy}(\tau)$ between $x(t)$ and $y(t)$. Give the conditions under which $\gamma_{xy}(\tau) = \alpha\gamma_{xx}(\tau)$, and then find the power p_y of $y(t)$, when this condition is fulfilled.

15.9. Starting with the continuous time-average definition of the autocorrelation function (ACF), prove that the Fourier Transform of the ACF is the power spectral density (PSD) of the signal.

15.10. The block schematic of a LTI digital system is shown in Figure P15.8. Here $x[n]$ and $y[n]$ are the input and output, respectively of the system and $h[n]$ is the system impulse response. The system is described by the difference equation, $y[n] = x[n] - ax[n-1]$. Here, a is a real scalar variable.

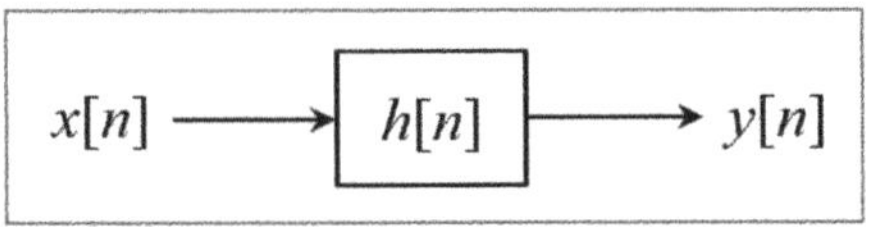

Figure P15.8.

It is assumed that $x[n]$ is real and zero mean wide-sense stationary (WSS) random process with variance σ_x^2 and its autocorrelation function (ACF), $\gamma_{xx}[k]$, for delay k is defined as, $\gamma_{xx}[k] = E\left[x[n]x[n-k]\right]$, where $E[.]$ denotes the expectation operation.

(a) Show that the variance σ_y^2 of the output is given by,

$$\sigma_y^2 = \sigma_x^2\left[1 + a^2 - 2a\rho_{xx}[1]\right],$$

where $\rho_{xx}[1] = \frac{\gamma_{xx}[1]}{\gamma_{xx}[0]}$ is the normalized ACF.

(b) Show that σ_y^2 attains its minimum value of $\sigma_x^2\left[1 - \rho_{xx}^2[1]\right]$ for $a = \rho_{xx}[1]$ (**Hint:** minimize σ_y^2 with respect to a).

(c) What is the condition under which $\sigma_y^2 < \sigma_x^2$?

15.11. The block schematic of a LTI digital system is shown in Fig. P15.9. Here, $x[n]$ and $y[n]$ are the input and output, respectively of the system. $h[n]$ is the impulse response of the system.

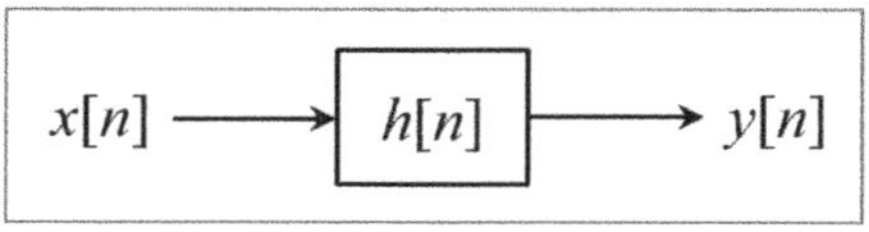

Figure P15.9.

The system is described by the difference equation,

$$y[n] = 0.8y[n-1] + x[n] + x[n-1].$$

$x[n]$ is a wide-sense stationary (WSS) random process with zero mean (i.e., $E[x[n]] = m_x = 0$) and its autocorrelation function (ACF), $\gamma_{xx}[k]$, for delay k is given by,

$$\gamma_{xx}[k] = \left(\frac{1}{2}\right)^{|k|}.$$

(a) Determine the system function, $H(z)$, and hence derive its impulse response, $h[n]$.

(b) Determine the power spectral density (PSD), $\Gamma_{yy}(\tilde{f})$, of the output $y[n]$. Here $\tilde{f}$ is the normalized frequency variable. (**Hint**: You may use "the theory of linear filtering of random signals" to derive the output PSD.)

(c) Derive an expression for the ACF, $\gamma_{yy}[k]$ of the output $y[n]$. (**Hint**: Use "Wiener-Khinchin theorem" to find $\gamma_{yy}[k]$ from $\Gamma_{yy}(\tilde{f})$. You may have to use partial fraction expansion of $\Gamma_{yy}(\tilde{f})$ here.)

(d) Determine the variance, σ_y^2, of the output $y[n]$. (**Hint**: Use properties of the ACF, $\gamma_{yy}[k]$, to determine the variance.)

15.12. Figure P15.10 shows a system with one input ($x(t)$) and one output ($y(t)$). As can be seen from the figure, it has two complex constants A and B. $v(t)$ in the figure is an intermediate variable. Assume that $x(t)$ is a stationary complex random process with mean zero and an autocorrelation function (ACF) $\gamma_{xx}(\tau)$ defined as $\gamma_{xx}(\tau) = E[x(t+\tau)x^*(t)]$. Here, * denotes the complex conjugation.

(a) Derive the expressions for the ACF ($\gamma_{yy}(\tau)$) of $y(t)$ and the cross-correlation function (CCF) $\gamma_{xv}(\tau)$ between $x(t)$ and $v(t)$ in terms of $\gamma_{xx}(\tau)$ and the complex constants A and B.

In the following (Fig. P15.11), the system in Fig. P15.10 has been modified to a two inputs ($x(t)$ and $w(t)$) two outputs ($y(t)$ and

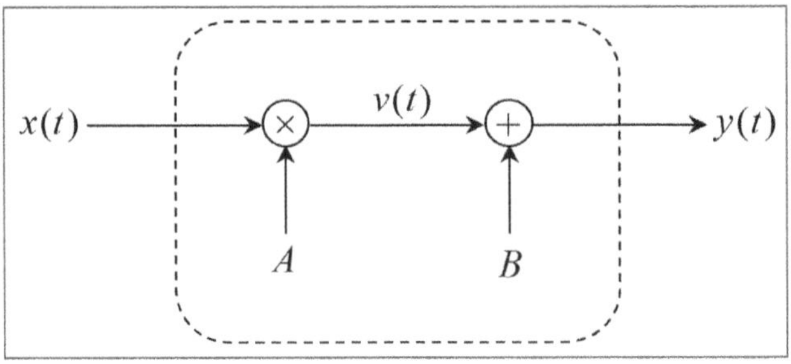

Figure P15.10.

$w(t)$) system, where $w(t)$ is another stationary complex random process with mean zero and ACF $\gamma_{ww}(\tau) = E[w(t+\tau)w^*(t)]$. The CCF between $x(t)$ and $w(t)$, $\gamma_{xw}(\tau) = E[x(t+\tau)w^*(t)]$ is also given. As can be seen in the figure, the summation and multiplication operations has been swapped within the system.

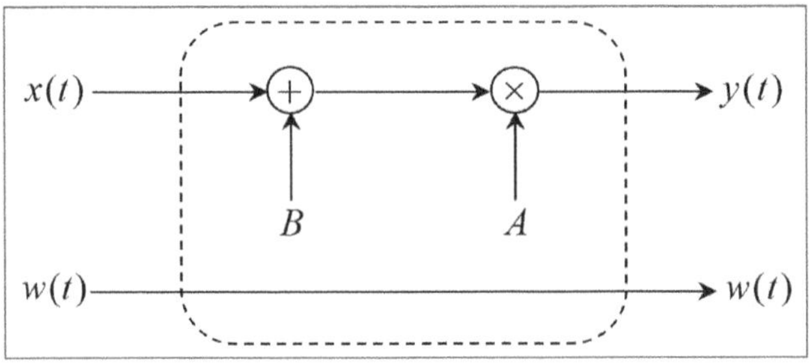

Figure P15.11.

Using the known parameters, $\gamma_{xx}(\tau)$, $\gamma_{ww}(\tau)$, $\gamma_{xw}(\tau)$, A, and B,

(b) Derive an expression for the ACF ($\gamma_{yy}(\tau)$) of $y(t)$.

(c) Derive expressions for the CCFs $\gamma_{yw}(\tau)$ between $y(t)$ and $w(t)$ and $\gamma_{wy}(\tau)$ between $w(t)$ and $y(t)$.

(d) Derive expressions for the CCFs $\gamma_{xy}(\tau)$ between $x(t)$ and $y(t)$ and $\gamma_{yx}(\tau)$ between $y(t)$ and $x(t)$.

15.13. Consider the block diagram representation of a LTI system as shown in Figure P2. Here $x[n]$ and $y[n]$ are the input and output, respectively of the system. $h[n]$ is the impulse response of the system. The difference equation relating the input and the output of the system is given by,

$$y[n] = ay[n-1] + x[n].$$

Here, a is a scalar variable.

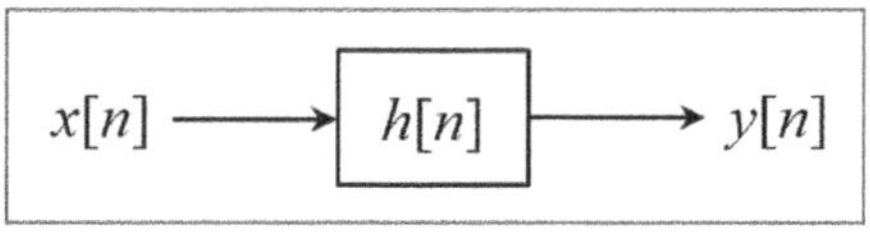

Figure P15.12.

(a) Determine the system function $H(z)$, which is the z-transform of $h[n]$ and suggest the range of values of a for which the system is stable.

(b) Determine the first three (3) samples (i.e., $h[0]$, $h[1]$, and $h[2]$) of $h[n]$ in terms of the variable a and hence determine the first three (3) values (i.e., $r_{hh}[0]$, $r_{hh}[1]$ and $r_{hh}[2]$) of the autocorrelation sequence $r_{hh}[k]$ of $h[n]$ in terms of a (*Hint*: By definition, $r_{hh}[k] = \sum_{n=-\infty}^{\infty} h[n]h[n+k]$; for $k = 0, \pm 1, \pm 2, \ldots$).

In the following, the value of a is chosen to be $\frac{1}{2}$ in the input-output relationship of the LTI system. The system is excited with an input $x[n] = \left(\frac{1}{4}\right)^n u[n]$, where $u[n]$ is a unit step function.

(c) Derive the expressions for the autocorrelation sequence $r_{xx}[k]$ of the input sequence $x[n]$ and the cross-correlation sequence $r_{xy}[k]$ between the input $x[n]$ and the output $y[n]$ of the system (*Hint*: We have $r_{xx}[k]$ and $R_{xx}(z)$ are z-transform pairs and $R_{xx}(z) = X(z)X(z^{-1})$ and $R_{xy}(z) = X(z)Y(z^{-1})$).

(d) Derive the expressions for the autocorrelation sequence $r_{yy}[k]$ of the output sequence $y[n]$ (*Hint*: $R_{yy}(z) = Y(z)Y(z^{-1})$).

15.14. Consider the block diagram representation of a linear time-invariant (LTI) system as shown in Fig. P15.13. Here $x(t)$ and $y(t)$ are the input and output, respectively of the system. $h(t)$ in the picture is the impulse response of the system. The system function $H(f)$ (Fourier transform of $h(t)$) is given by,

$$H(f) = \cos\left(\frac{\pi}{2f_c} f\right) \text{rect}\left(\frac{f}{2f_c}\right).$$

Here, t is the time variable, f is the frequency variable and f_c is a given specific frequency in Hz.

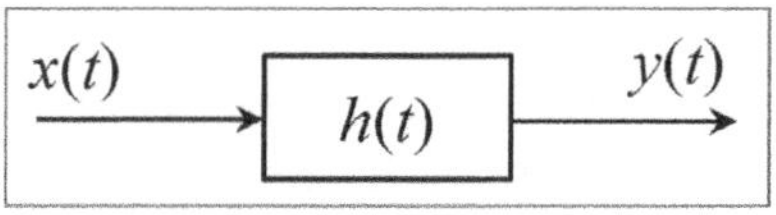

Figure P15.13.

(a) Derive an expression for the autocorrelation function (ACF), $r_{hh}(\tau)$ of the system. Here τ is the delay variable (*Hint*: Make use of the Wiener-Khinchin theorem that relates the power spectrum and the ACF).

(b) At the system input, a signal is applied with power spectral density $\Gamma_{xx}(f) = N_0 + K\delta(f)$. Here, N_0 and K are scalars and $\delta(f)$ is an impulse function. Find the ACF of the input signal, $\gamma_{xx}(\tau)$ and its mean value m_x.

(c) For the output signal $y(t)$, find the power p_y by using the statistical characteristics of the input and the system (*Hint*: Use the $r_{hh}(\tau)$ and $\gamma_{xx}(\tau)$ derived in Questions (a) and (b), respectively as well as the theory discussed in the linear filtering of random signals topic).

15.15. This problem involves the use of cross-correlation to detect a signal in noise and estimate the time delay in the signal. A signal $x[n]$ consists of a pulsed sinusoid corrupted by a stationary zero mean white noise sequence. That is,

$$x[n] = y[n - n_0] + w[n], \qquad 0 \leq n \leq N - 1$$

where $w[n]$ is the noise (white) with variance σ_w^2, N is the length of the data. The pulsed sinusoidal signal is given by,

$$y[n] = \begin{cases} A\cos(\omega_0 n), & 0 \leq n \leq M - 1 \\ 0, & \text{otherwise.} \end{cases}$$

Here, A is the amplitude of the sinusoid and M is the pulse duration. The length M is chosen in such a way that the pulsed sinusoid has integer number of periods of the cosine waveform in it. The frequency ω_0 of the sinusoid is known but the delay n_0, which is a positive integer, is unknown, and is to be determined by cross-correlating $x[n]$ with $y[n]$. Assume that $N > M + n_0$.

Let

$$r_{xy}[m] = \sum_{n=0}^{N-1} x[n]y[n - m],$$

denote the cross-correlation sequence between $x[n]$ and $y[n]$. In the absence of noise, this function exhibits a peak at delay $m = n_0$. Thus the time-delay n_0 is determined with no error. The presence of noise however can lead to errors in determining the unknown delay.

(a) For $m = n_0$, determine the expected value, $E\left(r_{xy}[n_0]\right)$. (*Note*: In the calculations, you may make use of the trigonometric identity $\cos^2(\omega_0 n) = \frac{1}{2}[1 + \cos(2\omega_0 n)]$ and that the double frequency term averages to zero. That is, $M \gg \frac{2\pi}{\omega_0}$.)

In the following question, we assume that the variance of $r_{xy}[n_0]$ (i.e., $\text{var}[r_{xy}[n_0]]$), due to the presence of noise is given by,

$$\text{var}[r_{xy}[n_0]] = \frac{MA^2}{2}\sigma_w^2.$$

(b) Determine the signal-to-noise ratio (S NR), defined as,

$$\text{SNR} = \frac{\left(E[r_{xy}[n_0]]\right)^2}{\text{var}[r_{xy}[n_0]]}.$$

(c) Discuss what is the effect of the pulse duration M on the SNR?

15.16. This problem considers a random process $y[n]$ that is generated by filtering a zero mean white noise $x[n]$, with a time-invariant filter as shown in Fig. P15.14. The sampling interval T_s of the digital signal is normalized to unity. The symbol z^{-1} in the figure represents a unit sample delay.

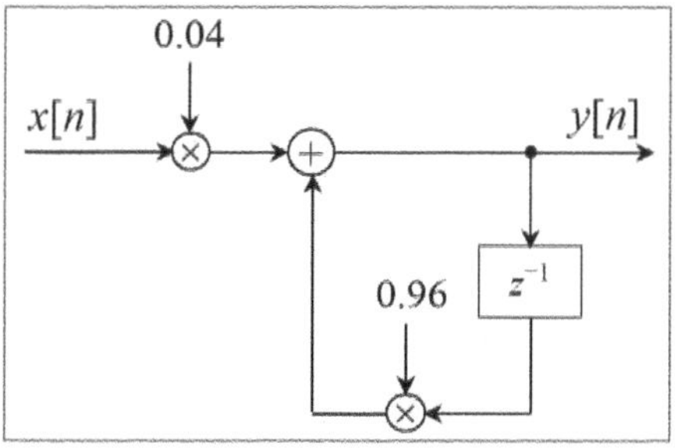

Figure P15.14.

(a) Write down the difference equation describing the filter shown in Fig. P15.14. Derive the transfer function $H(z)$ (in the z-domain) and the impulse response $h[n]$ of the filter.

(b) Sketch the magnitude response of the filter for $|f| \leq 1$. Is $H(z)$ a low-pass system (filter)?

In the following, we will assume that the variance of $x[n]$ is $\sigma_x^2 = 1$.

(c) Find the power spectrum $\Gamma_{yy}(\tilde{f})$ of the output signal $y[n]$. Then find the autocorrelation $\gamma_{yy}[k]$ of $y[n]$.

15.17. In a biomedical research laboratory, a researcher assumes the recorded electrical activities of the brain (EEG, electroencephalogram) to be a stationary random process and is denoted as $x(t)$.

(a) Starting with the continuous time-average definition of the auto-correlation function (ACF), prove that the Fourier Transform of the ACF is the power spectral density (PSD) of the signal, $x(t)$. Show all steps.

(b) The signal $x(t)$ is transmitted to another research laboratory through a channel. The received signal $y(t)$ is modelled as a scaled, shifted, and noisy version of $x(t)$ given as $y(t) = \alpha x(t - t_0) + \eta(t)$, where α is a scale factor, t_0 is the time delay and $\eta(t)$ is the noise. Assume that the noise process has zero mean and is statistically independent of $x(t)$, and that all process are stationary. Using the ensemble average definitions, derive expressions for the mean and the ACF of $y(t)$ in terms of the statistics of $x(t)$ and $\eta(t)$. Show all steps.

(c) If the researcher approximates $x(t)$ to be a sine wave of frequency ω_0 radians/second. That is, $x(t) = \sin(\omega_0 t)$, using the time-average definition of the ACF, derive an expression for the PSD of the signal. Show all steps.

15.18. A digital filter is described by the following difference equation

$$y[n] = \frac{1}{2}y[n-1] + x[n]$$

where $y[n]$ is the output and $x[n]$ is the input to the system. The filter is excited with an input $x[n] = \left(\frac{1}{4}\right)^n u[n]$, where $u[n]$ is the unit step sequence.

(a) Determine the autocorrelation sequences $r_{xx}[m]$, $r_{yy}[m]$, $r_{hh}[m]$, and the cross-correlation sequence $r_{xy}[m]$, where m is the delay.

15.19. The N-point DFT of a random sequence $x[n]$ is

$$X[k] = \sum_{n=0}^{N-1} x[n]e^{-j2\pi nk/N}, \qquad k = 0, 1, \ldots, N-1.$$

Assume that $E[x[n]] = 0$ and $E[x[n]x^*[n+m]] = \sigma_x^2\delta[m]$ [i.e., $x[n]$ is a white noise process].

(a) Determine the variance of the sequence $X[k]$.

(b) Determine the autocorrelation of the sequence $X[k]$.

DOI: 10.1561/9781680839791.ch16

Chapter 16

Modulation

As you may recall, in Chapters 2 to 5, we studied about signals and systems and their representation in time and frequency domains. In this chapter, we will use these concepts to study *modulation* which is a very important part of radar (communication systems) and medical ultrasound scanners (biomedical systems). In such systems, the strength of the received signal is related to the object that has reflected the transmitted pulse. Most importantly, *we want to illustrate how a strong foundation in signals and systems makes it very easy for us to understand the various theoretical and practical aspects of modulation.*

By studying this chapter, we should be able to receive clarity and understanding on the following.

- The reason for using modulation in communication/ultrasound systems.
- Different type of modulation schemes, and the advantages and disadvantages of these schemes.
- Important trade-offs underlying the design of communication/ultrasound systems.
- Principles of analog and digital modulation, envelope demodulation, coherent demodulation, matched filtering, etc.

16.1 Introduction

Let $x(t)$ be a real-valued message signal (e.g., speech signal) that we want to transmit. Then, *modulation is the process of impressing or loading $x(t)$ onto another signal $c(t)$ called as the 'carrier signal' which will carry $x(t)$ from the transmitter to the receiver.* The carrier signal is usually of the form

$$c(t) = A_c \cos(2\pi f_c t + \theta_c) \tag{16.1}$$

where A_c, f_c, and θ_c denote the amplitude, frequency and phase of the carrier. We can impress $x(t)$ onto $c(t)$ by making $x(t)$ modify any of these three parameters. As a result, we have three main categories of modulation schemes:

- *Amplitude Modulation (AM),*
- *Frequency Modulation (FM),* and
- *Phase Modulation (PM).*

The message signal is also known as *modulating signal.* The signal obtained by impressing the message onto the carrier is known as *modulated carrier.* The process of recovering the message signal from the modulated carrier is known as *demodulation.*

16.1.1 Base-Band and Pass-Band Signals

Let the one-sided bandwidth of $x(t)$ be B Hz. That is,[1]

$$|X(f)| = 0 \qquad \text{for} \quad |f| > B \tag{16.2}$$

where $X(f)$ is the Fourier transform (FT) of $x(t)$. Signals such as $x(t)$ whose spectra lie in the vicinity of $f = 0$ (e.g., speech, biomedical signals *etc.*) are called *base-band (or low-pass) signals.* On the other hand, suppose we create a signal $y(t)$ by multiplying $x(t)$ with $c(t)$. The spectrum of $y(t)$ will then consist of bands of width $2B$ around f_c and $-f_c$ (see Figure 16.4(a)–(b)).[2] Such signals whose spectra lie in bands away from $f = 0$ (e.g., signals transmitted by hand-phones) are called *pass-band (or band-pass) signals.* The process of modulation effectively converts base-band signals into pass-band signals.

---

1. The carrier frequency f_c is assumed to be much larger than the message bandwidth B, i.e., $f_c \gg B$.
2. If $y(t) = x(t)c(t)$ with $c(t)$ given by (16.1), we get

$$Y(f) = \frac{A_c}{2} e^{j\theta_c} X(f - f_c) + \frac{A_c}{2} e^{-j\theta_c} X(f + f_c). \tag{16.3}$$

16.1.2 Why Modulation?

Or why use a high frequency carrier?

Modulation of a carrier $c(t)$ by a message signal $x(t)$ is done to achieve one or more of the following objectives.

- Matching between Signal and Channel: To shift the frequency band of the message signal to the pass-band of the channel so that spectrum of the transmitted signal matches the characteristics of the channel.
- Antenna Size and Efficiency: For efficient transmission of signals, the antenna size should be comparable to the signal wavelength. Shifting the low-pass signals to higher frequencies using modulation allows antennas of practical sizes.
  - For example, consider the transmission of electromagnetic (EM) waves through space using antennas. The wavelength of an EM wave is given by $\lambda = \frac{c}{f} = \frac{3 \times 10^8}{f}$. It can be shown that the length of the antenna required to efficiently transmit the EM wave is $\approx \lambda$.
  - For voice signal at a frequency $f = 3\ kHz$, λ is 100000m.
  - If $f = 300\ MHz$, length of the antenna is 1m.
  - Higher the frequency of the carrier, the smaller the length of the antenna required. For example, a cellular (mobile) phone.
  - High frequency EM waves can propagate through atmosphere without much attenuation. For example, the space satellites use carrier frequency typically of the order of GHz.
- Transmission of Multiple Messages: To make provision for the transmission of multiple message signals, for example, by shifting the signals to adjacent frequency bands around the carrier frequency (i.e., frequency division multiplexing).
- Noise Immunity: To expand the signal bandwidth so as to increase immunity (resistance) to noise during transmission.
- Frequency Assignment: Using modulation, we can assign different frequency bands to different transmitting stations and/or programmes so that several stations and programmes can be simultaneously supported.

16.1.3 Choice of Modulation Scheme

The choice of modulation scheme depends on several factors such as,

- Characteristics (e.g., bandwidth) and number of message signals to be transmitted.

- Channel characteristics (e.g., type of transmission medium, channel disturbances).
- Performance requirements (e.g., received signal strength).
- Complexity and cost (e.g., complexity and cost of hardware, power, and bandwidth requirements).

The two most important resources in communication systems are *channel bandwidth and transmission power*. The *trade-off between cost and performance* in communication systems is mainly determined by these two resources.[3] The various modulation schemes require different combinations of bandwidth and power, and hence these schemes offer different trade-offs between cost and performance. Depending upon the kind of trade-off that is affordable in a particular application, we choose the type of modulation scheme.

We will now briefly discuss the principles, motivation, and main features of some of the modulation schemes.

16.2 Amplitude Modulation (AM)

In AM, we change the amplitude of the carrier $c(t)$ using the message signal $x(t)$. This is the most simplest to all modulation schemes to understand as well as to implement. We will explore a few types of AM schemes:

- conventional AM,
- AM with suppressed carrier (AM-SC),
- single side-band AM (SSB-AM), and
- quadrature AM (QAM).

These schemes provide different compromises among bandwidth, power, complexity and performance. Without loss of generality, we assume, that $A_c = 2$ and $\theta_c = 0$ in this section. That is, the carrier is given by

$$c(t) = 2\cos(2\pi f_c t). \tag{16.4}$$

16.2.1 Conventional AM (Conv-AM)

The Conv-AM modulated signal is given by

$$y(t) = 2\left[\alpha + x(t)\right]\cos(2\pi f_c t) \tag{16.5}$$

3. Complexity of the transmitter and receiver is another major aspect that influences the cost performance trade-off.

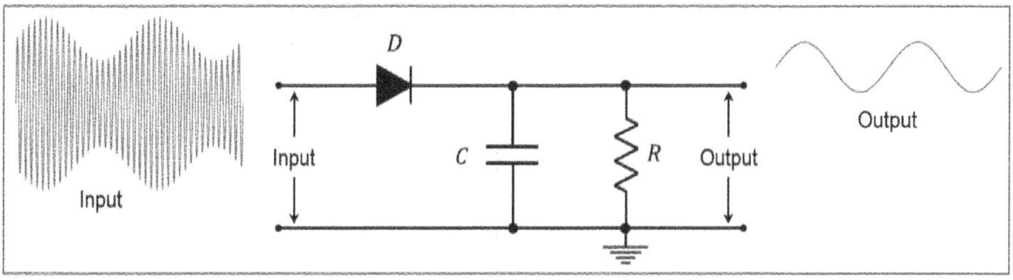

Figure 16.1. A simple envelope detector (demodulator) circuit.

where α is chosen such that

$$\alpha + x(t) > 0 \qquad \text{for all } t. \tag{16.6}$$

The purpose of the condition on α is to facilitate a simple mechanism, known as envelope detection, for demodulation.

§ *Let $x(t)$ be a periodic square wave signal, whose one period is given by*

$$x(t) = \begin{cases} +1 & \textit{for} \;\; 0 \leq t < 0.5T_0 \\ -1 & \textit{for} \;\; 0.5T_0 \leq t < T_0 \end{cases} \tag{16.7}$$

where $T_0 \gg 1/f_c$. Draw the Conv-AM signal $y(t)$ for (a) $\alpha = 0$ and (b) $\alpha = 1.5$. Comment on the amplitude envelopes of the modulated signal in the two cases.

16.2.1.1 Demodulation by Envelope Detection

Equation (16.6) ensures that we can recover $\alpha + x(t)$ from $y(t)$ by tracking the variations in the positive envelope (i.e., positive peak amplitude)[4] of $y(t)$. Since α is known, we can recover $x(t)$ from $\alpha + x(t)$. This is known as *envelope detection.* Thus, the receiver for Conv-AM is quite simple consisting of a half-wave rectifier followed by a low-pass filter (shown in Fig. 16.1).[5]

Figure 16.1 illustrates a simple envelope detection scheme. In this figure, D represents the rectifier, C is the capacitor and R is the resistor across which the output is taken. Input in this figure is an example of a modulated signal and the output is the extracted message signal. A (typical) modulating signal (Fig. 16.2(a)) and its envelope (Fig. 16.2(d)) is shown in Fig. 16.2. The original message signal

4. If Eq. (16.6) is not satisfied, then the positive envelope of $y(t)$ will be proportional to $|\alpha + x(t)|$ instead of $\alpha + x(t)$. Clearly, it is not possible to recover $x(t)$ from $|\alpha + x(t)|$.

5. For envelope detection to be accurate, the time constant τ of LPF must obey $\frac{1}{f_c} \ll \tau \ll \frac{1}{B}$ (i.e., small 'charging time constant' and large 'discharging time constant'). This helps the LPF to reject the high-frequency carrier and noise, while capturing the low-frequency message signal. The time constant gives an indication of the speed with which the circuit can respond to changes.

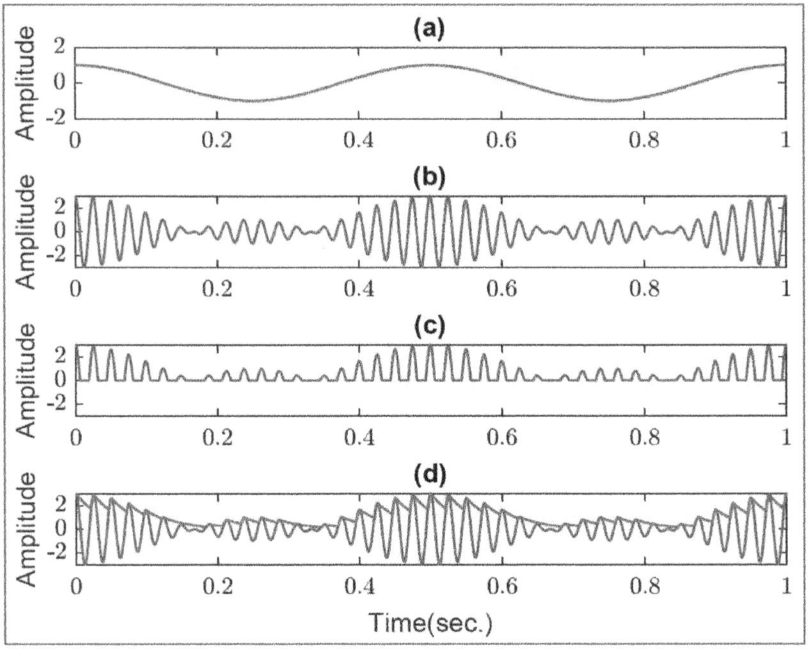

Figure 16.2. Signals associated with an envelope detector: (a) Modulating signal, $x(t)$, (b) Modulated signal, $y(t)$, (c) Signal at the output of the rectifier in the circuit, and (d) Detected signal (marked in red) at the output of the envelope detector plotted along with the modulated signal.

can be retrieved from the output (marked in red) signal by performing some signal processing techniques.

If the signal $x(t)$ is such that $-x_{max} < x(t) < x_{max}$, then Eq. (16.6) implies that we should choose α such that $\alpha > x_{max}$. Defining the *modulation index* for Conv-AM as

$$\beta_1 = \frac{x_{max}}{\alpha}, \tag{16.8}$$

we see that *envelope detection is possible only if* $0 < \beta_1 < 1$.

16.2.1.2 Spectra of Signals

The spectrum of $y(t)$ is given by[6]

$$\begin{aligned} Y(f) &= [\alpha\delta(f) + X(f)] \otimes [\delta(f - f_c) + \delta(f + f_c)] \\ &= \alpha\,[\delta(f - f_c) + \delta(f + f_c)] + [X(f - f_c) + X(f + f_c)]. \end{aligned} \tag{16.9}$$

6. Note: Recall that $\mathcal{F}[2\cos(2\pi f_c t)] = \delta(f - f_c) + \delta(f + f_c)$, $\mathcal{F}[\alpha] = \alpha\delta(f)$, and $\mathcal{F}[x(t)y(t)] = X(f) \otimes Y(f)$. Here, $\mathcal{F}$ denotes Fourier transform and $\otimes$ denotes convolution.

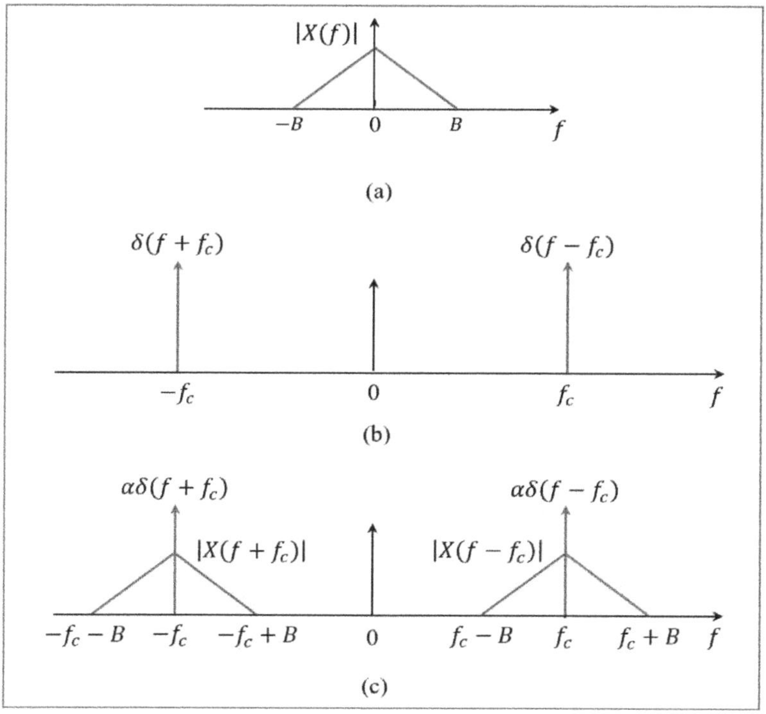

Figure 16.3. Spectra of signals in conventional AM: (a) message signal $x(t)$, (b) unmodulated carrier $c(t)$, and (c) modulated (transmitted) signal $y(t)$.

Figure 16.3 shows the spectra of $x(t)$, $c(t)$ and $y(t)$. Observe that the bands occupied by $y(t)$ are

$$-f_c - B < f < -f_c + B, \qquad f_c - B < f < f_c + B. \tag{16.10}$$

Thus, we see that transmission using Conv-AM requires double the bandwidth of the original message signal. Further, in addition to transmitting the message signal, we are also sending the carrier with power $2\alpha^2$, thus requiring extra transmission power.

16.2.1.3 Summary of Conv-AM

- The receiver is very simple (i.e., envelope detector), requiring no sophisticated approaches for demodulation.
- Requires extra power to transmit the carrier (i.e., $\alpha c(t)$ in Eq. (16.5)) to facilitate envelope detection.
- Requires double the bandwidth of the message signal.
- The extra power required to transmit the carrier increases with decrease in the modulation index (or, increase in α). But lower modulation index improves noise immunity.

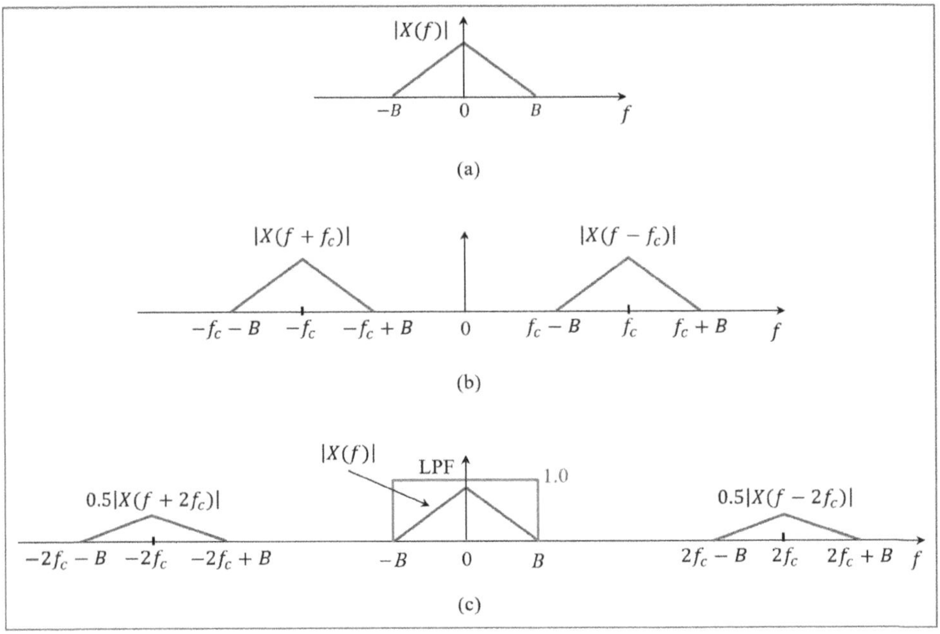

Figure 16.4. Spectra of signals in AM with suppressed carrier (AM-SC): (a) message signal $x(t)$, (b) modulated (transmitted) signal $y(t)$, and (c) LPF input $z(t)$. The LPF output is same as that in (a).

The different AM schemes we discuss below make the system more efficient in the use of bandwidth and power.

16.2.2 AM with Suppressed Carrier (AM-SC)

To save power, the AM-SC scheme chooses not to send the carrier component $\alpha c(t)$ along with the message in Eq. (16.5). The modulated signal and its spectrum in AM-SC are given by

$$y(t) = 2x(t)\cos(2\pi f_c t) \tag{16.11}$$

$$\begin{aligned} Y(f) &= X(f) \otimes [\delta(f - f_c) + \delta(f + f_c)] \\ &= [X(f - f_c) + X(f + f_c)]\,. \end{aligned} \tag{16.12}$$

Figure 16.4(b) shows this spectrum. As in Conv-AM, the AM-SC also needs double the bandwidth of the original message signal. However, as Eq. (16.12) shows, the AM-SC transmits only the message signal component and hence, unlike Conv-AM, it does not waste power to transmit the carrier component.

16.2.2.1 Coherent Demodulation

Unlike in Conv-AM, we cannot now recover $x(t)$ from the positive envelope of $y(t)$. But observe from Fig. 16.4 that if we multiply $y(t)$ with the carrier $\cos(2\pi f_c t)$, we will get a $X(f)$ in the base-band. That is,

$$\begin{aligned} z(t) &= y(t)\cos(2\pi f_c t) = 2x(t)\cos^2(2\pi f_c t) \\ &= x(t)\left[1+\cos(2\pi 2 f_c t)\right] \\ \Rightarrow \quad Z(f) &= X(f) + \frac{X(f-2f_c)+X(f+2f_c)}{2}. \end{aligned} \tag{16.13}$$

Figure 16.4(c) shows the spectrum $Z(f)$. Since the spectra $X(f - 2f_c)$ and $X(f + 2f_c)$ are situated around $\pm 2f_c$, we can eliminate these by filtering $z(t)$ with a LPF with cut-off frequency just above B Hz. Then, the LPF output will be same as $x(t)$. This is illustrated in Fig. 16.4(c).

For the above demodulation approach to work satisfactorily, *the phase and frequency of the carrier used for demodulation at the receiver must be exactly same as that used for modulation at the transmitter.* This is known as *coherent demodulation*. Any mismatch in phase and/or frequency of the carriers at transmitter and receiver will result in distortion in the recovered signal.[7] Therefore, it is very important to have special circuits at the receiver for generating the synchronous carrier. *This makes the receiver expensive in systems where coherent demodulation is necessary.*

§ *Use* $\cos(2\pi \tilde{f}_c t+\theta)$ *for demodulating the received AM-SC signal, where* $\tilde{f}_c = f_c + \Delta f$ *and* $\theta \neq 0$, *with* Δf *being an offset in the receiver carrier frequency. Using time domain as well as frequency-domain analysis, examine the effect of* Δf *and* θ *on demodulation.*

16.2.2.2 Compromise Approaches

To ease the problem of generating synchronous carrier, various approaches can be adopted.

- Approach 1: *Send a weak carrier* along with the AM-SC signal.
- Approach 2: *Send the carrier, once in a while, as pilot tones.*

The receiver could use the information about carrier to generate synchronous carrier. In this way, it is possible to strike an acceptable compromise between cost (i.e., extra transmission power) and receiver complexity.

7. Phase mismatch results in amplitude distortion and frequency mismatch results in frequency shift in the recovered message signal.

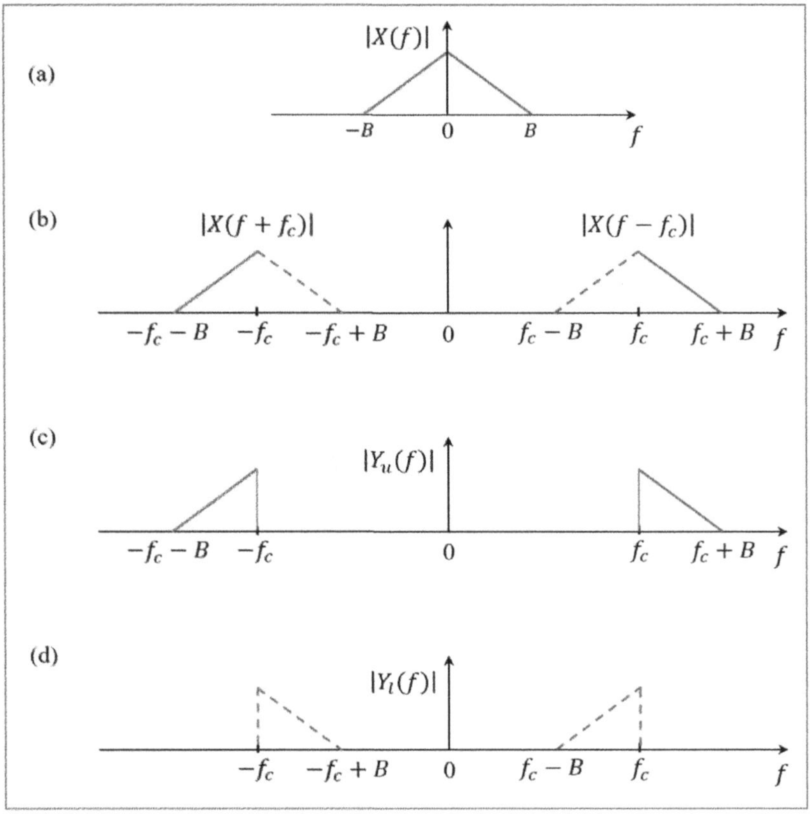

Figure 16.5. Spectra of signals in single side-band AM (SSB-AM): (a) message signal $x(t)$, (b) AM-SC modulated signal with both side-bands, (c) upper side-band signal $y_u(t)$, and (d) lower side-band signal $y_l(t)$.

16.2.2.3 Summary of AM-SC

- It transmits only the frequency-shifted message signal. Thus, compared to Conv-AM, the AM-SC scheme is power efficient.
- Requires double the bandwidth of the message signal.
- Receiver becomes complicated (and hence expensive) since it needs circuits for generating local carrier that is precisely synchronous with the carrier at the transmitter. It also must account for any changes in the carrier frequency/phase that arise due to the channel.
- The receiver complexity due to coherent demodulation can be minimized by resorting to the approaches of sending the carrier information using pilot tones or weak carrier.

16.2.3 Single Side-Band AM (SSB-AM)

The SSB-AM scheme is an improvement over AM-SC to save bandwidth. Observe from Fig. 16.5(b) (this is same as Fig. 16.4(b)) that the transmitted signal in AM-SC consists of two copies of the original message spectrum $X(f)$ shifted to $+f_c$ and $-f_c$. We note the following from Fig. 16.5(b).

- The spectrum shown in (blue) solid line (i.e., frequency bands $[-f_c - B, -f_c]$ and $[f_c, f_c + B]$ in Fig. 16.5(b)) contains complete information on the message $X(f)$. This part is shown separately in Fig. 16.5(c) and is called the *upper side-band.*
- The spectrum shown in (blue) dotted line (i.e., frequency bands $[-f_c, -f_c + B]$ and $[f_c - B, f_c]$ in Fig. 16.5(b)) also contains complete information on the message $X(f)$. This part is shown separately in Fig. 16.5(d) and is called the *lower side-band.*
- The upper and lower side-band signals contain exactly the same (and complete) information as the message signal $x(t)$.
- So, it is enough to transmit only one of the side-band signals. Thus, *the bandwidth required for SSB-AM transmission is same as the message bandwidth.*

The upper side-band signal $y_u(t)$ and lower side-band signal $y_l(t)$ can be generated as

$$y_u(t) = x(t)\cos(2\pi f_c t) - \hat{x}(t)\sin(2\pi f_c t) \tag{16.14}$$

$$y_l(t) = x(t)\cos(2\pi f_c t) + \hat{x}(t)\sin(2\pi f_c t) \tag{16.15}$$

where $\hat{x}(t)$ is -90° phase-shifted version of $x(t)$ (obtained by passing $x(t)$ through the so-called 'Hilbert transformer' filter). SSB-AM also uses coherent demodulation at the receiver.

16.2.3.1 Summary of SSB-AM

- Transmits only one side-band of the frequency-shifted message signal. Thus, SSB-AM is efficient in both power as well as bandwidth.
- The receiver is complicated due to the requirement to have coherent demodulation, just as in the case of AM-SC.
- SSB-AM systems require a 90° phase-shifter filter and/or band-pass filters with sharp cut-off characteristics (to filter the side-bands), which are costly to realize.

16.2.4 Quadrature AM (QAM)

While the SSB-AM scheme saves bandwidth by cutting-off one of the redundant side-bands, its implementation can be costly because of the special filters needed. On the other hand, the QAM approach addresses the bandwidth efficiency issue by not cutting-off a side-band but by sending two message signals $p(t)$ and $q(t)$ (instead of only one message) in the same bandwidth.

To ensure separability of the signals $p(t)$ and $q(t)$ at the receiver, they are modulated using two carriers[8] $\cos(2\pi f_c t)$ and $\sin(2\pi f_c t)$, respectively. The QAM modulated signal $y(t)$ and its spectrum are given by[9]

$$y(t) = 2p(t)\cos(2\pi f_c t) - 2q(t)\sin(2\pi f_c t) \tag{16.16}$$

$$\begin{aligned} Y(f) = &[P(f - f_c) + jQ(f - f_c)] \\ &+ [P(f + f_c) - jQ(f + f_c)] \end{aligned} \tag{16.17}$$

where $P(f)$ and $Q(f)$ are the spectra of $p(t)$ and $q(t)$, respectively.

16.2.4.1 QAM Demodulation

Coherent demodulation is used for recovering $p(t)$ and $q(t)$ from the received $y(t)$. Coherent demodulation using the carrier $\cos(2\pi f_c t)$ followed by low-pass filtering results in $p(t)$. This can be seen as follows.

$$\begin{aligned} \tilde{p}(t) &= y(t)\cos(2\pi f_c t) \\ &= 2p(t)\cos^2(2\pi f_c t) - 2q(t)\sin(2\pi f_c t)\cos(2\pi f_c t) \\ &= p(t) + [p(t)\cos(2\pi 2 f_c t) - q(t)\sin(2\pi 2 f_c t)]\,. \end{aligned} \tag{16.18}$$

The spectra of $p(t)\cos(2\pi 2 f_c t)$ and $q(t)\sin(2\pi 2 f_c t)$ are located around $\pm 2f_c$. Therefore, passing $\tilde{p}(t)$ through a LPF with bandwidth B will recover $p(t)$ and eliminate the components around $\pm 2f_c$. Similarly, coherent demodulation using $-\sin(2\pi f_c t)$ followed by low-pass filtering results in $q(t)$.

16.2.4.2 Complex Representation of QAM

The QAM modulation and demodulation operations can be expressed in equivalent complex representation as below. Noting that[10] $\cos(\theta) = \text{Re}\left\{e^{j\theta}\right\}$ and $-\sin(\theta) =$

---

8. Since $\cos(2\pi f_c t + 90°) = -\sin(2\pi f_c t)$, the carriers $\cos(2\pi f_c t)$ and $\sin(2\pi f_c t)$ are known as in-phase and quadrature carriers, respectively.

9. Note: Recall that $\mathcal{F}[2\sin(2\pi f_c t)] = [\delta(f - f_c) - \delta(f + f_c)]/j = -j\delta(f - f_c) + j\delta(f + f_c)$.

10. Here, $\text{Re}\{x\}$ denotes the 'real-part' of the complex quantity x.

$\mathrm{Re}\left\{je^{j\theta}\right\}$, we get

$$\begin{aligned}2p(t)\cos(2\pi f_c t) &= \mathrm{Re}\left\{2p(t)e^{j2\pi f_c t}\right\},\\ -2q(t)\sin(2\pi f_c t) &= \mathrm{Re}\left\{2jq(t)e^{j2\pi f_c t}\right\},\\ \Rightarrow \quad y(t) &= \mathrm{Re}\left\{2p(t)e^{j2\pi f_c t} + 2jq(t)e^{j2\pi f_c t}\right\},\\ &= \mathrm{Re}\left\{2v(t)e^{j2\pi f_c t}\right\},\end{aligned} \tag{16.19}$$

where $v(t) = p(t) + jq(t)$. Equation (16.19) implies that $p(t)$ and $q(t)$ can be QAM modulated by modulating the complex signal $p(t) + jq(t)$ using the complex carrier $e^{j2\pi f_c t}$ and taking real part of the resulting signal.

16.2.4.3 How about the Demodulation

Note that

$$\begin{aligned}\tilde{y}(t) = y(t)e^{-j2\pi f_c t} &= 2\mathrm{Re}\left\{v(t)e^{j2\pi f_c t}\right\}e^{-j2\pi f_c t}\\ &= \left[v(t)e^{j2\pi f_c t} + v^*(t)e^{-j2\pi f_c t}\right]e^{-j2\pi f_c t}\\ &= v(t) + v^*(t)e^{-j2\pi(2f_c)t} \qquad (16.20)\\ \Rightarrow \quad LPF\left\{\tilde{y}(t)\right\} &= v(t) = p(t) + jq(t). \qquad (16.21)\end{aligned}$$

Thus, QAM demodulation is equivalent to coherently demodulating $y(t)$ using the complex carrier $e^{-j2\pi f_c t}$ and low-pass filtering the result.

§ *Examine the spectra of signals at various stages in QAM modulator and demodulator.*

16.2.4.4 Summary of QAM

- Transmits two messages in the same bandwidth using in-phase and quadrature carriers. Thus, QAM is efficient in both power as well as bandwidth.
- The receiver is complicated due to the requirement to have coherent demodulation, just as in AM-SC and SSB-AM.
- In hardware complexity, QAM is simpler than SSB-AM.

16.3 Frequency Modulation (FM)

In FM, the message signal $x(t)$ is impressed onto the carrier by changing the *instantaneous frequency* of the carrier. FM has the advantage (compared to AM) of better resistance to noise, as noise affects amplitude much more seriously than frequency.

However, this advantage comes at the price of increased requirement on the channel bandwidth for transmitting FM signals.

The instantaneous frequency $f_i(t)$ and instantaneous phase $\theta_i(t)$ of the modulated carrier in FM are given by

$$f_i(t) = f_c + k_c x(t) \tag{16.22}$$

$$\Rightarrow \quad \theta_i(t) = 2\pi \int_{-\infty}^{t} f_i(\tau) d\tau \tag{16.23}$$

$$= 2\pi f_c t + 2\pi k_c \int_{-\infty}^{t} x(\tau) d\tau, \tag{16.24}$$

where k_c is known as *frequency sensitivity* of the frequency modulator and it determines the sensitivity of the instantaneous carrier frequency $f_i(t)$ to the message signal $x(t)$. The resulting modulated signal is given by

$$\begin{aligned} y(t) &= A_c \cos(\theta_i(t)) \\ &= A_c \cos\left(2\pi f_c t + 2\pi k_c \int_{-\infty}^{t} x(\tau) d\tau\right), \end{aligned} \tag{16.25}$$

where A_c is a constant amplitude.[11] Equations (16.22) and (16.25) show that *demodulation of a FM signal* involves determining the instantaneous frequency $f_i(t)$ of the FM signal and then subtracting the constant carrier frequency f_c from $f_i(t)$.

16.3.1 Bandwidth in FM Systems

From Eq. (16.22), we get the maximum deviation of the instantaneous frequency from f_c as

$$\Delta f_{max} = k_c x_{max} \qquad \text{where} \qquad x_{max} = \max\{|x(t)|\}. \tag{16.26}$$

Then, the *modulation index* in FM is defined as

$$\begin{aligned} \beta &= \frac{\text{Maximum Deviation in Carrier Frequency}}{\text{Bandwidth of Message Signal}} \\ &= \frac{\Delta f_{max}}{B} = \frac{k_c x_{max}}{B}, \end{aligned} \tag{16.27}$$

11. In *phase modulation* (PM), $x(t)$ is used to directly modify the phase of the carrier as $\theta_i(t) = 2\pi f_c t + k_p x(t)$ where k_p is the phase sensitivity parameter of the modulator. The resulting phase modulated signal is given by $y(t) = A_c \cos(\theta_i(t)) = A_c \cos\left(2\pi f_c t + k_p x(t)\right)$.

where B is the bandwidth of $x(t)$. Thus, we see that the frequency sensitivity parameter k_c has significant influence on the bandwidth of FM signals. The required transmission bandwidth increases with increase in k_c, and vice versa.

16.3.2 Narrow-Band and Wide-Band FM

With $\Delta\theta(t) = \theta_i(t) - 2\pi f_c t$, the FM signal given by Eq. (16.25) can be written as

$$y(t) = A_c \cos(\theta_i(t)) = A_c \cos(2\pi f_c t + \Delta\theta(t)) \tag{16.28}$$

$$= A_c \cos(2\pi f_c t)\cos(\Delta\theta(t)) - A_c \sin(2\pi f_c t)\sin(\Delta\theta(t)) \tag{16.29}$$

$$\Delta\theta(t) = 2\pi k_c \int_{-\infty}^{t} x(\tau)d\tau = \beta \frac{2\pi B}{x_{max}} \int_{-\infty}^{t} x(\tau)d\tau, \tag{16.30}$$

since $k_c = B\beta/x_{max}$. Depending on the maximum value of $|\Delta\theta(t)|$, we classify the FM into narrow-band FM and wide-band FM.

- Narrow-Band FM: This corresponds to the situation where $|\Delta\theta(t)| \ll 1$ for all t. This happens when the modulation index $\beta \ll 1$. Under this condition, the FM signal in Eq. (16.29) can be approximated as[12]

$$y(t) \approx A_c \cos(2\pi f_c t) - A_c \Delta\theta(t) \sin(2\pi f_c t) \tag{16.31}$$

  which is similar to Conv-AM signal with $\Delta\theta(t)$ as the message signal. Thus, bandwidth of narrow-band FM is similar to that of Conv-AM.
- Wide-Band FM: When the value of β is not very small (i.e., $|\Delta\theta(t)|$ is not small), then the spectrum of the resulting FM signal occupies a much wider band. This bandwidth expansion is the price that we pay for increased noise immunity in FM.[13]

In the next section, the matched filter methods for the demodulation operation is discussed in detail.

12. Note: $\cos(\theta) \approx 1$ and $\sin(\theta) \approx \theta$ when $|\theta| \ll 1$.

13. Clearly, narrow-band FM does not enjoy the noise immunity provided by wide-band FM.

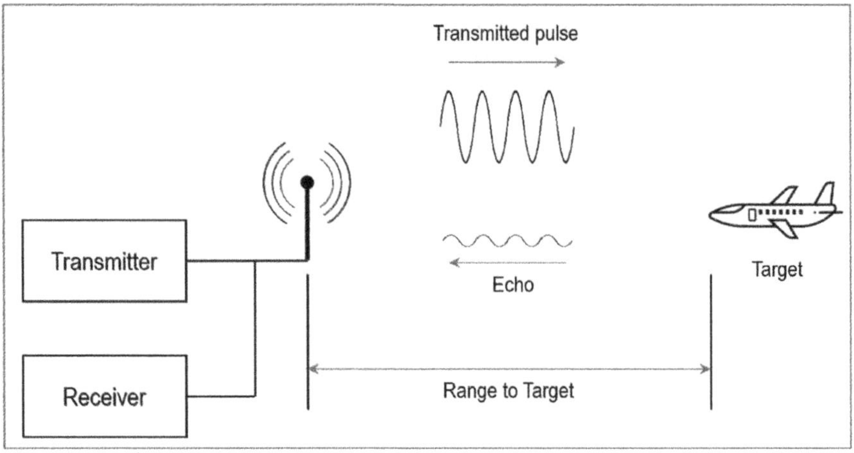

Figure 16.6. Illustration of a FM Radar scheme.

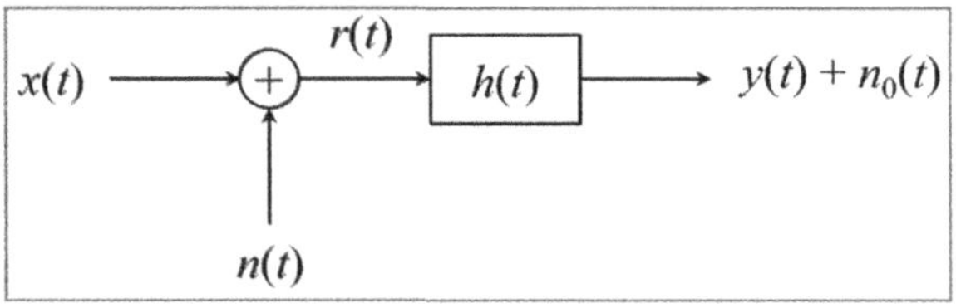

Figure 16.7. Matched filter scheme.

16.4 Detection (demodulation) by Matched Filter

Many systems are based on the transmission and reception (detection) of pulses. Radar and medical ultrasound scanners are examples of such systems. In such systems, the strength of the received signal is related to the object that has reflected the transmitted pulse. For example, a radar receives a stronger signal from bigger objects and weaker signals from smaller objects (Fig. 16.6). Such signals are often corrupted by noise, and it is therefore desirable to maximize the useful signals at the same time minimize the unwanted signals. A filter that is able to achieve the above objective is called the matched filter and in this section, the principle of matched filter scheme is discussed in detail.

Let us assume that the received signal $r(t)$ can be written as,

$$r(t) = x(t) + n(t) \tag{16.32}$$

where $x(t)$ is the desired response from the system - for example the echo from a single reflector, $n(t)$ is the noise. We also assume that there is no correlation between $x(t)$ and $n(t)$. We want now a linear and time-invariant (LTI) filter ($h(t)$ as shown in Fig. 16.7) which maximizes the ratio between the power of the desired signal and that of the noise at a given time instant t_m. This ratio is defined as,

$$\mathrm{SN}_p = \frac{|y(t_m)|^2}{E\{n_0^2(t_m)\}}, \tag{16.33}$$

where

$$y(t) = x(t) \otimes h(t) \tag{16.34}$$

is the filtered desired signal. $y(t_m)$ in Eq. (16.33) is the value of the signal at time instant t_m. $h(t)$ is the impulse response of the matched filter, $y(t)$ is the output of the matched filter without the noise, and $n_0(t) = h(t) \otimes n(t)$ is the filtered noise.

If we assume the noise to be stationary, we can replace t_m with t and the ratio in Eq. (16.33) can be expressed in the frequency domain as,

$$\frac{|y(t_m)|^2}{E\{n_0^2(t)\}} = \frac{|\int_{-\infty}^{+\infty} X(f)H(f)e^{j2\pi f t_m} df|^2}{\int_{-\infty}^{+\infty} \Gamma_{nn}(f)|H(f)|^2 df} \tag{16.35}$$

Since $n(t)$ is assumed stationary, its power is independent of time. $\Gamma_{nn}(f)$ is the power spectrum of the noise. If we assume that the noise is white, then we get,

$$\frac{|y(t_m)|^2}{E\{n_0^2(t)\}} = \frac{|\int_{-\infty}^{+\infty} X(f)H(f)e^{j2\pi f t_m} df|^2}{\sigma_n^2 \int_{-\infty}^{+\infty} |H(f)|^2 df} \tag{16.36}$$

where $\sigma_n^2 = \Gamma_{nn}(f)$ is the constant power density spectrum of the white noise. If we apply the *Schwartz's inequality*, we get

$$\left|\int_{-\infty}^{+\infty} X(f)H(f)e^{j2\pi f t_m} df\right|^2 \leq \int_{-\infty}^{+\infty} |H(f)|^2 df \int_{-\infty}^{+\infty} |X(f)e^{j2\pi f t_m}|^2 df \tag{16.37}$$

where the equality is achieved only for

$$H(f) = k_1 \left(X(f)e^{j2\pi f t_m}\right)^* = k_1 X^*(f)e^{-j2\pi f t_m}, \tag{16.38}$$

where, k_1 is a constant. The maximum of the ratio is therefore reached when Eq. (16.38) is fulfilled. In this case, we have (assuming $k_1 = 1$)

$$\begin{aligned}\frac{|y(t_m)|^2}{E\{n_0^2(t)\}} &\leq \frac{\int_{-\infty}^{+\infty} |H(f)|^2 df \int_{-\infty}^{+\infty} |X(f)e^{j2\pi f t_m}|^2 df}{\sigma_n^2 \int_{-\infty}^{+\infty} |H(f)|^2 df} \\ &\leq \frac{\int_{-\infty}^{+\infty} |X(f)|^2 df}{\sigma_n^2} = \frac{E}{\sigma_n^2}\end{aligned} \tag{16.39}$$

Here, $E = \int_{-\infty}^{+\infty} |X(f)|^2 df$ is the energy of the signal, $x(t)$. We used the fact that $\int |p|^2 df = \int p.p^* df$ and the factor $e^{j2\pi f t_m}$ is of no significance. At the moment of detection, the magnitude of the filter output for the desired signal is proportional to its energy, E.

The Fourier transform of the filter which maximizes the peak amplitude of the output signal relative to the noise is hence given by

$$H(f) = k_1 X^*(f) e^{-j2\pi f t_m} \tag{16.40}$$

which in the time domain is

$$h(t) = k_1 x(t_m - t). \tag{16.41}$$

The LTI filter is a time-reversed and delayed version of the original noise-free signal. The time delay t_m is chosen so that the filter is causal. This means that the filter impulse response ($h(t)$) is matched to the desired signal ($x(t)$), and hence the name '*Matched Filter*'.

16.5 Detection of Binary Signals in Noise

Detecting signals in the presence of noise is a common task. In radars, one wants to decide whether there is an air-plane at time instant t_m, and in telecommunications one wants to decide which symbol is being received.

For example consider a simplified case of the binary signal $x(t)$ as having the value either 0 or 1. When $x(t)$ has passed through the matched filter $h(t)$ we get the value K, when $x(t)$ has the value 1. In other words, at time instant t_m, we can have at the output of the matched filter either

$$y_1(t_m) = K + n_0(t_m) \tag{16.42}$$

or

$$y_1(t_m) = n_0(t_m) \tag{16.43}$$

when $x(t)$ has passed through a noisy channel and then filtered by the matched filter $h(t)$ (see Fig. 16.8). We must have a threshold to decide whether the received signal is 0 or 1. Let the value of the threshold be μ. This means that we assume $x(t_m)$ equal to 1 when $y_1(t_m) \geq \mu$ and is 0 when $y_1(t_m) < \mu$.

There are two possibilities for error because of the noise. Either the noise $n_0(t_m)$ can decrease the value of $y_1(t_m)$, so that it becomes less than μ although $x(t_m) = 1$, or $y_1(t_m) > \mu$ although $x(t_m) = 0$. The first error means that a pulse is not

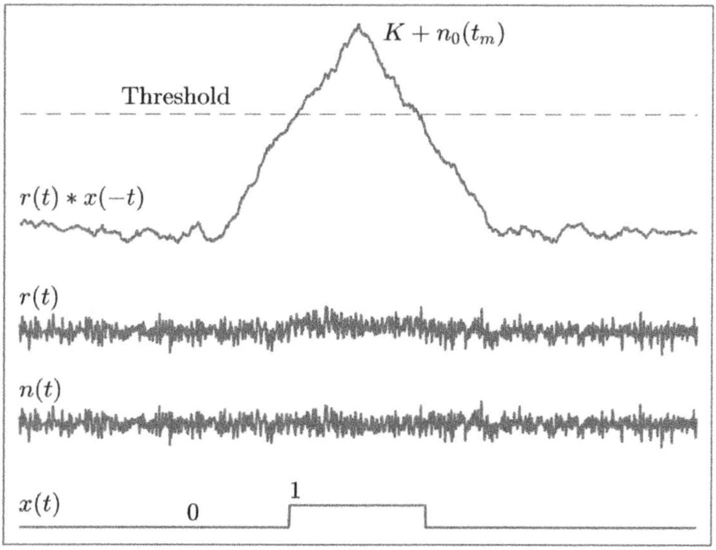

Figure 16.8. Matched filtering of binary signals: $x(t)$ is the transmitted message (binary) signal, $n(t)$ is the noise, which is getting added to $x(t)$ to generate the received signal, $r(t)$. $r(t) * x(-t)$ is the filtering (convolution) operation between the received signal and the matched filter (with impulse response, $h(t) = x(-t)$). It also can be seen as the cross-correlation between the received signal and the transmitted signal.

detected (*false negative*) and the second error means that we detect a signal which is actually not there (*false positive*).

An example with Gaussian noise is shown in Fig. 16.9. The hatched area illustrates how large the two possibilities are. It can be seen that there will be errors independent of the choice of μ. The total probability of error depends on how often 0 and K are present in the signal. If K is infrequent, then the false negative error will be small. The total probability of error is

$$P_f = P_0 P_{f_0} + P_K P_{f_K}, \tag{16.44}$$

where P_0 is the probability that $x(t_m) = 0$, and P_{f_0} is the probability that $n_0(t_m) > \mu$. Because there are only two symbols (0,1), we get

$$P_f = P_0 P_{f_0} + (1 - P_0) P_{f_K}. \tag{16.45}$$

If we have the same probability for 0 and K, then we get

$$P_f = \frac{1}{2}\left(P_{f_0} + P_{f_K}\right). \tag{16.46}$$

If the probability density function of the noise is symmetric about its mean value, then $P_{f_0} = P_{f_K}$, and the threshold $\mu = K/2$ since both types of the error have

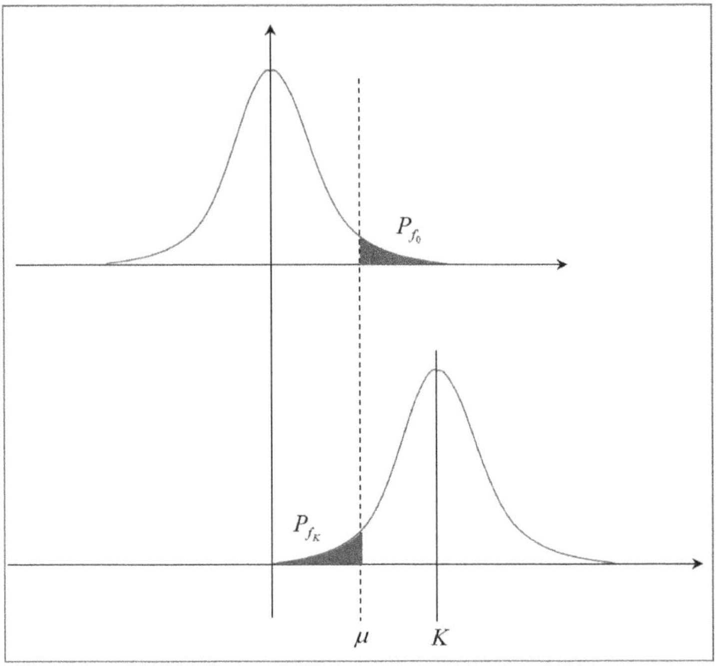

Figure 16.9. Probability density functions for the signal $y_1(t)$.

the same weight. The two probabilities P_{f0} and P_{fK}, for the case of Gaussian noise are given by,

$$P_f = P_{f0} = P_{fK} = \int_{\mu}^{\infty} \frac{1}{\sqrt{2\pi}\,\sigma_n} e^{\{\frac{-\xi^2}{2\sigma_n^2}\}} d\xi, \tag{16.47}$$

and are dependent on μ and σ_n, where σ_n is the standard deviation of the noise $n_0(t)$. The probability for error is shown as a function of $\frac{K}{2\sigma_n}$ in Fig. 16.10.

However, the probability of the two symbols is often not equal and the choice of a threshold is not easy.

The output of the matched filter at time instant t_m is a signal whose amplitude is proportional to the energy of the transmitted pulse. If there is no echo, for example a radar pulse reflected by an air-plane, then only noise is present. This is a situation equivalent to that is given in Eq. (16.43). Therefore, we have the same probability of error, and the same method for determining the threshold value can be used. Hence $\frac{K}{2\sigma_n}$ is the ratio between the energy of the received signal to twice the RMS value of the matched filtered noise.

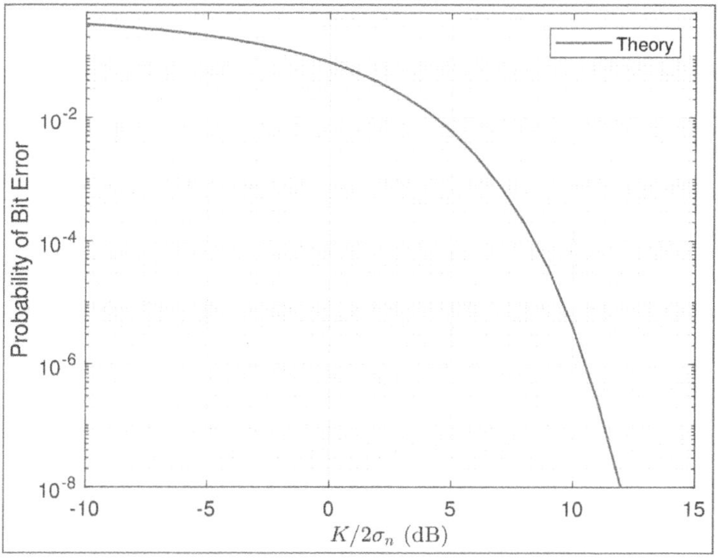

Figure 16.10. Probability of bit error with Gaussian noise, when the probability of the two bits (symbols) are equal.

Often in radar systems, P_K is much smaller than P_0 and P_K is usually unknown. Hence, the threshold value is usually determined empirically.

So far, we have discussed the analog modulation schemes. The remaining of this chapter is devoted for the discussion on digital modulation schemes.

16.6 Digital Modulation

Many applications require the transmission of a digital signal via a modem, network or digital broadcasting. This section deals with the topic of transmission of digital data (bits) stream. Just as we have different approaches to modulate analog signals, we need approaches that can convert the stream of bits into waveforms that are suitable for transmission. There are three basic modulation types:

- Amplitude (ASK - Amplitude Shift Keying),
- Frequency (FSK - Frequency Shift Keying), and
- Phase (PSK - Phase Shift Keying).

Let the sequence $\{b_1, b_2, b_3, b_4, b_5, \cdots\}$ denote a stream of bits at the rate of $\frac{1}{T_b}$ bits/second, i.e., T_b is the duration of one bit. Each bit b_n is either a '1' or a '0'. We first convert this bit sequence into analog signals and then modulate these signals using any of the approaches discussed above.

16.6.1 Amplitude Shift Keying (ASK)

This is the simplest modulation scheme for digital data. It is a special case of AM. For the given data, an analog signal is generated as

$$r(t) = \begin{cases} +1 & \text{for} \quad (n-1)T_b \le t < nT_b \quad \text{if} \quad b_n = 1 \\ -1 & \text{for} \quad (n-1)T_b \le t < nT_b \quad \text{if} \quad b_n = 0. \end{cases} \tag{16.48}$$

That is, we convert bit '1' into a positive pulse and bit '0' into a negative pulse, of width T_b each. Then, $r(t)$ is obtained by concatenating the pulses corresponding to the bits in the sequence. The waveform $r(t)$ is modulated using AM to result in the modulated signal

$$p(t) = r(t)A\sin(2\pi f_c t), \tag{16.49}$$

where f_c is the career frequency and A is its amplitude. Since $r(t)\sin(2\pi f_c t)$ is equivalent to switching the carrier amplitude to $+1$ or -1 during each bit interval depending on the value of the bit, this scheme is called "ASK". It is also known as "On-Off Keying (OOK)" if the pulse amplitude $r(t)$ is made zero when $b_n = 0$, which is equivalent to switching ON and OFF the carrier if the bit is '1' and '0', respectively.

The OOK signal corresponding to the binary string 10101011 is shown in Fig. 16.11. The top graph shows the ideal signal, and the bottom one shows the OOK signal.

When the signal is received after transmission, the binary data stream must be recreated. This can be done by applying a matched filter, and then determining whether 0 or 1 has been sent. If the probability for 0 and 1 are equal, the threshold value for detection is set to $E/2$, where E is the energy of the pulse $p(t)$. (K in Eq. (16.42) corresponds to E, hence the probability of error is given in Fig. 16.10.)

Example: The AM signal shown in the top graph in Fig. 16.11 is sent over a telephone channel with infinite bandwidth. The channel adds noise with Gaussian PDF and a constant power density of 10^{-4} W/Hz with a bandwidth of $f_n = 5\ kHz$. Find the error probability.

The energy of the signal after matched filtering is

$$\begin{aligned} E &= \int_0^{4/f_c} A^2 \sin^2(2\pi f_c t)dt \\ &= \frac{A^2}{2\pi f_c}\left[\frac{x}{2} - \frac{\sin 2x}{4}\right]_0^{8\pi} = \frac{2A^2}{f_c}. \end{aligned} \tag{16.50}$$

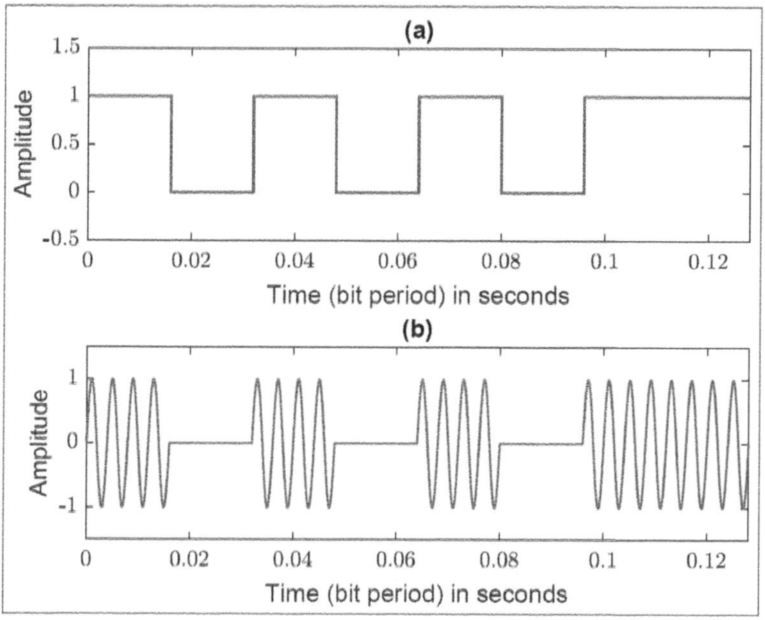

Figure 16.11. ASK signal for the binary sequence 10101011 sent with $f_c = 250$ Hz and $T = 4/f_c = 16$ ms. (a) Original digital signal (bit stream) and (b) ASK signal with two amplitudes.

As we know, the matched filter is a time-reversed version of the sent pulse. Its amplitude transfer function has a maximum value (amplification) exactly at the frequency of the transmitted pulse ($f = f_c$). The transfer function is concentrated at $f = f_c$ and the amplification decreases for higher values of f. If we assume that the amplification is negligible for $f > f_c$, we can find the power of the noise through Parseval's equation:

$$\begin{aligned} p_n &= \int_{-\infty}^{\infty} |H(f)|^2 \Gamma_{nn}(f) df = \Gamma_{nn}(0) \int_{-F_n}^{F_n} |H(f)|^2 df \\ &\approx \Gamma_{nn}(0) \int_{-\infty}^{\infty} |H(f)|^2 df = \Gamma_{nn}(0) \int_{-\infty}^{\infty} h^2(t) dt = \Gamma_{nn}(0) \frac{2A^2}{f_c} \end{aligned} \tag{16.51}$$

where $\Gamma_{nn}(f)$ is the power density of the noise. The equation assumes that the noise is white and overestimates its noise. The ratio $\frac{E}{2\sigma_n}$ is

$$\sigma_n = \sqrt{p_n} = \sqrt{\Gamma_{nn}(0) \frac{2A^2}{f_c}} \tag{16.52}$$

$$\frac{E}{2\sigma_n} = \frac{\frac{2A^2}{f_c}}{2\sqrt{\Gamma_{nn}(0) \frac{2A^2}{f_c}}} = \sqrt{\frac{A^2}{2\Gamma_{nn}(0) f_c}} \tag{16.53}$$

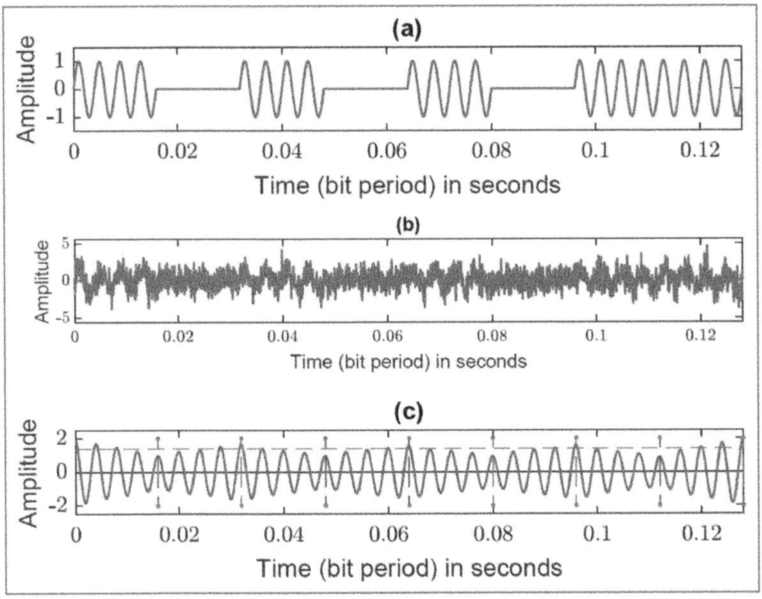

Figure 16.12. ASK signal for the binary sequence 10101011 sent with $f_c = 250$ Hz and $T = 4/f_c = 16$ ms. (a) The top graph shows the ideal ASK signal, (b) The ASK signal with noise added, and (c) The signal after passing through a matched filter. The horizontal line in this graph depicts the threshold value of $E/2$, and the vertical line depict the detection moments.

Substituting the given values in the above equation, we get

$$\frac{E}{2\sigma_n} = \sqrt{\frac{1^2}{2 \times 10^{-4} \times 1200}} = 6.2dB. \tag{16.54}$$

The probability for error is 2×10^{-2} (from Fig. 16.10). The various signals are shown in Fig. 16.12. for the binary sequence 10101011. If the amplitude of the pulse is increased from 1 to 2 volts, we get a ratio of 12dB and the probability for detection error falls to 8×10^{-5}, which is a significant improvement.

The probability of error is also dependent on whether the detection is performed at the right moment. It is therefore important that the receiver and transmitter are synchronized.

If the noise is white, then a simple relation for calculating the error probability can be derived. It gives insight in to the influence of the noise and the filter. If the noise is white, the power density is constant and the power of the noise after filtering is

$$p_n = \Gamma_{nn}(0)E \tag{16.55}$$

where E is the energy of the pulse $p(t)$. We get the following ratio:

$$\frac{E}{2\sigma_n} = \frac{E}{2\sqrt{\Gamma_{nn}(0)E}} = \sqrt{\frac{E}{4\Gamma_{nn}(0)}}. \tag{16.56}$$

In other words, the probability of error is determined by the ratio between the energy of the transmitted pulse and the power density of the noise introduced by the transmission channel.

16.6.2 Frequency Shift Keying (FSK)

Digital information can be transmitted using pulses with different centre frequencies. We sent $p_1(t)$ for the symbol '1' and $p_0(t)$ for the symbol '0' as follows:

$$p_1(t) = \begin{cases} A\sin(2\pi f_1 t) & \text{for } (n-1)T_b \leq t < nT_b \text{ if } b_n = 1 \\ 0 & \text{otherwise} \end{cases} \tag{16.57}$$

$$p_0(t) = \begin{cases} A\sin(2\pi f_0 t) & \text{for } (n-1)T_b \leq t < nT_b \text{ if } b_n = 0 \\ 0 & \text{otherwise} \end{cases} \tag{16.58}$$

where $f_1 T_b$ and $f_0 T_b$ are integers.

The binary string 10101011 is sent as shown in Fig. 16.13. We shift between frequency values and the method therefore is called the FSK. The individual FSK signals $p_0(t)$ and $p_1(t)$, corresponding to bits 0 and 1 are shown in Fig. 16.14.

Two matched filters are used for the detection: one for each pulse type. At the moment of detection, the value of the signal from the filters is proportional to the value of the energy of the transmitted pulse. Energy of the two pulses are equal because they are equally long. Let this be E. One possible detection method is shown in Fig. 16.15. The received signal goes through the two matched filters whose outputs get subtracted. This processing step can be written as

$$\begin{aligned} y(t) &= x(t) \otimes h_1(t) - x(t) \otimes h_2(t) \\ &= x(t) \otimes (h_1(t) - h_2(t)). \end{aligned} \tag{16.59}$$

Here, we neglected the presence of noise. At the moment of detection (t_m), we get:

$$y(t_m) = \int_{-\infty}^{\infty} x(\tau)(h_1(t_m - \tau) - h_2(t_m - \tau))d\tau \tag{16.60}$$

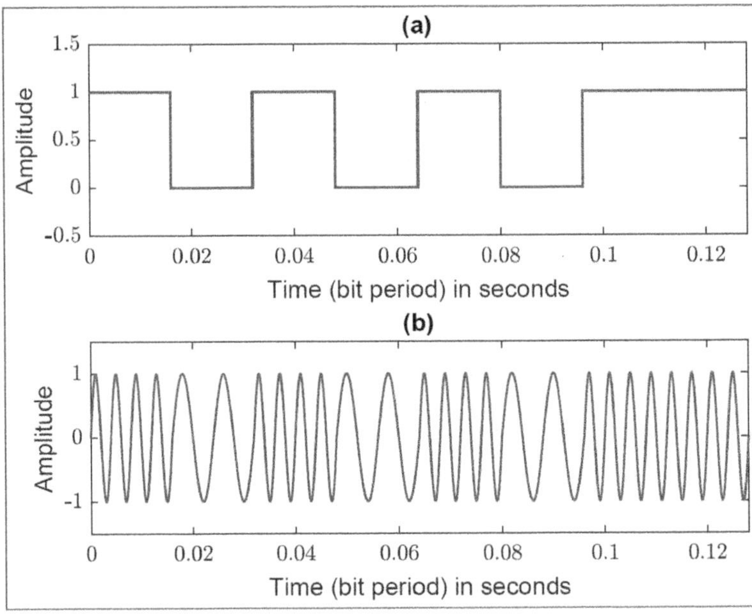

Figure 16.13. FSK signal for the binary sequence 10101011 sent with $f_1 = 4$ Hz, $f_0 = 2$ Hz and $T = 16$ ms. (a) The original (transmitted) signal (bit stream) and (b) The FSK signal with two frequencies for bits 1 and 0.

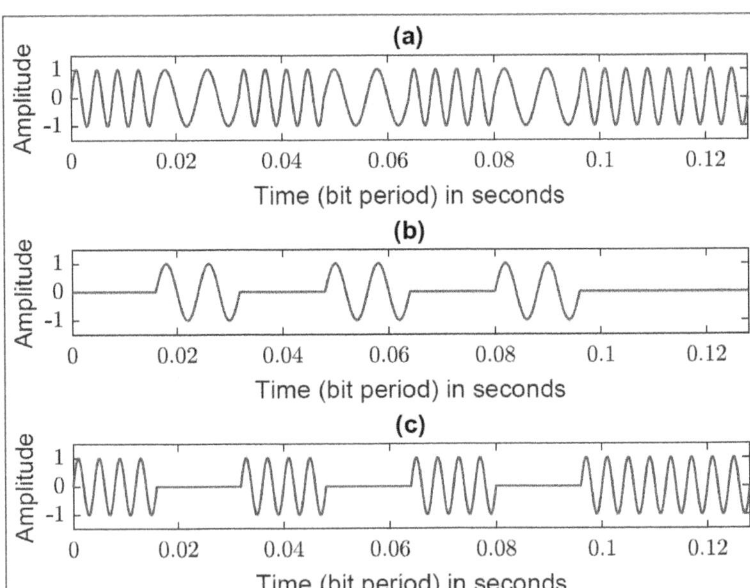

Figure 16.14. FSK signal for the binary sequence 10101011 sent with $f_1 = 4$ Hz, $f_0 = 2$ Hz and $T = 16$ ms. (a) The FSK signal with two frequencies for bits 1 and 0, (b) The FSK signal for bit 0 alone, and (c) The FSK signal for bit 1 alone.

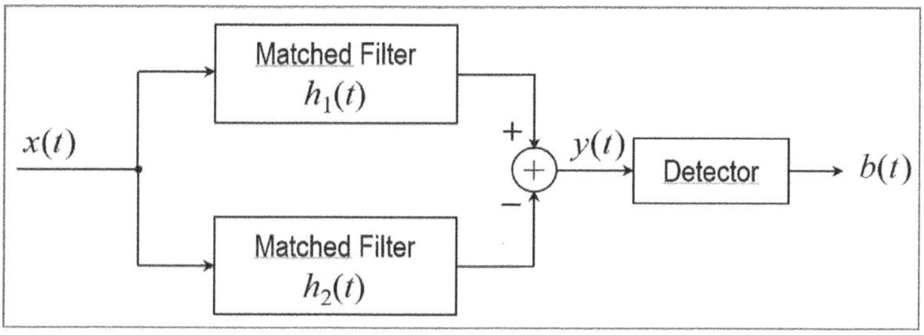

Figure 16.15. Detector used for FSK modulation.

where $h_1(t_m - t) = p_1(t_m - t_m + t) = p_1(t)$ (See Eq. (16.41)) and $h_2(t_m - t) = p_0(t_m - t_m + t) = p_0(t)$. At the moment of detection $t_m = 0$, we get a signal

$$y(0) = \int_0^{T_b} A\sin(2\pi f_1\tau)(A\sin(2\pi f_1\tau) - A\sin(2\pi f_0\tau))d\tau, \tag{16.61}$$

when '1' is received and the signal is noise-free. This can be re-written as

$$\begin{aligned} y(0) &= E - \int_0^{T_b} A\sin(2\pi f_1\tau)A\sin(2\pi f_0\tau)d\tau \\ &= E - \frac{A^2}{2}\int_0^{T_b}(\cos(2\pi(f_1 - f_0)\tau) - \cos(2\pi(f_1 + f_0)\tau))d\tau \\ &= E - \frac{A^2}{2}\int_0^{T_b}(\cos(2\pi 2\triangle f\tau) - \cos(2\pi(f_0 + \triangle f)\tau))d\tau. \end{aligned} \tag{16.62}$$

The value of the last integral will be 0, if we integrate over an integer number of periods of a sinusoidal signal with frequency $\triangle f = \frac{f_1 - f_0}{2}$. This is fulfilled when $2T_b\triangle f = k$, where k is an integer number, and the two signals ($A\sin(2\pi f_1 t)$ and $A\sin(2\pi f_0 t)$) are said to be orthogonal.

A similar derivation can be carried out for the symbol '0'. We get,

$$\begin{aligned} y(0) &= \int_0^{T_b} A\sin(2\pi f_0\tau)(A\sin(2\pi f_1\tau) - A\sin(2\pi f_0\tau))d\tau \\ &= -E \end{aligned} \tag{16.63}$$

The value that we get out of this detector is either E or $-E$ depending on which pulse has been sent, when the two signals are orthogonal. The threshold value can be set to zero.

The noise present in the signal is also being filtered and the power of the remaining noise (at the output) is

$$p_n = \int_{-\infty}^{\infty} |H_1(f) - H_2(f)|^2 \Gamma_{nn}(f) df, \tag{16.64}$$

where $\Gamma_{nn}(f)$ is the power density of the noise and $H_1(f)$ and $H_2(f)$ are the transfer functions of the two matched filters. If we assume that the noise has constant power in the pass-band of the two filters, we get

$$p_n = \Gamma_{nn}(0) \int_{-\infty}^{\infty} |H_1(f) - H_2(f)|^2 df. \tag{16.65}$$

The last term in the above equation can be written as

$$\begin{aligned} &= \int_{-\infty}^{\infty} (H_1(f) - H_2(f))(H_1^*(f) - H_2^*(f))df \\ &= \int_{-\infty}^{\infty} \left(|H_1(f)|^2 + |H_2(f)|^2 - H_1(f)H_2^*(f) - H_1^*(f)H_2(f)\right) df. \end{aligned} \tag{16.66}$$

The last two terms are equal to the signal $p_1(t)$ convolved with a time-reversed version of the signal $p_0(t)$, and the result is 0, since the two signals are orthogonal. We get hereby,

$$p_n = \Gamma_{nn}(0) \int_{-\infty}^{\infty} [|H_1(f)|^2 + |H_2(f)|^2] df = 2\Gamma_{nn}(0)E. \tag{16.67}$$

This means that the power of the noise is doubled by the subtraction.

The error probability can be found as in the case of ASK from Fig. 16.10. The threshold is set to 0. Hence, only noise samples with magnitude higher than E will cause an error. The power of the noise is doubled by the subtraction and the probability will be dependent on the ratio $E/\sqrt{2}\sigma_n$. The ratio which determines the probability of error can be shown to be

$$\sqrt{\frac{E^2}{2\sigma_n}} = \sqrt{\frac{E^2}{2\Gamma_{nn}(0)E}} = \sqrt{\frac{E}{2\Gamma_{nn}(0)}}. \tag{16.68}$$

In other words, we get 3dB improvement by using FSK rather than ASK.

16.6.3 Phase Shift Keying (PSK)

PSK is based on a shift in the phase of the signal, which is determined by the symbol (bit) being transmitted. When only two symbols are sent, the phase is shifted

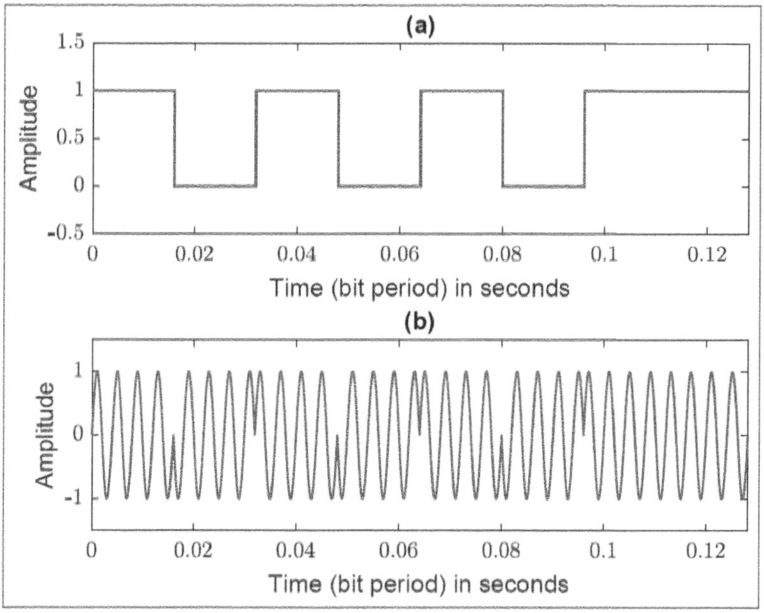

Figure 16.16. PSK signal for the binary sequence 10101011 sent at $f_c = 250$ Hz and $T = 2/f_c = 16$ ms. (a) Original digital signal (bit stream), and (b) The corresponding PSK signal.

between π and 0. The signal corresponding to the binary sequence 10101011 is shown in Fig. 16.16. The individual PSK signals corresponding to bits 0 and 1 are shown in Fig. 16.17.

A matched filter is used to recover the signal. The output of the filter is proportional to E when the received symbol is '1' and $-E$ when the received symbol is '0'. The detection threshold is set to 0, if the two symbols are equally probable. The power of the noise at the output of the filter is $p_n = \sigma_n^2 = \Gamma_{nn}(0)E$. The amplitude of the noise must be larger than E to cause a false detection. The ratio which determines the probability of error is

$$\frac{E}{\sigma_n} = \sqrt{\frac{E^2}{\Gamma_{nn}(0)E}} = \sqrt{\frac{E}{\Gamma_{nn}(0)}}, \tag{16.69}$$

which represents a 6dB improvement compared to ASK. This means that for the case given in the *Example*, we can transmit signals with an amplitude of 1 volt and achieve error probability of 8×10^{-5} by using PSK instead of ASK (2×10^{-2}).

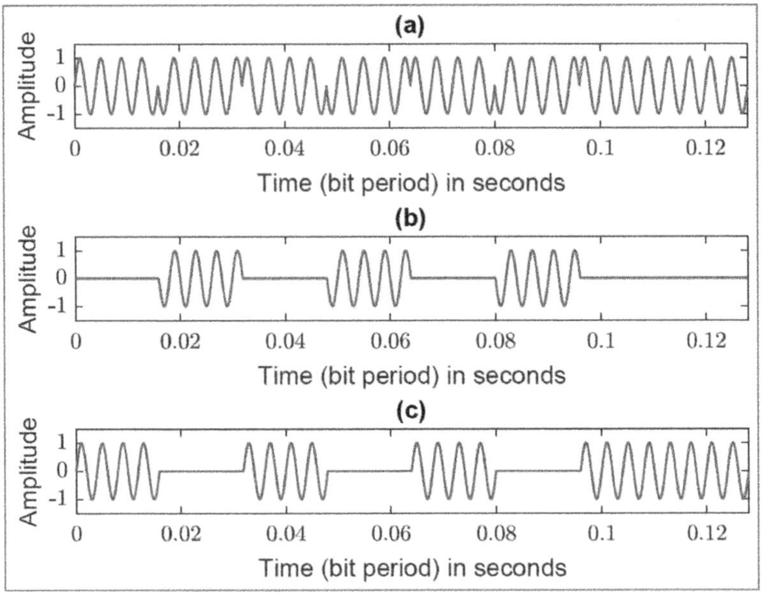

Figure 16.17. Individual PSK signals for the bits 1 and 0 for the binary sequence 10101011. (a) Ideal PSK signal, (b) The PSK signal for bit 0, and (c) The PSK signal for bit 1.

16.7 2-Bit Quadrature Amplitude Modulation (QAM)

The principles described here can be extended to n-bit QAM as well. Recall that the motivation for the development of QAM is to save bandwidth. In the same way, the 2-bit QAM is an approach to send more data bits (compared to ASK) through a channel of given bandwidth. This is explained below.

- The sequence of bits $\{b_1, b_2, b_3, b_4, b_5, \cdots\}$ is first converted into a sequence of symbols $\{s_1, s_2, s_3, s_4, \cdots\}$ where the symbols are defined by grouping 2 bits each as $s_1 = \{b_1, b_2\}$, $s_2 = \{b_3, b_4\}$, $s_3 = \{b_5, b_6\}$, $s_4 = \{b_7, b_8\}$, and so on. Thus, the time-duration of each symbol is $T_s = 2T_b$.
- The sequence of symbols is then converted into two analog signals as

$$r(t) = \begin{cases} +1 \text{ for } (n-1)T_s \le t < nT_s \text{ if } s_{n,1} = 1 \\ -1 \text{ for } (n-1)T_s \le t < nT_s \text{ if } s_{n,1} = 0 \end{cases} \tag{16.70}$$

$$q(t) = \begin{cases} +1 \text{ for } (n-1)T_s \le t < nT_s \text{ if } s_{n,2} = 1 \\ -1 \text{ for } (n-1)T_s \le t < nT_s \text{ if } s_{n,2} = 0 \end{cases} \tag{16.71}$$

where $s_{n,1}$ and $s_{n,2}$ are the first and second bits, respectively, in the symbol $s_n = \{s_{n,1},\ s_{n,2}\}$. That is, each symbol results in two pulses of widths T_s each.

- The resulting waveforms $r(t)$ and $q(t)$ are modulated using QAM and the modulated signal is given by

$$y(t) = 2r(t)\cos(2\pi f_c t) - 2q(t)\sin(2\pi f_c t). \tag{16.72}$$

Since $T_s = 2T_b$, we get $1/T_s = 0.5/T_b$. That is, the symbol-rate (i.e., number of symbols/second) is only half of the bit rate. For example, a bit rate of 10000 bits/second results in a symbol-rate of 5000 symbols/second. Therefore, the channel bandwidth we need to send the symbols using 2-bit QAM is only half of the bandwidth that we need to send the bits using ASK. Thus, the *QAM approach can be used to achieve high data-rate transmission in bandwidth constrained channels.*

16.8 Concluding Remarks

We can make the following concluding remarks based on the material presented in this chapter.

- Modulation is needed for matching the signal to the channel, and to support the transmission of several messages over a given channel.
- Modulation of a message onto a carrier is done by changing either the amplitude, frequency or phase of the carrier according to the message signal.
- While conventional AM is the most simple approach, it is inefficient in the use of power and bandwidth.
- AM-SC, SSB-AM, and QAM are efficient versions of AM. These approaches need coherent demodulation which makes the receiver complicated.
- FM has the advantage of noise immunity, which is obtained at the expense of increased bandwidth.
- The principle of matched filter for detection is described in detail.
- A demo Matlab programme, to show the application of Hilbert transform in communication is given in Appendix A.

Exercises

16.1. A message signal $m(t) = 20\cos(2\pi t)$, modulates a carrier $c(t) = 50\cos(100\pi t)$ to give a single-sideband wave. Determine the modulated waves when

(a) only the upper sideband is transmitted,
(b) only the lower sideband is transmitted.

16.2. An FM wave is described by $s(t) = 100\cos[\pi 2000000t + 0.005\cos(\pi 1000t)]$.

(a) Determine the instantaneous frequency of $s(t)$.
(b) What is the maximum frequency deviation?
(c) What is the modulating message signal?
(d) What is the modulation index?

16.3. Sketch the digitally modulated waveforms for an input binary sequence **1011010011**, where the carrier frequency f_c equals the bit rate, $1/T_b$, and T_b is the bit interval, using

(a) Amplitude Shift Keying (ASK)
(b) 2-bit Quadrature Amplitude Modulation (QAM)
Note: Assume that the pair of bits **00**, **10**, **11**, and **01** are represented by phase shifts equal to 0, $\pi/2$, π, and $3\pi/2$.

16.4. In creating a wireless home stereo system, an engineer is faced with the problem of sending two signals, $x_L(t)$ and $x_R(t)$, for the left and right audio channels respectively, to another room. A senior engineer suggests using a modulated signal, $y(t)$, whose amplitude is modulated by $x_L(t)$ using conventional AM and whose frequency is modulated by $x_R(t)$ using FM at a carrier frequency of f_0 Hz.

(a) Write down the form for $y(t)$. To evaluate the system, a test tone at 1 kHz is applied to both channels of the system. Write down an expression from which the spectrum of $y(t)$ can be calculated. Is this expression mathematically tractable?
(b) Suggest how $x_L(t)$ can be demodulated from $y(t)$ of Question (a). Is coherent demodulation essential in the recovery of $x_L(t)$ and $x_R(t)$ from $y(t)$? Which of the two channel signals will have a higher quality in the presence of noise and multi-path propagation? Explain your answer. If it is desired that both channels should have the same quality in using the basic modulation scheme outlined, suggest how this can be done.

16.5. An audio signal $s(t)$ with the spectrum $S(\Omega)$ (shown in Fig. P16.1) is to be transmitted in a communication scheme. In order to save the bandwidth, the signal spectrum is assumed to be 0 for $|\Omega| > \Omega_g = 2\pi \times 4$ kHz.

(a) Give the spectrum of the modulated signal $s_M(t)$ if the signal $s(t)$ is modulated by $\cos(\Omega_0 t)$ as shown in Fig. P16.2. Here Ω_0 is the modulating frequency.

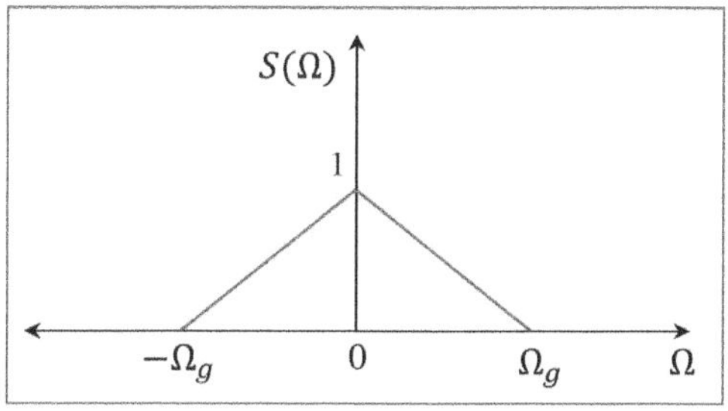

Figure P16.1.

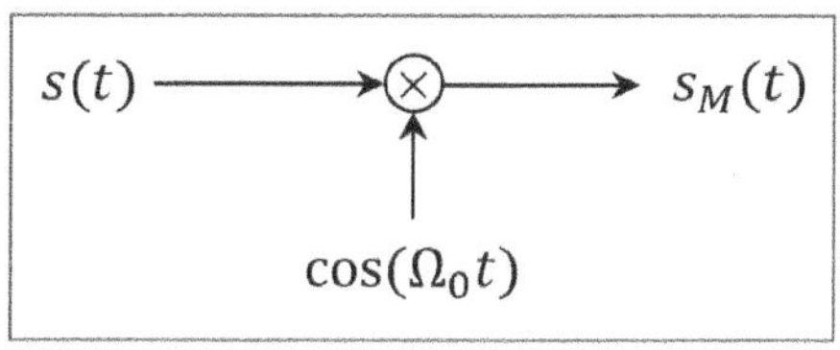

Figure P16.2.

(b) How much bandwidth does $S_M(\Omega)$ use compared to $S(\Omega)$? (*Note*: Represent it as a ratio of the bandwidth of $S_M(\Omega)$ to the bandwidth of $S(\Omega)$)

(c) To overcome the disadvantage observed in Question (b), the signal is transmitted according to the modulation scheme shown in Fig. P16.3. This is the single sideband (SSB) modulation scheme which takes advantage of the redundancy present in the spectrum of the real signal $s(t)$. Draw the spectrum of $S_{EM}(\Omega)$ and determine the bandwidth saving (if any) in the SSB scheme compared to the scheme in Question (a).

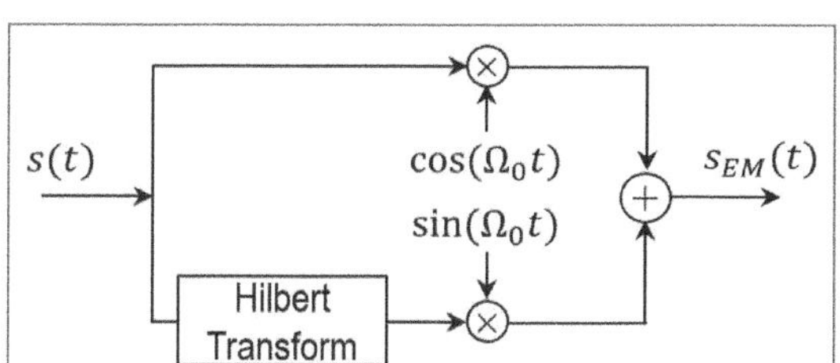

Figure P16.3.

DOI: 10.1561/9781680839791.ch17

Chapter 17

Power Spectrum Estimation

In statistical signal processing and physics, the spectral density, power spectral density (PSD), or energy spectral density (ESD), is a positive real function of a frequency variable associated with a stationary stochastic process, or a deterministic function of time. They have dimensions of power per Hz, or energy per Hz, respectively. It is often called simply the spectrum of the signal. Intuitively, the spectral density captures the frequency content of a stochastic process and helps identify the periodicity in the process.

Power spectrum estimation methods can be divided generally into two classes: classical methods or non-parametric methods (based on the periodogram) and modern methods or parametric methods (based on models). In this chapter, we are going to discuss the following methods of power spectrum estimation:

- Classical methods or Non-parametric methods
  - Periodogram approach,
  - Bartlett Method, Blackman & Tukey method, and
  - Welch Method.
- Modern methods OR Parametric methods
  - AR (autoregressive),
  - MA (moving average), and
  - ARMA (autoregressive and moving average)

Also, topics such as, DFT for spectrum estimation, windowing for spectrum estimation, time and frequency resolutions and spectral smoothing, are also dealt with in detail.

One of the major problems in spectrum estimation is the finite length of the available data, which will be discussed in the following section.

17.1 Estimation of Spectra from Finite Data

The availability of data only for a finite length is a major limitation on the quality of the power spectrum estimate. When dealing with statistically stationary signals, longer data record results in better estimate of the spectra. For non-stationary signals, the length of the data selected is determined by the rapidity of the time variations of signal statistics.

One of the problems that we encounter with the classical power estimation methods using a finite-length data is the *distortion* of the spectrum from the *actual*. This problem occurs in the computation of spectra for both deterministic as well as random signal. First, we discuss the effects on a finite data of deterministic signal, which is easier to observe and then consider random signals.

17.1.1 Computation of the ESD

Let us consider the computation of the spectrum of a deterministic signal from a finite sequence of data. The sequence $x[n]$ is usually the result of sampling a continuous time signal $x_a(t)$ at some uniform sampling rate f_s. Our objective is to obtain an estimate of the true spectrum from a finite $x[n]$.

Recall that if $x_a(t)$ is a finite energy signal, that is,

$$E = \int_{-\infty}^{\infty} |x_a(t)|^2 dt \quad < \infty \tag{17.1}$$

then its Fourier transform exists and is given by,

$$X_a(f) = \int_{-\infty}^{\infty} x_a(t) e^{-j2\pi f t} dt. \tag{17.2}$$

From Parseval's theorem, we have

$$E = \int_{-\infty}^{\infty} |x_a(t)|^2 dt = \int_{-\infty}^{\infty} |X_a(f)|^2 df. \tag{17.3}$$

The quantity $|X_a(f)|^2$ represents the distribution of the signal energy as a function of the frequency, and hence is called the energy spectrum density (ESD) of the signal, denoted as $S_{xx}(f)$, which is given as,

$$S_{xx}(f) = |X_a(f)|^2. \tag{17.4}$$

The total energy (E) in the signal is simply the integral of $S_{xx}(f)$ over all f.

It is also interesting to note that $S_{xx}(f)$ can be viewed as the Fourier transform (FT) of another function $r_{xx}(\tau)$, the *autocorrelation function* of the finite energy signal $x_a(t)$, defined as

$$r_{xx}(\tau) = \int_{-\infty}^{\infty} x_a^*(t)x_a(t+\tau)d\tau. \tag{17.5}$$

Indeed, it easily follows that,

$$\int_{-\infty}^{\infty} r_{xx}(\tau)e^{-j2\pi f\tau}d\tau = S_{xx}(f) = |X_a(f)|^2 \tag{17.6}$$

so that $r_{xx}(\tau)$ and $S_{xx}(f)$ are a Fourier transform pairs.

Now suppose that we compute the ESD of the signal $x_a(t)$ from its samples $(x[n], \ -\infty < n < \infty)$ taken at the rate f_s ($f_s > 2B$, B is the bandwidth of the signal in Hz). We have the FT,

$$X(\omega) = \sum_{n=-\infty}^{\infty} x[n]e^{-j\omega n} \tag{17.7}$$

or, equivalently,

$$X(f) = \sum_{n=-\infty}^{\infty} x[n]e^{-j2\pi fnT_s}. \tag{17.8}$$

Recall that $X(f)$ can be expressed in terms of $X_a(f)$ as

$$X(f) = f_s \sum_{k=-\infty}^{\infty} X_a(f-kf_s). \tag{17.9}$$

In the absence of aliasing, within the fundamental frequency range $|f| \leq f_s/2$, we have

$$X(f) = f_s X_a(f), \quad |f| \leq f_s/2. \tag{17.10}$$

Hence the spectrum of the sampled signal is identical to the spectrum of the analog signal. As a result, the ESD of the sampled signal is

$$S_{xx}(f) = |X(f)|^2 = f_s^2|X_a(f)|^2. \tag{17.11}$$

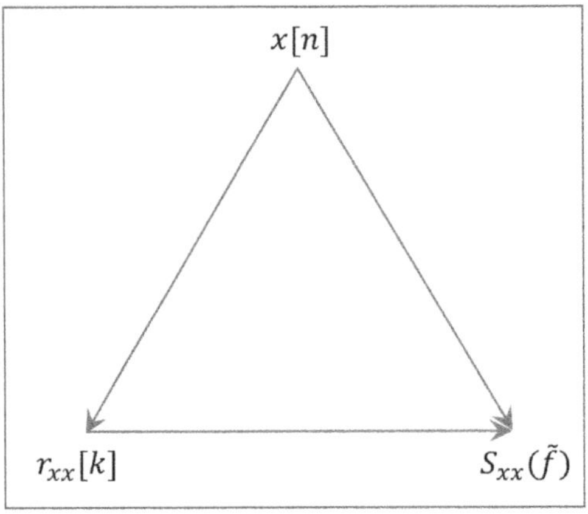

Figure 17.1. Two ways of estimation of the ESD.

We can proceed further by noting that the autocorrelation of the sampled signal defined by,

$$r_{xx}[k] = \sum_{n=-\infty}^{\infty} x^*[n]x[n+k] \tag{17.12}$$

has the Fourier transform (Wiener-Khinchin theorem)

$$S_{xx}(\tilde{f}) = \sum_{k=-\infty}^{\infty} r_{xx}[k]e^{-j2\pi k\tilde{f}}, \tag{17.13}$$

where $\tilde{f} = \frac{f}{f_s}$, f_s is the sampling frequency. Hence the ESD can be obtained by the Fourier transform of the autocorrelation of sequence $x[n]$. So, there are two distinct methods for computing the ESD of the signal $x_a(t)$ from its samples $x[n]$ (See Fig. 17.1). One is the direct method, where we compute the Fourier transform of $x[n]$ and then

$$S_{xx}(\tilde{f}) = |X(\tilde{f})|^2 = \left| \sum_{n=-\infty}^{\infty} x[n]e^{-j2\pi \tilde{f}n} \right|^2. \tag{17.14}$$

The second approach is the indirect method, which requires two steps. First the estimation of the autocorrelation $r_{xx}[k]$ using $x[n]$ and then the Fourier transform of $r_{xx}[k]$ to obtain the ESD (Eq. (17.13)).

In practice, however, we have only finite duration sequence $x[n]$, $0 \leq n \leq N-1$, for computing the spectrum of the signal. In effect, limiting the duration of the sequence is equivalent to multiplying $x[n]$ by a rectangular window of length

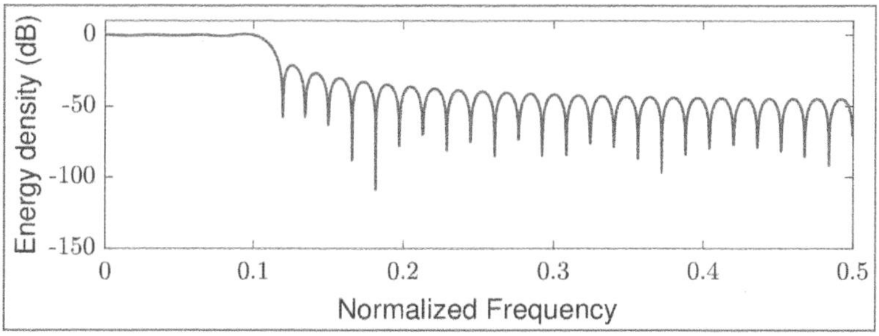

Figure 17.2. Spectrum obtained by convolving an $M = 61$ rectangular window with the ideal low pass spectrum in the example.

N. Thus we have,

$$\tilde{x}[n] = x[n]w[n] = \begin{cases} x[n] & \text{for } 0 \le n \le N-1 \\ 0 & \text{otherwise.} \end{cases}$$

Recall that *multiplication of two sequences in the time domain is equivalent to the convolution of their spectra in the frequency domain.* Hence,

$$\tilde{X}(\tilde{f}) = X(\tilde{f}) \otimes W(\tilde{f}) = \int_{-1/2}^{1/2} X(\alpha)W(\tilde{f} - \alpha)d\alpha. \tag{17.15}$$

Recall from our discussion on filter design (Section 10.3.2) that convolution of the window function $W(\tilde{f})$ with $X(\tilde{f})$ smoothes the spectrum, provided that the spectrum of $W(\tilde{f})$ is relatively narrow compared to $X(\tilde{f})$. But this condition implies that the window $w[n]$ be sufficiently long (i.e., N must be large). Moreover, even if $W(\tilde{f})$ is narrow compared to $X(\tilde{f})$, its convolution with the side-lobes of $W(\tilde{f})$ results in side-lobe energy (spectral leakage) in $\tilde{X}(\tilde{f})$, in frequency bands where the actual spectrum is zero.

Example: A signal with spectrum

$$X(\tilde{f}) = \begin{cases} 1 & \text{for } |\tilde{f}| \le 0.1 \\ 0 & \text{otherwise.} \end{cases}$$

is convolved with a rectangular window of length $N = 61$. The spectra $\tilde{X}(\tilde{f})$ is shown in Fig. 17.2. The width of the main lobe of the rectangular window is $\triangle\omega = 4\pi/61$ or $\triangle\tilde{f} = 2/61$, which is narrow compared to $X(\tilde{f})$. The energy leakage due to the high side lobe levels as well as broadening (smearing) of $X(\tilde{f})$ outside the range ($|\tilde{f}| \le 0.1$) is clearly seen.

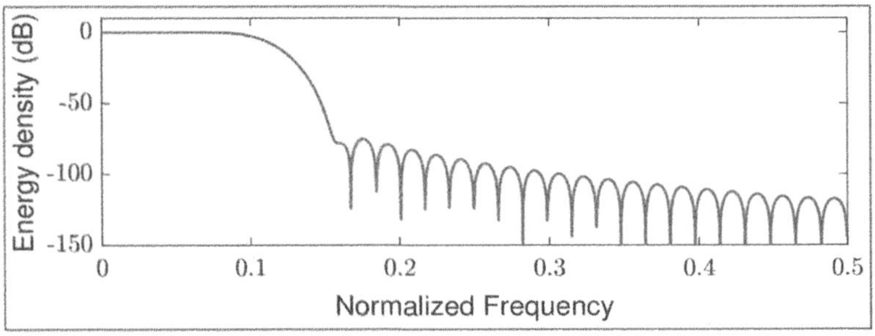

Figure 17.3. Spectrum obtained by convolving an $M = 61$ Blackman window with the ideal low pass spectrum in the example.

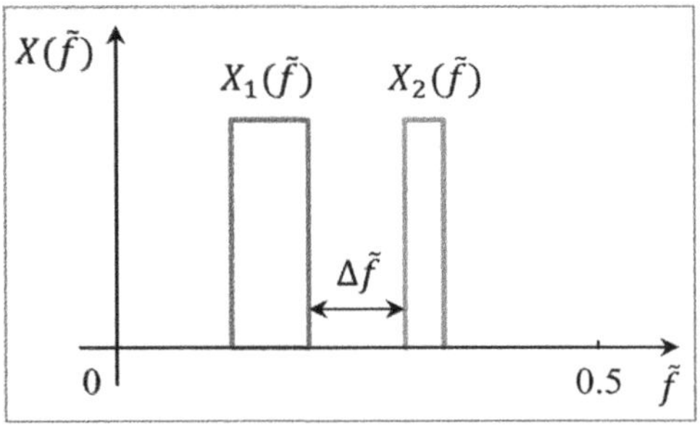

Figure 17.4. Two narrow-band signal spectra.

If we use another window (for example, the Blackman window) instead of the rectangular window, the spectrum $\tilde{X}(\tilde{f})$ will be as sown in Fig. 17.3. Here, the leakage has been reduced considerably, however, the spectrum has been widened by almost 50%.

So, the spectral leakage can be reduced by choosing windows that have low side-lobes. Though such window functions reduce the side-lobes, they result in an increase in smoothing or broadening the spectral characteristics of $X(\tilde{f})$. This will be a problem in cases where we wish to resolve signals with closely spaced frequency components. For example, the signals with spectral characteristic $X(\tilde{f}) = X_1(\tilde{f}) + X_2(\tilde{f})$, as shown in Fig. 17.4 cannot be resolved as two separate signals unless the width of the window function is significantly narrower than the frequency separation $\Delta\tilde{f}$.

It is clear that the ESD of the windowed sequence $\tilde{x}[n]$ is an approximation of the desired spectrum of the sequence $x[n]$. The spectral density obtained from $\tilde{x}[n]$ is

$$S_{\tilde{x}\tilde{x}}(\tilde{f}) = |\tilde{X}(\tilde{f})|^2 = \left| \sum_{n=0}^{N-1} \tilde{x}[n] e^{-j2\pi \tilde{f} n} \right|^2. \tag{17.16}$$

The spectrum given by Eq. (17.16) can be computed numerically at a set of N frequency points by means of the DFT. Thus (for $k = 0, 1, \ldots, N-1$),

$$\tilde{X}[k] = \sum_{n=0}^{N-1} \tilde{x}[n] e^{-j2\pi\left(\frac{k}{N}\right)n}. \tag{17.17}$$

Then

$$|\tilde{X}[k]|^2 = S_{\tilde{x}\tilde{x}}(\tilde{f})|_{\tilde{f}=k/N} = S_{\tilde{x}\tilde{x}}\left(\frac{k}{N}\right) \tag{17.18}$$

and hence

$$S_{\tilde{x}\tilde{x}}\left(\frac{k}{N}\right) = \left|\sum_{n=0}^{N-1} \tilde{x}[n] e^{-j2\pi\left(\frac{k}{N}\right)n}\right|^2, \tag{17.19}$$

which is a distorted version of the true spectrum $S_{xx}\left(\frac{k}{N}\right)$.

In the next section, we are going to discuss the power spectrum estimation of stationary random/stochastic signals.

17.2 Estimation of the Autocorrelation and Power Spectrum of Random Signals: The Periodogram

The finite energy signals considered in the previous section has Fourier transform possible and are characterized by the ESD. On the other hand, *the stationary random process do not have finite energy and hence do not possess a Fourier transform.* Such signals have finite average power and hence are characterized by the power spectral density (PSD). If $x(t)$ is a stationary random process, its autocorrelation function is,

$$\gamma_{xx}(\tau) = E[x^*(t)x(t+\tau)]. \tag{17.20}$$

The PSD is the Fourier transform of the autocorrelation function[1]

$$\Gamma_{xx}(f) = \int_{-\infty}^{\infty} \gamma_{xx}(\tau) e^{-j2\pi f\tau} d\tau. \tag{17.21}$$

In practice, we may have only one realization of the random process from which we have to estimate the power spectrum of the process. Also, we do not know $\gamma_{xx}(\tau)$ and as a result, Eq. (17.21) cannot be computed to obtain $\Gamma_{xx}(f)$. On

1. By Wiener-Khintchin theorem.

the other hand, from one realization of the random process, we can compute the *time-averaged* autocorrelation function,

$$\tilde{\gamma}_{xx}(\tau) = \frac{1}{2T_0}\int_{-T_0}^{T_0} x^*(t)x(t+\tau)dt, \tag{17.22}$$

where $2T_0$ is the observation interval. If the stationary random process is *ergodic* in the mean and autocorrelation function, then

$$\begin{aligned}\gamma_{xx}(\tau) &= \lim_{T_0\to\infty} \tilde{\gamma}_{xx}(\tau)\\ &= \lim_{T_0\to\infty} \frac{1}{2T_0}\int_{-T_0}^{T_0} x^*(t)x(t+\tau)dt.\end{aligned} \tag{17.23}$$

Furthermore, the Fourier transform of $\tilde{\gamma}_{xx}(\tau)$ provides an estimate $\tilde{\Gamma}_{xx}(f)$ of the PSD, that is,

$$\begin{aligned}\tilde{\Gamma}_{xx}(f) &= \int_{-T_0}^{T_0} \tilde{\gamma}_{xx}(\tau)e^{-j2\pi f\tau}d\tau\\ &= \frac{1}{2T_0}\int_{-T_0}^{T_0}\left[\int_{-T_0}^{T_0} x^*(t)x(t+\tau)dt\right]e^{-j2\pi f\tau}d\tau\\ &= \frac{1}{2T_0}\left|\int_{-T_0}^{T_0} x(t)e^{-j2\pi ft}dt\right|^2.\end{aligned} \tag{17.24}$$

The actual PSD is the expected value of $\tilde{\Gamma}_{xx}(f)$ in the limiting case,

$$\Gamma_{xx}(f) = \lim_{T_0\to\infty} E\left[\frac{1}{2T_0}\left|\int_{-T_0}^{T_0} x(t)e^{-j2\pi ft}dt\right|^2\right]. \tag{17.25}$$

From Eq. (17.22) and Eq. (17.24), we see that there are two possible approaches to compute $\tilde{\Gamma}_{xx}(f)$, the direct method as given by Eq. (17.24) and the indirect method, in which first obtain $\tilde{\gamma}_{xx}(\tau)$ (Eq. (17.22)) and then compute its Fourier transform to find the PSD.

We shall now consider the estimation of the power spectrum from samples $x[n]$ $(0 \le n \le N-1)$ of a single realization of the random process. We assume that the signal is sampled at $f_s > 2B$, where B is the highest frequency contained in the PSD of the random process.

The time-averaged autocorrelation sequence is (unbiased estimate)

$$\tilde{\gamma}'_{xx}[m] = \frac{1}{N-m}\sum_{n=0}^{N-m-1} x^*[n]x[n+m], \quad m = 0, 1, \ldots, N-1 \tag{17.26}$$

and for negative values of m, we have $\tilde{\gamma}'_{xx}[m] = [\tilde{\gamma}'_{xx}[-m]]^*$. Then we compute the Fourier transform

$$\tilde{\Gamma}'_{xx}(\tilde{f}) = \sum_{m=-N+1}^{N-1} \tilde{\gamma}'_{xx}[m]e^{-j2\pi \tilde{f} m}. \tag{17.27}$$

The normalization factor $N - m$ in Eq. (17.26) results in an estimate with mean value

$$\begin{aligned} E[\tilde{\gamma}'_{xx}[m]] &= \frac{1}{N-m} \sum_{n=0}^{N-m-1} E[x^*[n]x[n+m]] \\ &= \gamma_{xx}[m], \end{aligned} \tag{17.28}$$

where $\gamma_{xx}[m]$ is the *true* (statistical) autocorrelation of the sequence $x[n]$. Hence, $\tilde{\gamma}'_{xx}[m]$ is an unbiased estimate and its variance is approximately,[2]

$$\text{var}[\tilde{\gamma}'_{xx}[m]] \approx \frac{N}{[n-m]^2} \sum_{n=-\infty}^{\infty} \Big[|\gamma_{xx}[n]|^2 + \gamma^*_{xx}[n-m]\gamma_{xx}[n+m] \Big]. \tag{17.29}$$

Clearly, $\lim_{N\to\infty} \text{var}[\tilde{\gamma}'_{xx}[m]] = 0$ provided that $\sum_{n=-\infty}^{\infty} |\gamma_{xx}[n]|^2 < \infty$. Since $E[\tilde{\gamma}'_{xx}[m]] = \gamma_{xx}[m]$ and the variance of the estimate converges to zero as $N \to \infty$, the estimate $\tilde{\gamma}'_{xx}[m]$ is said to be *consistent.*

For large values of m, the estimate $\tilde{\gamma}'_{xx}[m]$ in Eq. (17.26) has a large variance, especially when m approaches N. Because in this case, only fewer number of data points are available for the estimate. As an alternative to Eq. (17.26), one could use

$$\tilde{\gamma}_{xx}[m] = \begin{cases} \frac{1}{N} \sum_{n=0}^{N-m-1} x^*[n]x[n+m] & \text{for } 0 \le m \le N-1 \\ \\ \frac{1}{N} \sum_{n=|m|}^{N-1} x^*[n]x[n+m] & \text{for } m = -1, -2, \ldots, 1-N \end{cases} \tag{17.30}$$

2. This result is given by Jenkins and Watts (*Spectral Analysis and its Applications*, Holden-Day San Francisco, 1968).

which *has a bias of* $\frac{|m|\gamma_{xx}[m]}{N}$, since its mean value is

$$E[\tilde{\gamma}_{xx}[m]] = \frac{1}{N}\sum_{n=0}^{N-m-1} E[x^*[n]x[n+m]]$$
$$= \frac{N-|m|}{N}\gamma_{xx}[m] = \left(1 - \frac{|m|}{N}\right)\gamma_{xx}[m]. \tag{17.31}$$

However, this estimate has a smaller variance given approximately as (because $\frac{1}{N} \leq \frac{N}{[n-m]^2}$)

$$\text{var}[\tilde{\gamma}_{xx}[m]] \approx \frac{1}{N}\sum_{n=-\infty}^{\infty}\left[|\gamma_{xx}[n]|^2 + \gamma_{xx}^*[n-m]\gamma_{xx}[n+m]\right]. \tag{17.32}$$

We observe that $\tilde{\gamma}_{xx}[m]$ is *asymptotically unbiased*, that is $\lim_{N\to\infty} E[\tilde{\gamma}_{xx}[m]] = \gamma_{xx}[m]$ and its variance converges to zero as $N \to \infty$. Hence, the estimate $\tilde{\gamma}_{xx}[m]$ is also a *consistent estimate* of $\gamma_{xx}[m]$.

We shall use the estimate $\tilde{\gamma}_{xx}[m]$ (Eq. (17.30)) in our treatment of power spectrum estimation and the estimate of the PSD is

$$\tilde{\Gamma}_{xx}(\tilde{f}) = \sum_{m=-(N-1)}^{N-1} \tilde{\gamma}_{xx}[m]e^{-j2\pi\tilde{f}m}. \tag{17.33}$$

Substituting for $\tilde{\gamma}_{xx}[m]$ from Eq. (17.30), Eq. (17.33) becomes

$$\tilde{\Gamma}_{xx}(\tilde{f}) = \frac{1}{N}\left|\sum_{n=0}^{N-1} x[n]e^{-j2\pi\tilde{f}n}\right|^2 = \frac{1}{N}|X(\tilde{f})|^2, \tag{17.34}$$

where $X(\tilde{f})$ is the Fourier transform of $x[n]$. This well-known form of the PSD estimate is called the *periodogram.*[3]

An illustration of the periodogram approach to the PSD estimation of a data $x[n]$ of length N samples is provided in Fig. 17.5.

From Eq. (17.33), the average value of the periodogram estimate $\tilde{\Gamma}_{xx}(\tilde{f})$ is

$$E[\tilde{\Gamma}_{xx}(\tilde{f})] = E\left[\sum_{m=-(N-1)}^{N-1} \tilde{\gamma}_{xx}[m]e^{-j2\pi\tilde{f}m}\right]$$

3. Introduced by Schuster (1898) to detect and measure “hidden periodicities” of the data.

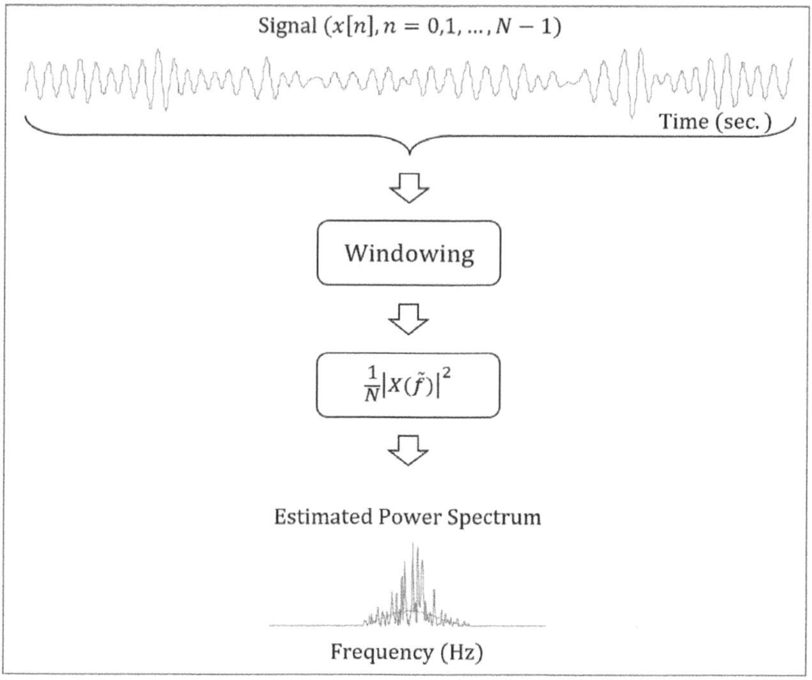

Figure 17.5. Periodogram approach to the estimation of PSD.

$$= \sum_{m=-(N-1)}^{N-1} E[\tilde{\gamma}_{xx}[m]]e^{-j2\pi \tilde{f} m}$$

$$= \sum_{m=-(N-1)}^{N-1} \left(1 - \frac{|m|}{N}\right) \gamma_{xx}[m]e^{-j2\pi \tilde{f} m} \tag{17.35}$$

The interpretation that we give to Eq. (17.35) is that the mean of the estimated spectrum is the Fourier transform of the windowed autocorrelation function,

$$\tilde{\tilde{\gamma}}_{xx}[m] = \left(1 - \frac{|m|}{N}\right) \gamma_{xx}[m], \tag{17.36}$$

where the window function is the triangular (Bartlett) window. Hence

$$E[\tilde{\Gamma}_{xx}(\tilde{f})] = \sum_{m=-\infty}^{\infty} \tilde{\tilde{\gamma}}_{xx}[m]e^{-j2\pi \tilde{f} m} = \int_{-1/2}^{1/2} \Gamma_{xx}(\alpha) W_B(\tilde{f} - \alpha) d\alpha \tag{17.37}$$

where $W_B(\tilde{f})$ is the spectral characteristics of the Bartlett window. The relation in Eq. (17.37) shows that the mean of the estimated spectrum is the convolution of the true PSD $\Gamma_{xx}(\tilde{f})$ with $W_B(\tilde{f})$. Hence, the mean of the estimated power spectrum is a smoothed version of the true spectrum and it suffers the spectral leakage problems.

We observe that the estimated spectrum is asymptotically unbiased, that is,

$$\lim_{N\to\infty} E\left[\sum_{m=-(N-1)}^{N-1} \tilde{\gamma}_{xx}[m]e^{-j2\pi\tilde{f}m}\right] = \sum_{m=-\infty}^{\infty} \gamma_{xx}[m]e^{-j2\pi\tilde{f}m} = \Gamma_{xx}(\tilde{f})$$

However, in general, the variance of the estimate $\tilde{\Gamma}_{xx}(\tilde{f})$ does not decay to zero as $N \to \infty$. For example, when the data sequence is Gaussian random process, the variance is easily shown to be [5],

$$var[\tilde{\Gamma}_{xx}(\tilde{f})] = \Gamma_{xx}^2(\tilde{f})\left[1 + \left(\frac{\sin 2\pi\tilde{f}N}{N\sin 2\pi\tilde{f}}\right)^2\right], \tag{17.38}$$

which in the limit as $N \to \infty$, becomes,

$$\lim_{N\to\infty} var[\tilde{\Gamma}_{xx}(\tilde{f})] = \Gamma_{xx}^2(\tilde{f}). \tag{17.39}$$

Hence, we conclude that the periodogram is not a consistent estimate of the true PSD (i.e., it does not converge to the true PSD).

17.2.1 Summary of the Periodogram based PSD Estimation

- The estimated autocorrelation $\tilde{\gamma}_{xx}[m]$ is a consistent estimate of the true autocorrelation function.
- $\tilde{\Gamma}_{xx}(\tilde{f})$, the periodogram which is the Fourier transform of $\tilde{\gamma}_{xx}[m]$ is not a consistent estimate of the true PSD.
- It is observed that $\tilde{\Gamma}_{xx}(\tilde{f})$ is an asymptotically unbiased estimate of $\Gamma_{xx}(\tilde{f})$, but for a finite duration sequence, the mean value of $\tilde{\Gamma}_{xx}(\tilde{f})$ contains a bias, which from Eq. (17.37) is evident as a distortion of the true power spectrum.
- The estimated spectrum suffers from the smoothing effects and the leakage problem thus limiting the ability to resolve closely spaced frequencies.

17.2.2 Use of DFT in Power Spectrum Estimation

The estimated ESD, $S_{\tilde{x}\tilde{x}}(\tilde{f})$ and the periodogram, $\tilde{\Gamma}_{xx}(\tilde{f})$, provided in Eqs. (17.16) and (17.34), respectively can be computed by the use of DFT and FFT algorithms. If we have N data points, the DFT computation provides samples of the periodogram

$$\tilde{\Gamma}_{xx}\left(\frac{k}{N}\right) = \frac{1}{N}\left|\sum_{n=0}^{N-1} x[n]e^{-j2\pi\left(\frac{k}{N}\right)n}\right|^2, \quad k = 0, 1, \ldots, N-1 \qquad (17.40)$$

at frequencies $\tilde{f}_k = \frac{k}{N}$.

In practice, however, such a sparse sampling of the spectrum does not provide a very good representation of the continuous spectrum estimate $\tilde{\Gamma}_{xx}(\tilde{f})$. By padding zeros, the effective length of the sequence is increased and the spectrum at other frequencies are also can be estimated. Thus, if we increase the length of the sequence to L points by zero padding, the L-point DFT is

$$\tilde{\Gamma}_{xx}\left(\frac{k}{L}\right) = \frac{1}{N}\left|\sum_{n=0}^{N-1} x[n]e^{-j2\pi\left(\frac{k}{L}\right)n}\right|^2, \quad k = 0, 1, \ldots, L-1. \qquad (17.41)$$

It may be noted that increasing the data length by zero padding ($L > N$) and computing DFT does not improve the frequency resolution[4] in the spectral estimate. It helps us to interpolate the spectral values at more frequency points.

Example: A sequence of $N = 16$ samples is obtained by sampling an analog signal consisting of two frequency components. The resulting discrete-time signal is

$$x[n] = \sin(2\pi(0.145)n) + \cos(2\pi(0.145 + \triangle\tilde{f})n), \quad n = 0, 1, \ldots, 15, \qquad (17.42)$$

where $\triangle\tilde{f}$ is the frequency separation. Evaluate the PSD, $\tilde{\Gamma}_{xx}(\tilde{f}) = \frac{1}{N}|X(\tilde{f})|^2$ at the frequencies $\tilde{f}_k = k/L, \ k = 0, 1, \ldots, L-1$ for $L = 16, 32, 64$, and 256 for values of $\triangle\tilde{f} = 0.06$ and 0.01.

The evaluated PSD for the given cases are illustrated in Figs.17.6 and 17.7. In Fig. 17.6, as you can see, the frequency separation between the two sinusoids is $\triangle\tilde{f} = 0.06$ and hence the DFT spectrum clearly show two peaks closer to the actual frequencies of the sinusoids. Increasing the number of points in the DFT does not improve the frequency resolvability, however, the spectrum becomes more smoother. This is pretty clear in Fig. 17.7, where the frequency separation between the two sinusoids is only $\triangle\tilde{f} = 0.01$. Here, the DFT spectrum fails to resolve the two frequencies even though the DFT lengths has been increased to 256. The spectrum, however, becomes smoother as in the first case.

4. Frequency resolution in the spectral estimate is determined by the data length N.

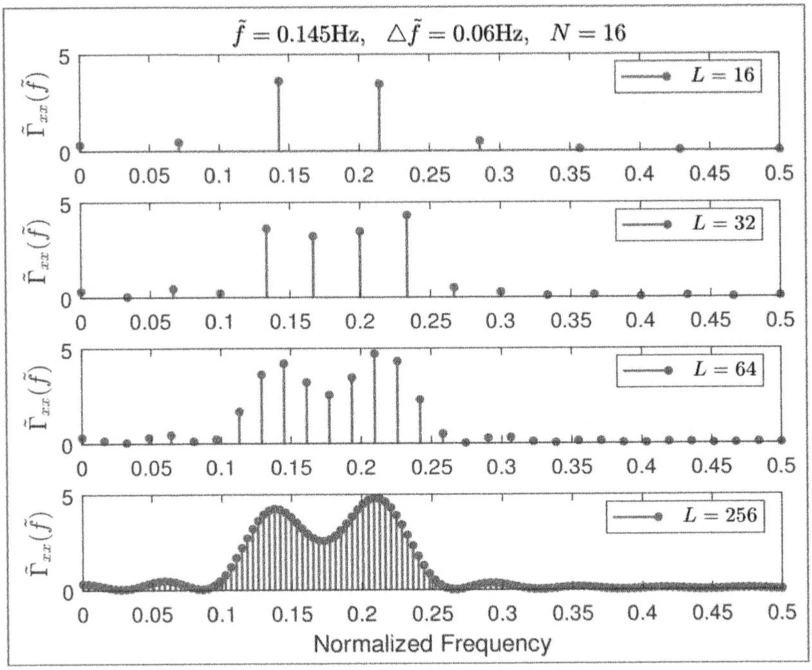

Figure 17.6. Spectra of two sinusoids with frequency separation $\triangle\tilde{f} = 0.06$.

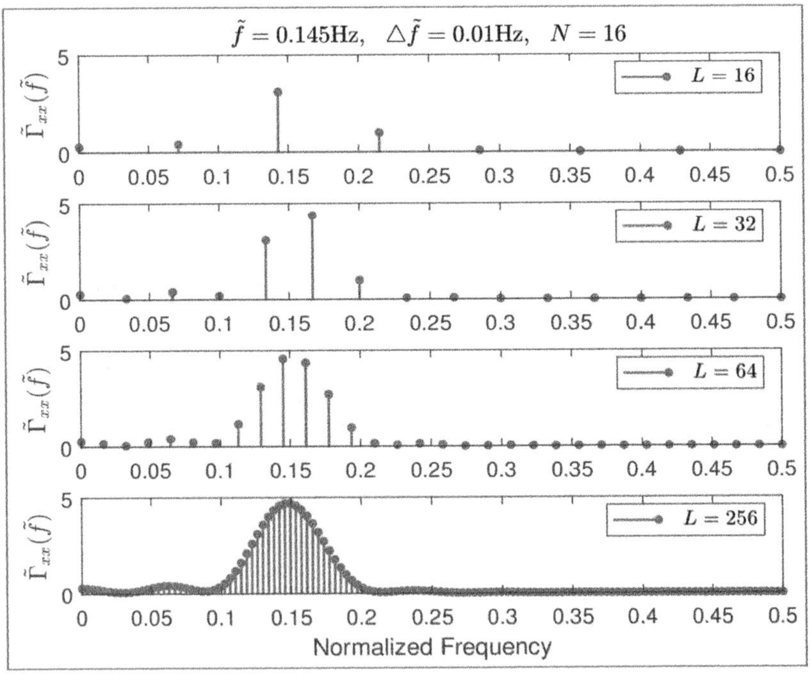

Figure 17.7. Spectra of two sinusoids with frequency separation $\triangle\tilde{f} = 0.01$.

17.3 Non-parametric Methods for Power Spectrum Estimation

These methods are the classical methods developed by Bartlett (1948), Blackman and Tukey (1958), and Welch (1967). These methods make no assumption about how the data were generated and hence the name *non-parametric.*

Since the estimates are based entirely on finite record of data (N), the frequency resolution of these methods is, at best, equal to the spectral width of the rectangular window of length N, which is $\approx 1/N$ at the $-3dB$ points.

First, we describe the estimates and then derive the mean and variance of each.

17.3.1 The Bartlett Method: Averaging the Periodograms

For reducing the variance in the periodogram approach, Bartlett suggested this method and it involves three steps.

- The N-point sequence is subdivided into K non-overlapping segments of length M. That is,

$$x_i[n] = x[n + iM], \tag{17.43}$$

where, $i = 0, 1, \ldots, K-1$ and $n = 0, 1, \ldots, M-1$.

- Compute the periodogram for each segment.

$$\tilde{\Gamma}_{xx}^{(i)}(\tilde{f}) = \frac{1}{M}\left|\sum_{n=0}^{M-1} x_i[n]e^{-j2\pi \tilde{f} n}\right|^2, \quad i = 0, 1, \ldots, K-1 \tag{17.44}$$

- Average the periodograms for K segments to obtain the Bartlett power spectral estimate

$$\tilde{\Gamma}_{xx}^{B}(\tilde{f}) = \frac{1}{K}\sum_{i=0}^{K-1} \tilde{\Gamma}_{xx}^{(i)}(\tilde{f}). \tag{17.45}$$

An illustration of these steps in Bartlett method of power spectrum estimation is provided in Fig. 17.8.

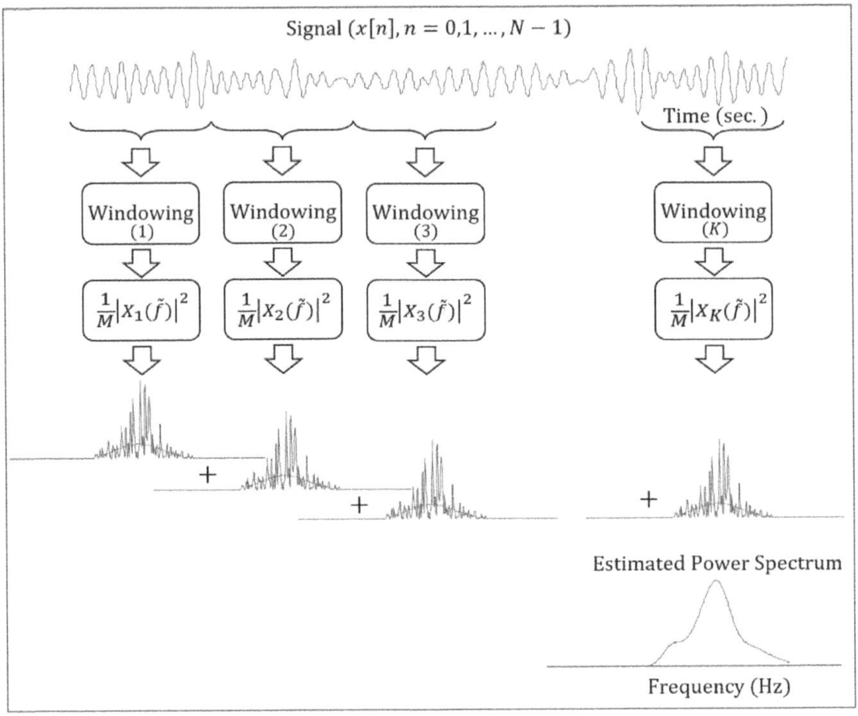

Figure 17.8. Bartlett method to the estimation of PSD.

Now, the statistical properties of the power spectrum estimates can be easily estimated.

Mean:

$$E[\tilde{\Gamma}^{B}_{xx}(\tilde{f})] = \frac{1}{K}\sum_{i=0}^{K-1} E[\tilde{\Gamma}^{(i)}_{xx}(\tilde{f})] \tag{17.46}$$

$$= E[\tilde{\Gamma}^{(i)}_{xx}(\tilde{f})]. \tag{17.47}$$

From Eq. (17.35) and Eq. (17.37), we have the expected value of the single periodogram as

$$E[\tilde{\Gamma}^{(i)}_{xx}(\tilde{f})] = \sum_{m=-(M-1)}^{M-1}\left(1-\frac{|m|}{M}\right)\gamma_{xx}[m]e^{-j2\pi\tilde{f}m}$$

$$= \frac{1}{M}\int_{-1/2}^{1/2}\Gamma_{xx}(\alpha)\left(\frac{\sin\pi(\tilde{f}-\alpha)M}{\sin\pi(\tilde{f}-\alpha)}\right)^2 d\alpha \tag{17.48}$$

where, $W_B(\tilde{f}) = \frac{1}{M}\left(\frac{\sin \pi \tilde{f} M}{\sin \pi \tilde{f}}\right)^2$ is the frequency characteristics of the Bartlett window.[5]

From Eq. (17.48), we observe that the true spectrum is now convolved with $W_B(\tilde{f})$. The effect of reducing the length of the data from N to $M(=\frac{N}{K})$ results in a window whose spectral width has been increased by a factor of K. As a result, the frequency resolution has been reduced by a factor K. However, the advantage is the reduced variance as shown below.

Variance:

$$var[\tilde{\Gamma}^B_{xx}(\tilde{f})] = \frac{1}{K^2}\sum_{i=0}^{K-1} var[\tilde{\Gamma}^{(i)}_{xx}(\tilde{f})]$$
$$= \frac{1}{K} var[\tilde{\Gamma}^{(i)}_{xx}(\tilde{f})]. \quad (17.49)$$

If we use Eq. (17.38) in Eq. (17.49), we get,

$$var[\tilde{\Gamma}^B_{xx}(\tilde{f})] = \frac{1}{K}\Gamma^2_{xx}(\tilde{f})\left[1 + \left(\frac{\sin 2\pi \tilde{f} M}{M \sin 2\pi \tilde{f}}\right)^2\right]. \quad (17.50)$$

This means that the variance of the Bartlett power spectrum estimate is reduced by the factor K compared to the periodogram approach.

17.3.2 The Welch Method: Averaging Modified Periodogram

Welch made two modifications to the Bartlett method:

- Made the data segments to overlap, which can be represented as (for $n = 0, 1, \ldots, M-1$ and $i = 0, 1, \ldots, L-1$):

$$x_i[n] = x[n + iD] \quad (17.51)$$

  where iD is the starting point of the ith sequence. For example, if $D = M/2$, there is 50% overlap between successive data segments.
- Window the data segments prior to computing the periodogram. The result is a "*modified*" periodogram.

---

5. The Bartlett window is given by

$$w_B[n] = \begin{cases} 1 - \frac{|m|}{M} & \text{for } |m| \leq M-1 \\ 0 & \text{otherwise} \end{cases}$$

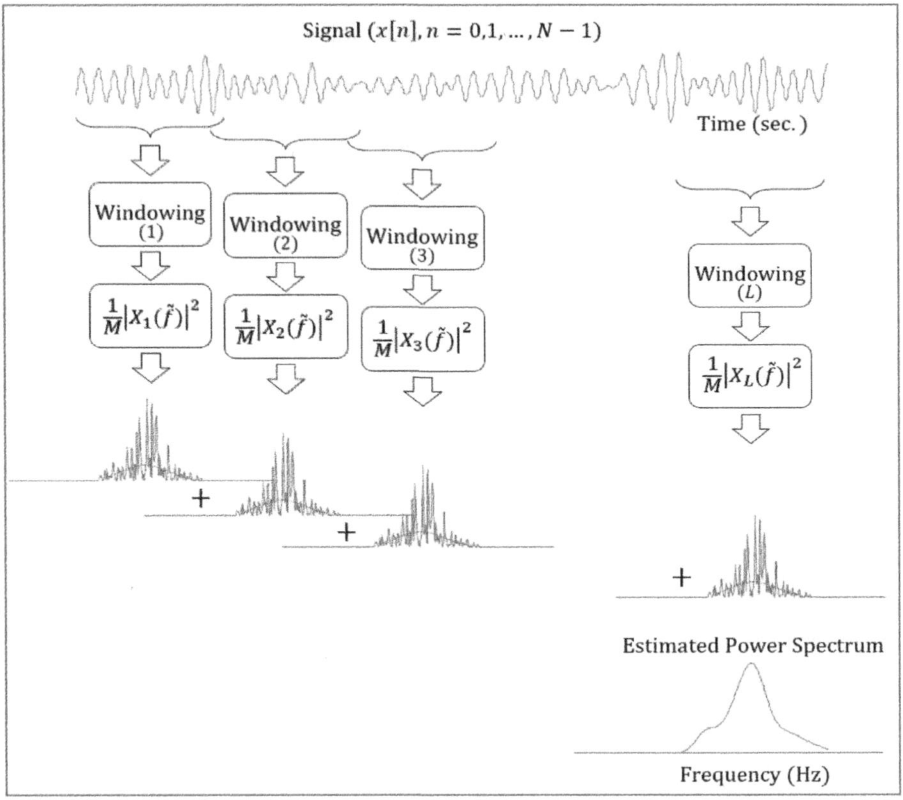

Figure 17.9. Welch's method to the estimation of PSD.

This is illustrated in Fig. 17.9.

The Welch power spectrum estimate is

$$\tilde{\Gamma}_{xx}^{W}(\tilde{f}) = \frac{1}{L}\sum_{i=0}^{L-1} \tilde{\tilde{\Gamma}}_{xx}^{(i)}(\tilde{f}) \tag{17.52}$$

where,

$$\tilde{\tilde{\Gamma}}_{xx}^{(i)}(\tilde{f}) = \frac{1}{MU}\left|\sum_{n=0}^{M-1} x_i[n]w[n]e^{-j2\pi \tilde{f} n}\right|^2, \quad i = 0, 1, \ldots, L-1 \tag{17.53}$$

where U is a normalization factor for the power in the window function and is selected as $U = \frac{1}{M}\sum_{n=0}^{M-1} w^2[n]$.

Mean:

$$E[\tilde{\Gamma}^{W}_{xx}(\tilde{f})] = \frac{1}{L}\sum_{i=0}^{L-1} E[\tilde{\tilde{\Gamma}}^{(i)}_{xx}(\tilde{f})] = E[\tilde{\tilde{\Gamma}}^{(i)}_{xx}(\tilde{f})]. \tag{17.54}$$

But the expected value of the modified periodogram is

$$E[\tilde{\tilde{\Gamma}}^{(i)}_{xx}(\tilde{f})] = \frac{1}{MU}\sum_{n=0}^{M-1}\sum_{m=0}^{M-1} w[n]w[m]E[x_i[n]x_i^*[m]]e^{-j2\pi\tilde{f}[n-m]} = \frac{1}{MU}\sum_{n=0}^{M-1}\sum_{m=0}^{M-1} w[n]w[m]\gamma_{xx}[n-m]e^{-j2\pi\tilde{f}[n-m]} \tag{17.55}$$

Using $\gamma_{xx}[n] = \int_{-1/2}^{1/2}\Gamma_{xx}(\alpha)e^{j2\pi\alpha n}d\alpha$ in Eq. (17.55), we get,

$$E[\tilde{\tilde{\Gamma}}^{(i)}_{xx}(\tilde{f})] = \frac{1}{MU}\int_{-1/2}^{1/2}\Gamma_{xx}(\alpha)\,[bb]\,d\alpha = \int_{-1/2}^{1/2}\Gamma_{xx}(\alpha)W(\tilde{f}-\alpha)d\alpha \tag{17.56}$$

where,

$$bb = \sum_{n=0}^{M-1}\sum_{m=0}^{M-1} w[n]w[m]e^{-j2\pi[n-m](\tilde{f}-\alpha)} \tag{17.57}$$

and by definition $W(\tilde{f}) = \frac{1}{MU}\left|\sum_{n=0}^{M-1} w[n]e^{-j2\pi\tilde{f}n}\right|^2$. The normalization factor U ensures that $\int_{-1/2}^{1/2} W(\tilde{f})d\tilde{f} = 1$.

Variance:

$$var[\tilde{\Gamma}^{W}_{xx}(\tilde{f})] = \frac{1}{L^2}\sum_{i=0}^{L-1}\sum_{j=0}^{L-1} E[\tilde{\tilde{\Gamma}}^{(i)}_{xx}(\tilde{f})\tilde{\tilde{\Gamma}}^{(j)}_{xx}(\tilde{f})] - \{E[\tilde{\Gamma}^{W}_{xx}(\tilde{f})]\}^2 \tag{17.58}$$

In the case of no overlap between successive data segments (i.e., $L = K$), Welch has shown that

$$var[\tilde{\Gamma}^{W}_{xx}(\tilde{f})] = \frac{1}{L}var[\tilde{\tilde{\Gamma}}^{(i)}_{xx}(\tilde{f})] \approx \frac{1}{L}\Gamma^2_{xx}(\tilde{f}). \tag{17.59}$$

In the case of 50% overlap between the data segments ($L = 2K$), the variance of the Welch power spectrum estimate with the Bartlett window is shown to be [5]

$$var[\tilde{\Gamma}_{xx}^{W}(\tilde{f})] \approx \frac{9}{8L}\Gamma_{xx}^{2}(\tilde{f}). \tag{17.60}$$

17.3.3 The Blackman and Tukey Method: Smoothing the Periodogram

In this method, the sample autocorrelation sequence is windowed[6] first and then Fourier transformed to give the estimate of the power spectrum. Thus, the Blackman-Tukey (BT) estimate is

$$\tilde{\Gamma}_{xx}^{BT}(\tilde{f}) = \sum_{m=-(M-1)}^{M-1} \tilde{\gamma}_{xx}[m]w[m]e^{-j2\pi\tilde{f}m} \tag{17.61}$$

where the window function $w[n]$ has length $2M - 1$ and zero for $|m| \geq M$. With this definition of $w[n]$, the limits on the sum in Eq. (17.61) can be extended to $(-\infty, \infty)$. Hence the frequency domain equivalent expression for Eq. (17.61) is the convolution integral

$$\tilde{\Gamma}_{xx}^{BT}(\tilde{f}) = \int_{-1/2}^{1/2} \tilde{\Gamma}_{xx}(\alpha)W(\tilde{f} - \alpha)d\alpha \tag{17.62}$$

where $\tilde{\Gamma}_{xx}(\tilde{f})$ is the periodogram. It is clear from Eq. (17.62) that the effect of windowing the autocorrelation is to *smooth the periodogram estimate*, thus decreasing the variance in the estimate at the cost of reducing the resolution. An illustration of the proposed method is provided in Fig. 17.10.

The window function $w[n]$ should be symmetric (even) about $n = 0$ to ensure that the estimate of the power spectrum is real. Furthermore, it is desirable to select the window spectrum to be non-negative, that is, $W(\tilde{f}) \geq 0, \quad |\tilde{f}| \leq 1/2$. This condition ensures that the estimated spectrum $\tilde{\Gamma}_{xx}^{BT}(\tilde{f}) \geq 0$ for $|\tilde{f}| \leq 1/2$, which is a desirable property of any spectral estimate.

Mean:

$$E[\tilde{\Gamma}_{xx}^{BT}(\tilde{f})] = \int_{-1/2}^{1/2} E[\tilde{\Gamma}_{xx}(\alpha)]W(\tilde{f} - \alpha)d\alpha. \tag{17.63}$$

6. The rationale for windowing $\tilde{\gamma}_{xx}[m]$ is that, for large lags, the estimation are less reliable because a smaller number $(N - m)$ of data points enter the estimate. For $m \to N$, the variance of $\tilde{\gamma}_{xx}[m]$ is very high and hence the estimates should be given less weighting.

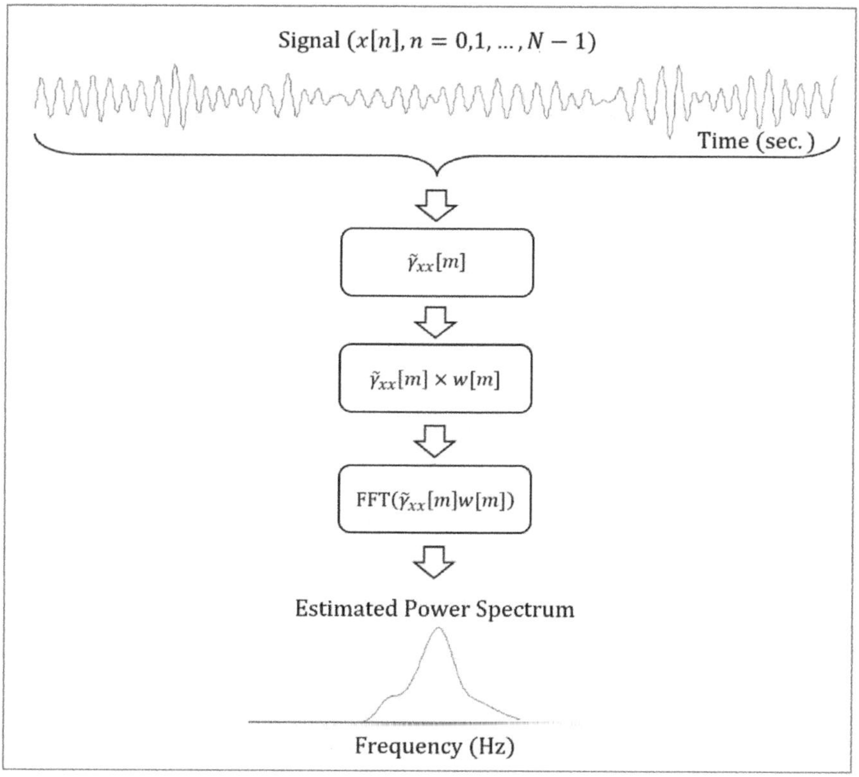

Figure 17.10. Blackman and Tukey method to the estimation of PSD.

Substituting for $E[\tilde{\Gamma}_{xx}(\alpha)]$ (from Eq. (17.37)) in the above equation, we get the double convolution integral

$$E[\tilde{\Gamma}_{xx}^{BT}(\tilde{f})] = \int_{-1/2}^{1/2}\int_{-1/2}^{1/2} \Gamma_{xx}(\theta)W_B(\alpha-\theta)W(\tilde{f}-\alpha)d\alpha d\theta. \qquad (17.64)$$

Equivalently, in the time domain, the expected value is

$$\begin{aligned} E[\tilde{\Gamma}_{xx}^{BT}(\tilde{f})] &= \sum_{m=-(M-1)}^{M-1} E[\tilde{\gamma}_{xx}[m]]w[m]e^{-j2\pi\tilde{f}m} \\ &= \sum_{m=-(M-1)}^{M-1} \gamma_{xx}[m]w_B[m]w[m]e^{-j2\pi\tilde{f}m} \end{aligned} \qquad (17.65)$$

where $w_B[n]$ is the Bartlett window. Clearly, the length of $w[n]$ should be such that $M \ll N$. That is, $w[n]$ should be narrower than $w_B[n]$ to provide additional smoothing of the periodogram. Under this condition, Eq. (17.64) becomes,[7]

$$E[\tilde{\Gamma}_{xx}^{BT}(\tilde{f})] \approx \int_{-1/2}^{1/2} \Gamma_{xx}(\theta) W(\tilde{f} - \theta) d\theta. \tag{17.66}$$

The mean of the PSD estimate using Blackman and Tukey method is a smoothed PSD.

Variance: The approximate expression for the variance is given by [5],

$$\begin{aligned} var[\tilde{\Gamma}_{xx}^{BT}(\tilde{f})] &\approx \Gamma_{xx}^2(f) \left[\frac{1}{N} \int_{-1/2}^{1/2} W^2(\theta) d\theta \right] \\ &\approx \Gamma_{xx}^2(f) \left[\frac{1}{N} \sum_{m=-(M-1)}^{M-1} w^2[m] \right]. \end{aligned} \tag{17.67}$$

The following table gives an overview of the mean and variance of the power estimates using the different non-parametric methods discussed in this section.

Method	Mean	Variance
Periodogram	$\int_{-0.5}^{0.5} \Gamma_{xx}(\alpha) W_B(\tilde{f} - \alpha) d\alpha$	$\Gamma_{xx}^2(\tilde{f}) \left[1 + \left(\frac{\sin 2\pi \tilde{f} N}{N \sin 2\pi \tilde{f}} \right)^2 \right]$
Bartlett	$\frac{1}{M} \int_{-0.5}^{0.5} \Gamma_{xx}(\alpha) W_B(\tilde{f} - \alpha) d\alpha$	$\frac{1}{K} \Gamma_{xx}^2(\tilde{f}) \left[1 + \left(\frac{\sin 2\pi \tilde{f} N}{N \sin 2\pi \tilde{f}} \right)^2 \right]$
Welch	$\int_{-0.5}^{0.5} \Gamma_{xx}(\alpha) W(\tilde{f} - \alpha) d\alpha$	$\approx \frac{1}{L} \Gamma_{xx}^2(\tilde{f})$ (no overlap)
Blackman & Tukey	$\int_{-0.5}^{0.5} \Gamma_{xx}(\alpha) W(\tilde{f} - \alpha) d\alpha$	$\Gamma_{xx}^2(\tilde{f}) \left[\frac{1}{N} \int_{-0.5}^{0.5} W^2(\theta) d\theta \right]$

17.3.4 Performance Characteristics of Non-parametric Power Spectrum Estimates

The quality of the Bartlett, Welch, and Blackman and Tukey power spectral estimates are compared by using the ratio of the square of mean of the power spectrum estimates to its variance as shown in the equation below:

$$Q_A = \frac{\{E[\tilde{\Gamma}_{xx}^A(\tilde{f})]\}^2}{var[\tilde{\Gamma}_{xx}^A(\tilde{f})]} \tag{17.68}$$

7. Since $\int_{-1/2}^{1/2} W_B(\alpha - \theta) W(\tilde{f} - \alpha) d\alpha = \int_{-1/2}^{1/2} W_B(\alpha) W(\tilde{f} - \theta - \alpha) d\alpha \approx W(\tilde{f} - \theta)$.

where, $A = B,\ W$ or BT for the three power spectrum estimates. Table below shows this comparison. It also shows the FFT based computational requirement for each of these methods. Here N is the length of the data and $\triangle \tilde{f}$ is the frequency resolution (measured at the $-3dB$ point).

Asymptotically, the periodogram is characterized by the quality factor

$$Q_P = \frac{\Gamma_{xx}^2(f)}{\Gamma_{xx}^2(f)} = 1. \tag{17.69}$$

It is fixed (= 1) and independent of data length N (another indication of poor quality of power spectrum estimation using periodogram).

Methods	Quality Factor	Computational Requirements (FFT)
Bartlett	$1.11N\triangle\tilde{f}$	$\frac{N}{2}\log_2\frac{0.9}{\triangle\tilde{f}}$
Welch (50% overlap)	$1.39N\triangle\tilde{f}$	$N\log_2\frac{5.12}{\triangle\tilde{f}}$
Blackman-Tukey	$2.34N\triangle\tilde{f}$	$N\log_2\frac{1.28}{\triangle\tilde{f}}$

We end the non-parametric methods of power spectrum estimation here. In the next section, the some of the modern (parametric) methods of power spectrum estimation are discussed.

17.4 Parametric Methods for Power Spectrum Estimation

The non-parametric methods for the power estimation are relatively simple, well understood, and easy to compute using DFT or FFT. However, they have the following limitations:

- They require long data records to get better frequency resolution.
- They suffer from spectral leakage due to windowing. This often masks the weak signals that are present in the data.
- The inherent assumption that the autocorrelation estimate $\tilde{\gamma}_{xx}[m]$ is zero for $m \geq N$ (where N is the data length) is unrealistic. This assumption limits the frequency resolution and the quality of the power spectrum estimate.
- The inherent assumption that the data are periodic with period N in the periodogram estimate is unrealistic.

The parametric methods do not require such assumptions. In fact, these methods extrapolate $\tilde{\gamma}_{xx}[m]$ for $m \geq N$. Extrapolation is possible if we have some a priori information on how the data were generated. Hence, a model for the signal generation is constructed using a number of parameters from the observed data. The model and the estimated parameters are then used to compute the PSD. The modeling approach eliminates the need for windowing and hence avoid the spectral leakage which provides a better frequency resolution compared to the FFT approach. This is especially better for time-varying or transient process where the data length (N) is very short and limited.

The parametric methods are based on modeling the data sequence $x[n]$ as the output of a linear system characterized by a rational system function of the form

$$H(z) = \frac{B(z)}{A(z)} = \frac{\sum_{k=0}^{q} b_k z^{-k}}{1 + \sum_{k=1}^{p} a_k z^{-k}}. \tag{17.70}$$

The corresponding difference equation is

$$x[n] = -\sum_{k=1}^{p} a_k x[n-k] + \sum_{k=0}^{q} b_k w[n-k], \tag{17.71}$$

where $w[n]$ is the input sequence to the system and $x[n]$ is the output of the system which is observable.

If $x[n]$ is characterized as stationary random process, then $w[n]$ is also assumed to be stationary random process and hence the PSD of the data is

$$\Gamma_{xx}(\tilde{f}) = |H(\tilde{f})|^2 \Gamma_{ww}(\tilde{f}), \tag{17.72}$$

where $\Gamma_{ww}(\tilde{f})$ is the PSD of $w[n]$ and $H(\tilde{f})$ is the frequency response of the model.

Since our objective is to estimate $\Gamma_{xx}(\tilde{f})$, it is convenient to assume that $w[n]$ is zero mean white noise sequence with autocorrelation $\gamma_{ww}[m] = \sigma_w^2 \delta[m]$, where $\sigma_w^2 = E[|w[n]|^2]$ is the variance. Then,

$$\Gamma_{xx}(\tilde{f}) = \sigma_w^2 |H(\tilde{f})|^2 = \sigma_w^2 \frac{|B(\tilde{f})|^2}{|A(\tilde{f})|^2}. \tag{17.73}$$

So, in the model-based (Parametric) approach, the spectrum estimation procedure consists of two steps. First, from the data sequence $\{x[n]\}$, $0 \leq n \leq N-1$, estimate the parameters $\{a_k\}$ and $\{b_k\}$ of the model. Then, from these estimates, compute the power spectrum according to Eq. (17.73).

It may be noted that the random process $x[n]$ generated by the pole-zero model in Eq. (17.70) or Eq. (17.71) is called an ARMA process of order (p, q) and is denoted as ARMA(p, q). If $q = 0$ and $b_0 = 1$, the resulting system model has a system function $H(z) = 1/A(z)$ and its output $x[n]$ is called an AR(p) process.

The third possible model is obtained by setting $A(z) = 1$, so that $H(z) = B(z)$. The output of this system is called a MA(q) process.

Before describing the methods for estimating the parameters in AR(p), MA(q), and ARMA(p, q) models, it is useful to establish the relationship between the model parameters and the autocorrelation sequence $\gamma_{xx}[m]$. In addition, we relate the AR model parameters to the coefficients in a linear predictor (LP) for the process $x[n]$.

17.4.1 Relationships Between the Autocorrelation and the Model Parameters

For the ARMA(p, q) process, the relationship is given by [5],

$$\gamma_{xx}[m] = \begin{cases} -\sum\limits_{k=1}^{p} a_k \gamma_{xx}[m-k] & \text{for } m > q \\ -\sum\limits_{k=1}^{p} a_k \gamma_{xx}[m-k] + \sigma_w^2 \sum_{k=0}^{q-m} h[k] b_{k+m} & \text{for } 0 \leq m \leq q \\ \gamma_{xx}^*[-m] & \text{for } m < 0. \end{cases} \tag{17.74}$$

From Eq. (17.74), for $m > q$, the set of linear equations for determining the model parameters $\{a_k\}$ are

$$(AA) \begin{pmatrix} a_1 \\ a_2 \\ \vdots \\ a_p \end{pmatrix} = - \begin{pmatrix} \gamma_{xx}[q+1] \\ \gamma_{xx}[q+2] \\ \vdots \\ \gamma_{xx}[q+p] \end{pmatrix} \tag{17.75}$$

where

$$AA = \begin{pmatrix} \gamma_{xx}[q] & \gamma_{xx}[q-1] & \cdots & \gamma_{xx}[q-p+1] \\ \gamma_{xx}[q+1] & \gamma_{xx}[q] & \cdots & \gamma_{xx}(q-p+2) \\ \vdots & \vdots & \vdots & \vdots \\ \gamma_{xx}[q+p-1] & \gamma_{xx}[q+p-2] & \cdots & \gamma_{xx}[q] \end{pmatrix}$$

An interpretation of the relationship in Eq. (17.75) is that the values of the autocorrelation $\gamma_{xx}[m]$ for $m > q$ are uniquely determined from the pole parameters $\{a_k\}$ and the values of $\gamma_{xx}[m]$ for $0 \leq m \leq p$. Hence, the linear system model automatically extends the values of $\gamma_{xx}[m]$ for $m > p$.

The $\{a_k\}$'s obtained from Eq. (17.75) will not help in determining the MA parameters $\{b_k\}$ as it requires the knowledge of the impulse response $h[n]$.

If we adopt an AR(p) model for the observed data, the relationship between the AR parameters and the autocorrelation is obtained form Eq. (17.74) by setting

$q = 0$. Thus we get,

$$\gamma_{xx}[m] = \begin{cases} -\sum_{k=1}^{p} a_k \gamma_{xx}[m-k] & \text{for } m > 0 \\ -\sum_{k=1}^{p} a_k \gamma_{xx}[m-k] + \sigma_w^2 & \text{for } m = 0 \\ \gamma_{xx}^*[-m] & \text{for } m < 0. \end{cases} \tag{17.76}$$

In this case, the AR parameters are obtained from the solution of the *Yule-Walker* or *normal equations* (shown below).

$$\begin{pmatrix} \gamma_{xx}[0] & \gamma_{xx}[-1] & \dots & \gamma_{xx}[-p+1] \\ \gamma_{xx}[1] & \gamma_{xx}[0] & \dots & \gamma_{xx}[-p+2] \\ \vdots & \vdots & \vdots & \vdots \\ \gamma_{xx}[p-1] & \gamma_{xx}[p-2] & \dots & \gamma_{xx}[0] \end{pmatrix} \begin{pmatrix} a_1 \\ a_2 \\ \vdots \\ a_p \end{pmatrix} = - \begin{pmatrix} \gamma_{xx}[1] \\ \gamma_{xx}[2] \\ \vdots \\ \gamma_{xx}[p] \end{pmatrix} \tag{17.77}$$

and the variance can be obtained from (Eq. 17.76, $m = 0$)

$$\sigma_w^2 = \gamma_{xx}[0] + \sum_{k=1}^{p} a_k \gamma_{xx}[-k] \tag{17.78}$$

Usually, Eqs. (17.77) and (17.78) are combined into a single matrix equation as

$$\begin{pmatrix} \gamma_{xx}[0] & \gamma_{xx}[-1] & \dots & \gamma_{xx}[-p] \\ \gamma_{xx}[1] & \gamma_{xx}[0] & \dots & \gamma_{xx}[-p+1] \\ \vdots & \vdots & \vdots & \vdots \\ \gamma_{xx}[p] & \gamma_{xx}[p-1] & \dots & \gamma_{xx}[0] \end{pmatrix} \begin{pmatrix} 1 \\ a_1 \\ \vdots \\ a_p \end{pmatrix} = \begin{pmatrix} \sigma_w^2 \\ 0 \\ \vdots \\ 0 \end{pmatrix} \tag{17.79}$$

It may be noted that the correlation matrix in Eq. (17.79) is Toeplitz,[8] and it can be efficiently inverted using the *Levinson-Durbin* algorithm.[9]

Thus all the system parameters in the AR(p) model are easily determined from the knowledge of the autocorrelation sequence $\gamma_{xx}[m]$ for $0 \le m \le p$. Furthermore, Eq. (17.76) can be used to extend the autocorrelation sequences beyond $m > p$, once the $\{a_k\}$ are obtained.

8. For a Toeplitz matrix, we have $A_{i,j} = A_{i-1,j-1}$, where $A_{i,j}$ is the $(i, j)^{th}$ element of A.

9. This is a computationally efficient recursive algorithm for estimating the model parameters.

For the MA(q) model for the observed data, the relationship between the autocorrelation and $\{b_k\}$ are given by [5]

$$\gamma_{xx}[m] = \begin{cases} \sigma_w^2 \sum_{k=0}^{q} b_k b_{k+m} & \text{for } 0 \le m \le q \\ 0 & \text{for } m > q \\ \gamma_{xx}^*[-m] & \text{for } m < 0. \end{cases} \tag{17.80}$$

17.4.2 AR Method of Power Spectrum Estimation

In this method, first estimate the AR parameters and then the power spectrum can be estimated.

17.4.2.1 Yule-Walker Method of AR Power Spectrum Estimation

In this method, the biased estimate of the autocorrelation (Eq. (17.81)) is used to ensure that the matrix is positive semi-definite and hence to ensure stable AR model.

$$\tilde{\gamma}_{xx}[m] = \frac{1}{N} \sum_{n=0}^{N-m-1} x^*[n]x[n+m], \quad m \ge 0. \tag{17.81}$$

The Levinson-Durbin algorithm [10] can be used to recursively estimate the model parameters. The corresponding power spectrum estimate is

$$\tilde{\Gamma}_{xx}^{YW}(\tilde{f}) = \frac{\hat{\sigma}_{wp}^2}{|1 + \sum_{k=1}^{p} \hat{a}_p[k] e^{-j2\pi \tilde{f} k}|^2} \tag{17.82}$$

where $\hat{a}_p[k]$ are the estimates of the AR parameters obtained from the Levinson-Durbin recursion and

$$\hat{\sigma}_{wp}^2 = \hat{E}_p^f = \tilde{\gamma}_{xx}[0] \prod_{k=1}^{p} \left[1 - |\hat{a}_p[k]|^2\right] \tag{17.83}$$

is the estimated minimum mean-square value for the pth order predictor.

17.4.2.2 The Burg Method of AR Power Spectrum Estimation

The major advantages of Burg's method are high frequency resolution, stability, and computational efficiency [10]. This method is based on the minimization of the forward and backward errors in linear predictors, with the constraint that the AR parameters satisfy the Levinson-Durbin recursion.

From the available data $x[n],\ n = 0, 1, \ldots, N-1$, let us consider the forward and backward linear prediction estimates of order m, given as

$$\hat{x}[n] = -\sum_{k=1}^{m} a_m[k]x[n-k] \tag{17.84}$$

$$\hat{x}[n-m] = -\sum_{k=1}^{m} a_m^*[k]x[n+k-m] \tag{17.85}$$

where $a_m[k], 0 \leq k \leq m-1, m = 0, 1, \ldots, p$, are the predictor coefficients. The corresponding forward and backward prediction errors are $f_m[n] = x[n] - \hat{x}[n]$, and $g_m[n] = x[n-m] - \hat{x}[n-m]$, respectively. The least-squares error that is to be minimized to estimate the prediction coefficients subject to the constraint that it satisfies the Levinson-Durbin recursion is

$$\varepsilon_m = \sum_{n=m}^{N-1} \left[|f_m[n]|^2 + |g_m[n]|^2 \right]. \tag{17.86}$$

The Levinson-Durbin recursion is given by (for $1 \leq k \leq m-1$ and $1 \leq m \leq p$),

$$a_m[k] = a_{m-1}[k] + K_m a_{m-1}^*[m-k] \tag{17.87}$$

where $K_m = a_m[m]$ is the mth reflection coefficient in the lattice filter realization of the predictor. The minimization of ε_m with respect to the complex-valued K_m results in (for $m = 1, 2, \ldots, p$),

$$\hat{K}_m = \frac{-\sum_{n=m}^{N-1} f_{m-1}[n] g_{m-1}^*[n-1]}{\frac{1}{2}\sum_{n=m}^{N-1} \left[|f_{m-1}[n]|^2 + |g_{m-1}[n-1]|^2 \right]}. \tag{17.88}$$

The term in the numerator of Eq. (17.88) is an estimate of the cross-correlation between the forward and backward prediction errors. With the normalization factor in the denominator, it is apparent that $K_m < 1$, so that the all-pole model obtained is stable.

We note that the denominator in Eq. (17.88) is simply the least-squares estimate of the forward and backward errors, E_{m-1}^f and E_{m-1}^b, respectively. Hence Eq. (17.88) can be expressed as (for $m = 1, 2, \ldots, p$)

$$\hat{K}_m = \frac{-\sum_{n=m}^{N-1} f_{m-1}[n] g_{m-1}^*[n-1]}{\frac{1}{2}\left[\hat{E}_{m-1}^f + \hat{E}_{m-1}^b \right]}, \tag{17.89}$$

where $\hat{E}^f_{m-1} + \hat{E}^b_{m-1} = \hat{E}_m$ is an estimate of the total squared error E_m. It was shown that the denominator term in Eq. (17.89) can be computed in a recursive fashion according to [10]

$$\hat{E}_m = (1 - |\hat{K}_m|^2)\hat{E}_{m-1} - |f_{m-1}[m-1]|^2 - |g_{m-1}[m-2]|^2 \qquad (17.90)$$

To summarize, the Burg algorithm computes the reflection coefficients in the equivalent lattice structure as specified by Eqs. (17.89) and (17.90), and the Levinson-Durbin algorithm is used to obtain the AR parameters. The power spectrum estimate is,

$$\tilde{\Gamma}^{BU}_{xx}(\tilde{f}) = \frac{\hat{E}_p}{|1 + \sum_{k=1}^{p} \hat{a}_p[k]e^{-j2\pi \tilde{f}k}|^2}. \qquad (17.91)$$

17.4.2.3 Selection of AR Model Order (p)

How do we choose the value of p? Too low an order result in a highly smoothed spectrum and too high a value introduce spurious peaks in the spectrum. The mean-square value of the residual error ($\hat{\sigma}^2_{wp}$) can be used as a measure to choose an optimum model order. The characteristics of this error is that it decreases as the model order increases. We can monitor the rate of decrease and decide to terminate the process when the rate of decrease becomes relatively slow.

Some of the popular methods are:

- Akaike's *final prediction error (FPE)* criteria, where the order is selected to minimize the performance index

$$FPE(p) = \hat{\sigma}^2_{wp}\left(\frac{N+p+1}{N-p-1}\right), \qquad (17.92)$$

where $\hat{\sigma}^2_{wp}$ is the estimated variance of the linear prediction (one-step) error.
- *Akaike's Information Criteria* (AIC), based on selecting the order that minimizes

$$AIC(p) = \ln \hat{\sigma}^2_{wp} + \frac{2p}{N}. \qquad (17.93)$$

Note that $\hat{\sigma}^2_{wp}$ decreases and therefore $\ln \hat{\sigma}^2_{wp}$ also decreases as p increases. However, $\frac{2p}{N}$ increases in p. Hence, a minimum value is obtained for some p.
- Rissanen's *Minimum Description Length (MDL)* criteria, where MDL is defined as

$$MDL(p) = N \ln \hat{\sigma}^2_{wp} + p \ln N. \qquad (17.94)$$

- Parzen's *Criterion Autoregressive Transfer (CAT)*, defined as

$$CAT(p) = \left(\frac{1}{N}\sum_{k=1}^{p}\frac{1}{\bar{\sigma}_{wk}^{2}}\right) - \frac{1}{\hat{\sigma}_{wp}^{2}}, \tag{17.95}$$

where

$$\bar{\sigma}_{wk}^{2} = \frac{N}{N-k}\hat{\sigma}_{wk}^{2}. \tag{17.96}$$

The order p corresponding to the minimum of $CAT(p)$ is to be chosen as the AR order.

17.4.3 MA Model for Power Spectrum Estimation

The parameters in an MA(q) model are related to the statistical autocorrelation $\gamma_{xx}[m]$ by Eq. (17.80). However,

$$B(z)B(z^{-1}) = D(z) = \sum_{m=-q}^{q} d_m z^{-m} \tag{17.97}$$

where the coefficients $\{d_m\}$ are related to the MA parameters by [5]

$$d_m = \sum_{k=0}^{q-|m|} b_k b_{k+m}, \quad |m| \le q. \tag{17.98}$$

Clearly, then (from Eq. (17.80)),

$$\gamma_{xx}[m] = \begin{cases} \sigma_w^2 d_m & \text{for } |m| \le q \\ 0 & \text{for } |m| > q \end{cases} \tag{17.99}$$

and the power spectrum for the MA(q) process is

$$\Gamma_{xx}^{MA}(\tilde{f}) = \sum_{m=-q}^{q} \gamma_{xx}[m]e^{-j2\pi\tilde{f}m}. \tag{17.100}$$

It is clear from these equations that we do not have to solve for the MA parameters $\{b_k\}$ to estimate the power spectrum. The estimate $\tilde{\gamma}_{xx}[m]$ for $|m| \le q$ is enough. From such estimates, we can compute the MA power spectrum, given by

$$\tilde{\Gamma}_{xx}^{MA}(\tilde{f}) = \sum_{m=-q}^{q} \tilde{\gamma}_{xx}[m]e^{-j2\pi\tilde{f}m}. \tag{17.101}$$

§ *An alternative method for determining $\{b_k\}$ is based on a high order AR approximation to MA process.*

The order q of the MA model may be determined empirically by AIC as described in the previous subsection. The criteria is to choose the order which gives the minimum of

$$AIC[q] = \ln \hat{\sigma}_{wq}^2 + \frac{2q}{N}, \tag{17.102}$$

where $\hat{\sigma}_{wq}^2$ is an estimate of the variance of the white noise.

17.4.4 ARMA Model for Power Spectrum Estimation

The ARMA model is particularly appropriate when the signal has been corrupted by noise. For example, suppose that the data $x[n]$ is generated by an AR system, where the system output is corrupted by white noise. The z-transform of the auto-correlation of the resultant signal can be expressed as

$$\begin{aligned} \Gamma_{xx}(z) &= \frac{\sigma_w^2}{A(z)A(z^{-1})} + \sigma_n^2 \\ &= \frac{\sigma_w^2 + \sigma_n^2 A(z)A(z^{-1})}{A(z)A(z^{-1})} \end{aligned} \tag{17.103}$$

where σ_n^2 is the variance of the additive noise. Therefore the process $x[n]$ is ARMA(p, p), where p is the order of the AR process.

From Eq. (17.75), for lags $|m| > q$, the equation involves only the AR parameters. Using the estimates of $\gamma_{xx}[m]$, we can solve Eq. (17.77) to obtain $\hat{a}_k$. For high-order models, however, this approach is likely to provide poor estimates of the parameters because of the poor estimates of the autocorrelation values at larger lags. Hence, this method is not recommended.

A more reliable method is to construct an over-determined set of linear equations for $m > q$, and to use the method of least-squares [14]. To elaborate, suppose the autocorrelation ($\gamma_{xx}[m]$) can be accurately estimated up to lag $M > p + q$. Then, we can write the following set of linear equations:

$$(AAA) \begin{pmatrix} a_1 \\ a_2 \\ \vdots \\ a_p \end{pmatrix} = - \begin{pmatrix} \gamma_{xx}[q+1] \\ \gamma_{xx}[q+2] \\ \vdots \\ \gamma_{xx}[m] \end{pmatrix} \tag{17.104}$$

where

$$AAA = \begin{pmatrix} \gamma_{xx}[q] & \gamma_{xx}[q-1] & \dots & \gamma_{xx}[q-p+1] \\ \gamma_{xx}[q+1] & \gamma_{xx}[q] & \dots & \gamma_{xx}(q-p+2) \\ \vdots & \vdots & \vdots & \vdots \\ \gamma_{xx}[M-1] & \gamma_{xx}[M-2] & \dots & \gamma_{xx}[M-p] \end{pmatrix}$$

or equivalently,

$$\mathbf{R}_{xx}\mathbf{a} = -\mathbf{r}_{xx} \tag{17.105}$$

where $\mathbf{R}_{xx}$ is the matrix of autocorrelations, $\mathbf{r}_{xx}$ is the vector of autocorrelations, and $\mathbf{a}$ is the vector of AR parameters.

Since the autocorrelation matrix $\mathbf{R}_{xx}$ is of dimension $(M-q) \times p$, and $M - q > p$, we can use the least-squares criterion to solve for the parameter vector $\mathbf{a}$. The result is,

$$\hat{\mathbf{a}} = -(\mathbf{R}_{xx}^T\mathbf{R}_{xx})^{-1}\mathbf{R}_{xx}^T\mathbf{r}_{xx} \tag{17.106}$$

This procedure is called the *least-squares modified Yule-Walker method.* A weighting factor can also be applied to the autocorrelation sequence to deemphasize the less reliable estimates for large lags.

Once the parameters for the AR part of the model have been estimated as indicated above, we have the system

$$\hat{A}(z) = 1 + \sum_{k=1}^{p} \hat{a}_k z^{-k}. \tag{17.107}$$

The sequence $x[n]$ can now be filtered by the FIR filter $\hat{A}(z)$ to yield the sequence

$$v[n] = x[n] + \sum_{k=1}^{p} \hat{a}_k x[n-k], \quad n = 0, 1, \dots, N-1. \tag{17.108}$$

The cascade of the ARMA(p, q) model with $\hat{A}(z)$ is approximately the MA(q) process generated by $B(z)$. Hence we can apply the MA estimate given in the preceding section to obtain the MA spectrum. To be specific, the filtered sequence $v[n]$ for $p \leq n \leq N-1$ is used to form the estimated correlation sequences $\tilde{\gamma}_{vv}[m]$, from which we obtain the MA spectrum

$$\tilde{\Gamma}_{vv}^{MA}(\tilde{f}) = \sum_{m=-q}^{q} r_{vv}[m]e^{-j2\pi\tilde{f}m}. \tag{17.109}$$

Finally, the estimated ARMA power spectrum is

$$\tilde{\Gamma}_{xx}^{ARMA}(\tilde{f}) = \frac{\tilde{\Gamma}_{vv}^{MA}(\tilde{f})}{|1 + \sum_{k=1}^{p} \hat{a}_k e^{-j2\pi \tilde{f} k}|^2}. \tag{17.110}$$

The order (p, q) of the ARMA model can be selected by using the following AIC index:

$$AIC(p, q) = \ln \hat{\sigma}_{wpq}^2 + \frac{2(p+q)}{N} \tag{17.111}$$

where $\hat{\sigma}_{wpq}^2$ is an estimate of the variance of the input error.

17.5 Discrete-Time White Noise

In Chapter 15, we studied about continuous-time white noise. In this section, we shall study about discrete-time white noise. *A discrete-time signal* $x[n]$ *is said to be a white noise if its power spectrum is flat over the complete frequency range.* That is, the power spectrum is given by

$$\Gamma_{xx}(\tilde{f}) = \sigma_x^2 \qquad -0.5 \leq \tilde{f} < 0.5. \tag{17.112}$$

Here, $\tilde{f}$ denotes the normalized frequency. Since $x[n]$ is a discrete-time signal, *its power spectrum* $\Gamma_{xx}(\tilde{f})$ *is periodic with period equal to 1* (i.e., normalized sampling frequency). Consequently, the power of $x[n]$ is given by,

$$p_x = \int_{-0.5}^{0.5} \Gamma_{xx}(\tilde{f}) d\tilde{f} = \sigma_x^2. \tag{17.113}$$

Taking the inverse DTFT of the power spectrum, we get the autocorrelation of $x[n]$ as

$$\gamma_{xx}[m] = \overline{x[n]x[n+m]} = \sigma_x^2 \delta[m] \tag{17.114}$$

This implies that

$$\gamma_{xx}[m] = \begin{cases} \sigma_x^2 & \text{if } m = 0 \\ 0 & \text{if } m \neq 0. \end{cases} \tag{17.115}$$

where m is an integer. Thus, the samples of a discrete-time white noise are completely uncorrelated with each other. Equation (17.112) requires that the mean of

$x[n]$ must be zero. Therefore, for discrete-time white noise, we have

$$\text{Power spectrum,}\quad \Gamma_{xx}(\tilde{f}) = \sigma_x^2 \qquad \text{for all} \quad \tilde{f} \tag{17.116}$$

$$\text{Autocorrelation,}\quad \gamma_{xx}[m] = \sigma_x^2 \delta[m] \tag{17.117}$$

$$\text{Mean,}\quad m_x = 0 \tag{17.118}$$

$$\text{Power,}\quad p_x = \overline{|x[n]|^2} = \sigma_x^2 \tag{17.119}$$

$$\text{Variance,}\quad \sigma_x^2 = \overline{|x[n] - m_x|^2} = p_x. \tag{17.120}$$

Thus, while the continuous-time white noise is an idealistic signal that does not exist in practice (nor can it be physically generated), the discrete-time white noise can be generated.[10]

17.6 DFT for Spectrum Estimation

In Chapter 12, we introduced the discrete Fourier transform (DFT) as a practical tool for estimating the spectra of signals. Recall the following from our study on DFT.

- The *DFT computes the spectrum at discrete frequencies, which are equally spaced, based on finite data records*. The spacing between the frequencies are given by $1/N$ where N is the length of the data record.
- The *DFT treats the finite data record* $\{x[0], x[1], x[2], \cdots, x[N-1]\}$ *as one period of a periodic signal with period N.*
- The *DFT spectrum, $X[k]$, is equivalent to sampling the DTFT spectrum, $X(f)$, of the finite data record* $\{x[0], x(1), x(2), \cdots, x[N-1]\}$ *at frequencies given by* $0,\ 1/N,\ 2/N,\ \cdots,\ (N-1)/N$.
- The *DFT of a sinusoidal signal will indicate the actual frequencies of the sinusoids only if the normalized frequencies of the sinusoids fall exactly on the frequencies at which the DFT is evaluated.*

Thus, when we use DFT to do spectrum estimation, we must know how the periodicity assumption of DFT and the value of N affect the estimated spectrum. Therefore, in the following sections, the objective is *to throw light on the various trade-offs that result from the periodicity assumption in DFT and the choice of N.*

10. Let the power of a discrete-time white noise be σ_x^2 and let the underlying sampling frequency be f_s. Then, in terms of un-normalized frequency, the power spectral density of the discrete-time white noise is given by σ_x^2/f_s

By studying these sections, we should be able to receive clarity and understanding on the following.

- Why windows are required in spectrum estimation? Which are the typical window functions?
- How does the choice of the window function and the value of N control the trade-off between spectral resolution and spectral leakage?
- When should leakage be given priority over resolution?
- How does the choice of N affect frequency and time resolutions?
- Issues that underlie spectral smoothing.

17.7 Windows and Spectrum Estimation

Let $x[n]$ denote the samples of a continuous-time signal of very large duration. In a particular experiment, we capture N samples of this signal and are given by

$$y[n] = x[n] \qquad \text{for} \quad n = 0, 1, \ldots, N-1. \tag{17.121}$$

We can also express the measured signal as

$$y[n] = x[n]w[n] \tag{17.122}$$

where

$$w[n] = \begin{cases} 1 & \text{for } n = 0, 1, \ldots, N-1 \\ 0 & \text{otherwise} \end{cases} \tag{17.123}$$

is known as a *rectangular window of duration N*. Thus, we find that

$$y[n] = \begin{cases} x[n] & \text{for } n = 0, 1, \ldots, N-1 \\ 0 & \text{otherwise} \end{cases} \tag{17.124}$$

Thus, *the window function makes $y[n]$ to be zero outside the time-span of the window.*

17.7.1 Periodicity Assumption of DFT

Using Eq. (17.122), we can express the spectrum (DTFT) of $y[n]$ as

$$Y(\tilde{f}) = X(\tilde{f}) \odot W(\tilde{f}) = \int_{-0.5}^{0.5} X(u)W(\tilde{f} - u)du. \tag{17.125}$$

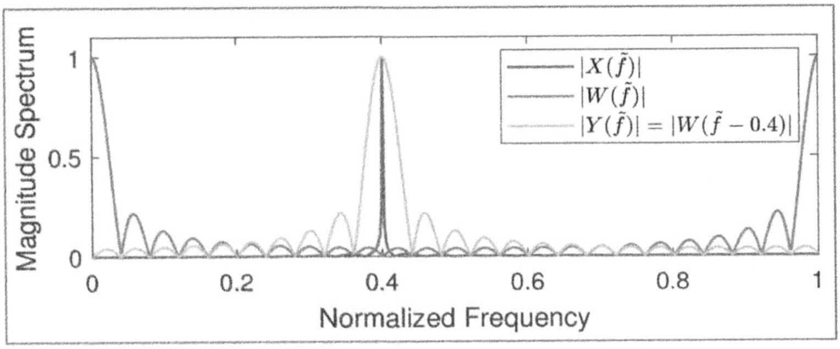

Figure 17.11. Illustration of windowing on the computed spectrum of a complex sinusoid at normalized frequency 0.4. (a) $X(\tilde{f})$ is the actual spectrum of the complex sinusoid, (b) $W(\tilde{f})$ is the spectrum of the rectangular window for length $N = 25$, and (c) $Y(\tilde{f}) = W(\tilde{f} - 0.4)$ is the computed spectrum of the complex sinusoid based on $N = 25$ samples. All the spectra are computed using the expression of DTFT.

For the sake of illustration, we choose $x[n] = e^{j2\pi \tilde{f}_1 n}$, i.e., a complex sinusoid at normalized frequency $\tilde{f}_1$. Then, we have

$$x[n] = e^{j2\pi \tilde{f}_1 n} \quad \text{and} \quad X(\tilde{f}) = \delta(\tilde{f} - \tilde{f}_1). \tag{17.126a}$$

Using Eq. (17.126a) in Eq. (17.125), we get

$$Y(\tilde{f}) = \int_{-0.5}^{0.5} \delta(u - \tilde{f}_1) W(\tilde{f} - u) du = W(\tilde{f} - \tilde{f}_1). \tag{17.126b}$$

Thus, *because of the finite data effect, the spectrum of the complex sinusoid computed based on N samples is given by the shifted spectrum of the window*. That is, instead of an impulse sitting at $\tilde{f} = \tilde{f}_1$ as the signal spectrum, the finite data effect causes the spectrum to take the shape of the window spectrum. This is illustrated in Figure 17.11 for $\tilde{f}_1 = 0.4$ and a rectangular window of length $N = 25$.

The spectrum of the rectangular window is given by

$$\begin{aligned} W(\tilde{f}) &= \sum_{n=-\infty}^{\infty} w[n] e^{j2\pi \tilde{f} n} = \sum_{n=0}^{N-1} w[n] e^{j2\pi \tilde{f} n} \\ &= \frac{1 - e^{j2\pi \tilde{f} N}}{1 - e^{j2\pi \tilde{f}}} = e^{j\pi (N-1)\tilde{f}} \frac{\sin(\pi \tilde{f} N)}{\sin(\pi \tilde{f})}. \end{aligned} \tag{17.127}$$

Therefore, the peak of $|W(\tilde{f})|$ occurs at $\tilde{f} = 0$ and the zero-crossings of $|W(\tilde{f})|$ occur at $\tilde{f} = \frac{l}{N}$ for non-zero $l = \pm 1, \pm 2, \ldots$.

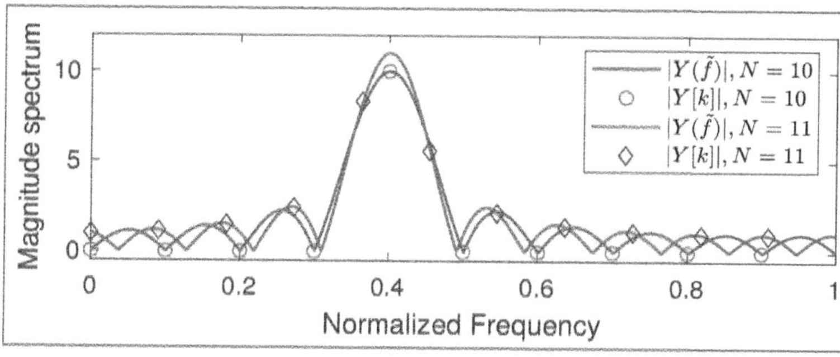

Figure 17.12. Illustration of the effect of periodicity and choice of N on the DFT spectrum for the spectrum of a complex sinusoid at normalized frequency 0.4. (a) DTFT $Y(\tilde{f})$ computed for $N = 10$, (b) DFT Y_k computed for $N = 10$, (c) DTFT $Y(\tilde{f})$ computed for $N = 11$, (d) DFT Y_k computed for $N = 11$.

Since DFT is obtained by sampling the DTFT at $\tilde{f} = \frac{k}{N}$, we get

$$\begin{aligned} Y[k] &= Y\left(\frac{k}{N}\right) = W\left(\frac{k}{N} - \tilde{f}_1\right) \\ &= e^{j\pi(N-1)\left(\frac{k}{N}-\tilde{f}_1\right)} \frac{\sin(\pi(k - N\tilde{f}_1))}{\sin(\pi(k - N\tilde{f}_1)/N)}. \end{aligned} \tag{17.128}$$

Therefore, *if N is chosen such that $N\tilde{f}_1$ is an integer (in the range $0, 1, \ldots, N-1$)*, then we will find that

$$|Y[k]| = \begin{cases} W[0] & \text{for } \; k = N\tilde{f}_1 \\ 0 & \text{for } \; k \neq N\tilde{f}_1. \end{cases} \tag{17.129}$$

For the complex sinusoid at $\tilde{f}_1 = 0.4$ and $N = 10$, we have $N\tilde{f}_1 = 4$. Therefore, the DFT spectrum will contain a peak at $k = 4$ and will be zero for all other values of $k = 0, 1, 2, 3, 5, 6, 7, 8, 9$. This is shown in Fig. 17.12 using the red 'o'.

On the other hand, *if we choose N such that $N\tilde{f}_1$ is not an integer in the range* $\{0, 1, \ldots, N-1\}$ (e.g., $N = 11$), then the DFT will be given by

$$|Y[k]| = \left|\frac{\sin(\pi(k - N\tilde{f}_1))}{\sin(\pi(k - N\tilde{f}_1)/N)}\right| \qquad \text{for all } \; k \tag{17.130}$$

and it will be non-zero for most values of k. This is shown in Figure 17.12 using the blue '◇' for the corresponding frequencies $k/N, k = 0, 1, \ldots, N-1$.

Based on this, we have the following remarks.

- *The choice of N determines the frequencies at which DTFT is evaluated and hence the shape of the resulting DFT spectrum.*

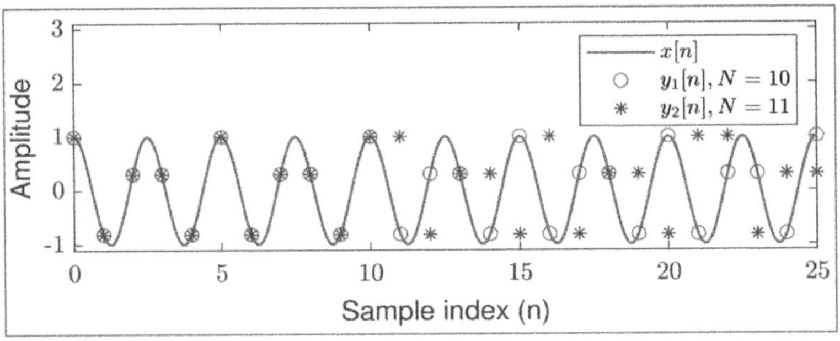

Figure 17.13. Illustration of the effect of periodicity and choice of N in the time-domain signal for a complex sinusoid at normalized frequency 0.4. (a) $y_1[n]$ is the periodic extension of $y[n]$ for $N = 10$, (b) $y_2[n]$ is the periodic extension of $y[n]$ for $N = 11$.

- *When N is chosen such that the periodicity assumption is satisfied, the DFT spectrum will correspond to the actual spectrum (see Eq. (17.129)).*
- *When N is chosen such that the periodicity assumption is not satisfied, the DFT spectrum will behave according to the sampled spectrum of the window given by Eq. (17.130).*

Figure 17.13 shows the periodic signals generated using the finite-date records of $y[n]$. The signal $y_1[n]$ is obtained by periodically extending $\{y[0], y[1], \cdots, y[9]\}$, i.e., with $N = 10$. The signal $y_2[n]$ is obtained by periodically extending $\{y[0], y[1], \cdots, y[10]\}$, i.e., with $N = 11$. Observe that $y_1[n]$ follows the original sinusoid while $y_2[n]$ does not resemble a sinusoid. Consequently, the DFT spectrum resulting from $N = 11$ will not resemble that of a sinusoid. Thus, we can see the agreement between Figs.17.12 and 17.13.

17.7.2 Spectral Leakage

Observe from Eq. (17.126b), i.e. $Y(\tilde{f}) = W(\tilde{f} - \tilde{f}_1)$, and Figure 17.11 that the computed spectrum, $Y(\tilde{f})$, of the sinusoid based on N samples has the shape of the window spectrum. That is, *the finite data effect causes the impulse spectrum of the sinusoid to spread to all the frequencies.* This effect is known as *spectral leakage*. The amount by which the impulse at $\tilde{f} = \tilde{f}_1$ leaks to the other frequencies depend upon the shape of the window spectrum $W(\tilde{f})$. For example, the leaked spectrum at frequency $\tilde{f} = 0.46$ is given by $W(0.46 - \tilde{f}_1)$.

Clearly, we would prefer to minimize the leakage. This can be achieved by choosing the window function such that its side-lobe levels are low. Figure 17.14 shows the spectra for typical window functions with $N = 10$. We can observe the following from Fig. 17.14.

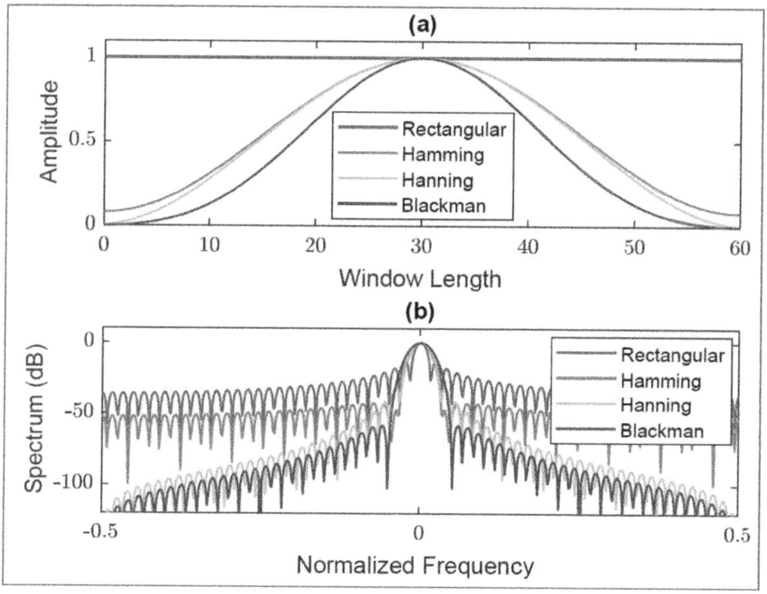

Figure 17.14. Time domain and frequency domain characteristics of different window functions: (a) Rectangular, (b) Hamming, (c) Hanning, and (d) Blackman.

- The rectangular window has the minimum main-lobe width. But it also has the highest side-lobe height.
- The Blackman window has the lowest side-lobe height. But it also has the widest main-lobe.
- The main-lobe width and side-lobe height of Hamming and Hanning windows lie intermediate to that of rectangular and Blackman windows.

Figure 17.14(a) shows the time-domain plots of these window functions. Observe that *while rectangular window causes the sharpest truncation at the boundaries, the Blackman window results in the smoothest truncation at the boundaries.* This is why, the spectra of these two windows have the opposite characteristics as highlighted above.

§ *Based on the side-lobe levels of the different windows shown in Fig. 17.14, we see that the rectangular window results in maximum spectral leakage, while the Blackman window results in minimum spectral leakage.*

17.7.3 Spectral Resolution

Spectral resolution refers to the ability of the spectral estimator to resolve closely spaced spectral lines (frequencies). For example, consider the following signal

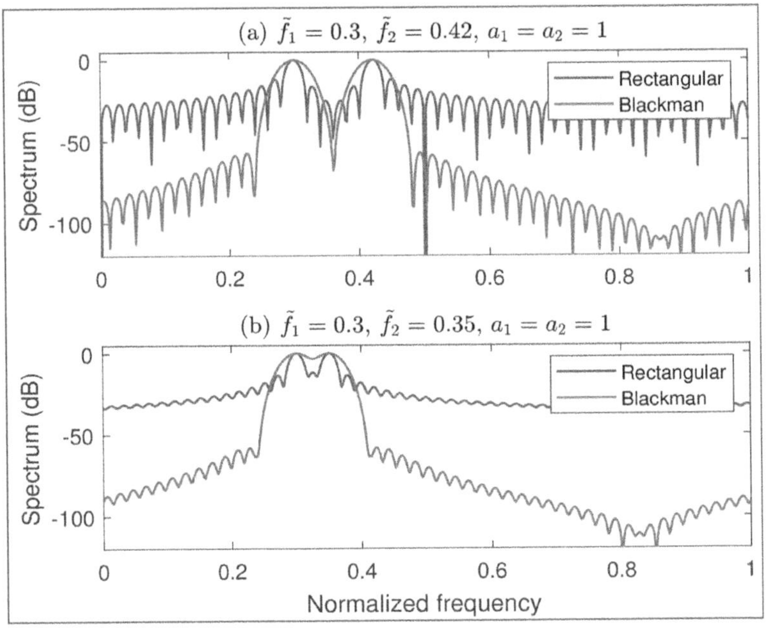

Figure 17.15. Spectra of the sum of two complex sinusoids (equal powers) using rectangular and Blackman windows with $N = 50$, $A_1 = 1$ and $A_2 = 0.1$. (a) $(\tilde{f}_1, \tilde{f}_2) = (0.3, 0.42)$, (b) $(\tilde{f}_1, \tilde{f}_2) = (0.3, 0.35)$.

obtained by summing two complex sinusoids of amplitudes A_1 and A_2 and frequencies $\tilde{f}_1$ and $\tilde{f}_2$:

$$x[n] = A_1 e^{j2\pi \tilde{f}_1 n} + A_2 e^{j2\pi \tilde{f}_2 n}. \tag{17.131}$$

When we use a spectral estimator to compute the spectrum of $x[n]$, we expect to see two spectral lines corresponding to the two sinusoids. But will the spectral estimator be able to show these two spectral lines even if the frequencies $\tilde{f}_1$ and $\tilde{f}_2$ are very close? To answer this, we need to study the resolution capability of the underlying spectral estimator.

Figure 17.15 shows the spectra of $x[n]$ for two sets of values of the frequencies. Figure 17.15(a) corresponds to $\tilde{f}_1 = 0.3$ and $\tilde{f}_2 = 0.42$, and Figure 17.15(b) corresponds to $\tilde{f}_1 = 0.3$ and $\tilde{f}_2 = 0.35$. The amplitudes of the sinusoids are set to be equal: $A_1 = 1 = A_2$. The selection of frequencies correspond to wide separation of spectral lines in Case (a) and closely spaced spectral lines in Case (b). We use rectangular and Blackman windows (with $N = 50$) to illustrate the difference in the resolution capability for different windows.

Observe that while the rectangular and Blackman windows are able to resolve the two spectral lines clearly in Case (a), the Blackman window fails to resolve the spectral lines in Case (b). This is because the main-lobe of Blackman window is

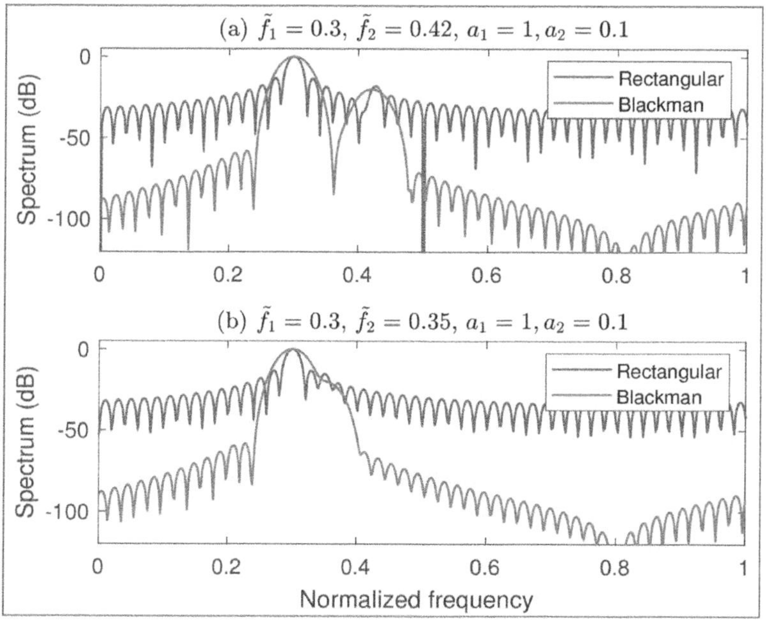

Figure 17.16. Spectra of the sum of two complex sinusoids (unequal powers) using rectangular and Blackman windows with $N = 50$, $A_1 = 1$ and $A_2 = 0.1$. (a) $(\tilde{f}_1, \tilde{f}_2) = (0.3, 0.42)$, (b) $(\tilde{f}_1, \tilde{f}_2) = (0.3, 0.35)$.

too wide to be able to resolve the closely spaced sinusoids in Case (b). Because rectangular window has narrow main-lobe, it is able to resolve the lines in Case (b).

Figure 17.16 shows the spectra of $x[n]$ when the amplitudes are chosen as $A_1 = 1$ and $A_2 = 0.1$. The frequencies and N are the same as that in Fig. 17.15. Observe that, compared to Fig. 17.15, the resolution provided by rectangular window in Fig. 17.16 is poor compared to that of Blackman window. This can be very clearly seen for Case (a). This arises because of the high spectral leakage caused by rectangular window compared to the Blackman window. Thus, when we have signals that consist of a mixture of strong and weak spectral components, we should give importance to minimizing spectral leakage. Otherwise, the weak component may get suppressed by the leakage arising from the strong component. Consequently, we prefer to use Blackman or Hamming type windows in such situations, compared to the rectangular window.

The phenomenon of spectral leakage and its effect on spectral resolution, as manifested in Figs.17.15 and 17.16, can be seen clearly from the following. Using Eq. (17.126b), we can write the DTFT of $x[n]$ (given in Eq. (17.131)) as

$$\begin{aligned} X(\tilde{f}) &= \left[A_1\delta(\tilde{f} - \tilde{f}_1) + A_2\delta(\tilde{f} - \tilde{f}_2)\right] \odot W(\tilde{f}) \\ &= A_1 W(\tilde{f} - \tilde{f}_1) + A_2 W(\tilde{f} - \tilde{f}_2). \end{aligned} \tag{17.132}$$

Therefore, we get

$$X(\tilde{f}_1) = A_1 W(0) + A_2 W(\tilde{f}_1 - \tilde{f}_2) \tag{17.133}$$

$$X(\tilde{f}_2) = A_1 W(\tilde{f}_2 - \tilde{f}_1) + A_2 W(0). \tag{17.134}$$

Thus, we see that the spectrum at $\tilde{f} = \tilde{f}_1$ is influenced not only by the main-lobe associated with the sinusoid with frequency $\tilde{f}_1$ (i.e., $A_1 W[0]$), but also by the side-lobe associated with the sinusoid of frequency $\tilde{f}_2$ (i.e., $A_2 W(\tilde{f}_1 - \tilde{f}_2)$). A similar phenomenon happens for the spectrum evaluated at frequency $\tilde{f}_2$ (see Eq. (17.134)), and for all the frequencies. This is illustrated in Fig. 17.17 which shows the spectra $W(\tilde{f} - \tilde{f}_1)$ and $W(\tilde{f} - \tilde{f}_2)$, with $(\tilde{f}_1, \tilde{f}_2) = (0.30, 0.42)$ and $A_1 = A_2 = 1$.

If the side-lobe level of a window is very low (see Fig. 17.17(b)), then the spectrum estimated at each frequency will be dominated by the contribution from the main-lobe, leading to a better spectrum estimate. On the other hand, if the side-lobe level is high (see Fig. 17.17(a)), then the spectrum estimated at each frequency will be significantly influenced by the contribution from the side-lobe, leading to a poor spectrum estimate.

If the main-lobe is very wide (see Fig. 17.17(b)), then a similar effect as in Eq. (17.132) takes place, resulting in the failure to resolve closely spaced spectral lines. Further understanding of leakage and resolution can be gained from Fig. 17.17 if we hold the frequency of the first sinusoid fixed at $\tilde{f}_1 = 0.3$ and vary the frequency ($\tilde{f}_2$) of the second sinusoid. For each value of $\tilde{f}_2$, examine how the spectra $W(\tilde{f} - 0.3)$ and $W(\tilde{f} - \tilde{f}_2)$ are adding together to result in the final spectrum $X(\tilde{f})$.

17.7.4 Windowing for Steady Spectrum

Observe from Fig. 17.14(a) that the non-rectangular windows *weigh down* the signal samples which are near the edges of the window. Therefore, the shape of the resulting periodic repetition of the signal will be less sensitive to small variations in the value of N, compared to that in the case of rectangular window. We illustrate this in Fig. 17.18 by showing the periodic repetitions of the windowed signal (real sinusoid) for $N = 11$ and $N = 13$. Observe that the signal in the case of Blackman window is very similar for the two values of N, whereas the signals look very different in the case of rectangular window. This will affect the shape of the spectra resulting from these two windows, as illustrated in Fig. 17.19.

The choice of the window affects the variations in the shape of the spectrum (e.g., height of the peak) with small changes in the signal frequencies. For example, if we only change the frequency of one of the sinusoids in a signal consisting of the sum of two sinusoids, we expect the spectral shape to remain more or less steady,

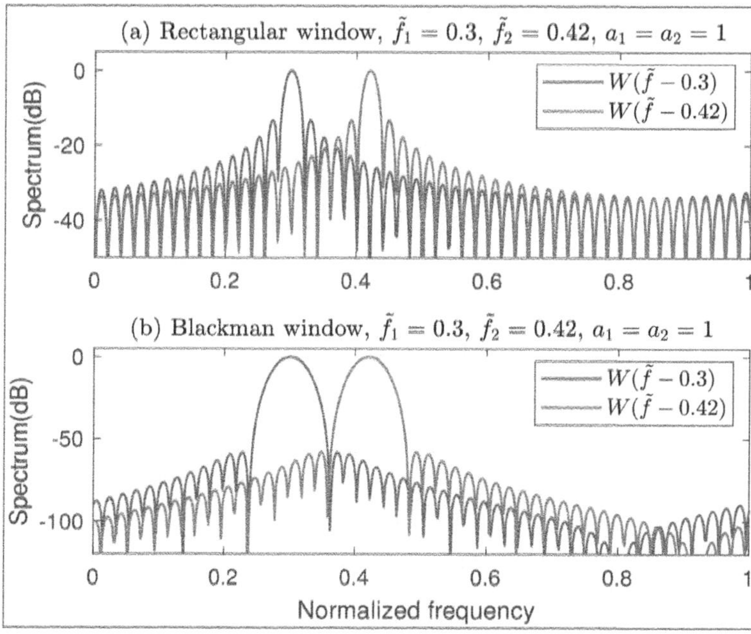

Figure 17.17. Spectra ($N = 50$) of two complex sinusoids (equal powers) at frequencies 0.3 and 0.42 (i.e., $W(\tilde{f} - 0.3)$ and $W(\tilde{f} - 0.42)$ given in Eq. (17.132)). (a) Rectangular window and (b) Blackman window.

except for changes in the positions of the spectral peaks. But, as the figures show, the spectral peaks vary significantly in the case of rectangular window, even with small changes in one of the frequencies. On the other hand, the spectra resulting from the Blackman window remain quite steady.

The jittery nature of the spectrum in the case of rectangular window is a result of the high spectral leakage resulting from its high side-lobe levels. Through the high side-lobes, any small change in the signal in one frequency gets reflected strongly throughout the frequency range.

17.8 Time and Frequency Resolutions

We saw in Section 17.7 that *the frequency resolution of the DFT spectrum (with rectangular window) is* $\frac{1}{N}$. That is, DFT will not be able to resolve two sinusoids if their frequencies are spaced at less than $\frac{1}{N}$. Hence, *the larger the value of* N, *the better will be the frequency resolution, and vice versa.*

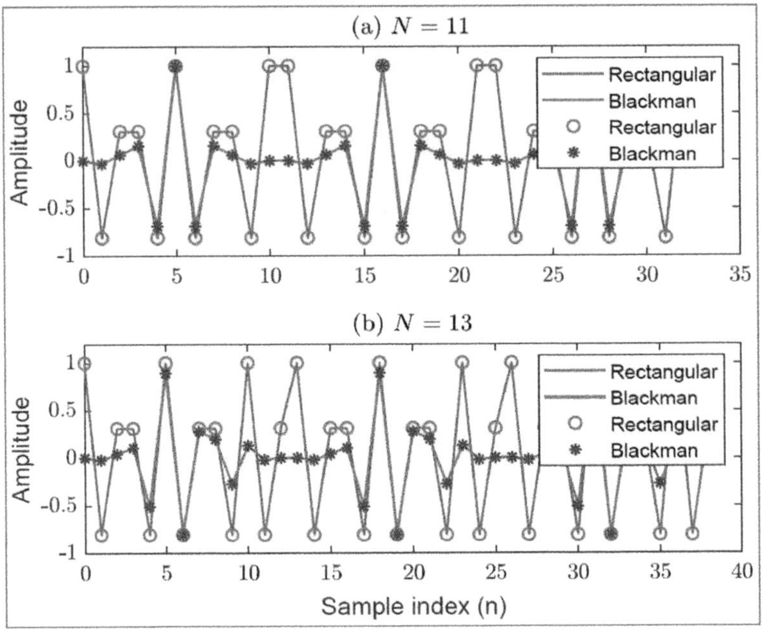

Figure 17.18. Periodic repetitions of the windowed real sinusoid with rectangular and Blackman windows. (a) $N = 11$, (b) $N = 13$.

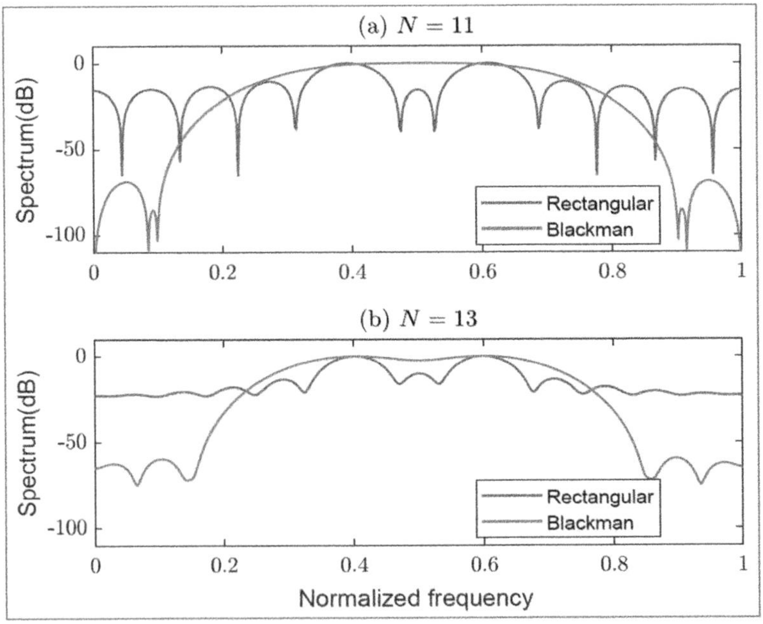

Figure 17.19. Spectra of the windowed real sinusoid with rectangular and Blackman windows. (a) $N = 11$, (b) $N = 13$.

On the other hand, *large values of N will result in reducing the time resolution of the DFT*. That is, choosing N to be large will make it difficult for the DFT to reflect the changes in the time domain.

Suppose that the signal we want to study has frequency components which are slowly varying with time. We may use DFT to study how the frequencies are changing with time. So, we divide the complete signal record into several short records of length N each, and compute the DFT spectrum for each short record. Clearly, *to track the changes well, we should choose small values for* N. Otherwise, *with a large* N, *the net changes in frequencies over a duration of* N *samples may be significant and the successive DFT spectra will not be able to show clearly how the frequencies are changing*. In other words, *the time resolution of the DFT is limited to* N. Thus, *while large* N *is preferred for good frequency resolution, small* N *is preferred for good time resolution*. In practice, depending on the scenario under consideration, we should strike a compromise between time and frequency resolutions while choosing the value of N.

17.9 Spectral Smoothing

When we compute the spectrum of a noisy signal using DFT, the estimated spectrum tends to exhibit fluctuations due to the noise effect. In practice, we prefer to get a spectrum that is smoothed to remove the fluctuations due to noise (and any other random phenomena). This can be done by the technique of spectral smoothing as described below.

- Divide the available data record (say, L samples) into shorter records of length N each. Let $L = MN$, i.e., there are M segments of N samples each.
- Compute the N-point DFT of all the M records. Let $X_{m,k}$ be the DFT of the mth record, with $k = 0, 1, \ldots, N-1$, and $m = 1, 2, \ldots, M$.
- Average the DFT spectra of the M records to obtain the smoothed DFT spectrum. That is,

$$|X_k|^2 = \frac{1}{M}\sum_{m=1}^{M}\left|X_{m,k}\right|^2, \quad k = 0, 1, \ldots, N-1. \tag{17.135}$$

Clearly, *to get better smoothing effect, we should have large number of records (i.e., large* M*) to average the spectrum*. But, if the total number of samples is limited, *choosing a large* M *will make the length of each data record (i.e.,* N*) quite small, thus leading to poor frequency resolution*.

17.10 Concluding Remarks

This chapter is concluded with the following remarks:

- Various techniques for estimation of the power spectrum has been discussed. They include the classical (non-parametric) as well as parametric methods.
- DFT uses the rectangular window implicitly while computing the spectrum.
- Non-trivial windows such as Hamming and Blackman help to reduce spectral leakage (and minimizes fluctuations in the spectrum) due to their low side-lobes. On the other hand, these windows have very wide main-lobes.
- Narrow main-lobe is required for good frequency resolution. The rectangular window has the narrowest main-lobe among the various windows.
- Choice of the DFT size, N, determines the trade-off between frequency and time resolutions in DFT spectrum. With increase in N, frequency resolution improves and time resolution suffers, and vice versa.
- The technique of spectral smoothing can be reduced to minimize fluctuations due to noise in spectra, at the expense of frequency resolution.

Exercises

17.1. Suppose we have $N = 1000$ samples from a sample sequence of random process.

(a) Determine the frequency resolution of the Bartlett, Welch (50% overlap), and Blackman-Tukey methods for a quality factor $Q = 10$.

(b) Determine the record lengths (M) for the Bartlett, Welch (50% overlap), and Blackman-Tukey methods.

17.2. Determine the mean and the autocorrelation of the sequence $x[n]$, which is the output from an ARMA(1,1) process described by the following difference equation

$$x[n] = 0.5x[n-1] + w[n] - w[n-1],$$

where $w[n]$ is a white noise process with variance σ_w^2.

17.3. Determine the mean and the autocorrelation of the sequence $x[n]$, generated by the MA(2) process described by the following difference equation

$$x[n] = w[n] - 2w[n-1] + w[n-2],$$

where $w[n]$ is a white noise process with variance σ_w^2.

17.4. An AR(2) process is described by the difference equation

$$x[n] = 0.81x[n-2] + w[n],$$

where $w[n]$ is a white noise process with variance σ_w^2.

Suppose that $x[n]$ is corrupted by an additive white noise process $v[n]$ with variance σ_v^2. Thus we have,

$$y[n] = x[n] + v[n].$$

(a) Determine the difference equation for $y[n]$ and thus demonstrate that $y[n]$ is an ARMA(2,2) process. Determine the coefficients of the ARMA(2,2) process.

(b) Generalize the result in part (a) to an AR(p) process.

$$x[n] = -\sum_{k=1}^{p} a_k x[n-k] + w[n],$$

and

$$y[n] = x[n] + v[n].$$

17.5. (a) A student is using an educational software to help understand the relationship between signals and spectra. The software is used to generate a real signal consisting of two sinusoids. The programme then calculates and displays the spectrum of the signal generated as shown in Fig. P17.6. Explain why the spectrum is plotted over the positive frequencies only. Why is the spectrum not consisting of purely two impulses corresponding to the frequencies of the two sinusoids that were generated? There seems to be some kind of "noise" next to the two spectral lines corresponding to the sinusoids. Is this due to the noise of the computer? Is it possible to get rid of this noise and how?

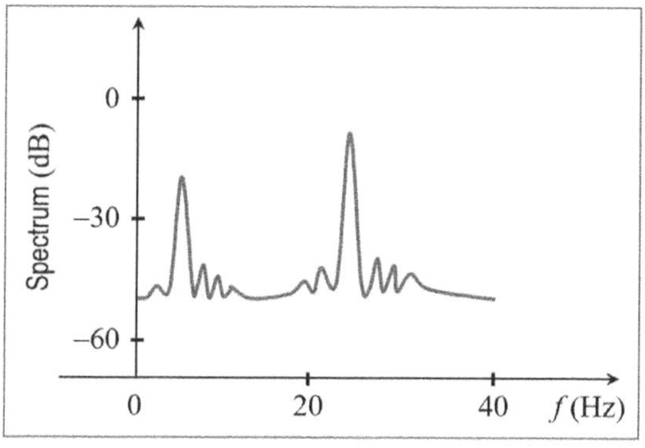

Figure P17.6.

(b) What is the most likely reason that the spectrum of Fig. P17.6 is plotted up to 40 Hz? Will the displayed spectrum change if different windowing functions are used? How? If the developer of the software would like the calculated spectrum to be as close as possible to the theoretical impulses, suggest an appropriate windowing function to use and state the restrictions on the sinusoids' frequencies that the student can choose.

17.6. (a) In a biomedical experiment, a signal $x_a(t)$ with spectrum as shown in Fig. P17.7 is being monitored. Typically, the spectrum consists of a background noise component, a fixed large component due to a known biological function, and more importantly, a slowly time-varying narrow-band component (at frequency f_C) due to another biological phenomenon under investigation. To detect and track f_C, the signal is sampled at 1 kHz, divided into non-overlapping blocks, appropriately windowed and then passed into a FFT processor. What should the block size be if f_C is to be tracked at the rate of 20 times/second? Describe the best scenario such that f_C can be estimated most accurately. What is the resolution for f_C under this scenario?

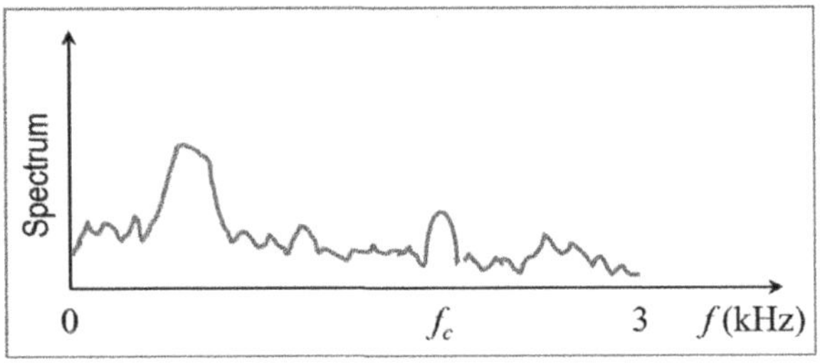

Figure P17.7.

(b) Discuss how the resolution and the stability of the detection of f_C in Question 17.6(a) will change as different windowing functions are used for the two situations when f_C is far away and close to the fixed high power spectral component. Which windowing function would you recommend and why?

17.7. An autoregressive process of order 2 (AR(2)) is described by the difference equation

$$x[n] = \frac{1}{2}x[n-1] + \frac{1}{4}x[n-2] + w[n],$$

where $w[n]$ is a white noise process with variance σ_w^2.

(a) Determine the mean and autocorrelation of the sequence $x[n]$.
Suppose that $x[n]$ is corrupted by an additive white noise process $v[n]$ with variance σ_v^2. Thus we have,

$$y[n] = x[n] + v[n].$$

(b) Show that $y[n]$ is an ARMA(2,2) process and determine the coefficients of the ARMA(2,2) process (ARMA here refers to autoregressive and moving average).

(c) Generalize the result in Question (b) to an AR(p) process ($x[n]$) and derive an expression for the z-transform of the autocorrelation of the process $y[n]$.

$$x[n] = -\sum_{k=1}^{p} a_k x[n-k] + w[n],$$

and

$$y[n] = x[n] + v[n].$$

Appendix A: Appendix

A.1 Hilbert Transform Demo

```
Ts = 0.05;      % Sample spacing
Fs = 1/Ts;      % Sampling frequency
t= -5:Ts:5;     % Sample times

g = sinc(t);  % Time signal

N = 2^13;       % Number of samples in FFT
G = fftshift(fft(g, N));
f = (-(N/2):(N/2-1))/N*Fs; % Frequency axis

figure(1); subplot(211);
plot(t,g); title('g(t) = sinc(t)');
subplot(212); plot(f,abs(G)); title('Amplitude of FFT of g(t)');

gplus = hilbert(g); % Hilbert transform is in imaginary part of gplus
ghat = imag(gplus);

Gplus = fftshift(fft(gplus, N));
Ghat = fftshift(fft(ghat, N));

%%

figure(2); subplot(211);
plot(t,ghat); title('Hilbert transform of g(t) = sinc(t)');
subplot(212); plot(f,abs(Ghat)); title('Amplitude of FFT');

%%
% Plot the pre-envelope
figure(3); plot(f,abs(Gplus));
title('Amplitude of FFT of gplus(t)');

%%
% Modulate by cosine function with frequency 5 Hz

xdsb = g .* cos(2*pi*5*t);
Xdsb = fftshift(fft(xdsb,N));
xssb = g .* cos(2*pi*5*t) - ghat .* sin(2*pi*5*t);
Xssb = fftshift(fft(xssb,N));
```

```
figure(4); subplot(211);
plot(t, xdsb); title('Double sideband (DSB) time signal');
subplot(212); plot(f, abs(Xdsb));
title('Amplitude of FFT of DSB signal');

%%

figure(5); subplot(211);
plot(t, xssb); title('Single sideband (SSB) time signal');
subplot(212); plot(f, abs(Xssb));
title('Amplitude of FFT of SSB signal');

% END
```

Appendix B: Formula Sheets

B.1 Trigonometric Formulas

B.1.1 Right-angle Triangle Definition

Assume that: $0 < \theta < \frac{\pi}{2}$ or $0^\circ < \theta < 90^\circ$

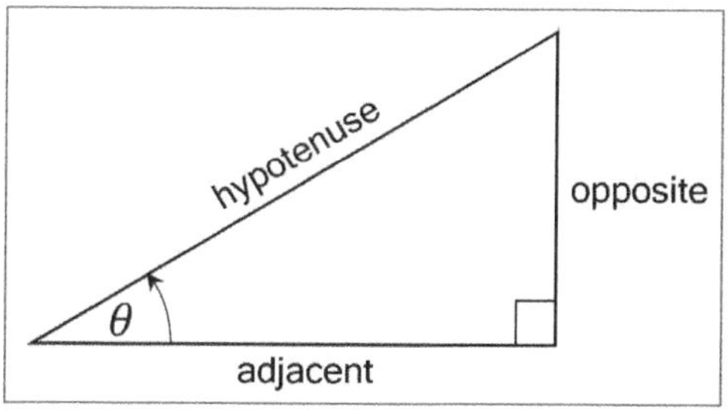

$$\sin(\theta) = \frac{opp}{hyp} \qquad \csc(\theta) = \frac{hyp}{opp}$$

$$\cos(\theta) = \frac{adj}{hyp} \qquad \sec(\theta) = \frac{hyp}{adj}$$

$$\tan(\theta) = \frac{opp}{adj} \qquad \cot(\theta) = \frac{adj}{opp}$$

B.1.2 Unit Circle Definition

Assume θ can be any angle,

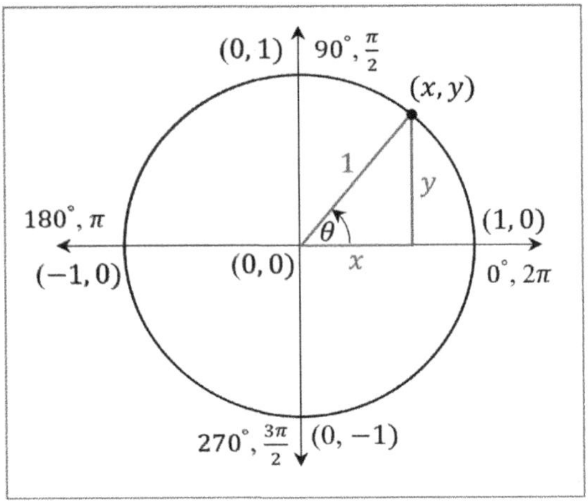

$$\sin(\theta) = \tfrac{y}{1} \quad \csc(\theta) = \tfrac{1}{y}$$

$$\cos(\theta) = \tfrac{x}{1} \quad \sec(\theta) = \tfrac{1}{x}$$

$$\tan(\theta) = \tfrac{y}{x} \quad \cot(\theta) = \tfrac{x}{1}$$

B.1.3 Domains of the Trigonometric Functions

$$\sin(\theta),\ \forall\ \theta \in (-\infty, \infty) \qquad \csc(\theta),\ \forall\ \theta \neq n\pi,\ n \in \mathbb{Z}$$

$$\cos(\theta),\ \forall\ \theta \in (-\infty, \infty) \qquad \sec(\theta),\ \forall\ \theta \neq \left(n + \tfrac{1}{2}\right)\pi,\ n \in \mathbb{Z}$$

$$\tan(\theta),\ \forall\ \theta \neq \left(n + \tfrac{1}{2}\right)\pi,\ n \in \mathbb{Z} \qquad \cot(\theta),\ \forall\ \theta \neq n\pi,\ n \in \mathbb{Z}$$

B.1.4 Ranges of the Trigonometric Functions

$$-1 \le \sin(\theta) \le 1 \qquad \csc(\theta) \ge 1 \text{ and } \csc(\theta) \le -1$$

$$-1 \le \cos(\theta) \le 1 \qquad \sec(\theta) \ge 1 \text{ and } \sec(\theta) \le -1$$

$$-\infty \le \tan(\theta) \le \infty \qquad -\infty \le \cot(\theta) \le \infty$$

B.2 Identities and Formulas

B.2.1 Tangent and Cotangent Identities

$$\tan(\theta) = \frac{\sin(\theta)}{\cos(\theta)} \qquad \cot(\theta) = \frac{\cos(\theta)}{\sin(\theta)}$$

B.2.2 Reciprocal Identities

$$\sin(\theta) = \frac{1}{\csc(\theta)} \qquad \csc(\theta) = \frac{1}{\sin(\theta)}$$

$$\cos(\theta) = \frac{1}{\sec(\theta)} \qquad \sec(\theta) = \frac{1}{\cos(\theta)}$$

$$\tan(\theta) = \frac{1}{\cot(\theta)} \qquad \cot(\theta) = \frac{1}{\tan(\theta)}$$

B.2.3 Pythagorean Identities

$$\sin^2(\theta) + \cos^2(\theta) = 1$$
$$\tan^2(\theta) + 1 = \sec^2(\theta)$$
$$1 + \cot^2(\theta) = \csc^2(\theta)$$

B.2.4 Even and Odd Formulas

$$\sin(-\theta) = -\sin(\theta) \qquad \csc(-\theta) = -\csc(\theta)$$
$$\cos(-\theta) = \cos(\theta) \qquad \sec(-\theta) = \sec(\theta)$$
$$\tan(-\theta) = -\tan(\theta) \qquad \cot(-\theta) = -\cot(\theta)$$

B.2.5 Periodic Formulas

If n is an integer,

$$\sin(\theta + 2\pi n) = \sin(\theta) \qquad \csc(\theta + 2\pi n) = \csc(\theta)$$
$$\cos(\theta + 2\pi n) = \cos(\theta) \qquad \sec(\theta + 2\pi n) = \sec(\theta)$$
$$\tan(\theta + \pi n) = \tan(\theta) \qquad \cot(\theta + \pi n) = \cot(\theta)$$

B.2.6 Double Angle Formulas

$$\sin(2\theta) = 2\sin(\theta)\cos(\theta)$$
$$\begin{aligned}\cos(2\theta) &= \cos^2(\theta) - \sin^2(\theta)\\ &= 2\cos^2(\theta) - 1\\ &= 1 - 2\sin^2(\theta)\end{aligned}$$
$$\tan(2\theta) = \frac{2\tan(\theta)}{1 - \tan^2(\theta)}$$

B.2.7 Degrees to Radians Formulas

If a is an angle in degrees and r is an angle in radians, then

$$\frac{\pi}{180^\circ} = \frac{r}{a} \quad \Rightarrow \quad r = \frac{\pi a}{180^\circ} \quad \text{and} \quad a = \frac{180^\circ r}{\pi}$$

B.2.8 Half-Angle Formulas

$$\sin(\theta) = \pm\sqrt{\frac{1 - \cos(2\theta)}{2}}$$

$$\cos(\theta) = \pm\sqrt{\frac{1 + \cos(2\theta)}{2}}$$

$$\tan(\theta) = \pm\sqrt{\frac{1 - \cos(2\theta)}{1 + \cos(2\theta)}}$$

B.2.9 Sum and Difference Formulas

$$\sin(\alpha \pm \beta) = \sin(\alpha)\cos(\beta) \pm \cos(\alpha)\sin(\beta)$$

$$\cos(\alpha \pm \beta) = \cos(\alpha)\cos(\beta) \mp \sin(\alpha)\sin(\beta)$$

$$\tan(\alpha \pm \beta) = \frac{\tan(\alpha) \pm \tan(\beta)}{1 \mp \tan(\alpha)\tan(\beta)}$$

B.2.10 Product to Sum Formulas

$$\sin(\alpha)\sin(\beta) = \frac{1}{2}\left[\cos(\alpha - \beta) - \cos(\alpha + \beta)\right]$$

$$\cos(\alpha)\cos(\beta) = \frac{1}{2}\left[\cos(\alpha - \beta) + \cos(\alpha + \beta)\right]$$

$$\sin(\alpha)\cos(\beta) = \frac{1}{2}\left[\sin(\alpha + \beta) + \sin(\alpha - \beta)\right]$$

$$\cos(\alpha)\sin(\beta) = \frac{1}{2}\left[\sin(\alpha + \beta) - \sin(\alpha - \beta)\right]$$

B.2.11 Sum to Product Formulas

$$\sin(\alpha) + \sin(\beta) = 2\sin\left(\frac{\alpha+\beta}{2}\right)\cos\left(\frac{\alpha-\beta}{2}\right)$$

$$\sin(\alpha) - \sin(\beta) = 2\cos\left(\frac{\alpha+\beta}{2}\right)\sin\left(\frac{\alpha-\beta}{2}\right)$$

$$\cos(\alpha) + \cos(\beta) = 2\cos\left(\frac{\alpha+\beta}{2}\right)\cos\left(\frac{\alpha-\beta}{2}\right)$$

$$\cos(\alpha) - \cos(\beta) = -2\sin\left(\frac{\alpha+\beta}{2}\right)\sin\left(\frac{\alpha-\beta}{2}\right)$$

B.2.12 Co-function Formulas

$$\sin\left(\tfrac{\pi}{2} - \theta\right) = \cos(\theta) \qquad \cos\left(\tfrac{\pi}{2} - \theta\right) = \sin(\theta)$$

$$\csc\left(\tfrac{\pi}{2} - \theta\right) = \sec(\theta) \qquad \sec\left(\tfrac{\pi}{2} - \theta\right) = \csc(\theta)$$

$$\tan\left(\tfrac{\pi}{2} - \theta\right) = \cot(\theta) \qquad \cot\left(\tfrac{\pi}{2} - \theta\right) = \tan(\theta)$$

B.3 Unit Circle

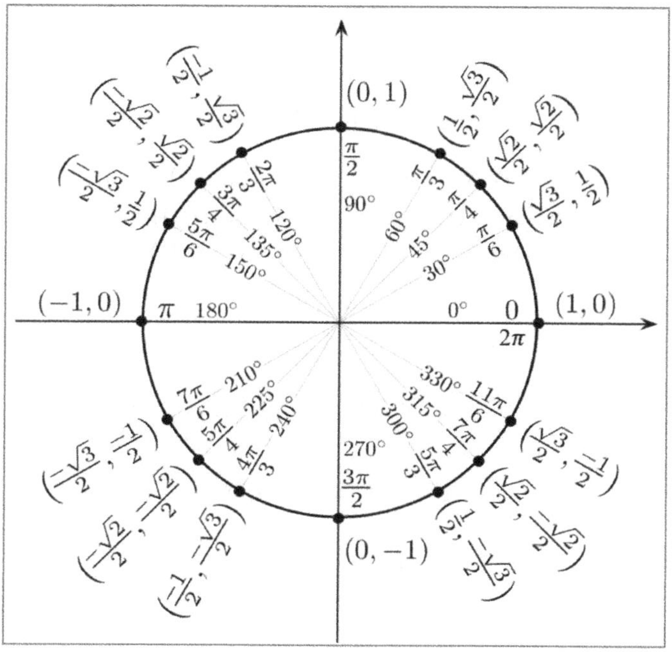

For any ordered pair (x, y) on the unit circle, $\cos(\theta) = x$ and $\sin(\theta) = y$.

Example:

$$\cos\left(\frac{4\pi}{3}\right) = -\frac{1}{2} \quad \text{and} \quad \sin\left(\frac{4\pi}{3}\right) = -\frac{\sqrt{3}}{2}$$

B.4 Inverse Trigonometric Functions

B.4.1 Definition

$$\theta = \sin^{-1}(x) \text{ is equivalent to } x = \sin(\theta)$$

$$\theta = \cos^{-1}(x) \text{ is equivalent to } x = \cos(\theta)$$

$$\theta = \tan^{-1}(x) \text{ is equivalent to } x = \tan(\theta)$$

B.4.2 Domain and Range

Function	Domain	Range
$\theta = \sin^{-1}(x)$	$-1 \le x \le 1$	$-\frac{\pi}{2} \le \theta \le \frac{\pi}{2}$
$\theta = \cos^{-1}(x)$	$-1 \le x \le 1$	$0 \le \theta \le \pi$
$\theta = \tan^{-1}(x)$	$-\infty \le x \le \infty$	$-\frac{\pi}{2} \le \theta \le \frac{\pi}{2}$

B.4.3 Inverse Properties

The properties hold for x in the domain and θ in the range.

$$\sin(\sin^{-1}(x)) = x \qquad \sin^{-1}(\sin(\theta)) = \theta$$
$$\cos(\cos^{-1}(x)) = x \qquad \cos^{-1}(\cos(\theta)) = \theta$$
$$\tan(\tan^{-1}(x)) = x \qquad \tan^{-1}(\tan(\theta)) = \theta$$

B.4.4 Other Notations

$$\sin^{-1}(x) = \arcsin(x)$$

$$\cos^{-1}(x) = \arccos(x)$$

$$\tan^{-1}(x) = \arctan(x)$$

B.5 Laws of Sines, Cosines, and Tangents

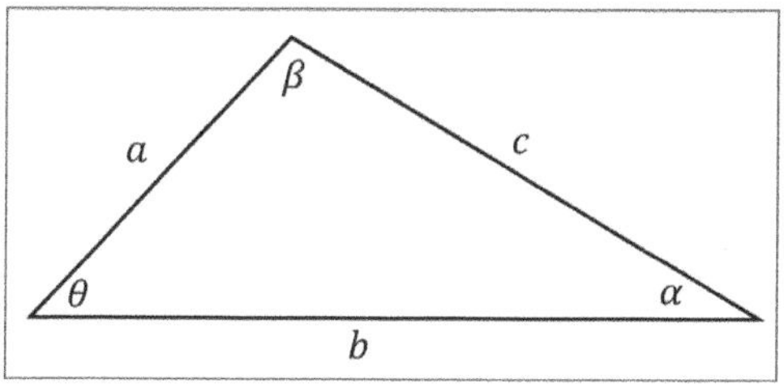

Law of Sines	Law of Cosines	Law of Tangents
$\frac{\sin(\alpha)}{a} = \frac{\sin(\beta)}{b} = \frac{\sin(\theta)}{c}$	$a^2 = b^2 + c^2 - 2bc\cos(\alpha)$	$\frac{a-b}{a+b} = \frac{\tan\frac{1}{2}(\alpha-\beta)}{\tan\frac{1}{2}(\alpha+\beta)}$
	$b^2 = a^2 + c^2 - 2ac\cos(\beta)$	$\frac{b-c}{b+c} = \frac{\tan\frac{1}{2}(\beta-\theta)}{\tan\frac{1}{2}(\beta+\theta)}$
	$c^2 = a^2 + b^2 - 2ab\cos(\theta)$	$\frac{a-c}{a+c} = \frac{\tan\frac{1}{2}(\alpha-\theta)}{\tan\frac{1}{2}(\alpha+\theta)}$

B.6 Complex Numbers

$$i = \sqrt{-1} \quad i^2 = -1 \quad i^3 = -i \quad i^4 = 1$$

$\sqrt{-a} = i\sqrt{a}, \quad a \geq 0.$

$(a + ib)(a - ib) = a^2 + b^2$

$(a + ib) + (c + id) = (a + c) + i(b + d)$

$|a + ib| = \sqrt{a^2 + b^2}$
Complex Modulus

$(a + ib) - (c + id) = (a - c) + i(b - d)$

$(a + ib)^* = a - ib$
Complex Conjugate

$(a + ib)(c + id) = (ac - bd) + i(ad + bc)$

$(a + ib)^*(a + ib)$
$= |a + ib|^2 = a^2 + b^2$

B.7 Sin, Cosine, and Tangent Functions

Figure below shows the functional relationship between sin function and the unit circle.

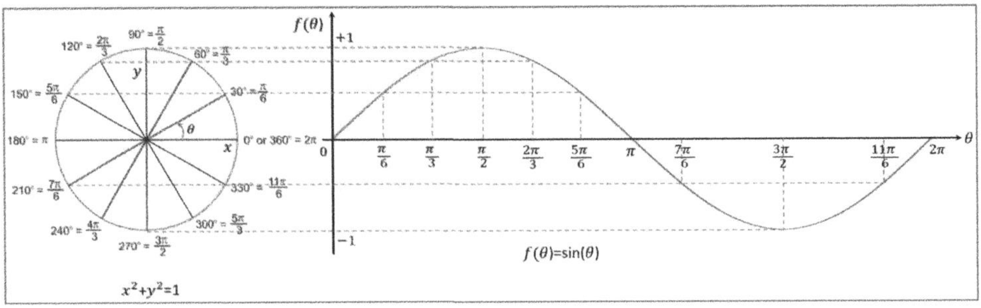

The following figures illustrates the sin, cos, and tan functions and the table shows the values of these functions at specific angle values.

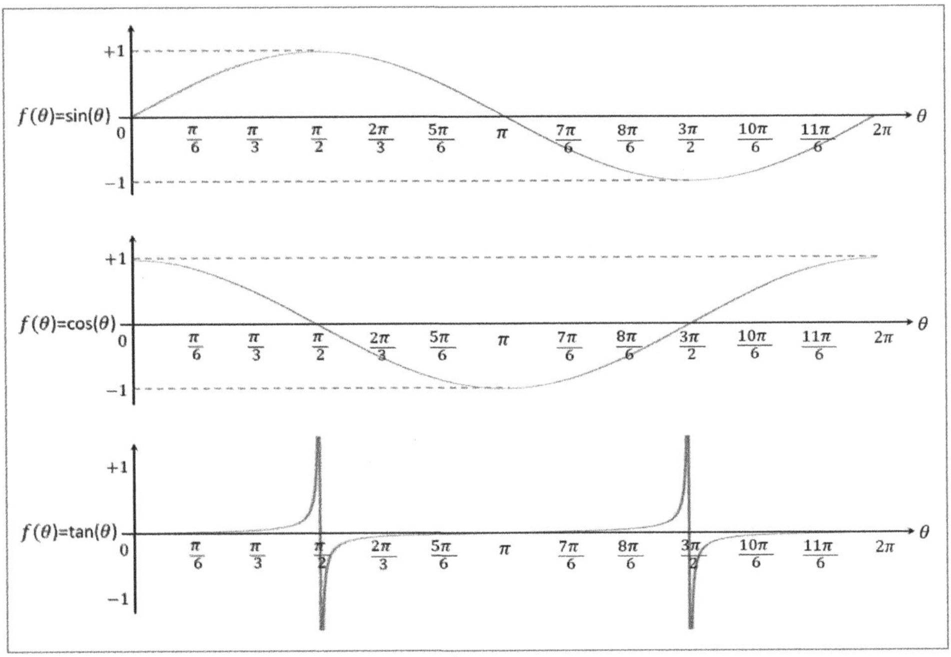

θ	$0, 2\pi$	$\frac{\pi}{6}$	$\frac{\pi}{3}$	$\frac{\pi}{2}$	$\frac{2\pi}{3}$	$\frac{5\pi}{6}$	π	$\frac{7\pi}{6}$	$\frac{8\pi}{6}$	$\frac{3\pi}{2}$	$\frac{10\pi}{6}$	$\frac{11\pi}{6}$
$\sin(\theta)$	0.0	$\frac{1}{2}$	$\frac{\sqrt{3}}{2}$	1.0	$\frac{\sqrt{3}}{2}$	$\frac{1}{2}$	0.0	$-\frac{1}{2}$	$-\frac{\sqrt{3}}{2}$	−1.0	$-\frac{\sqrt{3}}{2}$	$-\frac{1}{2}$
$\cos(\theta)$	1.0	$\frac{\sqrt{3}}{2}$	$\frac{1}{2}$	0.0	$-\frac{1}{2}$	$-\frac{\sqrt{3}}{2}$	−1.0	$-\frac{\sqrt{3}}{2}$	$-\frac{1}{2}$	0.0	$\frac{1}{2}$	$\frac{\sqrt{3}}{2}$
$\tan(\theta)$	0.0	$\frac{1}{\sqrt{3}}$	$\sqrt{3}$	∞	$-\sqrt{3}$	$-\frac{1}{\sqrt{3}}$	0.0	$\frac{1}{\sqrt{3}}$	$\sqrt{3}$	$-\infty$	$-\sqrt{3}$	$-\frac{1}{\sqrt{3}}$

B.8 Signal-to-noise Ratio

The relative amount of signal and noise present in a waveform is usually quantified by the signal-to-noise ratio (SNR). As the name implies, it is simply the ratio of signal to noise, both measured in root-mean-squared (RMS) amplitude and is often expressed in decibel (dB). SNR in dB is defined as,

$$\mathrm{SNR}_{dB} = 20\log_{10}\left(\frac{\text{Signal amplitude (RMS)}}{\text{Noise amplitude (RMS)}}\right), \tag{B.1}$$

$$= 10\log_{10}\left(\frac{\text{Signal power}}{\text{Noise power}}\right). \tag{B.2}$$

To convert from dB scale to linear scale,

$$\mathrm{SNR}_{\text{linear}} = 10^{\frac{\text{dB}}{20}}. \tag{B.3}$$

For example, a SNR of 20dB implies that the RMS value of signal is 10 times the RMS value of noise (i.e., $10^{\frac{20}{20}} = 10$). A SNR of +3dB indicates a ratio of $\sqrt{2} = 1.414$ (i.e.,$10^{\frac{3}{20}} = 1.414$). Similarly, a SNR of 0dB implies that the RMS values of the signal and noise are the same (i.e., $10^{\frac{0}{20}} = 1$).

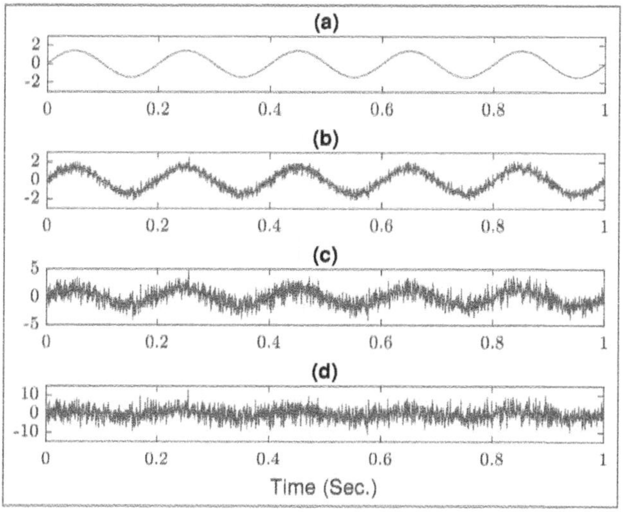

Figure B.1.

Figure B.1 shows a 5 Hz sinusoidal signal with various amounts of white noise added to it. Figure B.1(a) is the 5Hz sinusoid alone (or the SNR is ∞). The sinusoid with 10dB SNR is shown in Fig. B.1(b). Fig. B.1(c) illustrates the case with 0dB SNR and the last figure (Fig. B.1(d)) shows the signal at −10dB SNR.

B.9 z-Transform Properties

Let $x[n]$, $x_n[n]$, and $x_2[n]$ be three discrete-time signals with corresponding z-transforms, defined as,

$$X(z) = \sum_{n=-\infty}^{\infty} x[n]z^{-n}$$

$$X_1(z) = \sum_{n=-\infty}^{\infty} x_1[n]z^{-n}$$

$$X_2(z) = \sum_{n=-\infty}^{\infty} x_2[n]z^{-n}$$

For scalar constants a_1 and a_2, the following z-transform properties hold.

Property	Time Domain	z-Domain
Notation	$x[n]$ $x_1[n]$ $x_2[n]$	$X(z)$ $X_1(z)$ $X_2(z)$
Linearity	$a_1x_1[n] + a_2x_2[n]$	$a_1X_1(z) + a_2X_2(z)$
Time shifting	$x[n-k]$	$z^{-k}X(z)$
z-Scaling	$a_1^n x[n]$	$X(a_1^{-1}z)$
Time reversal	$x[-n]$	$X(z^{-1})$
Conjugation	$x^*[n]$	$X^*(z^*)$
z-Differentiation	$nx[n]$	$-z\frac{dX(z)}{dz}$
z-Differentiation	$n^k x[n]$	$(-1)^k z^k \frac{d^k X(z)}{dz^k}$
Convolution	$x_1[n] \otimes x_2[n]$	$X_1(z)X_2(z)$

B.10 Common z-Transform Pairs

The table below provides the z-transforms of some of the commonly used discrete-time functions. Here, $\omega_0 = 2\pi f_0$ is an arbitrary angular frequency.

Signal	z-Transform	ROC
$\delta[n]$	1	All z-domain
$u[n]$	$\frac{1}{1-z^{-1}}$	$\|z\| > 1$
$a^n u[n]$	$\frac{1}{1-az^{-1}}$	$\|z\| > \|a\|$
$na^n u[n]$	$\frac{az^{-1}}{(1-az^{-1})^2}$	$\|z\| > \|a\|$
$-a^n u[-n-1]$	$\frac{1}{1-az^{-1}}$	$\|z\| < \|a\|$
$-na^n u[-n-1]$	$\frac{az^{-1}}{(1-az^{-1})^2}$	$\|z\| < \|a\|$
$\cos(\omega_0 n)u[n]$	$\frac{1-z^{-1}\cos(\omega_0)}{1-2z^{-1}\cos(\omega_0)+z^{-2}}$	$\|z\| > 1$
$\sin(\omega_0 n)u[n]$	$\frac{z^{-1}\sin(\omega_0)}{1-2z^{-1}\cos(\omega_0)+z^{-2}}$	$\|z\| > 1$
$a^n \cos(\omega_0 n)u[n]$	$\frac{1-az^{-1}\cos(\omega_0)}{1-2az^{-1}\cos(\omega_0)+a^2z^{-2}}$	$\|z\| > \|a\|$
$a^n \sin(\omega_0 n)u[n]$	$\frac{az^{-1}\sin(\omega_0)}{1-2az^{-1}\cos(\omega_0)+a^2z^{-2}}$	$\|z\| > \|a\|$

B.11 Fourier Transform Theorems and Properties

Let $x(t)$ and $y(t)$ be two non-periodic signals with FTs $X(f)$ and $Y(f)$, respectively. That is,

$$x(t) = \int_{-\infty}^{\infty} X(f)e^{j2\pi ft}df, \qquad y(t) = \int_{-\infty}^{\infty} Y(f)e^{j2\pi ft}df$$

$$X(f) = \int_{-\infty}^{\infty} x(t)e^{-j2\pi ft}dt, \qquad Y(f) = \int_{-\infty}^{\infty} y(t)e^{-j2\pi ft}dt.$$

For real constants a, a_1, a_2, f_0, and t_0, the following properties are derived.

Property/Theorem	Time Domain	Frequency Domain
Notation	$x(t)$ $y(t)$	$X(f)$ $Y(f)$
Linearity	$a_1x(t) + a_2y(t)$	$a_1X(f) + a_2Y(f)$
Dilation	$x(at)$	$\frac{1}{\|a\|}X\left(\frac{f}{a}\right)$
Conjugation	$x^*(t)$	$X^*(-f)$

Property/Theorem	Time Domain	Frequency Domain
Duality	$X(t)$	$g(-f)$
Time shifting	$x(t-t_0)$	$e^{-j2\pi f t_0}X(f)$
Frequency shifting	$e^{j2\pi f_0 t}x(t)$	$X(f-f_0)$
Area under $X(f)$	$x(0)$	$\int_{-\infty}^{\infty} X(f)df$
Area under $x(t)$	$\int_{-\infty}^{\infty} x(t)dt$	$X(0)$
Time Differentiation	$\frac{dx(t)}{dt}$	$j2\pi f X(f)$
Time Integration	$\int_{-\infty}^{t} x(\tau)d\tau$	$\frac{1}{j2\pi f}X(f)$
Modulation Theorem	$x(t)y(t)$	$X(f)\otimes Y(f)$
Convolution Theorem	$x(t)\otimes y(t)$	$X(f)Y(f)$
Correlation Theorem	$x(t)\otimes y^*(-t)$	$X(f)Y^*(f)$
Energy Theorem	$\int_{-\infty}^{\infty} x^2(t)dt$	$\int_{-\infty}^{\infty} \lvert X(f)\rvert^2 df$

B.12 Continuous Time Fourier Series: Properties

Let $x(t)$ and $y(t)$ be two periodic signals with period, $T_p = \frac{1}{f_p}$, where f_p is the fundamental (cyclic) frequency of the signals with the corresponding angular frequency, $\Omega_p = 2\pi f_p$. The Fourier series relationships for both the signals can now be written as:

$$x(t) = \sum_{k=-\infty}^{+\infty} X_k e^{jk\Omega_p t}; \quad \text{where,} \quad X_k = \frac{1}{T_p}\int_{-\frac{T_p}{2}}^{\frac{T_p}{2}} x(t)e^{-jk\Omega_p t}dt$$

$$y(t) = \sum_{k=-\infty}^{+\infty} Y_k e^{jk\Omega_p t}; \quad \text{where,} \quad Y_k = \frac{1}{T_p}\int_{-\frac{T_p}{2}}^{\frac{T_p}{2}} y(t)e^{-jk\Omega_0 t}dt$$

Here, X_k and Y_k are the Fourier coefficients of $x(t)$ and $y(t)$, respectively. Now, for scalar quantities, α, β, and t_0, the following properties are hold.

Property	Periodic Signal	Fourier Coefficients
Notation	$x(t)$ $y(t)$	X_k Y_k
Linearity	$\alpha x(t) + \beta y(t)$	$\alpha X_k + \beta Y_k$
Time shifting	$x(t - t_0)$	$X_k e^{-jk\Omega_p t_0}$
Frequency shifting	$e^{jM\Omega_p t} x(t)$	X_{k-M}
Conjugation	$x^*(t)$	X^*_{-k}
Time reversal	$x(-t)$	X_{-k}
Periodic convolution	$\int_{T_p} x(\tau)y(t-\tau)d\tau$	$T_p X_k Y_k$
Multiplication	$x(t)y(t)$	$\sum_{l=-\infty}^{\infty} X_l Y_{k-l}$
Time Differentiation	$\frac{dx(t)}{dt}$	$jk\Omega_p X_k$
Time Integration	$\int_{-\infty}^{t} x(t)dt$	$\left(\frac{1}{jk\Omega_p}\right) X_k$
Conjugate Symmetry for Real Signals	$x(t)$ real	$X_k = X^*_{-k}$ $\Re e\{X_k\} = \Re e\{X_{-k}\}$ $\Im m\{X_k\} = -\Im m\{X_{-k}\}$ $\|X_k\| = \|X_{-k}\|$ $\angle X_k = -\angle X_{-k}$
Real and Even Signals	$x(t)$ real and even	X_k real and even
Real and Odd Signals	$x(t)$ real and odd	X_k purely imaginary and odd
Parseval's theorem	$\frac{1}{T_p}\int_{T_p} \|x(t)\|^2 dt$	$\sum_{k=-\infty}^{\infty} \|X_k\|^2$

B.13 Discrete Time Fourier Series: Properties

Let $x[n]$ and $y[n]$ be two periodic discrete-time signals with period, N, and fundamental frequency, $\omega_p = 2\pi f_p$, where $f_p = \frac{1}{N}$ is the fundamental (cyclic) frequency of the signals. Fourier series relationships for $x[n]$ and $y[n]$ can be written as:

$$x[n] = \sum_{k=0}^{N-1} X_k e^{jk\omega_p n}; \quad \text{where,} \quad X_k = \frac{1}{N}\sum_{n=0}^{N-1} x[n] e^{-jk\omega_p n}$$

$$y[n] = \sum_{k=0}^{N-1} Y_k e^{jk\omega_p n}; \quad \text{where,} \quad Y_k = \frac{1}{N}\sum_{n=0}^{N-1} y[n] e^{-jk\omega_p n}$$

Here, X_k and Y_k are the Fourier coefficients of $x[n]$ and $y[n]$, respectively.

Property	Periodic Signal	Fourier Coefficients
Notation	$x[n]$ $y[n]$	X_k Y_k
Linearity	$\alpha x[n] + \beta y[n]$	$\alpha X_k + \beta Y_k$
Time shifting	$x[n - n_0]$	$X_k e^{-jk\omega_p n_0}$
Frequency shifting	$e^{jM\omega_p n} x[n]$	X_{k-M}
Conjugation	$x^*[n]$	X^*_{-k}
Time reversal	$x[-n]$	X_{-k}
Periodic convolution	$\sum_{m=0}^{N-1} x[m]y[n-m]$	$N X_k Y_k$
Multiplication	$x[n]y[n]$	$\sum_{l=0}^{N-1} X_l Y_{k-l}$
First Difference	$x[n] - x[n-1]$	$\left(1 - e^{jk\omega_p}\right) X_k$
Running Sum	$\sum_{k=-\infty}^{n} x[k]$	$\left(\frac{1}{\left(1-e^{jk\omega_p}\right)}\right) X_k$
Conjugate Symmetry for Real Signals	$x[n]$ real	$X_k = X^*_{-k}$ $\Re e\{X_k\} = \Re e\{X_{-k}\}$ $\Im m\{X_k\} = -\Im m\{X_{-k}\}$ $\lvert X_k \rvert = \lvert X_{-k} \rvert$ $\angle X_k = -\angle X_{-k}$
Real and Even Signals	$x[n]$ real and even	X_k real and even
Real and Odd Signals	$x[n]$ real and odd	X_k purely imaginary and odd
Parseval's theorem	$\sum_{n=0}^{N-1} \lvert x[n] \rvert^2$	$\frac{1}{N} \sum_{k=0}^{N-1} \lvert X_k \rvert^2$

B.14 Uniform Random Variables

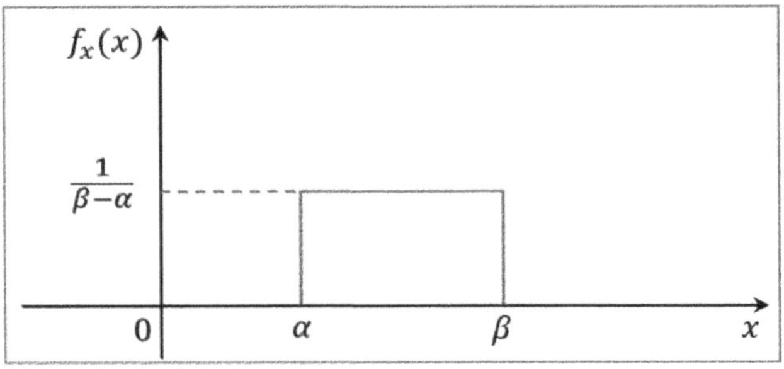

Figure B.2.

A random variable x is uniformly distributed in the interval $[\alpha, \beta]$ if its PDF is defined as

$$f_x(x) = \begin{cases} \frac{1}{\beta-\alpha} & \text{if} \quad \alpha \leq x \leq \beta \\ 0 & \text{otherwise} \end{cases} \tag{B.4}$$

Here, we have

- $$\frac{1}{\beta - \alpha} \int_{\alpha}^{\beta} dx = 1 \tag{B.5}$$

- $$m_x = E[x] = \int_{\alpha}^{\beta} x f_x(x) dx = \frac{1}{\beta - \alpha} \int_{\alpha}^{\beta} x dx = \frac{\alpha + \beta}{2} \tag{B.6}$$

- $$p_x = E[x^2] = \int_{\alpha}^{\beta} x^2 f_x(x) dx = \frac{1}{\beta - \alpha} \int_{\alpha}^{\beta} x^2 dx = \frac{\alpha^2 + \beta^2 + \alpha\beta}{3} \tag{B.7}$$

- $$\sigma_x^2 = \text{var}[x] = E[x^2] - \{E[x]\}^2 = \frac{(\beta - \alpha)^2}{12} \tag{B.8}$$

B.15 Normal (Gaussian) Random Variables

The PDF is given by,

$$f_x(x) = \frac{1}{\sigma_x \sqrt{2\pi}} \exp\left(-\frac{1}{2} \left(\frac{x - m_x}{\sigma_x} \right)^2 \right) \tag{B.9}$$

where

- m_x is the mean and σ_x is the standard deviation.
- The factor $\frac{1}{\sigma_x \sqrt{2\pi}}$ is a constant to make the area under the pdf to be unity.
- $f_x(x)$ is symmetric with respect to $x = m_x$ because the exponent is quadratic. This implies that when $m_x = 0$, the distribution is symmetric with respect to the y-axis (i.e., $x = 0$).
- The exponent term tends to zero faster as σ_x decreases.

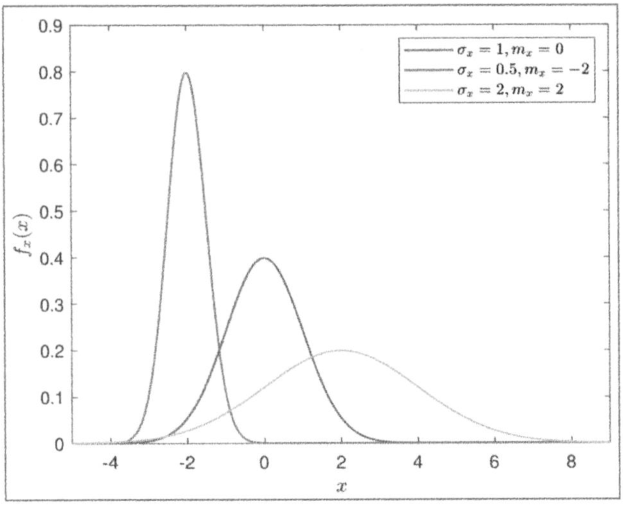

Figure B.3.

B.16 Central Limit Theorem (CLT)

If $x_1, x_2, \ldots, x_K$ be a sequence of independent and identically distributed (i.i.d) random variables each having mean m_x and variance σ_x^2. Then for K large, the distribution of $x_1 + x_2 + \ldots + x_K$ is approximately normal with mean Km_x and variance $K\sigma_x^2$.

It follows from the CLT that $\frac{x_1+x_2+\ldots+x_K-Km_x}{\sigma_x\sqrt{K}}$ is approximately a standard normal random variable. Thus for K large,

$$P\left\{\frac{x_1 + x_2 + \ldots + x_K - Km_x}{\sigma_x\sqrt{K}} < x\right\} \approx P\{z < x\} \tag{B.10}$$

This means that the distribution of an average tends to be Normal (Gaussian), even when the distribution from which the average is computed is decidedly non-normal.

Bibliography

[1] J.S. Bendat and A.G. Piersol, *Random Data: Analysis and Measurement Procedures*, 4th edition, John Wiley & Sons, New York, 2010.

[2] L. Ljung, *System Identification: Theory for the User*, Englewood Cliffs, NJ, Prentice-Hall, 1987.

[3] A. Papoulis, *Probability, Random variables, and Stochastic Processes*, 2nd Edition, McGraw-Hill, Inc., New York, 1984.

[4] A. Papoulis, *Signal Analysis*, 3rd Printing, McGraw-Hill, Inc., New York, 1987.

[5] J.G. Proakis and D.G. Manolakis, *Digital Signal Processing: Principles, Algorithms and Applications*, 4th Edition, Upper Saddle River, NJ, Prentice-Hall, 2007.

[6] G. Zelniker and F.J. Taylor, *Advanced Digital Signal Processing: Theory and Applications*, Marcel Dekker, Inc., New York, 1994.

[7] W.C. van Etten, *Introduction to Random Signals & Noise*, John Wiley & Sons, New York 2005.

[8] B.P. Lathi, *Signal Processing and Linear Systems*, Oxford University Press, 1998.

[9] P.J. Davis and P. Rabinowitz, *Methods of Numerical Integration*, 2nd Edition, Academic Press, New York, 1984.

[10] S.J. Orfanidis, *Optimum Signal Processing: An Introduction*, 2nd Edition, NY: Macmillan, New York, 1985.

[11] J. McClellan, R. Schafer, and M. Yoder, *DSP First: A Multimedia Approach*, Prentice Hall; Upper Saddle River, New Jersey, 1998.

[12] W. Rudin, *Real and Complex Analysis*, McGraw-Hill series in mathematics, McGraw-Hill, 2000.

[13] J.W. Cooley and J.W. Tukey, "An algorithm for the machine calculation of complex Fourier series", *Jl. of Math. Comp.*, Vol. 19, pp. 297–301, 1965.

[14] J.A. Cadzow, "Recursive digital filter synthesis via gradient based algorithms", *IEEE Transactions on Acoustics, Speech and Signal Processing*, Vol. 24, no. 4, pp. 349–355, 1976.

Index

Printed and bound by CPI Group (UK) Ltd, Croydon, CR0 4YY

15/09/2026

14962365-0007